MODERN PROBABILITY AND STATISTICS

QUEUEING THEORY

ALSO AVAILABLE IN MODERN PROBABILITY AND STATISTICS:

Generalized Poisson Models and their Applications in Insurance and Finance
V.E. Bening and V.Yu. Korolev

Robustness in Data Analysis: criteria and methods
G.L. Shevlyakov and N.O. Vilchevski

Asymptotic Theory of Testing Statistical Hypotheses: Efficient Statistics, Optimality, Power Loss and Deficiency
V.E. Bening

Selected Topics in Characteristic Functions
N.G. Ushakov

Chance and Stability. Stable Distributions and their Applcations
V.M. Zolotarev and V.V. Uchaikin

Normal Approximation: New Results, Methods and Problems
V.V. Senatov

Modern Theory of Summation of Random Variables
V.M. Zolotarev

MODERN PROBABILITY AND STATISTICS

Queueing Theory

P.P. Bocharov, C. D'Apice,
A.V. Pechinkin and S. Salerno

///VSP///

UTRECHT • BOSTON, 2004

VSP
an imprint of Brill Academic Publishers
P.O. Box 346
3700 AH Zeist
The Netherlands

Tel: +31 30 692 5790
Fax: +31 30 693 2081
vsppub@brill.nl
www.vsppub.com
www.brill.nl

First published in 2004

ISBN 90-6764-398-X

Printed in The Netherlands by Ridderprint bv, Ridderkerk.

Contents

Foreword

This book is the eighth in the series of monographs 'Modern Probability and Statistics' following the books

- V. M. ZOLOTAREV, *Modern Theory of Summation of Random Variables*;
- V. V. SENATOV, *Normal Approximation: New Results, Methods and Problems*;
- V. M. ZOLOTAREV AND V. V. UCHAIKIN, *Chance and Stability. Stable Distributions and their Applications*;
- N. G. USHAKOV, *Selected Topics in Characteristic Functions.*
- V. E. BENING, *Asymptotic Theory of Testing Statistical Hypotheses: Efficient Statistics, Optimality, Power Loss, and Deficiency.*
- G. L. SHEVLYAKOV AND N. O. VILCHEVSKI, *Robustness in Data Analysis: Criteria and Methods*;
- V. E. BENING AND V. YU. KOROLEV, *Generalized Poisson Models and their Applications in Insurance and Finance.*

The Russian school of probability theory and mathematical statistics made a universally recognised contribution to these sciences. Its potentialities are not only very far from being exhausted, but are still increasing. During last decade there appeared many remarkable results, methods and theories which undoubtedly deserve to be presented in monographic literature in order to make them widely known to specialists in probability theory, mathematical statistics and their applications.

However, due to recent political changes in Russia followed by some economic instability, for the time being, it is rather difficult to organise the publication of a scientific book in Russia now. Therefore, a considerable stock of knowledge accumulated during last years yet remains scattered over various scientific journals. To improve this situation somehow, together with the VSP publishing house and first of all, its director, Dr. Jan Reijer Groesbeek who with readiness took up the idea, we present this series of monographs.

The scope of the series can be seen from both the title of the series and the titles of the published and forthcoming books:

- J. Grandell, V. Korolev, R. Norberg, H.-P. Schmidli, Eds., *Selected Topics in Insurance Mathematics*;
- Yu. S. Khokhlov, *Generalizations of Stable Distributions: Structure and Limit Theorems*;
- A. V. Bulinski, *Limit Theorems for Associated Random Variables*;
- B. V. Gnedenko and A. N. Kolmogorov, *Limit Distributions for Sums of Independent Random Variables*, 2nd English edition.
- M. V. Mikhalevich and I. V. Sergienko, *Randomized Methods of Optimization through Preference Relations.*
- E. V. Morozov, *General Queueing Networks: the Method of Regenerative Decomposition*;
- A. N. Chuprunov, *Random Processes Observed at Random Times*;
- D. H. Mushtari, *Probabilities and Topologies on Linear Spaces*;
- V. G. Ushakov, *Priority Queueing Systems*;
- V. Yu. Korolev and V. M. Kruglov, *Random Sequences with Random Indices*;
- Yu. V. Prokhorov and A.P. Ushakova, *Reconstruction of Distribution Types*;
- L. Szeidl and V.M. Zolotarev, *Limit Theorems for Random Polynomials and Related Topics*;

as well as many others.

To provide high-qualified international examination of the proposed books, we invited well-known specialists to join the Editorial Board. All of them kindly agreed, so now the Editorial Board of the series is as follows:

We hope that the books of this series will be interesting and useful to both specialists in probability theory, mathematical statistics and those professionals who apply the methods and results of these sciences to solving practical problems. Of course, the choice of authors primarily from Russia is due only to the reasons mentioned above and by no means signifies that we prefer to keep to some national policy. We invite authors from all countries to contribute their books to this series.

V. Yu. Korolev,
V. M. Zolotarev,
Editors-in-Chief

Moscow, June 2003.

Introduction

Queueing theory is an applied mathematical discipline dealing with the performance of technical systems, which, in what follows, are referred to as *queueing systems* for processing of flows of *customers*. Random fluctuations in customer arrival and service processes play a pivotal role here.

To demonstrate the need for the queueing theory and the consequences that arise if the randomness is discarded in investigating the performance of queueing systems, let us consider a simple *server* at which customers arrive in a flow. Assume that prolonged observation has revealed that customers arrive at the server at a constant rate of six per hour. What then must be the server capacity to cope with this flow efficiently? The answer is self-evident: the server on the average must handle six customers per hour, i.e., turn out one customer every ten minutes. A thorough researcher would certainly provide a small margin—say, ten percent—for unforeseen events and suggest that the server must process one customer in nine minutes. Further increase of the performance would hardly yield any advantage since in this case the server would remain idle for a long time. Therefore, the server on the average must handle one customer in nine minutes; then customers will not line up at the server and the server will idle for six minutes in every hour.

In practice, however, the following fact is observed. Indeed, the server is free for 10% of time, but in many cases long queues appear at the server. For the above initial data, in particular, an eight-customer queue, on average, accumulates at the server if the input flow is Poisson and service is exponential (see Section 3.1). The reason for the accumulation of customers is the randomness in customer arrival and service processes.

It is easy to predict the further course of events. Since it is the random events that are to be blamed and since it is the probability theory that deals with random events, queueing systems must be studied using probabilistic techniques. Thus emerged another branch of the probability theory, the queueing theory, whose founder is believed to be an employee of the Copenhagen Telephone Company, the eminent Danish scientist A. K. Erlang. He was the first who used Markov processes with a discrete (finite or countable) state set to describe queueing systems. The use of discrete states is quite understandable, because telephone systems were the first to use the results of the queueing theory; at present, the application of the queueing theory embraces data, computer, and other systems and networks.

Queueing theory reached its peak in 1950–1970 when numerous papers and monographs devoted to various queueing problems had been published. In turn, queueing theory has stimulated the development of other fields in probability theory, e.g., the theory of random processes.

Later, interest in queueing theory somewhat waned for several reasons. Let us examine one of them—a mathematical one. On the one hand, a typical feature of queueing problems is that almost every queueing system requires its own special research tools and, on the other hand, the great interest in queueing theory has already yielded solutions to many problems that allow simple solutions—especially, in the computational sense. Moreover, analytical methods of studying queueing systems was eventually enhanced with a serious competing tool—simulation modelling.

Recently, interest in queueing theory has been revived not only as a result of new applied problems related, particularly, to the development and application of computers, but also due to the advent of new mathematical approaches to their solution. One such approach is the algorithmic approach, which is a consequence of the extensive use of computers, especially personal computers, in research and it provides solutions to queueing problems in the form of computer algorithms. The algorithmic approach, though inferior to the traditional analytical methods as regards clarity of results, applicability for optimisation, etc., has an indisputable advantage of being oriented towards developing applied software and tables, appreciated in applications much higher than even very elegant formulas.

Our aim in this book is to familiarise readers with certain tools of the queueing theory (including the latest techniques) and the results of their application. Undoubtedly, queueing problems are unwieldily diverse, and it is hardly worthwhile to describe even their formulations to the beginner. Therefore, we only consider the problems and approaches that, in our opinion, are frequently used and fruitful in queueing theory.

In mathematics, we often observe the following fact: the more recent the field of a science, the larger the amount of auxiliary information required to investigate it. This is also true of the queueing theory. Despite the apparent simplicity of formulations and even some solution methods, queueing theory makes use of numerous mathematical disciplines, of which probability theory is the pivotal. Therefore, the main body of the book is preceded by Chapter 1 of introductory nature devoted to preliminary topics from the probability theory. Prerequisites from other mathematical disciplines needed for understanding the book are fully covered by a standard university course of mathematics. Exceptional topics, whenever they appear in the text, are explained in detail. Note that Chapter 1 is of a collection of reference materials, and a prepared reader can start immediately from Chapter 2, turning to Chapter 1 whenever necessary.

Chapter 2 gives a formal description of queueing systems, their defining parameters, and performance indices. The main attention in this chapter is paid to the customer input flow because whether or not a model adequately describes a technical system largely depends on whether or not the flow is described correctly. Here, we present the Kendall classification of simple queueing systems and describe some probabilistic distributions widely used in the sequel.

Chapters 3 and 4 are devoted to the so-called Markov queueing systems: simplest models are described in Chapter 3 and more intricate ones in Chapter 4. They are investigated using a relatively simple mathematical apparatus. In particular, steady-state characteristics of queueing systems are determined from a system of equilibrium equations, which is a system of linear algebraic equations. Nonetheless, the problem of dimensionality, which is typical of such cases, is encountered here too. In queueing theory, this problem is surmounted by radically new methods, which as applied to particular queueing systems are illustrated in Chapter 4.

Chapter 5 gives a detailed analysis of a system, which in the Kendall classification is denoted by $M/G/1/\infty$. Precisely this model is a benchmark for testing and elaborating new methods of the queueing theory. On the basis of the results derived in Chapter 5, other 'traditional' queueing systems are investigated in Chapter 6.

Chapters 7 and 8 discuss in detail the $MAP/G/1/r \leq \infty$ queueing system with Markov input flow: the former deals with finite-buffer systems and the latter with infinite-buffer systems. Conceptually, these chapters have certain points in common with Chapter 5: a single-server queueing system with Markov arrivals in both these chapters is investigated by the same tools. In fact, the passage from the Poisson flow to a more general Markov flow has been shown to have no impact on the logic of the application of well-known approaches, but presupposes that the subject matter is expounded in matrix, rather than scalar, form. The Markov input flow is currently used in computer network modelling.

Chapter 9 is devoted to a more general $BMAP/SM/1/r \leq \infty$ system with batch Markov arrival process and semi-Markov service. Great attention is also paid to the $MAP/G/1/r$ queueing systems with special service disciplines such as priority systems, systems with a service discipline different from service according to the order of arrival, including priority service and service with retrial customers, and others. These disciplines may sometimes substantially enhance the performance of queueing systems with virtually no additional technical innovation.

Finally, the last chapter provides an introduction to queueing networks.

The bibliography includes the most popular textbooks, monographs, and some papers on queueing theory. Sources mentioned in the text and original papers have been used in preparing the manuscript of the book.

Dr. V. A. Naumov co-authored Sections 2.8 and 4.6.

The authors are deeply indebted to Dr. A. V. Kolchin who kindly agreed to edit this manuscript and wasted his summer vacation preparing it for publication. Many thanks are due to Prof. V. Yu. Korolev whose energy, in many aspects, made this publication possible.

1

Probabilistic apparatus of the queueing theory

The present chapter, which is of reference nature, describes in concise terms and without any proof the relatively general knowledge of the characteristic transformations of random variables and processes that are most frequently used by the queueing theory. Explanations are offered only for the results that are immediately used to derive and solve the equations describing the queueing system behaviour and provide a better insight into the physical picture of the phenomenon. The choice of material was dictated mostly by the fact that the general mathematical courses usually cover these divisions of the probability theory incompletely. The last section of this chapter defines the Kronecker matrix product which is used for compact notation of simultaneous equilibrium equations and the expressions of some characteristics of complicated queueing systems and describes the main properties of this product.

1.1. Characteristic transformations

The characteristic transformations of random variables include the characteristic function, Laplace–Stieltjes transform, and generating function. They define uniquely the distribution function of a random variable. Here, any random variable has a characteristic function, a nonnegative random variable has the Laplace–Stieltjes transform, and a nonnegative integer-valued random variable has the generating function. Since the queueing theory deals mostly with nonnegative and nonnegative integer-valued random variables, one makes use mostly of the Laplace–Stieltjes transform and generating function.

In this section we define also the Laplace transform which, notwithstanding the fact that it is not a characteristic transform of random variables, has the same basic properties and will be used below to study the transient characteristics of queueing system operation. We note that the characteristic transforms apply both to the random variables and distribution functions. Therefore, in what follows we also use the term 'characteristic transforms (characteristic function, Laplace–Stieltjes transform, and generating function) of the distribution function.' The characteristic transforms often allow one to represent in a simpler form the solutions of complicated equations for the desired queueing system characteristics.

Practical use of the characteristic transforms is hindered by the need for their inversion. In our opinion, the question here lies not in the excessive complexity of this problem which can be tackled to advantage by the modern computational mathematics and computers, but in the unwillingness of the majority of experts to take pains of bridging the almost shallow creek dividing the theory and practice of solving this problem.

1.1.1. Characteristic function

By the *characteristic function* of a random variable ξ is meant the function

$$\tilde{\alpha}(t) = \mathbf{E}e^{it\xi} = \int_{-\infty}^{\infty} e^{itx} dA(x),$$

where $A(x)$ is the distribution function of ξ and t is a real number.

The characteristic function has the following properties:

(1) $\tilde{\alpha}(t)$ is a continuous function, $\tilde{\alpha}(0) = 1$, $|\tilde{\alpha}(t)| \leq 1$, $-\infty < t < \infty$.

(2) If ξ_1 and ξ_2 are independent random variables with the characteristic functions $\tilde{\alpha}_1(t)$ and $\tilde{\alpha}_2(t)$, respectively, then their sum $\xi = \xi_1 + \xi_2$ has the characteristic function $\tilde{\alpha}(t) = \tilde{\alpha}_1(t)\tilde{\alpha}_2(t)$.

(3) If the random variable ξ has a (finite) moment $a^{(n)} = \mathbf{E}\xi^n$ of the nth order, then the characteristic function is n times differentiable and $\tilde{\alpha}^{(k)}(0) = i^k a^{(k)}$, $k \leq n$. For even n, the opposite is valid as well: if there exists a (finite) derivative $\tilde{\alpha}^{(n)}(0)$, then the random variable ξ has a (finite) moment of the nth order.

If the random variable ξ is continuous, then its characteristic function

$$\tilde{\alpha}(t) = \int_{-\infty}^{\infty} e^{itx} a(x)\, dx$$

coincides to within the factor $1/\sqrt{2\pi}$ with the Fourier transform of the distribution density $a(x) = A'(x)$. The main disadvantage of the characteristic function lies in the need for using the complex-valued function.

1.1.2. Laplace transform

By the *Laplace transform* of a function $p(x)$, $x \geq 0$, is meant the function

$$\pi(s) = \int_{0}^{\infty} e^{-sx} p(x)\, dx.$$

We will not emphasise the general conditions for existence of the Laplace transform. For our purposes, in most cases continuity and boundedness of $p(x)$ are sufficient for the Laplace transform to be defined for all $s > 0$.

The Laplace transform has the following properties:

(1) If $\pi(s)$ is the Laplace transform of a function $p(x)$, then the derivative $p'(x)$ has the Laplace transform $s\pi(s) - p(0)$.

(2) If $p(x)$ is the convolution of functions $p_1(x)$ and $p_2(x)$, that is,

$$p(x) = \int_0^x p_1(x-y)p_2(y)\,dy,$$

then its Laplace transform is of the form $\pi(s) = \pi_1(s)\pi_2(s)$, where $\pi_1(s)$ and $\pi_2(s)$ are the Laplace transforms of the functions $p_1(x)$ and $p_2(x)$ respectively.

(3) If there exists $\lim_{x\to\infty} p(x)$, then there exists $\lim_{s\to 0} s\pi(s) = \lim_{x\to\infty} p(x)$.

To invert the Laplace transform, one has to pass to the complex argument s. Then under the above assumptions the Laplace transform $\pi(s)$ is an analytic function in the domain $\Re s > 0$.

1.1.3. Laplace–Stieltjes transform

By the *Laplace–Stieltjes transform* of a nonnegative random variable ξ is meant the function

$$\alpha(s) = \mathbf{E}e^{-s\xi} = \int_0^\infty e^{-sx}\,dA(x), \qquad s \geq 0.$$

If ξ is a continuous random variable, then the Laplace–Stieltjes transform

$$\alpha(s) = \int_0^\infty e^{-sx} a(x)\,dx$$

coincides with the Laplace transform of its distribution density $a(x) = A'(x)$.

Similar to the characteristic function, the Laplace–Stieltjes transform has the following properties:

(1) $\alpha(s)$ is a positive continuous decreasing function, $\alpha(0) = 1$.

(2) If ξ_1 and ξ_2 are independent nonnegative random variables with the respective Laplace–Stieltjes transforms $\alpha_1(s)$ and $\alpha_2(s)$, then their sum $\xi = \xi_1 + \xi_2$ has the Laplace–Stieltjes transform $\alpha(s) = \alpha_1(s)\alpha_2(s)$.

(3) The random variable ξ has a (finite) moment $a^{(n)} = \mathbf{E}\xi^n$ of the nth order if and only if there exists a (finite) derivative $\alpha^{(n)}(0)$ and $a^{(n)} = (-1)^n\alpha^{(n)}(0)$.

The edge of the Laplace–Stieltjes transform on the characteristic function lies in the fact that, since it is a real function of a real argument, it is easier to deal with it than with the characteristic function.

As it is the case with the Laplace transform, one sometimes has to consider the Laplace–Stieltjes transform $\alpha(s)$ as a function of the complex variable s. Then $\alpha(s)$ is analytic on the half-plane $\Re s > 0$ and continuous up to the boundary $\Re s = 0$, and $|\alpha(s)| \leq 1$. We also note that in this case the value of the Laplace–Stieltjes transform $\alpha(s)$ at the point $s = it$ coincides with the value of the characteristic function $\tilde{\alpha}(t)$ at the point $-t$.

The formula

$$\alpha_{\mathrm{LST}}(s) = s\alpha_{\mathrm{LT}}(s)$$

relating the Laplace–Stieltjes transform $\alpha_{\mathrm{LST}}(s)$ of the random variable ξ with the Laplace transform of its distribution function

$$\alpha_{\mathrm{LT}}(s) = \int_0^\infty e^{-sx} A(x)\,dx$$

sometimes may be of advantage.

1.1.4. Generating function

By the *generating function* of a nonnegative integer random variable ξ with distribution

$$p_n = \mathbf{P}\{\xi = n\}, \qquad n = 0, 1, \ldots,$$

is meant the function

$$P(z) = \mathbf{E}z^{\xi} = \sum_{n=0}^{\infty} p_n z^n.$$

It is defined for all z from the interval $|z| \le 1$. We consider here, as usual, the real argument z, although sometimes we turn to the complex z in which case the generating function $P(z)$ is defined on the disk $|z| \le 1$. The Laplace–Stieltjes transform is obtained from the generating function if one assumes that $z = e^{-s}$.

We give now the properties of the generating function required in what follows:

(1) $P(z)$ is continuous in the segment $[-1, 1]$ (if $P(z)$ is considered as a function of the complex argument z, then it is analytic inside the disk $|z| < 1$ and continuous right up to the circle $|z| = 1$), $P(1) = 1$ and $|P(z)| \le 1$, $|z| \le 1$.

(2) If the random variable ξ is a sum of two independent nonnegative integers with the generating functions $P_1(z)$ and $P_2(z)$, respectively, then its generating function $P(z)$ is of the form $P(z) = P_1(z)P_2(z)$.

(3) The random variable ξ has a finite moment $a^{(n)} = \mathbf{E}\xi^n$ of the nth order if and only if there exists a (finite left-hand) derivative $P^{(n)}(1)$. Here,

$$P^{(n)}(1) = \mathbf{E}\xi(\xi - 1)\cdots(\xi - n + 1) = \mathbf{E}(\xi)_n;$$

the last quantity is called the *factorial moment* of the nth order. The initial and factorial moments can be expressed in terms of each other. For example,

$$a = \mathbf{E}\xi = \mathbf{E}(\xi)_1, \qquad a^{(2)} = \mathbf{E}\xi^2 = \mathbf{E}(\xi)_2 + \mathbf{E}(\xi)_1.$$

Introduction of the generating function is due to the fact that it is even simpler to handle than the Laplace–Stieltjes transform.

1.1.5. Inversion formulas

We conclude this section by presenting the inversion formulas which allow one to go from the characteristic transforms to the distributions of the random variables and making a little comment about their application.

The characteristic function $\tilde{\alpha}(t)$ is *inverted* using the formula

$$A(x_1) - A(x_2) = \frac{1}{2\pi i} \lim_{X\to\infty} \int_{-X}^{X} \frac{e^{-itx_2} - e^{-itx_1}}{t} \tilde{\alpha}(t)\,dt,$$

where x_1 and x_2 are arbitrary continuity points of the distribution function $A(x)$. It is worthwhile to note that numerical inversion of the characteristic function often encounters with difficulties involved in convergence of the integral. In this case, one can make use of the Berry–Esseen method (Feller, 1966). The *inversion formula of the Laplace transform* $\pi(s)$ is as follows:

$$p(x) = \frac{1}{2\pi i} \lim_{X\to\infty} \int_{\delta-iX}^{\delta+iX} e^{sx}\pi(s)\,ds,$$

where δ is an arbitrary positive number and integration is carried out on any (smooth) curve lying in the half-plane $\Re s > 0$ and connecting the points $\delta - iX$ and $\delta + iX$ (it is usually convenient to take the segment connecting these points as such curve).

The following theorems may be of utility for inversion of the Laplace transform.

THEOREM 1.1.1 (first expansion theorem). *Let there exist s_0 such that for $s > s_0$ the Laplace transform $\pi(s)$ is expandable into a series in the powers of $1/s$:*

$$\pi(s) = \sum_{n=0}^{\infty} \frac{c_n}{s^{n+1}}.$$

Then the inverse transform $p(x)$ is also representable as the converging (for any x) series

$$p(x) = \sum_{n=0}^{\infty} c_n \frac{x^n}{n!}.$$

THEOREM 1.1.2 (second expansion theorem). *Let the Laplace transform $\pi(s)$ be a rational function, that is, a function representable as a sum of the partial fractions*

$$\pi(s) = \sum_{j=1}^{m} \sum_{l=1}^{k_j} \frac{c_{jl}}{(s - s_j)^l}.$$

Then the inverse transform $p(x)$ is obtained using the formula

$$p(x) = \sum_{j=1}^{m} \sum_{l=1}^{k_j} c_{jl} \frac{x^{l-1}}{(l-1)!} e^{s_j x}.$$

THEOREM 1.1.3 (third expansion theorem). *Let $\pi(s)$ be a function which—with the exception of a finite number of singular points $s_1, \dots, s_n$—is analytic on the extended complex plane. Then*

$$p(x) = \sum_{k=1}^{n} \operatorname{res}\{e^{s_k x}\pi(s_k)\},$$

where $\operatorname{res}\{f(s_k)\}$ *is the residue of the function* $f(s)$ *at the singular point* s_k.

To invert *the Laplace–Stieltjes transform* $\alpha(s)$, one can use either any formula for inversion of the characteristic function or the formula of inversion of the Laplace transform (for example, if there exists a continuous bounded density $a(x)$).

The values of the probabilities $p_n = \mathbf{P}\{\xi = n\}$, $n = 0, 1, \dots,$ of the discrete random variable ξ are found by differentiating an appropriate number of times the generating function $P(z)$ at zero:

$$p_n = \frac{1}{n!}P^{(n)}(0).$$

In its turn, the integral Cauchy formula

$$P^{(n)}(0) = \frac{n!}{2\pi i}\int_L \frac{P(z)}{z^{n+1}}\,dz,$$

where L is a closed (smooth) curve bypassing the point 0 lying in the disk $|z| \le 1$, can often be helpful for numerical determination of the derivative $P^{(n)}(0)$. The unit circle $|z| = 1$ itself is usually taken as the loop L.

We do not dwell on the methods of inversion of the characteristic transforms and refer the interested reader to special publications.

1.2. Exponential and Poisson distributions

In this section we consider the properties of the exponential distribution. Special emphasis on the exponential distribution is due here to the 'memoryless' property with which the readers must be familiar from the probability theory. This property suggests that the exponential distribution is the 'parent' of the Markov processes which in turn underlie the queueing theory. We also recall the definition of the Poisson distribution which, as seen in what follows, is closely related to the exponential distribution.

1.2.1. Exponential distribution

The exponential distribution with parameter λ has the distribution function $G(x)$ and distribution density $g(x) = G'(x)$ of the form

$$G(x) = \begin{cases} 0, & x \le 0; \\ 1 - e^{-\lambda x}, & x > 0, \end{cases} \qquad g(x) = \begin{cases} 0, & x \le 0; \\ \lambda e^{-\lambda x}, & x > 0. \end{cases}$$

The expectation and variance of the random variable ξ distributed exponentially with the parameter λ obey the formulas

$$\mathbf{E}\xi = \int_0^\infty x\lambda e^{-\lambda x}\,dx = \frac{1}{\lambda}, \qquad \mathbf{Var}\,\xi = \int_0^\infty (x - \mathbf{E}\xi)^2 \lambda e^{-\lambda x}\,dx = \frac{1}{\lambda^2}.$$

LEMMA 1.2.1 (memoryless condition). *The random variable* ξ *is distributed exponentially if and only if the following memoryless condition is satisfied: for any* $y > 0$, *the conditional distribution*

$$G_y(x) = \mathbf{P}\{\xi - y < x \mid \xi > y\} = \begin{cases} 0, & x \le 0; \\ \dfrac{G(x+y) - G(y)}{1 - G(y)}, & x > 0, \end{cases}$$

coincides with the unconditional distribution $G(x) = \mathbf{P}\{\xi < x\}$, *that is,*

$$G_y(x) = G(x). \tag{1.2.1}$$

PROOF. Let us assume that $\bar{G}(x) = 1 - G(x)$, $\bar{G}_y(x) = 1 - G_y(x)$. Then condition (1.2.1) is equivalent to the condition $\bar{G}_y(x) = \bar{G}(x)$.

First, we prove that the last condition is necessary for $G(x)$ to be exponential. Indeed, by virtue of the definition of the conditional probability

$$\begin{aligned}\bar{G}_y(x) &= \mathbf{P}\{\xi > x + y \mid \xi > y\}\\ &= \frac{\mathbf{P}\{\xi > x + y, \xi > y\}}{\mathbf{P}\{\xi > y\}} = \frac{\mathbf{P}\{\xi > x + y\}}{\mathbf{P}\{\xi > y\}}, \qquad x > 0.\end{aligned}$$

Now, we recall the definition of the exponential law in order to obtain

$$\frac{\mathbf{P}\{\xi > x + y\}}{\mathbf{P}\{\xi > y\}} = \frac{e^{-\lambda(x+y)}}{e^{-\lambda y}} = e^{-\lambda x} = \bar{G}(x), \qquad x > 0,$$

hence,

$$\bar{G}_y(x) = \bar{G}(x),$$

which is what we set out to prove.

To prove that condition (1.2.1) is sufficient, we make use of the equality

$$\begin{aligned}\bar{G}(x + y) &= \mathbf{P}\{\xi > x + y\} = \mathbf{P}\{\xi > x + y \mid \xi > y\}\mathbf{P}\{\xi > y\}\\ &= \mathbf{P}\{\xi - y > x \mid \xi > y\}\mathbf{P}\{\xi > y\} = \bar{G}_y(x)\bar{G}(y).\end{aligned}$$

From condition (1.2.1) we obtain

$$\bar{G}(x + y) = \bar{G}(x)\bar{G}(y), \qquad x, y > 0. \tag{1.2.2}$$

By virtue of the properties of the distribution function, $\bar{G}(x)$ is a non-increasing function of x. All non-increasing solutions of equation (1.2.2) are known to be of the form

$$\bar{G}(x) = 0, \qquad x > 0,$$

or

$$\bar{G}(x) = e^{-\lambda x}, \quad x \geq 0, \quad 0 \leq \lambda < \infty.$$

It is reasonable to ignore the first solution. Moreover, in the second solution one must reject the case $\lambda = 0$. Therefore, we find that condition (1.2.1) is sufficient, which proves the lemma. □

Taking ξ as the time of customer service, the memoryless condition can be interpreted as follows: *the distribution* $G_y(x)$ *of the residual customer service time* $\xi - y$ *is independent of the time* y *during which it was already in service.*

In what follows, we employ a condition equivalent to the memoryless condition: if a customer was not served at the time y, then the *probability* $G_y(\Delta)$ *of completing service in the 'small' time interval* $(y, y + \Delta)$ *is independent of* y *and equal to* $\lambda\Delta$ *to within* $o(\Delta)$.

Indeed, for the memoryless case

$$G_y(\Delta) = G(\Delta) = 1 - e^{-\lambda\Delta} = 1 - [1 - \lambda\Delta + o(\Delta)] = \lambda\Delta + o(\Delta).$$

Vice versa, if

$$G_y(\Delta) = \lambda\Delta + o(\Delta),$$

then in view of the definition of $G_y(x)$ we obtain

$$G_y(\Delta) = \frac{G(y+\Delta) - G(y)}{1 - G(y)} = \lambda\Delta + o(\Delta).$$

Hence,

$$\frac{\bar{G}(y+\Delta) - \bar{G}(y)}{\Delta} = -\lambda\bar{G}(y) + o(1).$$

By letting Δ tend to zero, we obtain the differential equation

$$\bar{G}'(y) = -\lambda\bar{G}(y)$$

leading with regard for the initial condition $\bar{G}(0) = 1 - G(0) = 1$ to the equality $\bar{G}(y) = e^{-\lambda y}$, $y > 0$, which by virtue of Lemma 1.2.1 is equivalent to the memoryless condition.

Till the end of this subsection, we denote by ξ_i, $i = 1, \dots, m$, independent exponentially distributed random variables with parameters λ_i, $i = 1, \dots, m$, and by $\xi = \min_{1 \le i \le m} \xi_i$, the minimum of these random variables.

LEMMA 1.2.2 (minimum of exponentially distributed variables). *The random variable* ξ *has an exponential distribution with the parameter*

$$\lambda = \sum_{i=1}^{m} \lambda_i.$$

PROOF. Validity of the lemma follows from the chain of elementary inequalities and the definition of the exponential law:

$$\begin{aligned} \mathbf{P}\{\xi > x\} &= \mathbf{P}\{\min(\xi_1, \dots, \xi_m) > x\} = \mathbf{P}\{\xi_1 > x, \dots, \xi_m > x\} \\ &= \mathbf{P}\{\xi_1 > x\} \cdots \mathbf{P}\{\xi_m > x\} = e^{-\lambda_1 x} \cdots e^{-\lambda_m x} = e^{-\lambda x}, \qquad x > 0. \end{aligned}$$

□

LEMMA 1.2.3. *The following formula is valid:*

$$\mathbf{P}\{\xi = \xi_i\} = \frac{\lambda_i}{\lambda}, \qquad i = 1, \dots, m.$$

PROOF. According to the definitions of the exponential distribution and the distribution of the function of random variables, we obtain

$$\begin{aligned}
\mathbf{P}\{\xi = \xi_i\} &= \mathbf{P}\{\xi_i \le \xi_1, \dots, \xi_i \le \xi_{i-1}, \xi_i \le \xi_{i+1}, \dots, \xi_i \le \xi_m\} \\
&= \underset{\substack{0 \le x_i \le x_j \\ j \ne i}}{\int \cdots \int} \lambda_1 \cdots \lambda_m e^{-\lambda_1 x_1} \cdots e^{-\lambda_m x_m}\, dx_1 \cdots dx_m \\
&= \int_0^\infty \lambda_i e^{-\lambda_i x_i} \int_{x_i}^\infty \lambda_1 e^{-\lambda_1 x_1}\, dx_1 \cdots \int_{x_i}^\infty \lambda_{i-1} e^{-\lambda_{i-1} x_{i-1}}\, dx_{i-1} \\
&\quad \times \int_{x_i}^\infty \lambda_{i+1} e^{-\lambda_{i+1} x_{i+1}}\, dx_{i+1} \cdots \int_{x_i}^\infty \lambda_m e^{-\lambda_m x_m}\, dx_m\, dx_i \\
&= \int_0^\infty \lambda_i e^{-\lambda_i x_i} e^{-(\lambda_1 + \cdots + \lambda_{i-1} + \lambda_{i+1} + \cdots + \lambda_m) x_i}\, dx_i = \lambda_i \int_0^\infty e^{-\lambda x_i}\, dx_i = \frac{\lambda_i}{\lambda},
\end{aligned}$$

which proves the lemma. □

Now, we assume that $\eta_j = \xi_j - \xi_i$, $i, j = 1, \dots, m$.

LEMMA 1.2.4. *The following equality is valid:*

$$\begin{aligned}
&\mathbf{P}\{\eta_1 > x_1, \dots, \eta_{i-1} > x_{i-1}, \eta_{i+1} > x_{i+1}, \dots, \eta_m > x_m \mid \xi = \xi_i\} \\
&\quad = e^{-\lambda_1 x_1} \cdots e^{-\lambda_{i-1} x_{i-1}} e^{-\lambda_{i+1} x_{i+1}} \cdots e^{-\lambda_m x_m}, \quad x_j > 0, \quad j = 1, \dots, m, \quad j \ne i.
\end{aligned}$$

PROOF. We follow the way of proof of the preceding lemma. Indeed,

$$\begin{aligned}
&\mathbf{P}\{\eta_1 > x_1, \dots, \eta_{i-1} > x_{i-1}, \eta_{i+1} > x_{i+1}, \dots, \eta_m > x_m \mid \xi = \xi_i\} \\
&= \mathbf{P}\{\xi_i \le \xi_1 - x_1, \dots, \xi_i \le \xi_{i-1} - x_{i-1}, \xi_i \\
&\le \xi_{i+1} - x_{i+1}, \dots, \xi_i \le \xi_m - x_m\} (\mathbf{P}\{\xi = \xi_i\})^{-1} \\
&= \frac{1}{\mathbf{P}\{\xi = \xi_i\}} \underset{\substack{0 \le y_i \le y_j - x_j \\ j \ne i}}{\int \cdots \int} \lambda_1 \cdots \lambda_m e^{-\lambda_1 y_1} \cdots e^{-\lambda_m y_m}\, dy_1 \cdots dy_m \\
&= \frac{1}{\mathbf{P}\{\xi = \xi_i\}} \int_0^\infty \lambda_i e^{-\lambda_i y_i} \int_{y_i + x_1}^\infty \lambda_1 e^{-\lambda_1 y_1}\, dy_1 \cdots \int_{y_i + x_{i-1}}^\infty \lambda_{i-1} e^{-\lambda_{i-1} y_{i-1}}\, dy_{i-1} \\
&\quad \times \int_{y_i + x_{i+1}}^\infty \lambda_{i+1} e^{-\lambda_{i+1} y_{i+1}}\, dy_{i+1} \cdots \int_{y_i + x_m}^\infty \lambda_m e^{-\lambda_m y_m}\, dy_m\, dy_i \\
&= \frac{1}{\mathbf{P}\{\xi = \xi_i\}} e^{-(\lambda_1 x_1 + \cdots + \lambda_{i-1} x_{i-1} + \lambda_{i+1} x_{i+1} + \cdots + \lambda_m x_m)} \\
&\quad \times \int_0^\infty \lambda_i e^{-\lambda_i y_i} e^{-(\lambda_1 + \cdots + \lambda_{i-1} + \lambda_{i+1} + \cdots + \lambda_m) y_i}\, dy_i \\
&= \frac{1}{\mathbf{P}\{\xi = \xi_i\}} e^{-(\lambda_1 x_1 + \cdots + \lambda_{i-1} x_{i-1} + \lambda_{i+1} x_{i+1} + \cdots + \lambda_m x_m)} \lambda_i \int_0^\infty e^{-\lambda y_i}\, dy_i \\
&= e^{-\lambda_1 x_1} \cdots e^{-\lambda_{i-1} x_{i-1}} e^{-\lambda_{i+1} x_{i+1}} \cdots e^{-\lambda_m x_m},
\end{aligned}$$

which is the required result. □

Lemma 1.2.4 implies that if ξ_i is the minimal random variable among ξ_j, $j = 1, \ldots, m$, then subtracting it from the rest of the random variables we obtain new variables $\eta_j = \xi_j - \xi_i$, $j \neq i$, which, like ξ_j, $j \neq i$, are independent and exponentially distributed with parameters λ_j, $j \neq i$. We suggest that the readers prove the following generalisations of Lemmas 1.2.3 and 1.2.4 in order to be confident of their forces.

LEMMA 1.2.5. *The following equality is valid:*

$$\mathbf{P}\{\xi = \xi_i \mid \xi = x\} = \frac{\lambda_i}{\lambda}, \qquad i = 1, \ldots, m.$$

LEMMA 1.2.6. *The following relation is true:*

$$\begin{aligned}\mathbf{P}\{&\eta_1 > x_1, \ldots, \eta_{i-1} > x_{i-1}, \eta_{i+1} > x_{i+1}, \ldots, \eta_m > x_m \mid \xi = \xi_i, \xi = x\} \\ &= e^{-\lambda_1 x_1} \cdots e^{-\lambda_{i-1} x_{i-1}} e^{-\lambda_{i+1} x_{i+1}} \cdots e^{-\lambda_m x_m}, \quad x_j > 0, \quad j = 1, \ldots, m, \quad j \neq i.\end{aligned}$$

Lemma 1.2.5 means that even if we know the value x of ξ, the event $\{\xi = \xi_i\}$ has the probability λ_i / λ independent of x. Lemma 1.2.6 has a similar meaning.

1.2.2. Poisson distribution

The *Poisson distribution* with parameter λ is the distribution of a nonnegative integer random variable ν with the distribution series

$$p_i = \mathbf{P}\{\nu = i\} = \frac{\lambda^i}{i!} e^{-\lambda}, \qquad i \geq 0.$$

Then

$$\mathbf{E}\nu = \sum_{i=0}^{\infty} \frac{i\lambda^i}{i!} e^{-\lambda} = \lambda, \qquad \mathbf{Var}\,\nu = \sum_{i=0}^{\infty} (i - \lambda)^2 \frac{\lambda^i}{i!} e^{-\lambda} = \lambda,$$

that is, the parameter λ of the Poisson law is simultaneously the expectation and the variance of the random variable ν.

1.3. Renewal processes. Regenerative processes

Although the description of the renewal processes relies on the simplest notion of sequence of independent nonnegative random variables having identical distributions, the role of the renewal theory (the division of the probability theory dealing with the renewal processes) and especially of the key renewal theorem in the modern probability theory cannot be underestimated. As will be seen below, the elements of the renewal theory enable us to reduce the study of behaviour of many complex queueing systems to studying their behaviour in a chosen time interval.

To gain a better understanding of the presentation, we first consider the case of discrete (integer-valued) renewal process and then proceed to the general case, although both cases are usually united in one. This approach is justified also by the fact that in different problems we need both modifications of the renewal process, which in turn is related with choosing the method of study of the corresponding queueing system. The section is concluded by the regenerative processes which also prove to be useful for studying various queueing systems, although at first glance they seem to be an insignificant generalisation of the renewal processes.

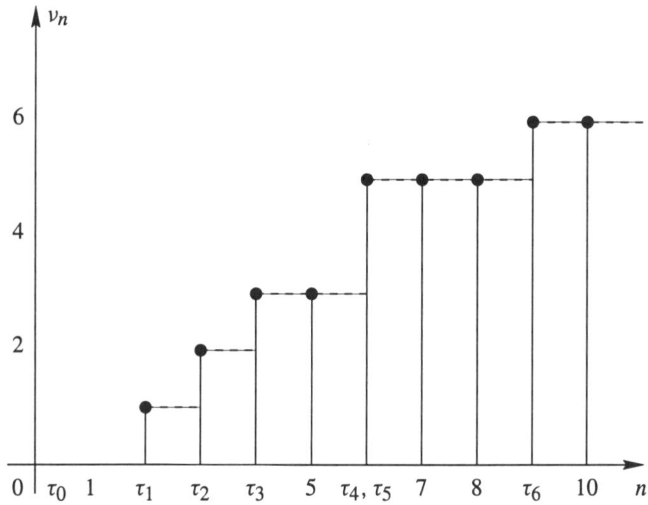

Figure 1.1.

1.3.1. Discrete renewal processes

Let $\{\xi_n,\ n \geq 1\}$ be a sequence of independent nonnegative integer random variables. We denote by $\{g_i^{(n)},\ i \geq 0\}$, $g_i^{(n)} = \mathbf{P}\{\xi_n = i\}$, the distribution series of the random variable ξ_n, $n \geq 1$. We assume that all ξ_n beginning from the second one are distributed identically to the distribution series $\{g_i,\ i \geq 0\}$, that is, $g_i^{(n)} = g_i$, $n \geq 2$. To avoid the trivial case, we assume additionally that $g_0 < 1$. Finally, we assume that there exists no integer l, $l \geq 2$, such that $g_i > 0$ only for $i = jl$, $j \geq 0$. The last assumption is not restrictive because if it is not satisfied, then we can change the 'time scale' and consider only the instants $n = jl$, $j \geq 0$.

We set $\tau_0 = 0$, $\tau_i = \sum_{j=1}^{i} \xi_j$, $i \geq 1$, the instant τ_i being called the *instant of the ith renewal*. The sequence $\{\tau_i,\ i \geq 0\}$ is nonnegative and nondecreasing. The random sequence $\{\nu_n,\ n \geq 0\}$, $\nu_n = \max_{i \geq 0}\{i\colon \tau_i \leq n\}$ is called the (general) *discrete renewal process* (Fig. 1.1). Obviously, ν_n is the number of renewals until the instant n inclusively.

A renewal process $\{\nu_n,\ n \geq 0\}$ is called the *simple* process if $g_i^{(1)} = g_i$, $i \geq 0$, that is, the distribution of the first instant of renewal $\tau_1 = \xi_1$ coincides with the distribution of the random variables ξ_n, $n \geq 2$.

A renewal process $\{\nu_n,\ n \geq 0\}$ is called the *stationary* process if the distribution series $\{g_i^{(1)},\ i \geq 0\}$ of the first instant of renewal $\tau_1 = \xi_1$ obeys the formula

$$g_0^{(1)} = 0, \quad g_i^{(1)} = \frac{1}{g} \sum_{j=i}^{\infty} g_j, \quad i \geq 1,$$

where $g = \mathbf{E}\xi_n = \sum_{i=1}^{\infty} i g_i$ is the mean value of ξ_n, that is, the time between the $(n-1)$th and nth instants of renewal, $n \geq 2$. It goes without saying that when defining the stationary renewal process we must assume that $g < \infty$.

The random variable ν_n has moments of any order. Moreover, for any renewal process $\{\nu_n,\ n \geq 0\}$ and each $n \geq 0$ there exists a number $C = C(n)$ such that $\mathbf{E}\nu_n^k \leq C^k k!$ for all $k \geq 0$.

By the *renewal function* H_n is meant the sequence $\{H_n,\ n \geq 0\}$, $H_n = \mathbf{E}\nu_n$ and by the *renewal series* h_n, the sequence $\{h_n,\ n \geq 0\}$, $h_0 = H_0$, $h_n = H_n - H_{n-1}$, $n \geq 1$. In other words, H_n is the mean number of renewals up to the instant n inclusive, and h_n is the mean number of renewals at the instant n. The renewal function H_n (or which is the same, the renewal series h_n) plays the main part in the renewal theory. In the formulas derived below, h_n can be treated as the probability that a renewal occurs at the nth instant.

The renewal series $\{h_n,\ n \geq 0\}$ satisfies the renewal equation

$$h_n = g_n^{(1)} + \sum_{i=0}^{n} h_i g_{n-i}, \qquad n \geq 0, \tag{1.3.1}$$

which is obtained from the formula of total probability as follows: at the nth instant, renewal occurs if and only if either the first renewal occurs with probability $g_n^{(1)}$ at the nth instant or the preceding renewal occurred with probability h_i at the ith instant, $i = 0, \ldots, n$, and the next one occurs with probability g_{n-i} at the nth instant.

To solve function (1.3.1), it is convenient to use the generating functions

$$G(z) = \sum_{i=0}^{\infty} g_i z^i, \quad G^{(1)}(z) = \sum_{i=0}^{\infty} g_i^{(1)} z^i, \quad H(z) = \sum_{i=0}^{\infty} h_i z^i.$$

We note that in contrast to the generating functions $G(z)$ and $G^{(1)}(z)$, the generating function $H(z)$ is defined only for $|z| < 1$. Then we obtain from (1.3.1)

$$H(z) = G^{(1)}(z) + H(z)G(z),$$

and, therefore,

$$H(z) = \frac{G^{(1)}(z)}{1 - G(z)}. \tag{1.3.2}$$

One can see from (1.3.2), in particular, that for the stationary renewal process

$$H(z) = \lambda \frac{z}{1 - z},$$

where $\lambda = 1/g$; by going to the original, we obtain

$$h_0 = 0, \quad h_n = \lambda, \quad n \geq 1.$$

Therefore, for the stationary renewal process the mean number of renewals h_n, $n \geq 1$, is independent of n and equal to $\lambda = 1/g$. The opposite assertion is valid as well: if $h_0 = 0$ and $h_n = \lambda = 1/g$, $n \geq 1$, then the renewal process is stationary. We note that it is reasonable to call the number λ the *renewal* (process) *intensity*. The asymptotic property of the renewal process is described by the following theorem.

THEOREM 1.3.1 (Blackwell theorem). *For the (general) renewal process* $h_n \xrightarrow[n\to\infty]{} \lambda$.

The following formulation of the Blackwell theorem usually proves beneficial for applications.

THEOREM 1.3.2 (Smith's key renewal theorem). *Let* $\{f_n,\ n \geq 0\}$ *be an arbitrary numerical sequence such that* $\sum_{n=0}^{\infty} |f_n| < \infty$. *Then*

$$\sum_{i=0}^{n} h_i f_{n-i} \xrightarrow[n\to\infty]{} \lambda \sum_{n=0}^{\infty} f_n.$$

Let us consider a corollary of the key renewal theorem. We denote by $\xi^+(n)$ the difference between the instant of the first-after-n renewal and n, that is, $\xi^+(n) = \tau_{\nu_n+1} - n$. It is reasonable to call $\xi^+(n)$ the *skip* over n or *residual waiting time* at the instant n. Let us find the distribution of $\xi^+(n)$. We take $f_i^+(n) = \mathbf{P}\{\xi^+(n) = i\}$, $i \geq 1$, and obtain from the formula of total probability that

$$f_i^+(n) = g_{n+i}^{(1)} + \sum_{j=0}^{n} h_j g_{n+i-j}, \quad i \geq 1.$$

The first addend in this equality tends to zero as $n \to \infty$, and the second addend, to

$$\lambda \sum_{j=0}^{\infty} g_{j+i} = \frac{1}{g} \sum_{j=i}^{\infty} g_j$$

by virtue of the key renewal theorem. Therefore, the limit distribution of the skip coincides with the distribution of the first instant of renewal of the stationary renewal process.

It is possible to demonstrate that in the stationary renewal process for any $n \geq 1$ the distribution of skip precisely coincides with the distribution of the instant of the first renewal. This means, in particular, that the stationary renewal process $\{\nu_n,\ n \geq 0\}$ is a sequence with stationary increments (the finite-dimensional distributions of the increments $\nu_{n+m} - \nu_m$ are independent of m), which justifies the use of the term 'stationary.' The essence of the key renewal theorem consists of that the (general) renewal process (for which $g < \infty$) tends with time to the stationary renewal process. If $g = \infty$, then $\lambda = 0$. Then $g_i(n) \xrightarrow[n\to\infty]{} 0$ for all $i \geq 0$ or, in other words, for $n \to \infty$ the skip $\xi^+(n)$ tends in probability to infinity.

The renewal processes $\{\nu_n,\ n \geq 0\}$ for which $\sum_{i=0}^{\infty} g_i^{(1)} \leq 1$, $\sum_{i=0}^{\infty} g_i < 1$ can be considered in the same way. The last inequality implies that with probability $1 - \sum_{i=0}^{\infty} g_i$ the nth renewal, $n \geq 1$, is the last one. The total number of renewals for such processes on $[0, \infty)$ is finite with probability 1.

1.3.2. Renewal processes (general case)

In actual fact, study of the general renewal processes does not differ from that of the discrete renewal processes.

Let $\{\xi_n,\ n \geq 1\}$ be a sequence of independent nonnegative random variables and $\mathbf{P}\{\xi_1 < x\} = G^{(1)}(x)$, $\mathbf{P}\{\xi_n < x\} = G(x)$, $n \geq 2$, that is, the variables ξ_n are distributed identically beginning from the second one. We assume that $G(0+) < 1$ and, additionally, there exists no l such that the random variables ξ_n, $n \geq 2$, can take the values jl only, $j \geq 0$. If the last assumption is not satisfied, then we arrive at the discrete renewal process.

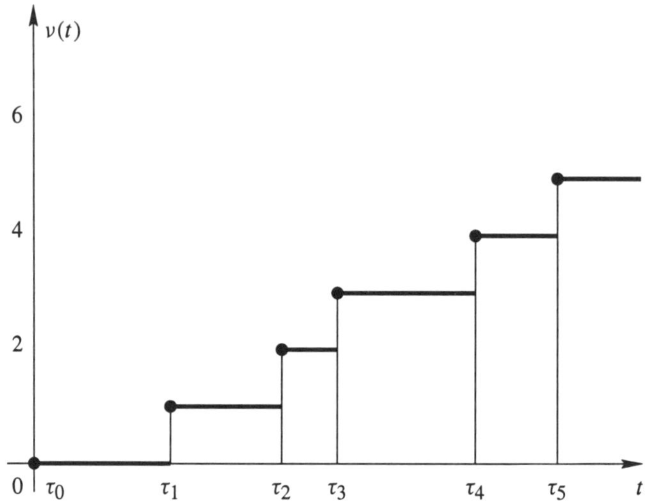

Figure 1.2.

As before, the instants $\tau_0 = 0$, $\tau_i = \sum_{j=1}^{i} \xi_j$, $i \geq 1$, is called the *instants of renewal*. The random process $\{\nu(t),\ t \geq 0\}$, $\nu(t) = \max_{i \geq 0}\{i : \tau_i \leq t\}$ is called the (general) *renewal process* (Fig.1.2).

The renewal process $\{\nu(t),\ t \geq 0\}$, is called *simple* if $G^{(1)}(x) = G(x)$ and *stationary* if $G^{(1)}(x) = \int_0^x [1 - G(y)]\, dy/g, x > 0$, where $g = \int_0^\infty x\, dG(x) = \int_0^\infty [1 - G(x)]\, dx = \mathbf{E}\xi_n$ is the mean time between the nth and $(n+1)$th renewals, $n \geq 1$.

The function $H(t) = \mathbf{E}\nu(t)$, $t \geq 0$, is called the *renewal function*. If there exists the derivative $h(t) = H'(t)$, it is called the *renewal density*. If there exists the renewal density $h(t)$, then the differential $dH(t) = h(t)\, dt$ has the sense of the probability that a renewal takes place at $(t, t + dt)$; $dH(t)$ admits a similar treatment (see the last subsection) also in the general case.

The renewal function $H(t)$ satisfies the *renewal equation*

$$H(t) = G^{(1)}(t) + \int_0^t G(t - x)\, dH(x). \tag{1.3.3}$$

Passing to the Laplace–Stieltjes transform

$$\varkappa(s) = \int_0^\infty e^{-st}\, dH(t), \qquad \gamma^{(1)}(s) = \int_0^\infty e^{-st}\, dG^{(1)}(t),$$

$$\gamma(s) = \int_0^\infty e^{-st}\, dG(t),$$

we obtain from (1.3.3)

$$\varkappa(s) = \gamma^{(1)}(s) + \gamma(s)\varkappa(s);$$

hence it follows that

$$\varkappa(s) = \frac{\gamma^{(1)}(s)}{1 - \gamma(s)}. \tag{1.3.4}$$

It follows from (1.3.4) that for the stationary renewal process

$$\varkappa(s) = \frac{\frac{1-\gamma(s)}{gs}}{1-\gamma(s)} = \frac{\lambda}{s},$$

where $\lambda = 1/g$ is the *renewal intensity*. Therefore,

$$H(x) = \lambda x,$$

that is, the renewal function $H(x)$ is linear with the coefficient $\lambda = 1/g$ equal to the renewal intensity. On the contrary, if $H(x) = \lambda x$, then the renewal process is stationary.

THEOREM 1.3.3 (Blackwell theorem). *For any x, the renewal function of the (general) renewal process satisfies the limit relation*

$$H(t+x) - H(t) \underset{t\to\infty}{\longrightarrow} \lambda x.$$

An equivalent formulation of the Blackwell theorem is as follows.

THEOREM 1.3.4 (Smith's key renewal theorem). *Let the function $f(x)$ satisfy the following conditions:*

(1) $|f(x)| \le f^+(x)$, *$f^+(x)$ being a monotone function and* $\int_0^\infty f^+(x)\,dx < \infty$.

(2) *$f(x)$ is monotone or continuous.*

Then

$$\int_0^t f(t-x)\,dH(x) \underset{t\to\infty}{\longrightarrow} \lambda \int_0^\infty f(x)\,dx.$$

In the general case, the key renewal theorem also implies that the renewal process $\{\nu(t),\ t \ge 0\}$ converges with time (for $g < \infty$) to the stationary process. The remaining remarks of the last subsection are applied to the general case without any further comments.

1.3.3. Regenerative processes

We confine our consideration only to the case of continuous time, although all the following presentation is completely applicable to the case of discrete time. We define now the regenerative process. Let there be a sequence $\{\xi_i,\ i \ge 1\}$ of nonnegative random variables ξ_i and a sequence $\{\eta_i(t),\ 0 \le t < \xi_{i+1},\ i \ge 0\}$ of random processes $\eta_i(t)$ defined in the intervals $[0, \xi_{i+1})$. We assume that the following conditions are satisfied:

(1) Beginning from the second pair, the pairs $\{(\xi_{i+1}, \eta_i(t)),\ 0 \le t < \xi_{i+1}\}$, $i \ge 0$, are independent and identically distributed.

(2) There exists no $t_0 > 0$ such that the random variables ξ_i, $i \ge 2$, can take only the values jt_0, $j \ge 0$.

(3) The distribution function $F(x) = \mathbf{P}\{\xi_i < x\}$, $i \ge 2$, satisfies the condition $F(0+) < 1$.

(4) The expectation $\mathbf{E}\xi_i = \int_0^\infty x\,dF(x)$, $i \ge 2$, is finite.

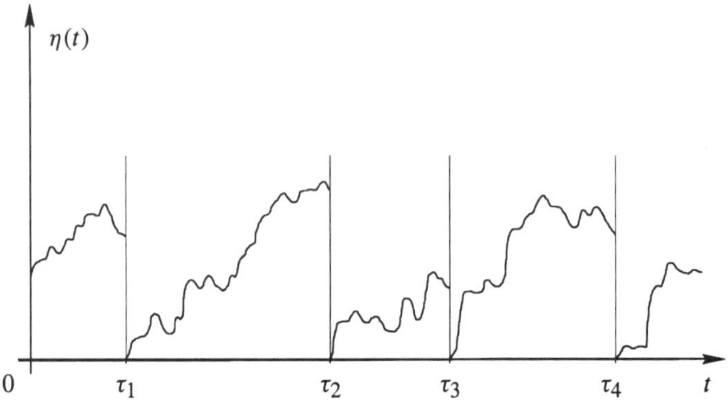

Figure 1.3.

We denote

$$\tau_0 = 0, \quad \tau_i = \sum_{j=1}^{i} \xi_j, \quad i \geq 1, \quad \nu(t) = \max_{i \geq 0}\{i : \tau_i \leq t\}.$$

Obviously, $\{\nu(t),\ t \geq 0\}$ is the (general) renewal process.

The regenerative process $\{\eta(t),\ t \geq 0\}$ is obtained by 'gluing' the processes $\{\eta_i(t),\ 0 \leq t < \xi_{i+1},\ i \geq 0\}$ (Fig. 1.3), the value of the process $\{\eta(t),\ t \geq 0\}$ coinciding with the value $\eta_i(t - \tau_i)$ of the process $\{\eta_i(t),\ t \geq 0\}$ in the interval $[\tau_i, \tau_{i+1})$. The instants τ_i of renewal of the process $\{\nu(t),\ t \geq 0\}$ is called the *regeneration instants* and the intervals $[\tau_i, \tau_{i+1})$, the *regeneration periods* of the process $\{\eta(t),\ t \geq 0\}$. At the instants τ_i, the process $\{\eta(t),\ t \geq 0\}$ 'forgets' completely its 'past,' which justifies its name 'regenerative.'

Let A be an arbitrary (measurable) subset of the set of states $\mathscr{X}$ of the process $\{\eta(t),\ t \geq 0\}$. We also assume that the following condition is satisfied:

(5) The function $P(A, t) = \mathbf{P}\{\eta_1(t) \in A, \xi_2 > t\}$ is continuous in t.

Let us determine the probability $\mathbf{P}\{\eta(t) \in A\}$ that at the instant t the process $\eta(t)$ is in the set A. By means of the formula of total probability, it can be expressed in terms of the renewal function $H(t)$ of the renewal process $\nu(t)$:

$$\mathbf{P}\{\eta(t) \in A\} = \mathbf{P}\{\eta_0(t) \in A, \xi_1 > t\} + \int_0^t P(A, t - x)\, dH(x).$$

Now, let us consider the limit behaviour of $\mathbf{P}\{\eta(t) \in A\}$ for $t \to \infty$. On the one hand, the renewal process $\nu(t)$ satisfies the key renewal theorem 1.3.4 by virtue of the above assumptions. On the other hand,

$$\mathbf{P}\{\eta_0(t) \in A, \xi_1 > t\} \leq \mathbf{P}\{\xi_1 > t\} \xrightarrow[t \to \infty]{} 0,$$

$$P(A, t) \leq \mathbf{P}\{\xi_i > t\} = 1 - F(t), \qquad i \geq 2,$$

$$\int_0^\infty [1 - F(t)]\, dt = \mathbf{E}\xi_i < \infty, \qquad i \geq 2.$$

Therefore, $P(A,t)$ as the function of t satisfies the condition of the key renewal theorem, and we arrive at the following result which is formulated as a theorem in view of its importance for further presentation.

THEOREM 1.3.5 (ergodicity theorem for the regenerative process). *If Conditions 1–5 are satisfied, then*

$$\mathbf{P}\{\eta(t) \in A\} \xrightarrow[t\to\infty]{} \lambda \int_0^\infty P(A,x)\,dx,$$

where $1/\lambda = \int_0^\infty x\,dF(x)$ *is the mean length of the* i*th regenerative period,* $i \geq 2$.

The quantity $\lambda \int_0^\infty P(A,t)\,dt$ which is representable by means of the notion of conditional expectation as

$$\lambda \int_0^\infty P(A,t)\,dt = \lambda \mathbf{E} \int_0^{\xi_2} \mathbf{P}\{\eta_1(t) \in A \mid \xi_2\}\,dt$$

is nothing but the probability $\mathbf{P}\{\eta(t) \in A\}$ averaged over one regenerative period $[\tau_i, \tau_{i+1})$, $i \geq 1$.

1.4. Markov chains

The Markov chain is a random discrete-time process whose values at different time instants are interdependent at the distance of one step. As will be seen in the following chapters, despite their relative simplicity, the Markov chains allow us to conduct a rather full (sometimes, exhaustive) study of the performance of individual queueing systems. We deal here only with Markov chains having the discrete, that is, finite or countable, state set $\mathcal{X}$.

The Markov chains used in what follows are aperiodic and irreducible. Therefore, we do not distract the readers by describing the influence produced on the Markov chains by the appearance of unessential and periodic states or several ergodic classes. We also recall that throughout this book consideration is given to the time-homogeneous Markov chains.

1.4.1. Definition and general properties of Markov chains with discrete state set

Let $\{\nu_n,\ n \geq 0\}$ be a random sequence with a discrete, i.e., finite, $\mathcal{X} = \{X_1, X_2, \dots, X_m\}$, or countable, $\mathcal{X} = \{X_1, X_2, \dots\}$ state set $\mathcal{X}$. We label any state $X_i \in \mathcal{X}$ with its ordinal i and define the set $\mathcal{I}$ of state labels which will be used below for convenience of notation. For the same reasons, we identify in this chapter the state X_i with its label i and the state set $\mathcal{X}$, with the set of labels $\mathcal{I}$. However, we stress that in the chapters to follow it will be convenient to return when studying the queueing systems (and the readers will understand why) to the general definition of the state set $\mathcal{X}$.

The sequence $\{\nu_n,\ n \geq 0\}$ is called the (homogeneous) *Markov chain* if it satisfies the Markov property: for any $n \geq 1$ and $i_0, i_1, \dots, i_{n-2}, i, j \in \mathcal{I}$,

$$\mathbf{P}\{\nu_n = j \mid \nu_0 = i_0, \dots, \nu_{n-2} = i_{n-2}, \nu_{n-1} = i\} = \mathbf{P}\{\nu_n = j \mid \nu_{n-1} = i\} = p_{ij}.$$

The Markov property can be briefly characterised as follows: *for a fixed 'present,' the 'future' and 'past' are independent.* In what follows, we apply this property not only to the random sequences with the discrete state set, but also to the random processes with arbitrary time and arbitrary state set.

The probability p_{ij} is called the *probability of one-step transition* from the ith to the state j of the Markov chain $\{\nu_n,\ n \geq 0\}$. The matrix

$$P = (p_{ij}) = \begin{pmatrix} p_{11} & p_{12} & \cdots \\ p_{21} & p_{22} & \cdots \\ \vdots & \vdots & \ddots \end{pmatrix}$$

is called the *transition probability matrix*. For the transition probability matrix of the Markov chain, the following relations are valid:

$$\sum_{j \in \mathcal{J}} p_{ij} = 1, \qquad i \in \mathcal{J}.$$

Any matrix with nonnegative elements which satisfies these relations is called the *stochastic matrix*. We note that if for the matrix P with nonnegative elements the relations

$$\sum_{j \in \mathcal{J}} p_{ij} \leq 1, \qquad i \in \mathcal{J},$$

hold and the strict inequality is satisfied at least for one i, then it is called the *semi-stochastic matrix*.

For the probabilities $p_{ij}^{(n)} = \mathbf{P}\{\nu_{n+m} = j \mid \nu_m = i\}, n \geq 2$, of n-step transition from the ith to the state j, the Kolmogorov–Chapman equation

$$p_{ij}^{(n+l)} = \sum_{k \in \mathcal{J}} p_{ik}^{(n)} p_{kj}^{(l)}$$

is valid. According to it, the probability $p_{ij}^{(n)}$ obeys the following formula:

$$p_{ij}^{(n)} = \sum_{k_1, \ldots, k_{n-1}} p_{ik_1} p_{k_1 k_2} \cdots p_{k_{n-1} j},$$

and the matrix $P^{(n)} = (p_{ij}^{(n)})$ of the probabilities of transition in n steps is representable as the the nth power of the matrix P, that is, $P^{(n)} = P^n$.

To determine completely the distribution of the Markov chain $\{\nu_n,\ n \geq 0\}$, one must define together with the transition probability matrix $P = (p_{ij})$ the initial distribution $p_i(0) = \mathbf{P}\{\nu_0 = i\}$, $i \in \mathcal{J}$, or, which is the same, the column vector $\boldsymbol{p}(0)$ of the initial probabilities. Then the state probabilities $p_i(n) = \mathbf{P}\{\nu_n = i\}$ are as follows:

$$p_i(n) = \sum_{j \in \mathcal{J}} p_j(0) p_{ji}^{(n)}, \qquad i \in \mathcal{J},$$

or in the matrix form,

$$\boldsymbol{p}^T(n) = \boldsymbol{p}^T(0) P^n.$$

The Markov chain $\{\nu_n,\ n \geq 0\}$ is called the *stationary chain* if $p_i(1) = p_i(0)$ for all $i \in \mathcal{I}$; the last equality implies the identity $p_i(n) = p_i(0)$, $i \in \mathcal{I}$, for all $n \geq 1$. In this case, the probabilities $p_i = p_i(0)$, $i \in \mathcal{I}$, are called the *stationary probabilities* of the states of the Markov chain $\{\nu_n,\ n \geq 0\}$. If $p_i > 0$, $i \in \mathcal{I}$, then $\{p_i,\ i \in \mathcal{I}\}$ is called also the *equilibrium distribution*. The stationary state probabilities satisfy the *simultaneous equilibrium equations*

$$p_i = \sum_{j \in \mathcal{I}} p_j p_{ji}, \qquad i \in \mathcal{I}, \tag{1.4.1}$$

or in the matrix form,

$$\boldsymbol{p}^T = \boldsymbol{p}^T P,$$

and the intrinsic *normalisation condition*

$$\sum_{i \in \mathcal{I}} p_i = 1. \tag{1.4.2}$$

The opposite is true as well: if some set $\{p_i,\ i \in \mathcal{I}\}$ of nonnegative numbers satisfies equilibrium equations (1.4.1) and the normalisation condition (1.4.2), then together with the initial distribution $p_i(0) = p_i$, $i \in \mathcal{I}$, the matrix of transition probabilities P generates a stationary Markov chain with the state set $\mathcal{I}$.

The states i and j are called *communicating* if there are $n_1 \geq 1$ and $n_2 \geq 1$ such that $p_{ij}^{(n_1)} > 0$ and $p_{ji}^{(n_2)} > 0$. The Markov chain $\{\nu_n,\ n \geq 0\}$ all of whose states are communicating is called the *irreducible chain*. We note that the transition probability matrix of the irreducible Markov chain is indecomposable.

Let i be an arbitrary state of the Markov chain $\{\nu_n,\ n \geq 0\}$. We consider the set $\mathcal{N}_i$ of all the instants $n \geq 1$ for which $p_{ii}^{(n)} > 0$, that is, the instants where one can return to the state i from the state i. We denote by l the greatest common divisor of all $n \in \mathcal{N}_i$. The state i is called the *periodic* state with period l if $l > 1$ and *non-periodic* (aperiodic) state if $l = 1$. Any two communicating states simultaneously are either non-periodic or periodic with the same period l. The Markov chain $\{\nu_n,\ n \geq 0\}$ all of whose states are non-periodic (for the irreducible chain, non-periodicity of only one chosen arbitrary state suffices to this end) is called the *aperiodic* Markov chain.

When studying the performance of different queueing systems, we are primarily interested in their limit behaviour in time because, as experience shows, with time the majority of physical systems rapidly enters the stationary (stable, equilibrium) state. The limit behaviour of the Markov chains is closely related with the so-called notion of ergodicity.

The Markov chain $\{\nu_n,\ n \geq 0\}$ is called *ergodic* if there exists a probability distribution $\{p_i,\ i \in \mathcal{I}\}$, $p_i > 0$, $i \in \mathcal{I}$, such that the transition probabilities $p_{ij}^{(n)}$ satisfy the limit relation

$$p_{ij}^{(n)} \xrightarrow[n \to \infty]{} p_j, \ i, j \in \mathcal{I}.$$

In this case, the distribution $\{p_i,\ i \in \mathcal{I}\}$ is called the *limit (final)* distribution of the chain $\{\nu_n,\ n \geq 0\}$.

As one can see from this definition, ergodicity of the Markov chain is defined only by its transition probability matrix $P = (p_{ij})$ and is independent of the initial distribution $\{p_i(0),\ i \in \mathcal{I}\}$. Therefore, it is possible to state that all the Markov chains with the same matrix P are ergodic and disregard the initial distribution $\{p_i(0),\ i \in \mathcal{I}\}$ when studying the Markov chain for ergodicity.

The limit probabilities p_i, $i \in \mathcal{I}$, of the ergodic Markov chain are necessarily the stationary probabilities of the same chain (or more precisely, of some stationary chain with the same state set $\mathcal{I}$ and the same transition probability matrix $P = (p_{ij})$ and the initial probabilities $p_i(0) = p_i, i \in \mathcal{I}$).

We note that the name 'ergodic' itself which was borrowed from the theory of stationary random processes* is explained by the following *property of the ergodic Markov chain.* Let us consider an arbitrary state i and determine the random variable $\zeta_n(i)$ which is equal to the total number of passages via the state i by the chain in n steps divided by n. Then $\zeta_n(i) \xrightarrow[n\to\infty]{} p_i$ with probability 1. Ergodicity, in particular, enables one to simulate the queueing systems using one sufficiently long realization instead of many realizations.

We conclude this section by brief discussion of the notion of the terminating Markov chain. When a finite Markov chain with the transition probability matrix $P = (p_{ij})_{i,j=1,\dots,m}$ evolves in time, it may happen that at the instant of current transition the chain with the probability $1 - \sum_{j=1}^{m} p_{ij}$, leaves the state i and terminates, that is, does not get anymore in any state. This Markov chain is called the *terminating Markov chain.* In this case, the matrix P is semi-stochastic. The inverse is true as well: some terminating Markov chain corresponds to any semi-stochastic matrix P.

We note that if for the terminating Markov chain the (semi-stochastic) matrix P of the transition probabilities is indecomposable, then for the matrix $I - P$ there exists an inverse matrix $(I - P)^{-1}$, and here

$$(I - P)^{-1} = I + P + P^2 + \cdots$$

1.4.2. Ergodicity of the Markov chain with a finite number of states

Thus, let the Markov chain $\{\nu_n,\ n \geq 0\}$ be defined on the finite state set $\mathcal{I} = \{1, 2, \dots, m\}$.

THEOREM 1.4.1 (ergodic theorem for a finite Markov chain). *Any irreducible non-periodic Markov chain $\{\nu_n,\ n \geq 0\}$ with a finite number of states $\mathcal{I}$ is ergodic. The limit probabilities p_i, $i = 1, \dots, m$, are determined as the unique solution of the equilibrium equations* (1.4.1) *with the normalisation condition* (1.4.2).

Therefore, the ergodic theorem 1.4.1 states that for any irreducible non-periodic Markov chain there exists a single stationary modification which the chain $\{\nu_n,\ n \geq 0\}$ itself converges to in the sense of convergence of the finite-dimensional distributions. The opposite, generally speaking, is not true. There may exist different stationary Markov chains with the same transition probability matrix P. These chains, however, are not irreducible.

In the case of irreducible periodic chain, there necessarily is a single stationary distribution $\{p_i,\ i = 1, \dots, m\}$, but the distribution $\{p_i(n),\ i = 1, \dots, m\}$ converges to the distribution $\{p_i,\ i = 1, \dots, m\}$, but not for any initial distribution $\{p_i(0),\ i = 1, \dots, m\}$.

*The notion of ergodicity that was presented here and is used by the Markov process theory, generally speaking, does not agree with the notion of ergodicity accepted by the theory of stationary random processes ((Gikhman and Skorokhod, 1966)). For example, any finite stationary irreducible periodic Markov chain is ergodic in the sense of the theory of stationary random processes, but is not ergodic in the above sense. However, the authors never encountered a queueing system describable by a random process which would have a stationary ergodic modification, but not be ergodic in the above sense.

1.4.3. Ergodicity of the Markov chain with a countable state set

In the case of the countable state set $\mathcal{I} = \{1, 2, \dots\}$, the situation with ergodicity of the Markov chain is essentially the same as in the case of finite state, although it is somewhat more complicated because the chain can 'go' to infinity.

Let at instant 0 the Markov chain $\{\nu_n,\ n \geq 0\}$ be in some state i. We denote by $\tau_1^{(i)}$ the instant of the first (after 0) return to the state i. The state i is called the *recurrent* state if $\mathbf{P}\{\tau_1^{(i)} < \infty\} = 1$ and *non-recurrent* state, otherwise. Recurrence of the state i means that after leaving the state i the Markov chain returns to it at least once (and therefore, an infinite number of times) with probability 1. The recurrent state i is called the *positively recurrent* state if $\mathbf{E}\tau_1^{(i)} < \infty$ and *zero-recurrent* state if $\mathbf{E}\tau_1^{(i)} = \infty$.

In the irreducible Markov chain with the countable state set, all states are simultaneously either non-recurrent, or zero-recurrent, or positively recurrent. Therefore, it is possible to regard the entire chain as *non-recurrent, zero-recurrent*, or *positive recurrent.* Recurrency of a chain is defined only by the transition probability matrix P and is independent of the initial distribution $\{p_i(0),\ i \geq 1\}$.

We note that the sequence $\{\tau_j^{(i)},\ j \geq 1\}$ of the instants of return to the state i generates a simple renewal process. Here, if the state i is non-recurrent, then $F^{(i)}(\infty) = \mathbf{P}\{\tau_1^{(i)} < \infty\} < 1$ and, consequently, the renewal process terminates in a finite number of steps. The sequence of the instants of reaching the state i for any initial distribution $\{p_i(0),\ i \geq 1\}$ admits a similar treatment (but in the form of the general renewal process). Now we can formulate the following result.

THEOREM 1.4.2 (Feller ergodic theorem). *Let $\{\nu_n,\ n \geq 0\}$ be an irreducible non-periodic Markov chain (with a countable state set). In this case,*

- *if the chain is positive recurrent, then it is ergodic, the limit (stationary) distribution $\{p_i,\ i \geq 1\}$ being defined as the unique solution of the equilibrium equations* (1.4.1) *satisfying the condition $\sum_{i=1}^{\infty} |p_i| < \infty$ and the normalisation condition* (1.4.2)*;*
- *if the chain is zero-recurrent, then $p_{ij}^{(n)} \xrightarrow[n\to\infty]{} 0,\ i, j \geq 1$; and*
- *if the chain is non-recurrent, then for any initial distribution $\{p_i(0),\ i \geq 1\}$ it returns with probability 1 to each state i only a finite number of times and, consequently, $p_{ij}^{(n)} \xrightarrow[n\to\infty]{} 0,\ i, j \geq 1$, as in the case of zero-recurrence.*

We note that by the Feller theorem it is reasonable to treat the case of non-recurrent Markov chain as its exit to infinity with probability 1. Use of the Feller theorem usually comes up against difficulties due to the fact that at least for a single state i one needs to find the distribution of the instant $\tau_1^{(i)}$ of first return. In some cases, one succeeds in getting a nontrivial solution of the equilibrium equations (1.4.1), sometimes as a recurrent procedure. Then another theorem may be used to check the Markov chain $\{\nu_n,\ n \geq 0\}$ for ergodicity.

THEOREM 1.4.3 (Foster ergodic theorem). *For an irreducible non-periodic Markov chain $\{\nu_n,\ n \geq 0\}$ to be ergodic, it is necessary and sufficient that there exists a nontrivial solution $\{p_i,\ i \geq 1\}$ of the equilibrium equations* (1.4.1) *such that $\sum_{i=1}^{\infty} |p_i| < \infty$. The solution $\{p_i,\ i \geq 1\}$ coincides with the limit (stationary) distribution within the normalising factor.*

Finally, we present another result which in many cases makes it easier to check the Markov chain for ergodicity.

THEOREM 1.4.4 (Moustafa criterion). *For an irreducible non-periodic Markov chain* $\{\nu_n,\ n \geq 0\}$ *to be ergodic, it suffices that there exist a number* $\varepsilon > 0$, *an integer* i_0, *and a set of nonnegative numbers* $x_1, x_2, \ldots$ *such that*

$$\sum_{j=1}^{\infty} p_{ij} x_j \leq x_i - \varepsilon, \qquad i \geq i_0, \tag{1.4.3}$$

$$\sum_{j=1}^{\infty} p_{ij} x_j < \infty, \qquad i < i_0. \tag{1.4.4}$$

1.4.4. Random walks

We define here another type of the Markov chains with a continuous, generally speaking, state set. Let $S_0, \xi_1, \xi_2, \ldots$ be a sequence of independent random variables, $\xi_1, \xi_2, \ldots$ being identically distributed with the distribution function $F(x)$. Then the sequence of partial sums $\{S_n,\ n \geq 0\}$,

$$S_n = S_0 + \sum_{i=1}^{n} \xi_i,\ n \geq 1,$$

is called the *random walk* (on the straight line).

The queueing theory encounters with Markov chains $\{\eta_n,\ n \geq 0\}$ behaving as the random walk $\{S_n,\ n \geq 0\}$ but only as long as $S_n \geq 0$. Their values η_n, however, cannot be smaller than zero. Then the Markov chain $\{\eta_n,\ n \geq 0\}$ obeys the recurrence relation

$$\eta_0 = S_0, \quad \eta_{n+1} = \max\{\eta_n + \xi_{n+1}, 0\}, \quad n \geq 0.$$

It is reasonable to call the so-defined Markov chain the random walk *with delaying barrier* at zero.

1.4.5. Numerical methods to solve the equilibrium equations

As we will see below, use of the methods of the queueing theory often comes up against equilibrium equations for specially constructed Markov chains. Therefore, we conclude the present section by touching upon the numerical algorithms to solve the simultaneous equilibrium equations.

The Markov chains for which the stationary probabilities will be found are assumed to be finite and ergodic[†], although, as will be seen below, the algorithms described are applicable to the special Markov chains with a countable number of states which often occur in the queueing theory.

In the case of an arbitrary Markov chain, that is, a Markov chain with matrix of transition probabilities of general type it is reasonable to solve the equilibrium equations by the Gaussian method. Practice, however, demonstrates that for a great number of states this

[†] In actual practice, it suffices to assume only that the Markov chains are finite and irreducible.

method leads to great errors to the extent of negative probabilities arising as the result of calculations.

Though being modifications of the Gaussian method and that of chasing, the algorithms presented here take into account the specific features of the equilibrium equations and work well. They allow one to solve on the existing personal computers the equilibrium equations for Markov chains with 10^4 and more states.

To grasp the probabilistic sense of the first algorithm, we present a finding of the theory of ergodic processes. Since it is often helpful in simplifying appreciably the study of queueing systems, we formulate auxiliary Lemma 1.4.1 and Corollary 1.4.1 for the general case of continuous time and also Corollary 1.4.2 for the ergodic Markov chains. In this section, we need only Corollary 1.4.2.

Let $\{\eta(t),\ t \geq 0\}$ be a random process with the state set $\mathscr{X}$ (and σ-algebra $\mathfrak{B}(\mathscr{X})$ on the set $\mathscr{X}$ relative to which the values of the process $\{\eta(t),\ t \geq 0\}$ are measurable). We refer to $\{\eta(t),\ t \geq 0\}$ as the generalised ergodic process if for any (measurable) subset $A \subset \mathscr{X}$ there exists with probability 1

$$\lim_{T\to\infty} \frac{1}{T} \int_0^T \chi_A(\eta(t))\,dt = \mu(A),$$

where $\chi_A(x)$ is the indicator of the set A, that is, the function equal to 1 if $x \in A$ and 0 if $x \notin A$, and $\mu(A)$ is a numerical function.

Let now $\{\eta(t),\ t \geq 0\}$ be a generalised ergodic process. We fix some (measurable) subset $X \subset \mathscr{X}$ such that $\mu(X) > 0$ and mark the instants t when $\eta(t) \notin X$ (Fig. 1.4a). We disregard the marked time instants and paste the remaining pieces of the process $\eta(t)$ to obtain a new process $\{\eta_X(t),\ t \geq 0\}$ whose state set coincides with X (Fig. 1.4b).

LEMMA 1.4.1. *The process $\{\eta_X(t),\ t \geq 0\}$ is a generalised ergodic process, and for any (measurable) set $A \subset X$,*

$$\lim_{T\to\infty} \frac{1}{T} \int_0^T \chi_A(\eta_X(t))\,dt = \mu_X(A) = \frac{\mu(A)}{\mu(X)}.$$

PROOF. We denote by $\tau(T)$ a random instant for the process $\{\eta(t),\ t \geq 0\}$ corresponding to the instant T for the process $\{\eta_X(t),\ t \geq 0\}$ (see Fig. 1.4). The instant $\tau(T)$ obeys the equation

$$\int_0^{\tau(T)} \chi_X(\eta(t))\,dt = T.$$

By virtue of generalised ergodicity of the process $\{\eta(t),\ t \geq 0\}$, with probability 1 there exists

$$\lim_{T\to\infty} \frac{T}{\tau(T)} = \mu(X),$$

or, which is the same,

$$\lim_{T\to\infty} \frac{\tau(T)}{T} = \frac{1}{\mu(X)}.$$

Again, using the generalised ergodicity of the process $\{\eta(t),\ t \geq 0\}$, we find that with probability 1 there exists

$$\lim_{T\to\infty} \frac{1}{\tau(T)} \int_0^T \chi_A(\eta_X(t))\,dt = \lim_{T\to\infty} \frac{1}{\tau(T)} \int_0^{\tau(T)} \chi_A(\eta(t))\,dt = \mu(A).$$

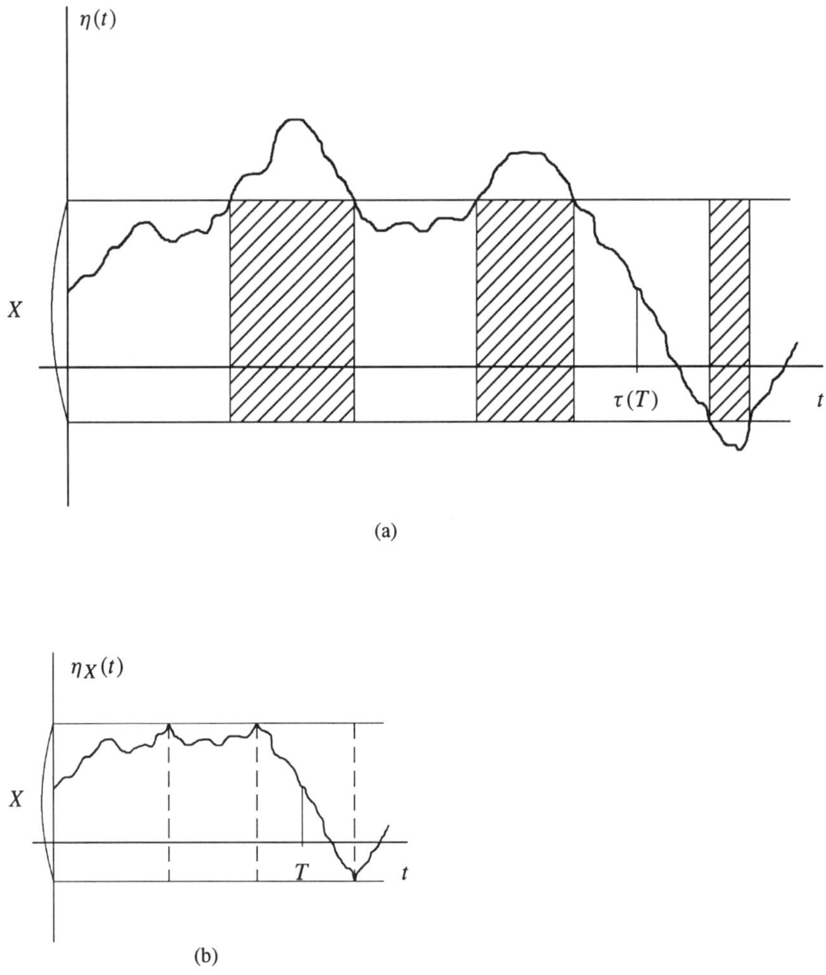

Figure 1.4.

Hence, we see that with probability 1 there exists

$$\lim_{T\to\infty}\frac{1}{T}\int_0^T \chi_A(\eta_X(t))\,dt = \lim_{T\to\infty}\frac{\tau(T)}{\tau(T)T}\int_0^{\tau(T)} \chi_A(\eta(t))\,dt = \frac{\mu(A)}{\mu(X)},$$

which proves the lemma. □

COROLLARY 1.4.1. *Let processes* $\{\eta(t), t \geq 0\}$ *and* $\{\eta_X(t), t \geq 0\}$ *satisfy Lemma* 1.4.1 *and, additionally, let there exist the limit distributions*

$$\mathbf{P}(A) = \lim_{t\to\infty}\mathbf{P}\{\eta(t) \in A\}$$

and

$$\mathbf{P}_X(A) = \lim_{t\to\infty} \mathbf{P}\{\eta_X(t) \in A\}.$$

Then for all $A \subset X$

$$\mathbf{P}_X(A) = \frac{\mathbf{P}(A)}{\mathbf{P}(X)}.$$

PROOF. Validity of this assertion immediately follows from the equalities

$$\begin{aligned}
\mu(A) &= \mathbf{E}\mu(A) = \mathbf{E} \lim_{T\to\infty} \frac{1}{T}\int_0^T \chi_A(\eta(t))\,dt \\
&= \lim_{T\to\infty} \frac{1}{T}\int_0^T \mathbf{E}\chi_A(\eta(t))\,dt = \lim_{T\to\infty} \frac{1}{T}\int_0^T \mathbf{P}\{\eta(t) \in A\}\,dt = \mathbf{P}(A), \\
\mu_X(A) &= \mathbf{E}\mu_X(A) = \mathbf{E} \lim_{T\to\infty} \frac{1}{T}\int_0^T \chi_A(\eta_X(t))\,dt \\
&= \lim_{T\to\infty} \frac{1}{T}\int_0^T \mathbf{E}\chi_A(\eta_X(t))\,dt = \lim_{T\to\infty} \frac{1}{T}\int_0^T \mathbf{P}\{\eta_X(t) \in A\}\,dt = \mathbf{P}_X(A).
\end{aligned}$$

□

Let now $\{\nu_n,\ n \geq 0\}$ be an ergodic Markov chain with the finite number of states $\mathcal{I}$. Let $X \subset \mathcal{I}$ be a (nonempty) subset of states. Let us consider a random sequence $\{\nu'_n,\ n \geq 0\}$ made up by the values of the Markov chain $\{\nu_n,\ n \geq 0\}$ at the successive instants of falling in the states of the subset X. Obviously, $\{\nu'_n,\ n \geq 0\}$ also is the ergodic Markov chain.

The following assertion is the counterpart of Corollary 1.4.1 for the ergodic Markov chains.

COROLLARY 1.4.2. *Let p'_i, $i \in X$, and p_i, $i \in \mathcal{I}$, be stationary probabilities, respectively, of Markov chains $\{\nu'_n,\ n \geq 0\}$ and $\{\nu_n,\ n \geq 0\}$. Then the equality $p'_i = cp_i$, where $c^{-1} = \sum_{i\in X} p_i$, is valid for all $i \in X$.*

The proof of Corollary 1.4.2 follows the same lines as that of Corollary 1.4.1, and we omit it.

Now, we dwell on two methods to solve the equilibrium equations that will be used repeatedly in what follows.

We recall that the equilibrium equations for the Markov chain $\{\nu_n,\ n \geq 0\}$ with the state set $\mathcal{I} = \{1, 2, \dots, m\}$ has the following matrix form:

$$\boldsymbol{p}^T P = \boldsymbol{p}^T,$$

where $\boldsymbol{p}^T = \{p_1, p_2, \dots, p_m\}$ is the vector of stationary probabilities of states, and

$$P = \begin{pmatrix} p_{11} & \cdots & p_{1,m-1} & p_{1m} \\ \vdots & \ddots & \vdots & \vdots \\ p_{m-1,1} & \cdots & p_{m-1,m-1} & p_{m-1,m} \\ p_{m1} & \cdots & p_{m,m-1} & p_{mm} \end{pmatrix}$$

is the matrix of transition probabilities of the Markov chain $\{\nu_n,\ n \geq 0\}$.

Elimination method: Probabilistic interpretation. Let us transform step by step the matrix P by eliminating successively the states of the original Markov chain. With this end in view, we introduce at the first step a new Markov chain $\{\nu_n^{(1)},\ n \geq 0\}$ with the state set $\mathcal{J}^{(1)} = \{1, 2, \ldots, m-1\}$ that was formed by the values of the Markov chain $\{\nu_n,\ n \geq 0\}$ at the successive instants of getting into the states of the subset $\mathcal{J}^{(1)}$. Let us assume that

$$p_{mm}^{(1)} = (1 - p_{mm})^{-1}. \tag{1.4.5}$$

Now, we define the probability $p_{mk}^{(1)}$, $k = 1, \ldots, m-1$, that at the instant of the first exit from the state m the Markov chain $\{\nu_n,\ n \geq 0\}$ falls into the state k:

$$p_{mk}^{(1)} = p_{mk} + p_{mm} p_{mk} + (p_{mm})^2 p_{mk} + \cdots = p_{mm}^{(1)} p_{mk}.$$

Finally, we define the transition probabilities $p_{kl}^{(1)}$, $k, l = 1, \ldots, m-1$, of the Markov chain $\{\nu_n^{(1)},\ n \geq 0\}$. Since the Markov chain $\{\nu_n,\ n \geq 0\}$ can either immediately get from the state k to the state l with probability p_{kl} or first get to the state m with probability p_{km} and then with the probability p_{ml} to the state l at the instant of the first exit from the state m, we obtain

$$p_{kl}^{(1)} = p_{kl} + p_{km} p_{ml}^{(1)}.$$

Let us assume that

$$\hat{P}^{(1)} = \begin{pmatrix} p_{11}^{(1)} & \cdots & p_{1,m-1}^{(1)} & p_{1m} \\ \vdots & \ddots & \vdots & \vdots \\ p_{m-1,1}^{(1)} & \cdots & p_{m-1,m-1}^{(1)} & p_{m-1,m} \\ p_{m1}^{(1)} & \cdots & p_{m,m-1}^{(1)} & p_{mm}^{(1)} \end{pmatrix},$$

One can readily see that the principal minor

$$P^{(1)} = \begin{pmatrix} p_{11}^{(1)} & \cdots & p_{1,m-2}^{(1)} & p_{1,m-1}^{(1)} \\ \vdots & \ddots & \vdots & \vdots \\ p_{m-1,1}^{(1)} & \cdots & p_{m-2,m-2}^{(1)} & p_{m-2,m-1}^{(1)} \\ p_{m-1,1}^{(1)} & \cdots & p_{m-1,m-2}^{(1)} & p_{m-1,m-1}^{(1)} \end{pmatrix}$$

of the matrix $\hat{P}^{(1)}$ is the matrix of transition probabilities of the Markov chain $\{\nu_n^{(1)},\ n \geq 0\}$ and also is a stochastic matrix.

At the next step, the same procedure is performed on the matrix $P^{(1)}$, and so on until the $(m-1)$th step where we obtain the one-element matrix $P^{(m-1)} = (1)$.

To avoid new notation, which is necessary for software realization of the algorithm in order to reduce the area of computer memory used, the successively obtained matrices $P^{(1)}$, $P^{(2)}$, and so on are denoted by P. We note that at the $(m-l)$th step, $l = 1, \ldots, m$, the elements p_{kl}, $k = 1, \ldots, l-1$, of the resulting matrix P are the probabilities of transitions from the state k to the state l for the Markov chain $\{\nu_n^{(m-l)},\ n \geq 0\}$ (with this notation, we assume that the original Markov chain $\{\nu_n,\ n \geq 0\}$ coincides with the Markov chain $\{\nu_n^{(0)},\ n \geq 0\}$), and the element $(p_{ll})^{-1}$ is the probability that this chain leaves the state l.

In view of the aforementioned, the stationary probabilities $p_k^{(m-2)}$, $k = 1, 2$, of the Markov chain $\{\nu_n^{(m-2)},\ n \geq 0\}$ satisfy the balance equation

$$p_1^{(m-2)} p_{12} = p_2^{(m-2)} (p_{22})^{-1}.$$

Hence, we assume that

$$q_1 = 1, \qquad q_2 = q_1 p_{12} p_{22}$$

and obtain

$$\begin{aligned} p_1^{(m-2)} &= p_1^{(m-2)} q_1, \\ p_2^{(m-2)} &= p_1^{(m-2)} q_2, \end{aligned} \tag{1.4.6}$$

where $p_1^{(m-2)}$ is the stationary probability of state 1 for the Markov chain $\{\nu_n^{(m-2)},\ n \geq 0\}$.

By virtue of Corollary 1.4.2, the stationary probabilities of the Markov chains $\{\nu_n^{(m-3)},\ n \geq 0\}$ and $\{\nu_n^{(m-2)},\ n \geq 0\}$ are the same to within a constant, that is,

$$\begin{aligned} p_1^{(m-3)} &= p_1^{(m-3)} q_1, \\ p_2^{(m-3)} &= p_1^{(m-3)} q_2. \end{aligned} \tag{1.4.7}$$

This formula differs from (1.4.6) only in the normalising factor $p_1^{(m-3)}$, the stationary probability of state 1 for the Markov chain $\{\nu_n^{(m-3)},\ n \geq 0\}$.

The stationary probabilities $p_k^{(m-3)}$, $k = 1, 2, 3$, of the Markov chain $\{\nu_n^{(m-3)},\ n \geq 0\}$ are related by

$$p_1^{(m-3)} p_{13} + p_2^{(m-3)} p_{23} = p_3^{(m-3)} (p_{33})^{-1},$$

from which, in view of (1.4.7), we find

$$p_3^{(m-3)} = p_1^{(m-3)} q_3,$$

where

$$q_3 = (q_1 p_{13} + q_2 p_{23}) p_{33}.$$

Continuing this procedure, we finally arrive at the following expression for the stationary probabilities $p_k = p_k^{(0)}$ of the original Markov chain $\{\nu_n,\ n \geq 0\}$:

$$p_k = p_0 q_k, \qquad k = 1, \dots, m.$$

Here, q_k obeys the recurrence relation

$$q_1 = 1, \quad q_k = \left(\sum_{l=1}^{k-1} q_l p_{lk} \right) p_{kk}, \quad k = 2, \dots, m, \tag{1.4.8}$$

and the probability p_0 is determined as usual from the normalisation condition

$$p_0 = \left(\sum_{k=1}^{m} q_k \right)^{-1}. \tag{1.4.9}$$

We note that the proposed algorithm can also yield pronounced errors caused by the single subtraction in (1.4.5). These errors, however, can be brought virtually to nothing if $p_{mm}^{(1)}$ is calculated not by (1.4.5) but by the formula

$$p_{mm}^{(1)} = \left(\sum_{k=1}^{m-1} p_{mk}\right)^{-1}. \tag{1.4.10}$$

Quasi-triangular transition matrix: Recursive solution of equilibrium equations. In what follows, we will often encounter Markov chains with the following transition probability matrix:

$$P = \begin{pmatrix} p_{11} & p_{12} & p_{13} & \cdots & p_{1,m-1} & p_{1m} \\ p_{21} & p_{22} & p_{23} & \cdots & p_{2,m-1} & p_{2m} \\ 0 & p_{32} & p_{33} & \cdots & p_{3,m-1} & p_{3m} \\ 0 & 0 & p_{43} & \cdots & p_{4,m-1} & p_{4m} \\ \vdots & \vdots & \vdots & \ddots & \vdots & \vdots \\ 0 & 0 & 0 & \cdots & p_{m-1,m-1} & p_{m-1,m} \\ 0 & 0 & 0 & \cdots & p_{m,m-1} & p_{m,m} \end{pmatrix}.$$

In this case, the following algorithm to calculate the stationary probabilities p_k, $k = 1, \dots, m$, can be suggested.

First, by using the leading equation of the equilibrium equations which in this case is of the form

$$p_1 p_{11} + p_2 p_{21} = p_1,$$

we express the stationary probability p_2 in terms of the stationary probability p_1:

$$p_2 = p_1 q_2,$$

where

$$q_2 = q_1(1 - p_{11})(p_{21})^{-1}.$$

Then by using the second equation

$$p_1 p_{12} + p_2 p_{22} + p_3 p_{32} = p_2$$

of the equilibrium equations, we determine the stationary probability p_3:

$$p_3 = p_1 q_3,$$

where

$$q_3 = (q_1 p_{12} + q_2(1 - p_{22}))(p_{32})^{-1}.$$

This procedure is continued to obtain

$$p_k = p_0 q_k, \qquad k = 1, \dots, m,$$

where q_k obeys the recurrence relation

$$q_1 = 1,$$

$$q_k = \left(\sum_{l=1}^{k-2} q_l p_{l,k-1} + q_{k-1}(1 - p_{k-1,k-1})\right)(p_{k,k-1})^{-1}, \qquad k = 2, \dots, m,$$

and p_0, as before, is defined by the normalisation condition (1.4.9).

The above algorithms can be extended to the matrix case, that is, the case where the elements p_{ij} themselves of the matrix P are matrices. We consider the matrix algorithm whenever necessary in order to gain a better insight into the matter.

1.5. Markov processes with discrete state set

Whereas the Markov chains with discrete (finite or countable) state set constitute, in general, only an auxiliary apparatus to study the queueing systems, the continuous-time Markov processes with discrete state set serve as the main tool for studying their entire class which even are called the Markov queueing systems. In what follows, we omit for brevity the term 'continuous-time' because we identify the discrete-time processes with the random sequences.

To avoid misunderstanding, we state immediately that the following description of the Markov processes with discrete state set is far from covering all possible cases. It is precisely this description that allows us to kill at once two birds with one stone: first, to reject *a priori* all processes that are 'exotic' for the queueing theory and second, to avoid using the rather complicated theory employed to study the general Markov processes with the discrete state set. The Markov processes with discrete state set used below have many properties in common with the Markov chains (with discrete state set). Therefore, they also are often called the Markov chains. We do not follow this tradition, however, because sometimes we encounter both the Markov processes and chains which must be discriminated. We also recall that we agreed to identify in this chapter the state set $\mathcal{X}$ with the set $\mathcal{I}$ of the labels of these states.

Behaviour of the Markov processes with discrete state set is defined by the state set $\mathcal{I}$ even to a greater extent than behaviour of the Markov chains. If $\mathcal{I}$ is finite, then there exists an exact analogy between the Markov processes and chains. Moreover, the processes are even somewhat simpler because they do without the notion of periodicity. In the case of countable state set $\mathcal{I}$, the situation aggravates because, along with the possibility of going to 'infinity' in 'infinite' time, an additional possibility for the process to go to 'infinity' in a finite time appears. It is fair to say, however, that the exit to 'infinity' in a finite time is not characteristic of the physical queueing systems. That is why in this section we confine ourselves to formulating the results which are required in what follows and discuss the possible anomalies only by way of a simple example of the so-called process of pure birth just to provide an insight to the origin of such deviations.

1.5.1. Infinitesimal matrix

Let $\{\eta(t),\ t \geq 0\}$ be a *continuous-time* random process with the *discrete state set* $\mathcal{I}$, where $\mathcal{I} = \{1, 2, \dots, m\}$ and $\mathcal{I} = \{1, 2, \dots\}$, respectively, for finite and countable state sets.

The process $\{\eta(t),\ t \geq 0\}$ is called the *Markov process* if for any assemblies of time instants $t_1, t_2, \dots, t_{n+1}$, $0 \leq t_1 < t_2 < \cdots < t_{n+1}$ and states $i_1, i_2, \dots, i_{n+1} \in \mathcal{I}$ the following equality holds:

$$\mathbf{P}\{\eta(t_{n+1}) = i_{n+1} \mid \eta(t_1) = i_1, \dots, \eta(t_n) = i_n\} == \mathbf{P}\{\eta(t_{n+1}) = i_{n+1} \mid \eta(t_n) = i_n\}.$$

The Markov process $\{\eta(t),\ t \geq 0\}$ is called the time-*homogeneous* process if $\mathbf{P}\{\eta(t+s) = j \mid \eta(s) = i\}$, $i, j \in \mathcal{I}$, is independent of s. In what follows, we always—except for specially qualified situations—consider only the homogeneous Markov processes.

We denote by $p_i(t) = \mathbf{P}\{\eta(t) = i\}$ the probability that at the instant t the process $\eta(t)$ is in the state i and by $p_{ij}(t) = \mathbf{P}\{\eta(t+s) = j \mid \eta(s) = i\}$, the probability of going in time t from the state i to the state j. We use below for $p_i(t)$, $i \in \mathcal{I}$, and $p_{ij}(t)$, $i, j \in \mathcal{I}$, the matrix notation $\boldsymbol{p}(t)$ and $P(t)$, where $\boldsymbol{p}(t)$ is a column vector and $P(t)$ is a matrix.

The functions $p_{ij}(t)$ and $p_i(t)$ satisfy the *Kolmogorov–Chapman equation*

$$p_{ij}(t+s) = \sum_{k \in \mathcal{I}} p_{ik}(t) p_{kj}(s), \qquad i, j \in \mathcal{I}, \tag{1.5.1}$$

$$p_i(t+s) = \sum_{j \in \mathcal{I}} p_j(t) p_{ji}(s), \qquad i \in \mathcal{I}. \tag{1.5.2}$$

or, in the matrix form,

$$P(t+s) = P(t)P(s),$$
$$\boldsymbol{p}^T(t+s) = \boldsymbol{p}^T(t)P(s).$$

Equation (1.5.2) is obtained as follows. For the process $\{\eta(t),\ t \geq 0\}$ to be at the instant $t+s$ in the state i, it is necessary that at the instant t it is in the state j and goes in the remaining time s from the state j to the state i. Using the total probability formula and taking into account the Markov property of the process $\{\eta(t),\ t \geq 0\}$, we arrive at (1.5.2). Equation (1.5.1) is obtained similarly.

We assume everywhere below that

$$p_{ij}(\Delta) \underset{\Delta \downarrow 0}{\longrightarrow} \delta_{ij}, \qquad i, j \in \mathcal{I}, \tag{1.5.3}$$

where δ_{ij} is the Kronecker symbol. This condition is intrinsic in applications.

As was shown, it follows from condition (1.5.3) that there exist the limits

$$a_{ij} = \lim_{\Delta \downarrow 0} \frac{p_{ij}(\Delta)}{\Delta}, \qquad i, j \in \mathcal{I}, \quad i \neq j,$$
$$-a_i = a_{ii} = \lim_{\Delta \downarrow 0} \frac{p_{ii}(\Delta) - 1}{\Delta}, \qquad i \in \mathcal{I}, \tag{1.5.4}$$

and

$$0 \leq a_i \leq \infty, \quad i \in \mathcal{I}, \quad 0 \leq a_{ij} < \infty, \quad i, j \in \mathcal{I}, \quad i \neq j,$$
$$\sum_{j \in \mathcal{I}} a_{ij} \leq 0, \qquad i \in \mathcal{I}. \tag{1.5.5}$$

The number a_{ij}, $i, j \in \mathcal{I}$, $i \neq j$, is called the *intensity of transition* from the state i to the state j, the number a_i, $i \in \mathcal{I}$, *intensity of exit* from the state i, and $A = (a_{ij})$, the *matrix of transition intensities* or *infinitesimal matrix*. These names are explained by the fact that for 'small' Δ the probabilities $p_{ij}(\Delta)$ are

$$p_{ij}(\Delta) = \delta_{ij} + a_{ij}\Delta + o(\Delta), \qquad i, j \in \mathcal{I}. \tag{1.5.6}$$

In what follows, we deal only with the processes for which $a_i < \infty$.

If, instead of inequality (1.5.5) the strict equality

$$\sum_{j \in \mathcal{I}} a_{ij} = 0, \qquad i \in \mathcal{I}, \tag{1.5.7}$$

is satisfied, then $\{\eta(t),\ t \geq 0\}$ is called the *conservative* (locally regular) process. For the countable $\mathcal{I}$ and, if not stated otherwise, also for the finite $\mathcal{I}$ we consider everywhere below only the conservative processes.

1.5.2. Constructive description of Markov processes

Let $\{\eta(t),\ t \geq 0\}$ be a conservative Markov process with the initial distribution $\{p_i(0),\ i \in \mathcal{I}\}$ and infinitesimal matrix $A = (a_{ij})$. We assume that $a_i \neq 0,\ i \in \mathcal{I}$, and take

$$q_{ij} = \begin{cases} \dfrac{a_{ij}}{a_i}, & i \neq j; \\ 0, & i = j. \end{cases}$$

It follows from (1.5.7) that $Q = (q_{ij})$ is the transition probability matrix of some Markov chain.

Physical interpretation of the Markov process $\{\eta(t),\ t \geq 0\}$ is as follows. First, we choose the initial state i with regard for the initial distribution $\{p_i(0),\ i \in \mathcal{I}\}$. The process $\{\eta(t),\ t \geq 0\}$ stays in the state i during a random time distributed exponentially with parameter a_i. Then with probability q_{ij} which is independent of the time of sojourn in the state i, it goes to the state j where it stays during time distributed exponentially with the parameter a_j and independent of when and how it got into the state j. Then it goes to the state k with the probability q_{jk}, and so on.

Let $\tau_0 = 0$, τ_n, $n \geq 1$, be the instant of the nth change of state and $\nu_n = \eta(\tau_n + 0)$ be the process state immediately after the nth change of state. Then $\{\nu_n,\ n \geq 0\}$ is the Markov chain with the initial distribution $\{p_i(0),\ i \in \mathcal{I}\}$ and the transition probability matrix $Q = (q_{ij})$. This chain is referred to as the *embedded Markov chain* (at the instants of all jumps) for the process $\{\eta(t),\ t \geq 0\}$. If $\nu_n = i$, then the time between the nth and $(n+1)$th instants of state changes is exponentially distributed with the parameter a_i and does not depend on how and in what time the process got into the state i.

When defining the process $\{\eta(t),\ t \geq 0\}$, we admitted the possibility of $a_i = 0$, which corresponds to the fact that upon getting into the state i the process never leaves it. Therefore, it is reasonable to call the state i for which $a_i = 0$ the *absorbing* state. We always assume below that the process can have at most one absorbing state. Additionally, occurrence of the absorbing state is specially stipulated and has a transparent physical ground.

Thus, the defining parameters of the Markov process $\{\eta(t),\ t \geq 0\}$ with the discrete state set $\mathcal{I}$ are as follows:

- initial distribution $\{p_i(0),\ i \in \mathcal{I}\}$;

- set of parameters $\{a_i,\ i \in \mathcal{S}\}$; and
- transition probability matrix $Q = (q_{ij})_{i,j\in\mathcal{S}}$, with $q_{ii} = 0$, of the embedded Markov chain $\{\nu_n,\ n \geq 0\}$.

But, as we saw above, a single matrix $A = (a_{ij})$, where $a_{ij} = a_i q_{ij}$, $i, j \in \mathcal{S}$, $i \neq j$, and $a_{ii} = -a_i$, $i \in \mathcal{S}$, can be defined instead of $\{a_i,\ i \in \mathcal{S}\}$ and $Q = (q_{ij})$.

The above constructive description of the (conservative) Markov process is incomplete. Indeed, let us assume that $\tau_\infty = \sup_{n\geq 0} \tau_n$. It may happen that $\mathbf{P}\{\tau_\infty < \infty\} > 0$, that is, the process $\{\eta(t),\ t \geq 0\}$ is subject to an infinite number of transitions in a finite time with a non-zero probability. In this case, we assume that if $\tau_\infty < t$, then the value of $\eta(t)$ at the instant t (and the next instants) is not defined or, stated differently, the process $\{\eta(t),\ t \geq 0\}$ terminated before t.

To avoid the possibility of going to 'infinity' in a finite time, we introduce the notion of regularity. The conservative process $\{\eta(t),\ t \geq 0\}$ is called the *regular* (stable, abrupt) process if $\tau_\infty = \infty$ with probability 1, that is, the process does not terminate in any finite time. The above notion of conservative property does not still guarantee regularity of the process $\{\eta(t),\ t \geq 0\}$. Regularity is based on the following theorem.

THEOREM 1.5.1 (regularity criterion). *For the process $\{\eta(t),\ t \geq 0\}$ to be regular, it suffices that*

- *either a_i is uniformly bounded, that is, $a_i < c < \infty$, $i \in \mathcal{S}$,*
- *or all the states of the embedded Markov chain $\{\nu_n,\ n \geq 0\}$ are recurrent.*

This theorem, in particular, implies that any conservative Markov process with a finite state set is regular.

We note that in the queueing theory—or at least in that its part that will considered below—only the regular processes appear which is due to the fact that, taking some liberty of speech, the number of changes of the states of the Markov process describing a queueing system does not differ dramatically from the number of customers arrived which in turn is finite with probability 1 in any finite time interval. Therefore, many textbooks on the queueing theory do without the notion of regularity at all.

1.5.3. Kolmogorov differential equations

Let $\{\eta(t),\ t \geq 0\}$ be a conservative Markov process. Since we admit the possibility of termination of the process $\{\eta(t),\ t \geq 0\}$ in a finite time, the conditions for normalisation of the probabilities $p_i(t)$ and $p_{ij}(t)$ by unity are replaced by weaker conditions:

$$\sum_{i\in\mathcal{S}} p_i(t) \leq 1, \tag{1.5.8}$$

$$\sum_{j\in\mathcal{S}} p_{ij}(t) \leq 1, \qquad i \in \mathcal{S}. \tag{1.5.9}$$

Relation (1.5.8) can change into a strict inequality only simultaneously for all $t > 0$. The same holds for (1.5.9) for each $i \in \mathcal{S}$. The process $\{\eta(t),\ t \geq 0\}$ is regular if and only if $\sum_{i\in\mathcal{S}} p_i(t) = 1$ for some $t > 0$.

For the probabilities $p_i(t)$, the *set of (forward) Kolmogorov differential equations*

$$p_i'(t) = \sum_{j\in\mathcal{I}} a_{ji} p_j(t), \qquad i \in \mathcal{I}, \tag{1.5.10}$$

holds, which admits the matrix form representation

$$\frac{d}{dt}\boldsymbol{p}^T(t) = \boldsymbol{p}^T(t)A.$$

The *initial conditions* for (1.5.10) are defined by the initial distribution of the process

$$p_i(0) = \mathbf{P}\{\eta(0) = i\}, \qquad i \in \mathcal{I}. \tag{1.5.11}$$

Since in the studies of the Markov queueing systems below the simultaneous equations (1.5.10) plays the key role, we give a sketch how to derive it. Let us consider the states of the process $\{\eta(t),\ t \geq 0\}$ at the instants t and $t + \Delta$, where Δ is a 'small' time increment. By putting down the Kolmogorov–Chapman equation (1.5.2) for the instants t and $t + \Delta$, we obtain

$$p_i(t + \Delta) = \sum_{j\in\mathcal{I}} p_j(t) p_{ji}(\Delta).$$

Upon subtraction of $p_i(t)$ from both sides of this equality and division by Δ, we get

$$\frac{p_i(t + \Delta) - p_i(t)}{\Delta} = \frac{p_{ii}(\Delta) - 1}{\Delta} p_i(t) + \sum_{j\in\mathcal{I}\setminus\{i\}} \frac{p_{ji}(\Delta)}{\Delta} p_j(t).$$

By letting now Δ go to zero and using the limit relation (1.5.4), we arrive at (1.5.10).

We note that to justify rigorously the above conclusion for the countable $\mathcal{I}$ one needs to prove legitimacy of the passage to the limit under the sum sign. It is a rather difficult problem, and the readers are referred to special publications (for example, (Chung, 1967)) for its solution.

From (1.5.1) and (1.5.4), we obtain in a similar way *the set of (forward) Kolmogorov differential equations* for $p_{ij}(t)$:

$$p_{ij}'(t) = \sum_{k\in\mathcal{I}} a_{kj} p_{ik}(t), \qquad i, j \in \mathcal{I}, \tag{1.5.12}$$

with the initial conditions

$$p_{ij}(0) = \delta_{ij}, \qquad i, j \in \mathcal{I}, \tag{1.5.13}$$

which by means of the matrix $P(t) = (p_{ij}(t))$ can be reduced to

$$P'(t) = P(t)A.$$

In the case of a finite state set $\mathcal{I}$, equations (1.5.10), as well as (1.5.12), are simultaneous linear differential equations of the first order with constant coefficients. They have a unique solution $p_i(t)$, $i \in \mathcal{I}$, (or $p_{ij}(t)$, $i, j \in \mathcal{I}$) satisfying the initial condition (1.5.11) (or (1.5.13)).

In the case of countable $\mathcal{I}$, (1.5.10), along with $p_i(t)$, can have other solutions satisfying the non-negativeness conditions $p_i(t) \geq 0$, normalisation condition (1.5.8), and the initial conditions (1.5.11). Physically, these solutions are related with the fact that, having attained 'infinity' in a finite time, the process $\{\eta(t),\ t \geq 0\}$ can return back along the same way (we recall that from the very beginning we agreed to disregard these cases). However, if the process $\{\eta(t),\ t \geq 0\}$ is regular, then they have no other solutions satisfying the above conditions, the normalisation condition (1.5.8) here becoming, as it should, the strict equality $\sum_{i \in \mathcal{I}} p_i(t) = 1$. The aforementioned applies to (1.5.12) as well.

Solution of simultaneous equations (1.5.10) and (1.5.12) is formally representable as

$$P(t) = e^{At},$$
$$\boldsymbol{p}^T(t) = \boldsymbol{p}^T(0)e^{At}, \tag{1.5.14}$$

where the matrix function

$$e^{At} = \sum_{i=0}^{\infty} \frac{t^i}{i!} A^i \tag{1.5.15}$$

is the so-called matrix exponent. The formal series (1.5.15) converges if, for example, a_i are uniformly bounded or the set $\mathcal{I}$ is finite.

To solve equations (1.5.10), it is often convenient to use the Laplace transform

$$\pi_i(s) = \int_0^{\infty} e^{-st} p_i(t)\, dt, \qquad i \in \mathcal{I}.$$

The properties of the Laplace transform (see Section 1.1) offer a means to obtain from (1.5.10) the simultaneous linear algebraic equations

$$s\pi_i(s) - p_i(0) = \sum_{i \in \mathcal{I}} a_{ji}\pi_j(s), \qquad i \in \mathcal{I}. \tag{1.5.16}$$

In a similar manner, (1.5.12) can be reduced to simultaneous linear algebraic equations.

1.5.4. Terminating Markov processes with finite state set

Let $\{\eta(t),\ t \geq 0\}$ be a Markov process with the finite state set $\mathcal{I} = \{1, 2, \dots, m\}$. We return to the case where it is not conservative, that is, inequality (1.5.5) is satisfied instead of equality (1.5.7). As in the case of the conservative process, we assume that $q_{ii} = 0$, $i = 1, \dots, m$, $q_{ij} = a_{ij}/a_i$, $i, j = 1, \dots, m$, $i \neq j$. Now, $Q = (q_{ij})$ is the semi-stochastic matrix of the probabilities of transitions of some terminating Markov chain.

Nevertheless, with one exception the constructive description given in Section 1.5.2 for the conservative process applies also to this case: at the time of the next transition the process can go out of the state i and get in no other state with probability $q_i = 1 - \sum_{j=1}^{m} q_{ij}$. It is reasonable to call it the *terminating* process, although here the nature of termination is another: the process terminates with a nonzero probability after a finite number of transitions.

For the state probabilities $p_i(t)$, $i = 1, \dots, m$, and the transition probabilities $p_{ij}(t)$, $i, j = 1, \dots, m$, of a terminating Markov process with finite state set $\mathcal{I}$, sets (1.5.10) and (1.5.12) of (forward) Kolmogorov differential equations with initial conditions (1.5.11) and (1.5.13) are valid as well, and their solutions are defined by formulas (1.5.14).

We denote by $p_0(t)$ the probability of termination of the process $\{\eta(t),\ t \geq 0\}$ before the instant t. The sum $\sum_{i=1}^{m} p_i(t) = 1 - p_0(t)$ is the probability that the process does not terminate before the instant t. Since it is representable in the matrix form as

$$\sum_{i=1}^{m} p_i(t) = \boldsymbol{p}^T(t)\mathbf{1},$$

where $\mathbf{1}$ is the column vector of unities of size m, from (1.5.14) we obtain

$$1 - p_0(t) = \boldsymbol{p}^T(0)e^{At}\mathbf{1}.$$

Hence, it follows, in particular, that

$$p_0(t) = 1 - \boldsymbol{p}^T(0)e^{At}\mathbf{1}.$$

This formula defines in the matrix-exponential form the probability $p_0(t)$ which is the distribution function of the time before the instant of termination of the process $\{\eta(t),\ t \geq 0\}$.

By adding a supplementary absorbing state denoted, say, by 0, the nonconservative process $\{\eta(t),\ t \geq 0\}$ can be readily reduced to the conservative process $\{\tilde{\eta}(t),\ t \geq 0\}$. Here, the matrix $\tilde{\tilde{A}} = (\tilde{a}_{ij})$, where $\tilde{a}_{ij} = a_{ij}$, $i, j = 1, \dots, m$, $\tilde{a}_{0i} = 0$, $i = 0, \dots, m$, $\tilde{a}_{i0} = -\sum_{j=1}^{m} a_{ij}$, $i = 1, \dots, m$, must be taken as the infinitesimal matrix of the process $\{\tilde{\eta}(t),\ t \geq 0\}$. Then the mere transition of the process $\{\tilde{\eta}(t),\ t \geq 0\}$ into the absorbing state 0 corresponds to termination of the process $\{\eta(t),\ t \geq 0\}$. We do, however, without it only for considerations of convenience in order not to describe each time the supplementary state.

The nonconservative (terminating) Markov process with the finite state set is a very convenient tool for describing the *PH*-distributions used in Chapter 4 to study the general Markov queueing systems.

Along with the notation $\{\eta(t),\ t \geq 0\}$, we denote below the terminating Markov process by $\{\eta(t),\ t \in [0, \tau)\}$, where τ is the instant of process termination.

The nonconservative Markov process $\{\eta(t),\ t \geq 0\}$ with the countable state set $\mathcal{J}$ can be described in a similar manner. In this case, the process can also be transformed into the conservative process $\{\tilde{\eta}(t),\ t \geq 0\}$ by adding a supplementary absorbing state. Again for considerations of convenience we do without the nonconservative processes because here the two types of termination give rise to another difficulty: exit of the process to 'infinity' in a finite time and its disappearance after the next transition.

1.5.5. Stationary Markov processes

A Markov process $\{\eta(t),\ t \geq 0\}$ is called the *stationary* process if the probabilities $p_i(t) = p_i$, $i \in \mathcal{J}$, are independent of t. The probabilities p_i, $i \in \mathcal{J}$, are called here the *stationary state probabilities* of the process $\{\eta(t),\ t \geq 0\}$. As in the case of the Markov chains, if $p_i > 0$, $i \in \mathcal{J}$, then $\{p_i,\ i \in \mathcal{J}\}$ also is called *the equilibrium distribution.*

The stationary state probabilities satisfy the equilibrium equations

$$0 = \sum_{j \in \mathcal{J}} a_{ji} p_j, \qquad i \in \mathcal{J}, \tag{1.5.17}$$

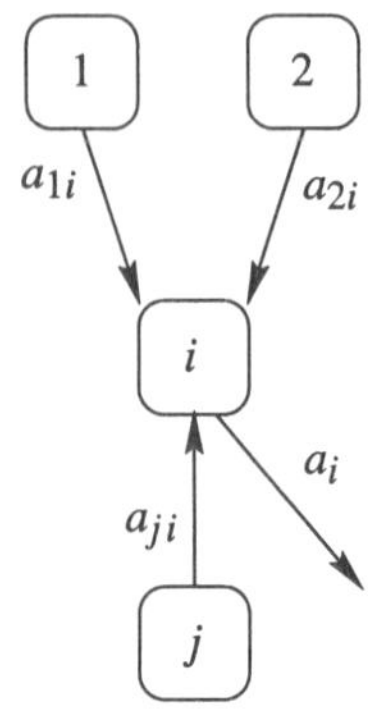

Figure 1.5.

obtained from (1.5.10) by substituting 0 for $p_i'(t)$, as well as the normalisation condition

$$\sum_{i\in\mathcal{J}} p_i = 1. \tag{1.5.18}$$

By virtue of the agreement that termination is possible, the necessary condition for the process $\{\eta(t),\ t \geq 0\}$ to be stationary is its regularity.

The inverse is true: if for the regular process $\{\eta(t),\ t \geq 0\}$ some set $\{p_i,\ i \in \mathcal{J}\}$ of nonnegative numbers satisfies the equilibrium equations (1.5.17) and the normalisation condition (1.5.18), then some stationary process with the state set $\mathcal{J}$ is generated by the infinitesimal matrix together with the initial distribution $p_i(0) = p_i$, $i \in \mathcal{J}$.

Since in what follows we often deal with the equilibrium equations (1.5.17), we dwell on the general principles of its construction. We represent all states i, $i \in \mathcal{J}$, of the process $\{\eta(t),\ t \geq 0\}$ as a planar graph (Fig. 1.5) and choose a state i. Then it would appear reasonable to treat $a_i p_i(t)$ as the *flow of probabilities leaving at the instant t the state i*. In turn, $a_{ji} p_j(t)$ is the flow of probabilities from the state j to the state i and $\sum_{j\in\mathcal{J}\setminus\{i\}} a_{ji} p_j(t)$ is the *total flow of probabilities entering the state i*. For the stationary process $\{\eta(t),\ t \geq 0\}$, these flows must be balanced, which leads to the equality

$$a_i p_i = \sum_{j\in\mathcal{J}\setminus\{i\}} a_{ji} p_j$$

which is the ith equation of the equilibrium equations (1.5.17). In the queueing theory, it is often called the *global balance equation* of the state i.

Along with the global balance equations, the queueing theory makes use of the local balance equations. Let the state set $\mathcal{J}$ be divided arbitrarily into two subsets $\mathcal{J}_1$ and $\mathcal{J}_2 = \mathcal{J} \setminus \mathcal{J}_1$ (Fig. 1.6). Then $\sum_{i\in\mathcal{J}_1} \sum_{j\in\mathcal{J}_2} a_{ij} p_i(t)$ is the total flow of probabilities from the subset $\mathcal{J}_1$ to the subset $\mathcal{J}_2$ and $\sum_{i\in\mathcal{J}_1} \sum_{j\in\mathcal{J}_2} a_{ji} p_j(t)$ is the flow of probabilities from the subset $\mathcal{J}_2$ to the subset $\mathcal{J}_1$. In the stationary case, we obtain by equating these flows the *local balance equation*

$$\sum_{i\in\mathcal{J}_1} \sum_{j\in\mathcal{J}_2} a_{ij} p_i = \sum_{i\in\mathcal{J}_1} \sum_{j\in\mathcal{J}_2} a_{ji} p_j$$

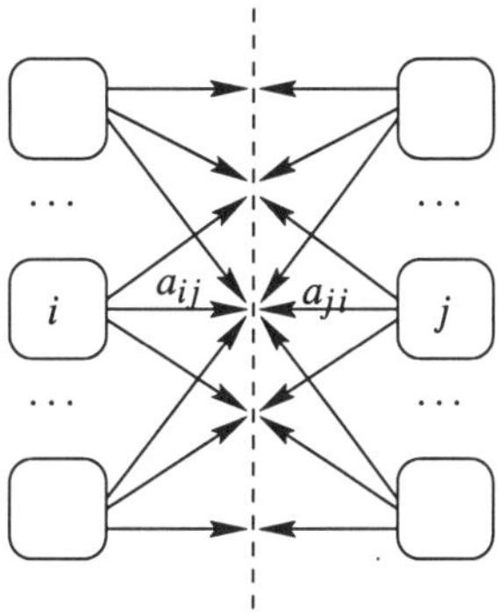

Figure 1.6.

between the subsets $\mathcal{S}_1$ and $\mathcal{S}_2$.

Finally, we mention another kind of balance between the states, the *partial balance* between the states i and j by which the equality

$$a_{ij} p_i = a_{ji} p_j$$

is meant. In contrast to its global and local counterparts, the partial balance is far from being always satisfied. However, if it is satisfied, then as a rule this leads to far-reaching consequences. In particular, the relation $p_i = a_{ji} p_j / a_{ij}$ between the stationary state probabilities p_i and p_j follows from the partial balance between the states i and j.

1.5.6. Ergodicity of the Markov process

The states i and j are called the *communicating states* if there are $t_1 > 0$ and $t_2 > 0$ such that $p_{ij}(t_1) > 0$ and $p_{ji}(t_2) > 0$. The states i and j are communicating if and only if $p_{ij}(t) > 0$ and $p_{ji}(t) > 0$ for all $t > 0$. The Markov process $\{\eta(t),\ t \geq 0\}$ is called *irreducible* if all its states are communicating. The Markov process is irreducible if and only if its embedded Markov chain is irreducible.

The Markov process $\{\eta(t),\ t \geq 0\}$ is called *ergodic* if there exists a probabilistic distribution $\{p_i,\ i \in \mathcal{S}\}$, $p_i > 0$, $i \in \mathcal{S}$, such that the transition probabilities $p_{ji}(t)$ satisfy the limit relation

$$p_{ji}(t) \xrightarrow[t\to\infty]{} p_i, \qquad i, j \in \mathcal{S}.$$

The distribution $\{p_i,\ i \in \mathcal{S}\}$ is called the *limit (final) distribution* of the process $\{\eta(t),\ t \geq 0\}$.

The limit probabilities p_i, $i \in \mathcal{S}$, of the ergodic Markov process are the stationary probabilities of some Markov process with the same infinitesimal matrix A and initial probabilities $p_i(0) = p_i$, $i \in \mathcal{S}$.

The ergodic Markov process has the following property similar to the property of the ergodic Markov chain. Let us consider the random variable $\zeta_T(i)$ which is equal to the total sojourn time of the process in the state i divided by the observation time T. Then $\zeta_T(i) \xrightarrow[T\to\infty]{} p_i$ with probability 1. Stated differently, the stationary probability p_i, $i \in \mathcal{S}$, of the ergodic Markov process can be treated as that part of time during which the process stays in the state i in a long time interval. Ergodicity of the Markov process, as well as of the chain, depends substantially on what (finite or countable) is the state set $\mathcal{J}$.

THEOREM 1.5.2 (ergodic theorem for a Markov process with finite state set). *Any irreducible Markov process* $\{\eta(t),\ t \geq 0\}$ *with finite state set* $\mathcal{J} = \{1, 2, \dots, m\}$ *is ergodic. Here the limit probabilities* p_i, $i = 1, \dots, m$, *are determined as the unique solution of the equilibrium equations* (1.5.17) *with the normalisation condition* (1.5.18).

THEOREM 1.5.3 (ergodic theorem for a Markov process with countable state set). *Let* $\{\eta(t),\ t \geq 0\}$ *be a regular irreducible Markov process with countable state set* $\mathcal{J} = \{1, 2, \dots\}$. *Then*

- *either* $p_{ij}(t) \xrightarrow[t\to\infty]{} 0$, $i, j \geq 1$,
- *or the process* $\{\eta(t),\ t \geq 0\}$ *is ergodic, the limit distribution* $\{p_i,\ i \geq 1\}$ *being determined as the unique solution of the equilibrium equations* (1.5.17) *satisfying the condition* $\sum_{i=1}^{\infty} |p_i| < \infty$ *and the normalisation condition* (1.5.18).

THEOREM 1.5.4 (Foster ergodic theorem). *For a regular irreducible Markov process* $\{\eta(t),\ t \geq 0\}$ *to be ergodic, it is necessary and sufficient that there exists a nontrivial solution* $\{p_i,\ i \geq 1\}$ *of the equilibrium equations* (1.5.17) *such that* $\sum_{i=1}^{\infty} |p_i| < \infty$; *this solution coincides with the limit distribution to within the normalising factor.*

We note that the notions of *non-recurrence, positive recurrence, and zero-recurrence* can be introduced for the Markov processes, as well as for the Markov chains. Again, zero-recurrence of the regular irreducible Markov process corresponds to the 'flight' of the process (in infinite time) to 'infinity' in probability, and non-recurrence, with probability 1. Positive recurrence of the irreducible Markov process amounts to its ergodicity.

1.5.7. Birth-and-death processes

We consider here one class of the Markov processes which describe the simplest Markov queueing systems. As we will see below, the equilibrium equations (1.5.17) can be solved for these processes explicitly, and the Kolmogorov differential equations (1.5.10) can be solved explicitly in terms of the Laplace transform.

Let the states of the Markov process $\{\eta(t),\ t \geq 0\}$ be enumerable so that (Fig. 1.7)

$$a_{ij} = \begin{cases} \lambda_i, & j = i + 1; \\ \mu_i, & j = i - 1; \\ -(\lambda_i + \mu_i), & j = i; \\ 0, & |j - i| \geq 2, \end{cases} \qquad i, j \in \mathcal{J}. \tag{1.5.19}$$

The process $\{\eta(t),\ t \geq 0\}$ is called the *birth-and-death process.*

We note that it is easier to treat the material if we enumerate the states of the birth-and-death process (and also some other processes with the discrete state set) beginning from 0 and not 1. Therefore, we assume that $\mathcal{J} = \{0, 1, \dots, m\}$ in the case of the finite set $\mathcal{J}$ and $\mathcal{J} = \{0, 1, \dots\}$ in the case of the countable set.

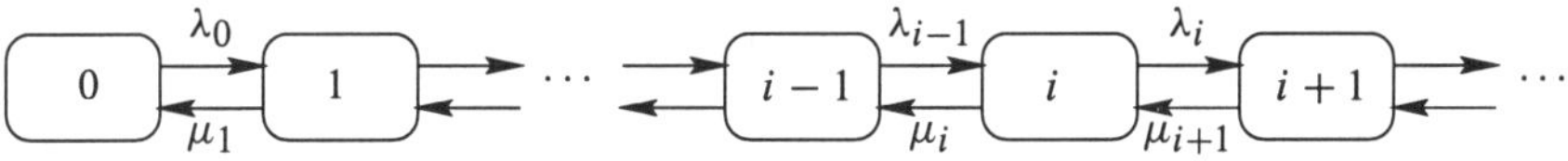

Figure 1.7.

The birth-and-death process has a very simple physical sense: direct transition from the state i can be done only to the adjacent states $(i+1)$ with intensity λ_i and $(i-1)$ with intensity μ_i. Additionally, since the state 0 has only one neighbouring state 1, transition from it is possible only to 1 and $a_{00} = -\lambda_0$. A similar remark is true for m in the case of the finite state set $\mathcal{I} = \{0, 1, \dots, m\}$ where $a_{mm} = -\mu_m$.

Everywhere below, unless otherwise specified, we assume that $\lambda_i > 0$ and $\mu_i > 0$ for all states i for which they are defined. Obviously, in this case the birth-and-death process are irreducible. The equilibrium equations of the process $\{\eta(t),\ t \geq 0\}$ are as follows:

$$\begin{aligned} -\lambda_0 p_0 + \mu_1 p_1 &= 0, \\ -(\lambda_i + \mu_i) p_i + \lambda_{i-1} p_{i-1} + \mu_{i+1} p_{i+1} &= 0, \qquad i \geq 1. \end{aligned} \tag{1.5.20}$$

In the case of the finite state set $\mathcal{I}$, one needs here to add an individual equation for p_m:

$$-\mu_m p_m + \lambda_{m-1} p_{m-1} = 0.$$

Simultaneous equations (1.5.20) are easily solved. Indeed, by rewriting it as

$$\begin{aligned} -\lambda_0 p_0 + \mu_1 p_1 &= 0, \\ -\lambda_i p_i + \mu_{i+1} p_{i+1} - (-\lambda_{i-1} p_{i-1} + \mu_i p_i) &= 0, \qquad i \geq 1, \end{aligned}$$

we obtain

$$-\lambda_{i-1} p_{i-1} + \mu_i p_i = 0, \qquad i \geq 1,$$

which yields the recurrence relation

$$p_i = \frac{\lambda_{i-1}}{\mu_i} p_{i-1} = \frac{\lambda_{i-1}}{\mu_i} \cdot \frac{\lambda_{i-2}}{\mu_{i-1}} p_{i-2} = \cdots, \qquad i \geq 1,$$

or finally,

$$p_i = p_0 \prod_{j=1}^{i} \frac{\lambda_{j-1}}{\mu_j}, \qquad i \geq 1. \tag{1.5.21}$$

In order to find p_0, we make use of the normalisation condition (1.5.18). In the case of countable $\mathcal{I}$, we obtain

$$p_0 = \left(1 + \sum_{i=1}^{\infty} \prod_{j=1}^{i} \frac{\lambda_{j-1}}{\mu_j} \right)^{-1}, \tag{1.5.22}$$

and in the case of finite $\mathcal{J}$,

$$p_0 = \left(1 + \sum_{i=1}^{m} \prod_{j=1}^{i} \frac{\lambda_{j-1}}{\mu_j}\right)^{-1}. \tag{1.5.23}$$

Therefore, in the case of the finite state set $\mathcal{J}$ the process $\{\eta(t),\ t \geq 0\}$ is ergodic and (1.5.21) and (1.5.23) determine its limit (stationary) distribution. In the case of countable $\mathcal{J}$, to check the process $\{\eta(t),\ t \geq 0\}$ for ergodicity, one must first to check it for regularity. Here, one can mostly use the regularity criterion (Theorem 1.5.1). Here, the transition probability matrix $Q = (q_{ij})$ of the embedded Markov chain $\{\nu_n,\ n \geq 0\}$ is as follows:

$$q_{01} = 1, \quad q_{0j} = 0, \quad j \neq 1,$$

$$q_{ij} = \begin{cases} \dfrac{\lambda_i}{\lambda_i + \mu_i}, & j = i+1; \\ \dfrac{\mu_i}{\lambda_i + \mu_i}, & j = i-1; \qquad i \geq 1, \quad j \geq 0. \\ 0, & |j-i| \neq 1, \end{cases}$$

Sometimes, the *sufficient Karlin–McGregor condition* (Karlin and McGregor, 1957b; Karlin and McGregor, 1957a)

$$\sum_{i=1}^{\infty} \prod_{j=1}^{i} \frac{\mu_j}{\lambda_j} = \infty \tag{1.5.24}$$

is of advantage for checking the birth-and-death process for regularity. If the process $\{\eta(t),\ t \geq 0\}$ is regular, then by the Foster theorem it follows from (1.5.21) and (1.5.22) that convergence of the series in the right-hand side of (1.5.22), that is, fulfilment of the condition

$$\sum_{i=1}^{\infty} \prod_{j=1}^{i} \frac{\lambda_{j-1}}{\mu_j} < \infty, \tag{1.5.25}$$

is the necessary and sufficient condition for existence of the limit (stationary) distribution $\{p_i,\ i \geq 0\}$. We note that simultaneous fulfilment of conditions (1.5.24) and (1.5.25) is not only *sufficient*, but also *necessary* for ergodicity of the birth-and-death process.

The transient distribution $\{p_i(t),\ i \in \mathcal{J}\}$ of the birth-and-death process $\{\eta(t),\ t \geq 0\}$ can also be found explicitly, but in terms of the Laplace transform. Indeed, simultaneous equations (1.5.16) for the Laplace transform $\pi_i(s)$ in this case are of the following form:

$$\begin{aligned} s\pi_0(s) - p_0(0) &= -\lambda_0\pi_0(s) + \mu_1\pi_1(s), \\ s\pi_i(s) - p_i(0) &= -(\lambda_i + \mu_i)\pi_i(s) + \lambda_{i-1}\pi_{i-1}(s) + \mu_{i+1}\pi_{i+1}(s), \qquad i \geq 1. \end{aligned} \tag{1.5.26}$$

Equations (1.5.26) can be solved recurrently, but we do not do so because in particular cases below they are solved with the use of special techniques.

As we have promised above, we conclude this section by demonstrating by way of a simple example what is the cause of irregularity of the process $\{\eta(t),\ t \geq 0\}$. Let the birth-and-death process $\{\eta(t),\ t \geq 0\}$ satisfy the condition $\mu_i = 0$, $i \geq 1$. Then transition from the state i is possible only to the state $(i+1)$, and it is reasonable to refer to this process as the *process of pure birth.*

We assume for simplicity that at the initial instant 0 the process $\{\eta(t),\ t \geq 0\}$ of pure birth with the countable state set is in the state 0, that is, $p_0(0) = 1,\ p_i(0) = 0,\ i \geq 1$. Equations (1.5.26) take the following form:

$$s\pi_0(s) - 1 = -\lambda_0\pi_0(s),$$
$$s\pi_i(s) = -\lambda_i\pi_i(s) + \lambda_{i-1}\pi_{i-1}(s), \qquad i \geq 1.$$

We solve it and find that

$$\pi_0(s) = \frac{1}{s+\lambda_0},$$
$$\pi_i(s) = \frac{1}{s+\lambda_0}\prod_{j=1}^{i}\frac{\lambda_{j-1}}{s+\lambda_j}, \qquad i \geq 1.$$

In particular, by setting $\pi_\infty(s) = 1/s - \sum_{i=0}^{\infty}\pi_i(s)$, we obtain

$$\pi_\infty(s) = \frac{1}{s}\prod_{j=0}^{\infty}\frac{\lambda_j}{s+\lambda_j}. \tag{1.5.27}$$

The function $\pi_\infty(s)$ is the Laplace transform of the probability $p_\infty(t)$ that the process $\{\eta(t),\ t \geq 0\}$ terminates before the instant t. Formula (1.5.27) can be obtained without use of (1.5.26). For this purpose, it suffices to note that the time of reaching 'infinity' by the pure birth process is the sum of an 'infinite number' of independent random variables that are exponentially distributed with the parameters λ_i, $i \geq 0$; now we obtain (1.5.27) by passing to the Laplace transform.

By means of (1.5.27), one can demonstrate that $p_\infty(t) = 0$ for all $t > 0$ if and only if

$$\sum_{i=0}^{\infty}\frac{1}{\lambda_i} = \infty; \tag{1.5.28}$$

otherwise, $p_\infty(t) > 0$ also for all $t > 0$ (see also (Feller, 1966)). Therefore, condition (1.5.28) is necessary and sufficient for the pure birth process to be regular. Condition (1.5.28) can be readily understood if one recalls that $1/\lambda_i$ is the mean time of sojourn in the state i. Then $\sum_{i=0}^{\infty} 1/\lambda_i$ is the mean time spent by the process in all states.

If the pure birth process has $\lambda_i = \lambda$, $i \geq 0$, then it is called the *Poisson process.*

1.5.8. Non-homogeneous Markov processes

Sometimes we have to depart from the time-homogeneous Markov processes with the discrete state set. True, the non-homogeneous processes to be used have a simple structure which does not differ very much from the structure of the homogeneous processes. The only difference between them lies in that the transition intensities a_{ij} can depend on time t: $a_{ij} = a_{ij}(t)$. For these processes, the set of Kolmogorov differential equations (1.5.10) holds true as well, the intensities a_{ij} being time-dependent. Since the non-homogeneous processes can appear only occasionally, we do not dwell on them now and consider the techniques of solving the Kolmogorov differential equations (1.5.10) separately for each particular case.

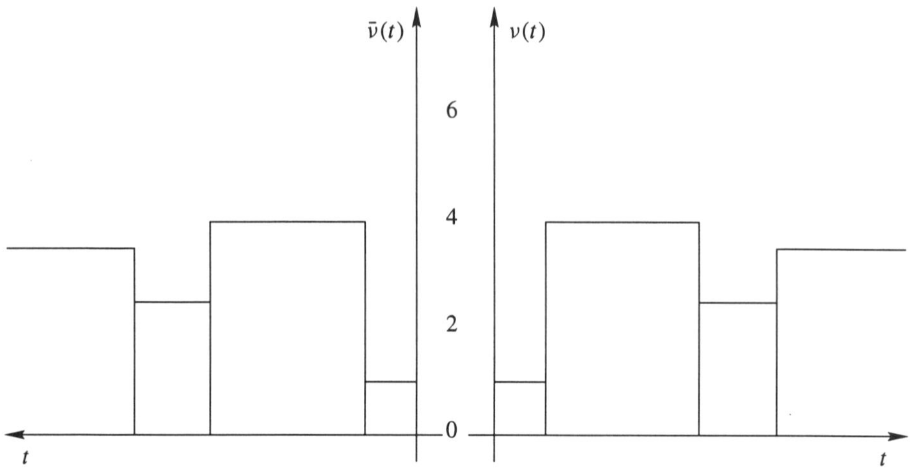

Figure 1.8.

1.5.9. Reversed Markov processes

The output flow of the queueing system is of special interest in the studies of queueing networks. The following method of reversed time is of great help here. First, we assume that some (regular and irreducible) Markov process $\{\eta(t),\ t \geq 0\}$ is defined in the time interval $[0, \infty)$. We assume for definiteness that $\eta(0) = 1$, that is, at the initial instant 0 the process is in the state 1 with probability 1. In the case of the countable state set $\mathcal{I}$ of the process $\{\eta(t),\ t \geq 0\}$, we also assume that a_{ji} are all bounded.

The *reversed* process $\{\tilde{\eta}(t),\ t \leq 0\}$ is obtained from the process $\{\eta(t),\ t \geq 0\}$ by substituting time $-t$ for t, that is, $\tilde{\eta}(t) = \eta(-t)$ (Fig. 1.8).

Since due to the fact that the process $\{\eta(t),\ t \geq 0\}$ is Markov, its 'past' and 'future' under fixed 'present' are independent and just change places while going from the process $\{\eta(t),\ t \geq 0\}$ to the process $\{\tilde{\eta}(t),\ t \leq 0\}$, and the process $\{\tilde{\eta}(t),\ t \leq 0\}$ is Markov as well.

Let us see what is the reversed process $\{\tilde{\eta}(t),\ t \leq 0\}$. Its state set coincides with the state set $\mathcal{I}$ of the process $\{\eta(t),\ t \geq 0\}$, that is, it is discrete. We prove that for the process $\{\tilde{\eta}(t),\ t \leq 0\}$ there exist for all $t < 0$ the intensities

$$\tilde{a}_{ij}(t) = \lim_{\Delta \to 0} \frac{\mathbf{P}\{\tilde{\eta}(t+\Delta) = j \mid \tilde{\eta}(t) = i\}}{\Delta}, \quad i, j \in \mathcal{I}, \quad j \neq i.$$

To this end, we make use of the definitions of conditional probability and reversed process. Then

$$\frac{\mathbf{P}\{\tilde{\eta}(t+\Delta) = j \mid \tilde{\eta}(t) = i\}}{\Delta} = \frac{\mathbf{P}\{\tilde{\eta}(t+\Delta) = j, \tilde{\eta}(t) = i\}}{\Delta \mathbf{P}\{\tilde{\eta}(t) = i\}}$$
$$= \frac{\mathbf{P}\{\eta(-t-\Delta) = j, \eta(-t) = i\}}{\Delta \mathbf{P}\{\eta(-t) = i\}}.$$

By replacing now $\mathbf{P}\{\eta(-t-\Delta) = j, \eta(-t) = i\}$ by the equivalent infinitesimal value

$p_j(-t-\Delta)a_{ji}\Delta \sim p_j(-t)a_{ji}\Delta$ and letting Δ go to zero, we obtain

$$\tilde{a}_{ij}(t) = \frac{a_{ji}p_j(-t)}{p_i(-t)}.$$

Similarly,

$$\begin{aligned}\frac{\mathbf{P}\{\tilde{\eta}(t+\Delta)=i \mid \tilde{\eta}(t)=i\}-1}{\Delta} &= \frac{\mathbf{P}\{\tilde{\eta}(t+\Delta)=i, \tilde{\eta}(t)=i\}-\mathbf{P}\{\tilde{\eta}(t)=i\}}{\Delta\mathbf{P}\{\tilde{\eta}(t)=i\}} \\ &= \frac{\mathbf{P}\{\eta(-t-\Delta)=i, \eta(-t)=i\}-\mathbf{P}\{\eta(-t)=i\}}{\Delta\mathbf{P}\{\eta(-t)=i\}} \\ &= -\frac{\mathbf{P}\{\eta(-t-\Delta)\neq i, \eta(-t)=i\}}{\Delta\mathbf{P}\{\eta(-t)=i\}} \\ &= -\frac{\sum_{j\neq i}\mathbf{P}\{\eta(-t-\Delta)=j, \eta(-t)=i\}}{\Delta\mathbf{P}\{\eta(-t)=i\}}.\end{aligned}$$

Again, replacing $\mathbf{P}\{\eta(-t-\Delta)=j, \eta(-t)=i\}$ by the equivalent infinitesimal values $p_j(-t-\Delta)a_{ji}\Delta \sim p_j(-t)a_{ji}\Delta$ and letting Δ tend to zero, we obtain

$$\tilde{a}_{ii}(t) = \lim_{\Delta\to 0}\frac{\mathbf{P}\{\tilde{\eta}(t+\Delta)=i \mid \tilde{\eta}(t)=i\}-1}{\Delta} = -\frac{\sum_{j\neq i}a_{ji}p_j(-t)}{p_i(-t)}, \qquad i\in\mathcal{J}.$$

Finally, we obtain

$$\tilde{a}_{ij}(t) = \begin{cases}\dfrac{a_{ji}p_j(-t)}{p_i(-t)}, & j\neq i;\\ -\dfrac{\sum_{j\neq i}a_{ji}p_j(-t)}{p_i(-t)}, & j=i,\end{cases} \qquad i,j\in\mathcal{J}. \tag{1.5.29}$$

We see that the reversed process $\{\tilde{\eta}(t),\ t\leq 0\}$ is a non-homogeneous Markov process described in Section 1.5.8. Its non-homogeneity becomes obvious if one recalls that it must fall in the state 1 at the instant 0.

Let now the process $\{\eta(t),\ -\infty<t<\infty\}$ defined in the time interval $(-\infty,\infty)$ be stationary with the distribution $\{p_i,\ i\in\mathcal{J}\}$. Then the reversed process $\{\tilde{\eta}(t), -\infty<t<\infty\}$ is also defined in the interval $(-\infty,\infty)$. That said above about the interval $(-\infty,0]$ applies to the interval $(-\infty,\infty)$. By virtue of (1.5.29), the reversed process is also homogeneous and stationary with the same distribution $\{p_i,\ i\in\mathcal{J}\}$, its infinitesimal matrix $\tilde{A}=(\tilde{a}_{ij})$ obeying the formula

$$\tilde{a}_{ij} = \begin{cases}\dfrac{a_{ji}p_j}{p_i}, & j\neq i;\\ -\dfrac{\sum_{j\neq i}a_{ji}p_j}{p_i}, & j=i,\end{cases} \qquad i,j\in\mathcal{J}.$$

1.6. Semi-Markov, linearwise, and piecewise-linear processes

Continuous-time Markov processes with discrete state set are insufficient for studying non-Markov queueing systems. Therefore, we describe below some other classes of random processes which can also be classified as Markov processes whose state sets are no more discrete. Although, in view of the definition below, the semi-Markov proccss and the Markov chain with the semi-Markov control have a discrete state set, they are not Markov; if it is desired to markovise them, then a continuous component must be added to them.

The processes below have a distinctive feature which is characteristic of all queueing systems: substantial changes in them can occur only at the discrete time instants or, according to a vivid expression of A. N. Kolmogorov, 'they are random processes with discrete interference of contingency.' The processes considered in Sections 1.3–1.5 belong to this type. Their internal point in common is the presence of the regenerative (renewal) points.

In this section, we confine our consideration only to the constructive description of the processes under study. We underline that, as before, we do not pursue general definitions, but as a rule impose too-strict constraints which would enable us to simplify substantially the description, but are satisfied in what follows.

1.6.1. Semi-Markov processes

A semi-Markov process is a natural generalisation of the continuous-time Markov process with discrete state set $\mathcal{I} = \{1, 2, \dots, m\}$ or $\mathcal{I} = \{1, 2, \dots\}$ to the case where the distribution of the time of sojourn in the state i can be other than exponential. Constructive description of the semi-Markov process $\{\nu(t),\ t \geq 0\}$ is as follows. Let there exist

- a distribution $\{p_i(0),\ i \in \mathcal{I}\}$ called the initial distribution of the process;
- a set of distribution functions $F_i(x)$, $i \in \mathcal{I}$, defining the time of sojourn of the process at the state i; and
- a set of nonnegative functions $q_{ij}(x)$, $i, j \in \mathcal{I}$, $\sum_{j \in \mathcal{I}} q_{ij}(x) = 1$, $i \in \mathcal{I}$, which are the probabilities of direct transitions from the state i to the state j provided that the process stays at the state i during the period x.

Behaviour of semi-Markov process $\{\nu(t),\ t \geq 0\}$ is defined as follows (Fig. 1.9). First, at the instant 0 the process is in the state i with probability $p_i(0)$. It stays in this state during a random time τ_1 distributed by the law $F_i(x)$. Then the process goes to the state j with probability $q_{ij}(x)$ which depends on the time x of sojourn in the state i and stays there during the random time $\tau_2 - \tau_1$ distributed by the law $F_j(x)$, no matter when and how the process got into the state j. Then with probability $q_{jk}(y)$ depending only on the time y of sojourn in the state j it goes to the state k, and so on. The value of the semi-Markov process $\nu(t)$ at the time t is the state i in which it is at the given instant.

When defining the semi-Markov process, we made another minor generalisation concerning the fact that, in contrast to the Markov process, the probability $q_{ii}(x)$ can also be not equal to zero, that is, transition from the state i into itself is possible.

If one considers the state of the process $\nu_n = \nu(\tau_n + 0)$ immediately after the nth transition, one gets the Markov chain $\{\nu_n,\ n \geq 0\}$ which can be reasonably called, as in the case of the Markov process, the *embedded Markov chain* of the semi-Markov process $\{\nu(t),\ t \geq 0\}$. By the formula of total probability, the transition probabilities $q_{ij} = \mathbf{P}\{\nu_{n+1} = j \mid \nu_n = i\}$ of the embedded Markov chain $\{\nu_n,\ n \geq 0\}$ are

$$q_{ij} = \int_0^\infty q_{ij}(x)\, dF_i(x).$$

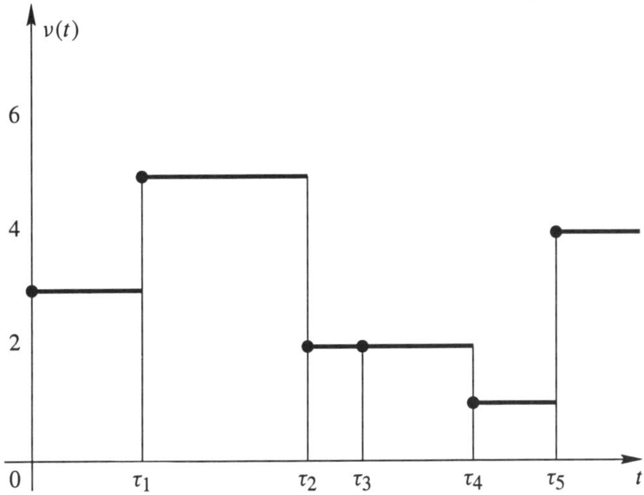

Figure 1.9.

In particular, the probabilities $q_{ij}(x)$ can be independent of x, and in this case it is reasonable that $q_{ij}(x) = q_{ij}$ for any $x \geq 0$, $i, j \in \mathcal{I}$.

As in the case of the Markov process, the so-defined semi-Markov process is not defined completely because exit to 'infinity' in a finite time is possible. Now we present conditions which are readily verifiable, on the one hand, and guarantee against anomalous situations, on the other hand. We take

$$F^+(x) = \sup_{i \in \mathcal{I}} F_i(x), \qquad F^-(x) = \inf_{i \in \mathcal{I}} F_i(x)$$

and assume that

(1) $F^+(0+) < 1$ and $F^-(x)$ is the distribution function, that is, $F^-(+\infty) = \lim_{x \to \infty} F^-(x) = 1$, and there exists a (finite) expectation $f^- = \int_0^\infty x\, dF^-(x)$.

The function $F^+(x)$ is, obviously, a distribution function, and $0 < f^+ = \int_0^\infty x\, dF^+(x) \leq f^- < \infty$ if Condition 1 is satisfied. Additionally, by denoting by $f_i = \int_0^\infty x\, dF_i(x)$ the mean time of sojourn in the state i for the process, we obtain $f^+ \leq f_i \leq f^-$, $i \in \mathcal{I}$.

We also assume that another technical condition is satisfied as well.

(2) There exists no $l > 0$ such that all $F_i(x)$ are the distribution functions of discrete random variables taking the values kl, $k \geq 0$ only.

We note that if the latter condition is not satisfied, then a semi-Markov discrete-time process results.

Let now $H_i(t)$ be the mean number of transitions to the state i in time t.

THEOREM 1.6.1. *Let Conditions* 1 *and* 2 *be satisfied for the process* $\{\nu(t),\ t \geq 0\}$ *and the embedded Markov chain* $\{\nu_n,\ n \geq 0\}$ *be irreducible and non-periodic. If the embedded Markov chain is ergodic, then for any* $s > 0$

$$H_i(t+s) - H_i(t) \xrightarrow[t\to\infty]{} \frac{sp_i^*}{\sum_{j\in\mathcal{I}} f_j p_j^*}, \qquad i \in \mathcal{I},$$

where p_i^* *are the limit (stationary) probabilities for the embedded Markov chain; otherwise,*

$$H_i(t+s) - H_i(t) \xrightarrow[t\to\infty]{} 0, \qquad i \in \mathcal{I}.$$

Theorem 1.6.1 is a generalisation of the Blackwell theorem to the semi-Markov processes. Since from this viewpoint the properties of the semi-Markov process are adequate to those of the renewal process, the semi-Markov process sometimes is called the *Markov renewal process.*

Theorem 1.6.1 can be formulated in a form similar to Smith theorem. Let us assume that $p_i(t) = \mathbf{P}\{\nu(t) = i\}, i \in \mathcal{I}$. The result below follows from Theorem 1.6.1.

THEOREM 1.6.2. *Let the hypotheses of Theorem* 1.6.1 *be satisfied for the process* $\{\nu(t),\ t \geq 0\}$. *If the embedded Markov chain* $\{\nu_n,\ n \geq 0\}$ *is ergodic, then*

$$p_i(t) \xrightarrow[t\to\infty]{} \frac{f_i p_i^*}{\sum_{j\in\mathcal{I}} f_j p_j^*}, \qquad i \in \mathcal{I};$$

otherwise,

$$p_i(t) \xrightarrow[t\to\infty]{} 0, \qquad i \in \mathcal{I}.$$

Theorem 1.6.2 relates the limit (stationary) probabilities of the states of the semi-Markov process and its embedded Markov chain.

1.6.2. Markov chains with the semi-Markov control

The semi-Markov process, unfortunately, is used in the queueing theory mostly as an auxiliary tool. To introduce it, one has to make additional, often complicated, constructions. The inverse passage from the semi-Markov process to the queueing system performance indices is also rather difficult. We are going to describe another class of random processes which can be directly used to analyse the queueing system. These processes differ from the semi-Markov process in that additional jumps obeying the laws of the Markov process are admitted in addition to the state changes controlled by the semi-Markov process. For ease of presentation we somewhat simplify the structure of the 'reference' semi-Markov process, though, and require that

(i) the state set be finite and

(ii) the transition probabilities $q_{ij}(t) = q_{ij}$ be time-independent.

As above, we define the process from its constructive description. Since we are interested here only in the stationary characteristics, we do not dwell on defining the initial state of the process.

Let us consider the random process $\{\eta(t),\ t \geq 0\} = \{(\nu(t), \zeta(t)),\ t \geq 0\}$ with a finite state set $\mathscr{X} \subset \mathscr{I} \times \mathscr{N}$, which is a subset of the set $\mathscr{I} \times \mathscr{N}$ of all possible pairs (i, n), $i \in \mathscr{I}$, $n \in \mathscr{N}$. Let at the instant τ_{k-1}, $k \geq 1$, the component $\zeta(t)$ of the process $\eta(t)$ change and take the value n. Then it remains equal to n until the instant τ_k, the difference $\tau_k - \tau_{k-1}$ being independent of the history of process and distributed by the law $F_n(x)$, $F_n(0+) = 0$. Further, during the time (τ_{k-1}, τ_k) when the component $\zeta(t)$ is in the state n, the component $\nu(t)$ is a (homogeneous) Markov process with the infinitesimal matrix $A_n = (a_n^{i,j})_{i,j \in \mathscr{I}_n}$, where $\mathscr{I}_n = \{i \in \mathscr{I} \mid (i, n) \in \mathscr{X}\}$ is the set of those values i which $\nu(t)$ can take when the component $\zeta(t)$ takes the value n. At the instant τ_k of change of state of the component $\zeta(t)$, the states of the component $\nu(t)$ change as well, the components $\nu(t)$ and $\zeta(t)$ taking with probability

$$q_{n,m}^{i,j} = \mathbf{P}\{\nu(\tau_k) = j, \zeta(\tau_k) = m \mid \nu(\tau_k - 0) = i, \zeta(\tau_k - 0) = n\}, \qquad (i, n), (j, m) \in \mathscr{X},$$

the values j and m, respectively, independently of behaviour of the process $\{\eta(t),\ t \geq 0\}$ until the instant τ_k. The so-defined process $\{\eta(t),\ t \geq 0\}$ is called the *Markov chain with the semi-Markov control.*

In this subsection we consider only the case where the Laplace–Stieltjes transform of the distribution function $F_n(x)$, $n \in \mathscr{N}$, is of rational form, that is, is a polynomials ratio. As we will see below, the stationary distributions then can be determined from simultaneous equations that are like equilibrium equations, but are not necessarily actual equilibrium ones. In the general case, the Markov chains with semi-Markov control can be considered either by constructing an embedded semi-Markov process or by the method of introducing an supplementary variable which will be considered in the next subsection.

Let us consider now two embedded Markov chains $\{\eta_k^+,\ k \geq 1\} = \{(\nu(\tau_k), \zeta(\tau_k)),\ k \geq 1\}$ and $\{\eta_k^-,\ k \geq 1\} = \{(\nu(\tau_k - 0), \zeta(\tau_k - 0)),\ k \geq 1\}$. We note that the set $\mathscr{X}^+$ of states of the Markov chain $\{\eta_k^+,\ k \geq 1\}$ does not necessarily coincide with the set $\mathscr{X}$ of states of the chain $\{\eta(t),\ t \geq 0\}$, because one often encounters chains $\{\eta(t),\ t \geq 0\}$ with zero probabilities of hitting some states from $\mathscr{X}$ immediately after the instants of changes of values of the component $\zeta(t)$; it is reasonable to ignore these states. We assume that the Markov chains $\{\eta(t),\ t \geq 0\}$, $\{\eta_k^+,\ k \geq 1\}$ and $\{\eta_k^-,\ k \geq 1\}$ are ergodic.

We take

$$p(j, m) = \lim_{t \to \infty} \mathbf{P}\{\eta(t) = (j, m)\}, \qquad p^{\pm}(j, m) = \lim_{k \to \infty} \mathbf{P}\{\eta_k^{\pm} = (j, m)\}.$$

Since we assumed that the Laplace–Stieltjes transform of the distribution function $F_n(x)$, $n \in \mathscr{N}$, are rational functions, there exists for $F_n(x)$ the matrix-exponential representation (Bocharov and Naumov, 1977)

$$F_n(x) = 1 - \boldsymbol{\gamma}_n^T e^{-G_n x} \mathbf{1}, \ x \geq 0, \ n \in \mathscr{N}, \tag{1.6.1}$$

where G_n is an $l_n \times l_n$ matrix whose eigen-values have positive real parts and $\boldsymbol{\gamma}_n$ is a vector of dimensionality l_n. We assume that $\boldsymbol{g}_n = G_n \mathbf{1}$. The following theorem is valid.

THEOREM 1.6.3. *There exists a unique set of vectors $\boldsymbol{x}(i, n) \in \mathbb{R}^{l_n}$, $(i, n) \in \mathscr{X}$, satisfy-*

ing the simultaneous equations

$$x^T(j,m)G_m = \sum_{i\in\mathcal{I}_m} x^T(i,m)a_m^{i,j} + \sum_{(i,n)\in\mathcal{X}} x^T(i,n)g_n q_{n,m}^{i,j}\gamma_m^T, \qquad (j,m)\in\mathcal{X}; \tag{1.6.2}$$

$$\sum_{(i,n)\in\mathcal{X}} x^T(i,n)\mathbf{1} = 1; \tag{1.6.3}$$

for all $(j,m)\in\mathcal{X}$

$$p^+(j,m) = \sum_{(i,n)\in\mathcal{X}} (x^T(i,n)g_n)^{-1} \sum_{(i,n)\in\mathcal{X}} x^T(i,n)g_n q_{n,m}^{i,j}, \tag{1.6.4}$$

$$p(i,n) = x^T(i,n)\mathbf{1}, \tag{1.6.5}$$

and the stationary probabilities $p^+(i,n)$ *and* $p^-(i,n)$ *are related by*

$$p^+(j,m) = \sum_{(i,n)\in\mathcal{X}} p^-(i,n)q_{n,m}^{i,j}. \tag{1.6.6}$$

Now, we consider the set $\mathcal{X}^2 = \mathcal{X}\times\mathcal{X}$ and fix some subset $\tilde{\mathcal{X}}\subset\mathcal{X}^2$. Let $0<\tau_{l_1}<\tau_{l_2}<\dots$ be a subsequence of the transition instants of the semi-Markov component of the process $\{\eta(t),\ t\ge 0\}$ such that $(\eta(\tau_{l_k}-0),\eta(\tau_{l_k}))\in\tilde{\mathcal{X}}$, and let $\tilde{\theta}_k^- = \eta(\tau_{l_k}-0)$, $\tilde{\theta}_k^+ = \eta(\tau_{l_k})$, $k\ge 1$. One can prove that the subsequences $\{\tilde{\theta}_k^+\ k\ge 1\}$ and $\{\tilde{\theta}_k^-,\ k\ge 1\}$ constitute homogeneous Markov chains which are assumed to be ergodic.

We set

$$\tilde{p}^\pm(i,n) = \lim_{k\to\infty} \mathbf{P}\{\tilde{\theta}_k^\pm = (i,n)\}, \qquad (i,n)\in\mathcal{X},$$

and introduce the set

$$\tilde{\mathcal{X}}_{(j,m)} = \{(i,n) \mid ((i,n),(j,m))\in\tilde{\mathcal{X}}\}.$$

The limit (stationary) probabilities $\tilde{p}^\pm(i,n)$ are defined by the following theorem.

THEOREM 1.6.4. *For all* $(j,m)\in\mathcal{X}$,

$$\tilde{p}^+(j,m) = \left(\sum_{(i,n),(k,l)\in\mathcal{X}} x^T(i,n)g_n q_{n,l}^{i,k}\right)^{-1} \sum_{(i,n)\in\tilde{\mathcal{X}}_{(j,m)}} x^T(i,n)g_n q_{n,m}^{i,j}, \tag{1.6.7}$$

and the stationary probabilities $\tilde{p}^+(j,m)$ *and* $\tilde{p}^-(i,n)$ *are related by*

$$\tilde{p}^+(j,m) = \sum_{(i,n)\in\tilde{\mathcal{X}}_{(j,m)}} \tilde{p}^-(i,n)q_{n,m}^{i,j}, \qquad (j,m)\in\mathcal{X}. \tag{1.6.8}$$

We conclude by saying that the type of processes considered above—Markov chains with semi-Markov control—is a special case of a wider class of semi-regenerative processes.

1.6.3. Linearwise Markov processes

The concept of a linearwise Markov process is as follows. A Markov chain with semi-Markov control from the last subsection is not a Markov process, but it becomes such a process by just attaching a supplementary continuous coordinate (variable) which follows the time which the component $\zeta(t)$ stays in the state n. Two cases are possible here: to take as the supplementary variable either the time $\xi^+(t)$ remaining for the component $\zeta(t)$ to be in the state n or the time $\xi^-(t)$ which the component already spent in this state. These two cases yield two linearwise process types.

Now, we proceed to describing constructively a linearwise Markov process of the first type and begin with the set of values $\mathscr{X}$. For further purposes, it is advisable to segregate the state 0 for which the continuous component is not defined. In the remaining cases, the state of the linearwise process $\{\eta(t),\ t \geq 0\}$ is defined by the pair $(\nu(t), \xi(t))$, where the state set $\mathscr{I}$ of the discrete component $\nu(t)$ is finite ($\mathscr{I} = \{1, 2, \dots, m\}$) or countable ($\mathscr{I} = \{1, 2, \dots\}$) and the state set of the continuous component $\xi(t)$ is the semi-axis $[0, \infty)$. Therefore, the set of values $\mathscr{X}$ of the linearwise process $\{\eta(t),\ t \geq 0\}$ is of the form $\mathscr{X} = \{0;\ (i, x),\ i \in \mathscr{I},\ x \geq 0\}$.

The defining parameters of the linearwise process $\{\eta(t),\ t \geq 0\}$ are as follows:

- the initial probabilities $p_0(0) = \mathbf{P}\{\eta(0) = 0\}$ that at the instant 0 the process is in the state 0 and $P_i(x, 0) = \mathbf{P}\{\nu(0) = i, \xi(0) < x\}, i \in \mathscr{I}$, that at the instant 0 the discrete component takes on the value i, and the continuous one, a value smaller than x;

- the distribution function $F_i(x)$, $i \in \mathscr{I}$, of the value of jump of the continuous component provided that at the instant of jump the discrete component had the value i;

- the probabilities q_{ij}, $i \in \mathscr{I}$, $j \in \mathscr{I} \cup \{0\}$, of transition of the discrete component from the state i to the state j at the instant when the continuous component vanishes;

- the intensities a_{ij}, $i, j \in \mathscr{I} \cup \{0\}$, of the additional transitions of the discrete component from the state i to the state j which obey the laws of the Markov process.

The following conditions which are readily verifiable will be always satisfied below:

(1) $a_{i0} = 0, i \in \mathscr{I}$.

(2) $\sum_{j \in \mathscr{I} \cup \{0\}} a_{ij} = 0, i \in \mathscr{I} \cup \{0\}$.

(3) $a_0 = -a_{00} > 0$.

(4) One can go in some time with a nonzero probability from the state with the value i, $i \in \mathscr{I}$, of the discrete component to a state with the value j, $j \in \mathscr{I} \cup \{0\}$ of the discrete component.

(5) Intensities $a_i = -a_{ii}$, $i \in \mathscr{I}$, are bounded uniformly.

(6) The function $F^+(x) = \sup_{i \in \mathscr{I}} F_i(x)$ satisfies the equality $F^+(0+) = 0$.

(7) The function $F^-(x) = \inf_{i \in \mathscr{I}} F_i(x)$ is a distribution function, that is, $F^-(+\infty) = 1$, with (finite) expectation $f^- = \int_0^\infty x\, dF^-(x)$.

Condition 1 implies that the additional transitions cannot drive the discrete component to the state 0. Condition 2 requires that the Markov process controlling the additional transitions must be conservative. Conditions 3 and 4 can be reasonably called irreducibility conditions. Conditions 5 and 6 guarantee that the process is regular. Finally, as will be seen below, Conditions 6 and 7 make process ergodicity to be defined by the ergodicity of its embedded Markov chain. We note that in the case of finite $\mathcal{I}$ Condition 5 is satisfied automatically and Conditions 6 and 7 are equivalent to the conditions $F_i(0+) = 0, i \in \mathcal{I}$, and conditions for existence of the (finite) expectations $f_i = \int_0^\infty x\,dF_i(x), i \in \mathcal{I}$.

The time behaviour of the linearwise process $\{\eta(t),\ t \geq 0\}$ is as follows. At the initial instant 0, either the value 0 of the process is chosen with probability $p_0(0)$ or the values i and x of the discrete and continuous components, respectively, are chosen according to the distribution $P_i(x, 0), i \in \mathcal{I}$. Further (Fig.. 1.10), the continuous component $\xi(t)$ decreases with unity rate until it goes to 0. The discrete component $\nu(t)$ can vary here: if at the instant t it had the value j, then with probability $a_{jk}\Delta + o(\Delta)$, which is independent of process behaviour until the instant t, it can change its value in a 'small' time Δ to k, $k \in \mathcal{I}$, the change itself producing no impact on the continuous component $\xi(t)$. At the instant when the continuous component $\xi(t)$ vanishes, the value of the discrete component changes from j to k, $k \in \mathcal{I} \cup \{0\}$ with probability q_{jk} which is also independent of the history. Then a new value of the continuous component $\xi(t)$ is determined using the distribution $F_k(x)$, and so on. If at the initial or some other instant when the continuous component vanishes the discrete component takes zero value, then the continuous component $\xi(t)$ is not determined and the process $\eta(t)$ stays in the state 0 during time exponentially distributed with parameter a_0 and then with probability a_{0i}/a_0 goes to the state with the discrete component equal to i.

If we consider the sequence $\nu_n = \nu(\tau_n + 0)$ of values of the discrete component immediately after the nth instant τ_n of vanishing of the continuous component, we obtain the Markov chain $\{\nu_n,\ n \geq 1\}$ which can be reasonably called the *embedded Markov chain* of the linearwise process $\{\eta(t),\ t \geq 0\}$. Here, the state set of the embedded Markov chain $\{\nu_n,\ n \geq 1\}$ comprises only those states i which the chain gets into at least for one $n > 0$ with non-zero probability.

The functions $p_0(t) = \mathbf{P}\{\eta(t) = 0\}$ and $P_i(x, t) = \mathbf{P}\{\nu(t) = i, \xi(t) < x\}$ describe the time behaviour of the linearwise process $\{\eta(t),\ t \geq 0\}$. Let us write *simultaneous equations* similar to the Kolmogorov simultaneous differential equations for $p_0(t)$ and $P_i(x, t)$. To this end, we consider the state of the process $\{\eta(t),\ t \geq 0\}$ at the instants t and $t + \Delta$, where Δ is a 'small' time increment. At the instant $t + \Delta$, the process is in a state with the zero discrete component if at the instant t either

- it was in a state with the discrete component equal to 0 (with probability $p_0(t)$) and in time Δ it did not change (with probability $1 - a_0\Delta + o(\Delta)$), or

- the discrete component was equal to i, $i \in \mathcal{I}$, and the continuous component was smaller than Δ, that is, $\nu(t) = i$ and $\xi(t) < \Delta$ (the probability of this event is $P_i(\Delta, t)$), and upon vanishing of the continuous component the process went with probability q_{i0} into the state 0.

The probability of the remaining possible transitions is $o(\Delta)$.

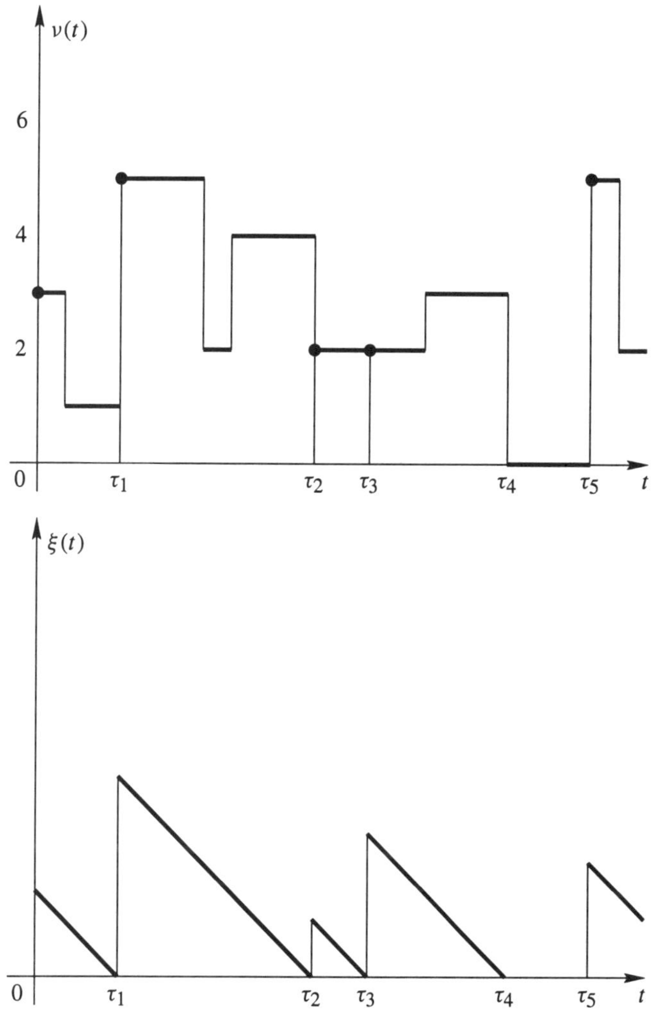

Figure 1.10.

Therefore, from the formula of total probability we obtain

$$p_0(t+\Delta) = p_0(t)(1-a_0\Delta) + \sum_{i\in\mathcal{I}} P_i(\Delta, t)q_{i0} + o(\Delta). \tag{1.6.9}$$

For the process $\{\eta(t),\ t \geq 0\}$ at the instant $t+\Delta$ to be in a state with the discrete component i, $i \in \mathcal{I}$, and the continuous component smaller than x, it is necessary that at the instant t either

- the discrete component is in the state i and the continuous one lies within the range from Δ to $x+\Delta$ (with probability $P_i(x+\Delta, t) - P_i(\Delta, t)$) and in the time Δ the discrete component did not change with probability $1 - a_i\Delta + o(\Delta)$; or

- the discrete component was in the state j, the continuous one was within the range from Δ to $x+\Delta$ (with probability $P_j(x+\Delta,t)-P_j(\Delta,t)$) and in the time Δ the discrete component changed its value from j to i (with probability $a_{ji}\Delta+o(\Delta)$); or
- the discrete component was in the state j, the continuous one was within the range from 0 to Δ (with probability $P_j(\Delta,t)$) and in the interval $(t,t+\Delta)$ where the continuous component vanished the discrete one changed its value to i (with probability q_{ji}) and the continuous component took a value smaller than x (with probability $F_i(x)$); or, finally,
- the process was in the state 0 (with probability $p_0(t)$), in the time Δ the discrete component changed its value to i (with the probability $a_{0i}\Delta+o(\Delta)$) and at the instant when the discrete component changed its value the continuous component took a value smaller than x (with probability $F_i(x)$).

The probability of the remaining possible transitions is $o(\Delta)$.

We use the formula of total probability to obtain

$$P_i(x,t+\Delta)=[P_i(x+\Delta,t)-P_i(\Delta,t)](1-a_i\Delta)+\sum_{j\in\mathcal{J}\setminus\{i\}}[P_j(x+\Delta,t)-P_j(\Delta,t)]a_{ji}\Delta$$
$$+\sum_{j\in\mathcal{J}}P_j(\Delta,t)q_{ji}F_i(x)+p_0(t)a_{0i}\Delta F_i(x)+o(\Delta),\qquad i\in\mathcal{J}.\quad(1.6.10)$$

Now we subtract $p_0(t)$ and $P_i(x+\Delta,t)$, respectively, from both sides of equalities (1.6.9) and (1.6.10) and divide by Δ:

$$\frac{p_0(t+\Delta)-p_0(t)}{\Delta}=-a_0p_0(t)+\frac{1}{\Delta}\sum_{i\in\mathcal{J}}P_i(\Delta,t)q_{i0}+o(1),$$
$$\frac{P_i(x,t+\Delta)-P_i(x+\Delta,t)}{\Delta}=\frac{P_i(x,t+\Delta)-P_i(x,t)-P_i(x+\Delta,t)+P_i(x,t)}{\Delta}$$
$$=-a_iP_i(x+\Delta,t)-\frac{1}{\Delta}P_i(\Delta,t)$$
$$+\sum_{j\in\mathcal{J}\setminus\{i\}}P_j(x+\Delta,t)a_{ji}+\frac{1}{\Delta}\sum_{j\in\mathcal{J}}P_j(\Delta,t)q_{ji}F_i(x)$$
$$+p_0(t)a_{0i}F_i(x)+o(1),\qquad i\in\mathcal{J}.$$

We let Δ go to zero and obtain the simultaneous differential equations

$$p_0'(t)=-a_0p_0(t)+\sum_{i\in\mathcal{J}}\frac{\partial}{\partial x}P_i(x,t)\Big|_{x=0}q_{i0},$$
$$\frac{\partial}{\partial t}P_i(x,t)-\frac{\partial}{\partial x}P_i(x,t)=\sum_{j\in\mathcal{J}}P_j(x,t)a_{ji}-\frac{\partial}{\partial x}P_i(x,t)\Big|_{x=0}$$
$$+\sum_{j\in\mathcal{J}}\frac{\partial}{\partial x}P_j(x,t)\Big|_{x=0}q_{ji}F_i(x)$$
$$+p_0(t)a_{0i}F_i(x),\qquad i\in\mathcal{J}.\quad(1.6.11)$$

We derived simultaneous equations (1.6.11) under the assumption that the derivatives involved in the equations do exist, although it is ungrounded and, moreover, can be invalid. The general theory of Markov processes gets out of the situation by going to the so-called conjugate space. We, however, do not want to get bogged down in these intricacies and deal only with equations (1.6.11). There are several excuses for doing so, but the main reason lies in the fact that if we find a solution of these equations—explicit or in terms of the Laplace–Stieltjes transform—which satisfies the probabilistic properties (non-negativeness and normalisation to unity) and the initial conditions $p_0(0)$ and $P_i(x,0)$, $i \in \mathcal{I}$, then it is the desired distribution.

The functions $p_0(t)$ and $P_i(x,t)$, $i \in \mathcal{I}$, tend with t to the limits which either are 0 simultaneously for all i and x or are positive and satisfy the normalisation condition. In the latter case, $p_0 = \lim_{t\to\infty} p_0(t)$ and $P_i(x) = \lim_{t\to\infty} P_i(x,t)$, $i \in \mathcal{I}$, which are called the *limit* (and also *stationary*) state probabilities of the linearwise process $\{\eta(t),\ t \geq 0\}$, are determined from the equations

$$\begin{aligned} 0 &= -a_0 p_0 + \sum_{i\in\mathcal{I}} P_i'(0) q_{i0}, \\ -P_i'(x) &= \sum_{j\in\mathcal{I}} P_j(x) a_{ji} - P_i'(0) + \sum_{j\in\mathcal{I}} P_j'(0) q_{ji} F_i(x) + p_0 a_{0i} F_i(x), \qquad i \in \mathcal{I}, \end{aligned} \tag{1.6.12}$$

resulting from (1.6.11) if it is made independent of time. In this case, the process $\{\eta(t),\ t \geq 0\}$ is said to be *ergodic*. Understandably, the normalisation condition

$$p_0 + \sum_{i\in\mathcal{I}} P_i(\infty) = 1$$

must be taken into consideration when solving (1.6.12). We observe that, in contrast to (1.6.11), all derivatives involved in (1.6.12) exist.

THEOREM 1.6.5 (ergodicity of a linearwise Markov process of the first type). *For the linearwise Markov process $\{\eta(t),\ t \geq 0\}$ to be ergodic (provided that Conditions 1–7 are satisfied), it is necessary and sufficient for the embedded Markov chain $\{\nu_n,\ n \geq 1\}$ to be positively recurrent.*

It follows from Theorem 1.6.5 (provided that Conditions 1–7 are satisfied), that:

- if the embedded Markov chain is non-periodic (and it will be always the case in our following studies), then ergodicity of the linearwise process amounts to ergodicity of this chain, and
- any linearwise process with the finite state set of the discrete component is ergodic.

The *linearwise Markov process of the second type* is described for the special case of $F_i(x) = F(x)$, $i \in \mathcal{I}$. In all other respects, the linearwise process of the second type is defined by the same parameters. It is clear that Conditions 6 and 7, which were postulated when defining the linearwise process of the first type, are satisfied immediately if the conditions $F(0+) = 0$ and $\int_0^\infty x\,dF(x) < \infty$ are satisfied.

The difference between the linearwise process of the second type from that of the first type lies in the fact that at the instant when the discrete component $\nu(t)$ changes its value from i to j, $i, j \in \mathcal{J}$, so that the continuous component changes as well, the continuous component $\xi(t)$ vanishes and then increases (but does not decrease) with unity rate. More precisely, if at the instant t the continuous component has the value x and the discrete component has the value i, $i \in \mathcal{J}$, then in the time $(t, t+\Delta)$ the discrete component changes with probability $[F(x+\Delta) - F(x)]/[1 - F(x)]$ and takes the value j with probability q_{ij}, $j \in \mathcal{J} \cup \{0\}$, and if $j \neq 0$, then continuous component vanishes. As for the linearwise process of the first type, the discrete component can change from i, $i \in \mathcal{J} \cup \{0\}$, to j, $j \in \mathcal{J}$, with probability $a_{ij}\Delta + o(\Delta)$ in a 'small' time Δ; if $i \neq 0$, then this change does not influence the continuous component which continues to increase with the unity rate, and if $i = 0$, then the continuous component takes the zero value and also begins to increase with unity rate. Again, if the discrete component is equal to 0, then the continuous one is cannot be determined (Fig. 1.11).

The values $\nu_n = \nu(\tau_n + 0)$ of the discrete component immediately after the instants τ_n when the continuous component vanishes constitute the embedded Markov chain $\{\nu_n, n \geq 1\}$.

We denote, as above, $p_0(t) = \mathbf{P}\{\nu(t) = 0\}$ and $P_i(x,t) = \mathbf{P}\{\nu(t) = i, \xi(t) < x\}$, $i \in \mathcal{J}$. We assume that there exist the derivatives $p_i(x,t) = \partial P_i(x,t)/\partial t$. We derive equations satisfied by $p_0(t)$ and $p_i(x,t)$, $i \in \mathcal{J}$. To this end, we consider the instants t and $t + \Delta$, where Δ is a 'small' time increment.

At the instant $t + \Delta$, the process is in a state with the discrete component i, $i \in \mathcal{J}$, and the continuous component $x + \Delta$ if at the instant t either

- the discrete component had the value i and the continuous one, x (with the probability density $p_i(x,t)$), and in the time Δ the discrete component did not change (with probability $1 - a_i\Delta + o(\Delta)$) and the continuous component did not vanish (with probability $[1 - F(x+\Delta)]/[1 - F(x)]$), or
- the discrete component had the value j, $j \in \mathcal{J}$, and the continuous one, x (with the probability density $p_j(x,t)$), and in the time Δ the discrete component changed its value to i (with the probability $a_{ji}\Delta + o(\Delta)$) and the continuous component did not vanish (with probability $[1 - F(x+\Delta)]/[1 - F(x)]$).

The probability of the remaining possible transitions is $o(\Delta)$.

By means of the formula of total probability, we obtain

$$p_i(x+\Delta, t+\Delta) = p_i(x,t)(1 - a_i\Delta)\frac{1 - F(x+\Delta)}{1 - F(x)} + \sum_{j \in \mathcal{J}\setminus\{i\}} p_j(x,t) a_{ji}\Delta \frac{1 - F(x+\Delta)}{1 - F(x)} + o(\Delta).$$

By taking $p_i(x,t) = [1 - F(x)]r_i(x,t)$, subtracting $p_i(x,t)$ from both sides of the equality, and dividing by Δ, we obtain

$$\frac{r_i(x+\Delta, t+\Delta) - r_i(x,t)}{\Delta} = \frac{r_i(x+\Delta, t+\Delta) - r_i(x+\Delta, t) + r_i(x+\Delta, t) - r_i(x,t)}{\Delta}$$
$$= a_i r_i(x,t) + \sum_{j \in \mathcal{J}\setminus\{i\}} a_{ji} r_j(x,t) + o(1).$$

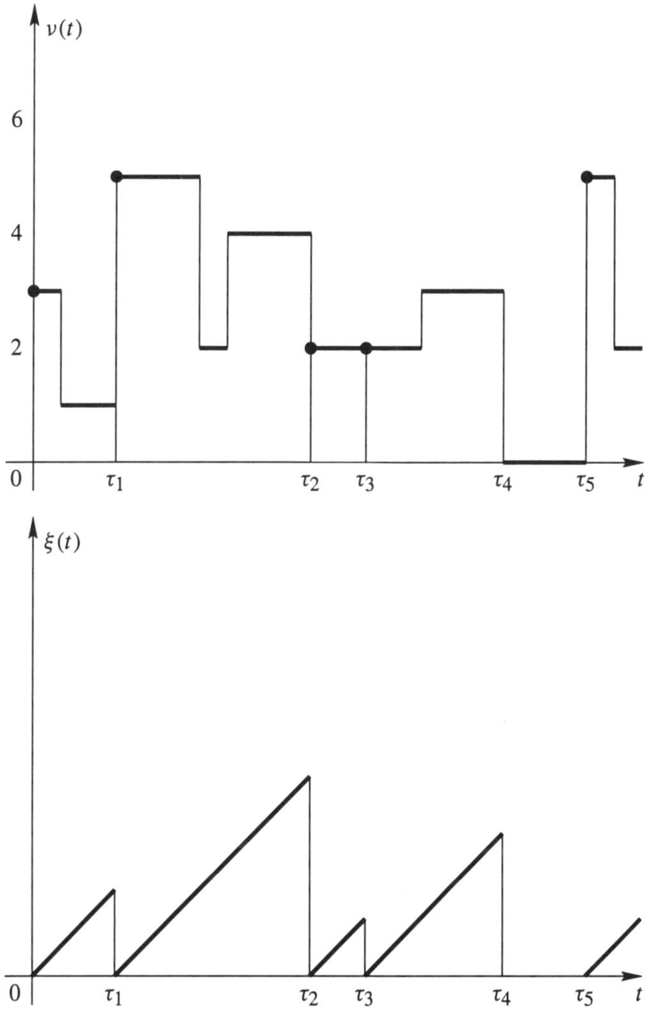

Figure 1.11.

Now, we let Δ tend to zero and obtain the equality

$$\frac{\partial}{\partial t}r_i(x,t) + \frac{\partial}{\partial x}r_i(x,t) = \sum_{j\in\mathcal{I}} a_{ji}r_j(x,t). \tag{1.6.13}$$

For the process to be at the state 0 at the instant $t+\Delta$, it is required that at the time t either

- it is in the state 0 (with probability $p_0(t)$) and in the time Δ its state does not change (with probability $1 - a_0\Delta + o(\Delta)$), or
- it is in the state (i, x), $i \in \mathcal{I}$, (with probability density $p_i(x,t)$) and in the time Δ the

continuous component vanishes (with probability $[F(x+\Delta)-F(x)]/[1-F(x)]$) and the discrete one takes the valuc 0 (with probability q_{i0}).

The probability of the remaining possible transitions is $o(\Delta)$.

By means of the formula of total probability, we obtain

$$p_0(t+\Delta)=p_0(t)(1-a_0\Delta)+\sum_{i\in\mathcal{I}}\int_0^\infty p_i(x,t)\frac{F(x+\Delta)-F(x)}{1-F(x)}\,dx q_{i0}+o(\Delta).$$

We recall that $p_i(x,t)=[1-F(x)]r_i(x,t)$, subtract $p_0(t)$ from both sides of the equality, and divide by Δ to obtain

$$\frac{p_0(t+\Delta)-p_0(t)}{\Delta}=-a_0p_0(t)+\sum_{i\in\mathcal{I}}\int_0^\infty r_i(x,t)\frac{F(x+\Delta)-F(x)}{\Delta}\,dx q_{i0}+o(1).$$

As was shown in (Sevastyanov, 1957), if $r_i(x,t)$ is a continuous bounded function of the argument x, then the integral in the right-hand side of the last equality tends to $\int_0^\infty r_i(x,t)\,dF(x)$ for $\Delta\to 0$. Therefore, by letting Δ tend to zero, we obtain the final equation

$$p_0'(t)=-a_0p_0(t)+\sum_{i\in\mathcal{I}}\int_0^\infty r_i(x,t)\,dF(x)q_{i0}. \tag{1.6.14}$$

Finally, we write the relation for $p_i(0,t)$. To this end, we note that at the instant $t+\Delta$ the process is in the states with the discrete component equal to i, $i\in\mathcal{I}$, and the continuous one lying within the limits from 0 to Δ if at the instant t either

- the process was in the state 0 (with probability $p_0(t)$) and in the time Δ the discrete component changed its value to i (with probability $a_{0i}\Delta+o(\Delta)$), or
- the process was in a state with the values of the discrete component equal to j, $j\in\mathcal{I}$, and the continuous one, x (with the probability density $p_j(x,t)$) and in the time Δ the continuous component vanished (with probability $[F(x+\Delta)-F(x)]/[1-F(x)]$) and the discrete component changed its value to i (with probability q_{ji}).

Since the remaining transitions have the probability $o(\Delta)$, we find, using the formula of total probability, that

$$P_i(\Delta,t)=p_0(t)a_{0i}\Delta+\sum_{j\in\mathcal{I}}\int_0^\infty p_j(x,t)\frac{F(x+\Delta)-F(x)}{1-F(x)}\,dx q_{ji}+o(\Delta).$$

Reasoning as above and taking into account that $\lim_{\Delta\to 0}P_i(\Delta,t)/\Delta=p_i(0,t)=r_i(0,t)$, we obtain

$$r_i(0,t)=p_0(t)a_{0i}+\sum_{j\in\mathcal{I}}\int_0^\infty r_j(x,t)\,dF(x)q_{ji}. \tag{1.6.15}$$

Therefore, we see that the functions $p_0(t)$, $r_i(x,t)$, $i\in\mathcal{I}$, satisfy the linear differential equations (1.6.13)–(1.6.15). It is clear that the initial distribution $\{p_0(0);\ P_i(x,0), i\in\mathcal{I}\}$

must be defined in order to solve equations (1.6.13)–(1.6.15). As for existence of the corresponding derivatives and solution of equations (1.6.13)–(1.6.15), all the remarks concerning the linearwise process of the first type are applicable.

The *limit (stationary)* probabilities $p_0 = \lim_{t\to\infty} p_0(t)$ and $P_i(x) = \lim_{t\to\infty} P_i(x,t)$, $i \in \mathcal{I}$, are determined by the simultaneous equations

$$\begin{aligned} 0 &= -a_0 p_0 + \sum_{j\in\mathcal{I}} \int_0^\infty r_i(x)\, dF(x) q_{i0}, \\ r_i'(x) &= \sum_{j\in\mathcal{I}} a_{ji} r_j(x), \qquad & i \in \mathcal{I}, \\ r_i(0) &= p_0 a_{oi} + \sum_{j\in\mathcal{I}} \int_0^\infty r_j(x)\, dF(x) q_{ji}, \qquad & i \in \mathcal{I}, \end{aligned} \tag{1.6.16}$$

obtained from equations (1.6.13)–(1.6.15) by equating the time derivatives to zero. In these equations, $P'(x) = p_i(x) = [1 - F(x)]r_i(x)$. The following theorem is a counterpart of Theorem 1.6.5.

THEOREM 1.6.6 (ergodicity of a linearwise Markov process of the second type). *For the linearwise process $\{\eta(t),\ t \geq 0\}$ to be ergodic (provided that Conditions 1–7 are satisfied), it is necessary and sufficient that the embedded Markov chain $\{\nu_n,\ n \geq 1\}$ be positive recurrent.*

Here, ergodicity of the linearwise process amounts to ergodicity of the embedded Markov chain provided that this chain is non-periodic. We observe that the equations (1.6.16) and (1.6.12) resemble one another. As a rule, equations (1.6.16) are solved somewhat easier because the initial conditions constitute individual equations. Therefore, if some queueing system admits a description in terms of linearwise processes of both types, then the second type is preferable.

The ergodic linearwise processes $\{\eta(t),\ t \geq 0\}$ (of both types), as well as the ergodic Markov chain and Markov process with the discrete state set, have the following property. Let $\mathcal{X}_1$ be some (measurable) subset in the state set $\mathcal{X}$. We assume for simplicity that the discrete component equal to i and the continuous component with value in the interval $[0, x)$ correspond to the subset $\mathcal{X}_1$. We consider the random variable $\zeta_T(\mathcal{X}_1)$ which is equal to the fraction of time in the interval $[0, T]$ during which the process stays in the subset $\mathcal{X}_1$, that is, in the states with the discrete component equal to i and continuous component varying from 0 to x. Then the relation $\zeta_T(\mathcal{X}_1) \to P_i(x)$ is satisfied for $T \to \infty$ with probability 1.

1.6.4. Piecewise-linear Markov processes

In some cases (while studying the $M/G/n/0$ queueing system, in particular), more than one supplementary variable must be added to obtain the Markov process. This construction often can be realized using the piecewise-linear Markov process. As in the case of the linearwise process, two types of the piecewise-linear process can be introduced. We present here only a brief characterisation of the piecewise-linear process of the first type. The piecewise-linear process, as well as the linearwise process, comprises the discrete component taking the values i, $i \in \mathcal{I} \cup \{0\}$, and more than one continuous component varying with constant but, generally, different rates.

We assign to each state i, $i \in \mathcal{I}$, some positive integer $|i|$ ($|0| = 0$ is assigned to the state 0) which defines the number of continuous variables that must be added to markovise the process. The value of the piecewise-linear process $\{\eta(t),\ t \geq 0\}$ is a set consisting of the discrete component $\nu(t)$ which takes values in the set $\{i,\ i \in \mathcal{I}\}$ and of $|\nu(t)|$ nonnegative continuous components $\xi_1(t), \dots, \xi_{|\nu(t)|}(t)$.

If at the instant t the piecewise-linear process $\{\eta(t),\ t \geq 0\}$ is in the state $(i, x_1, \dots, x_{|i|})$ and $x_1 > 0, \dots, x_{|i|} > 0$, then the values $\xi_j(t)$, $j = 1, \dots, |i|$, of the continuous components at this instant decrease with constant rates d_{ij}.

The process $\{\eta(t),\ t \geq 0\}$ varies at the instants when one of the continuous components (let it be $\xi_k(t)$) vanishes. Then the discrete component $\nu(t)$ goes from the state i to the state j with probability $q_{ij}^{(k)}$; it is required here that $|j| = |i| - 1$ (the number of continuous component is reduced by one). The vanished coordinate $\xi_k(t)$ is eliminated from the notation, the rest of them continuing to decrease with rates d_{jl}, $l = 1, \dots, |j|$.

Additionally, changes are possible at arbitrary instants according to the Markov law. If at the instant t the discrete component $\nu(t)$ had the value i, $i \in \mathcal{I}$, then it can go to the state j in a 'small' time Δ with probability $a_{ij}\Delta + o(\Delta)$; in so doing, $|j| \geq |i|$. If $|j| > |i|$, then $|j| - |i|$ new coordinates $\xi_{|i|+1}(t), \dots, \xi_{|j|}(t)$ which are independent of the behaviour of the process prior to the instant t and distributed by the law

$$F(x_1, \dots, x_{|j|-|i|}) = \mathbf{P}\{\xi_{|i|+1}(t) < x_1, \dots, \xi_{|j|}(t) < x_{|j|-|i|}\}$$

are added to the existing continuous coordinates $\xi_1(t), \dots, \xi_{|i|}(t)$.

For complete description of the piecewise-linear process, it is required to define the initial distribution. Additionally, for correct description of the process, it is required to select the defining parameters so that two continuous coordinates cannot vanish simultaneously. As for the Markov processes of other types, the piecewise-linear processes in what follows always exhibits 'good' properties—ergodicity, in particular. We demonstrate how to cope with the piecewise-linear processes by way of particular examples.

1.7. Kronecker matrix product

Let A and B be the respective rectangular $m \times n$ and $p \times q$ matrices. By the *Kronecker product* $A \otimes B$ of the matrix A and B is meant the following $mp \times nq$ block matrix:

$$A \otimes B = (a_{ij}B)_{i,j=1,\dots,n} = \begin{pmatrix} a_{11}B & a_{12}B & \cdots & a_{1n}B \\ a_{21}B & a_{22}B & \cdots & a_{21}B \\ \vdots & \vdots & \ddots & \vdots \\ a_{n1}B & a_{n2}B & \cdots & a_{nn}B \end{pmatrix}.$$

With regard for the fact that this matrix operation is not prevalent, we list its most important properties (see, for example, (Bellman, 1960; Graham, 1981)).

(1) $(\alpha A) \otimes B = A \otimes (\alpha B) = \alpha(A \otimes B)$, where α is a number.

(2) $(A + B) \otimes C = A \otimes C + B \otimes C$.

(3) $A \otimes (B + C) = A \otimes B + A \otimes C$.

(4) $A \otimes (B \otimes C) = (A \otimes B) \otimes C$.

(5) $(A \otimes B)^T = A^T \otimes B^T$.

(6) If the products AB and CD are valid, then

$$(AB) \otimes (CD) = (A \otimes C)(B \otimes D).$$

Properties 1–6 hold true for any matrix for which the corresponding operations make sense. The following properties apply to the square matrices.

(7) Let A and B be square matrices of the orders m and n, respectively, and I_n and I_m be the identity matrices of the same orders. Then

$$(A \otimes B) = (A \otimes I_n)(I_m \otimes B) = (I_m \otimes B)(A \otimes I_n).$$

(8) If A and B are nondegenerate matrices, then

$$(A \otimes B)^{-1} = A^{-1} \otimes B^{-1}.$$

(9) Let λ and $\boldsymbol{x}$ be the eigenvalue and the eigenvector of the matrix A of the order m, and μ and $\boldsymbol{y}$ be the eigenvalue and the eigenvector of the matrix B of the order n. Then

- the Kronecker product $\boldsymbol{x} \otimes \boldsymbol{y}$ is the eigenvector of the matrix $A \otimes B$ corresponding to the eigen-value $\lambda\mu$,
- $\boldsymbol{x} \otimes \mathbf{1}_n + \mathbf{1}_m \otimes \boldsymbol{y}$ is the eigenvector of the matrix $A \otimes I_n + I_m \otimes B$ corresponding to the eigen-value $\lambda + \mu$ (here, $\mathbf{1}_k$ is the vector consisting of k units), and
- all mn eigen-values and their corresponding eigenvectors of the matrices $A \otimes B$ and $A \otimes I_n + I_m \otimes B$ have the aforementioned form.

The matrix $A \otimes I_n + I_m \otimes B$ is often called the *Kronecker sum* of the matrices A and B and denoted by $A \oplus B$. This operation has the following property:

(10) If $C = A \oplus B$, then $e^C = e^A \otimes e^B$.

2

Defining parameters of queueing systems

The Introduction already gave the reader to understand the diversity of physical systems that are studied by the queueing theory. It goes without saying that prior to using the mathematical methods one must state in formal mathematical terms the problem of calculating the indices characterising the performance of the queueing systems. The present chapter deals precisely with this issue.

We can extract the following components that are common to all queueing systems:

INPUT FLOW of customers, that is, the process of arrival of the customers at the system;

SYSTEM STRUCTURE, that is, number and types of the servers, as well as the capacities of the buffers located before all and/or individual servers;

TIMES OF CUSTOMER SERVICE by the servers, that is, the actual times (disregarding the times of waiting and service interrupt) during which the customers undergo complete service and depart from the system; and

SERVICE DISCIPLINE, that is, the process of allocating the customers to the servers, generating the queues, picking the customers from the queues, and so on.

Now, we can proceed to finding the *performance indices*, that is, user-defined characteristics of service which show to what extent the queueing system copes with the tasks incumbent on it.

2.1. Input flow

The input flow is distinguished for the random instants of customer arrivals at the system and—for more complicated systems—also by the types of the customers arrived at these instants. In this section we will consider description of the input flows and dwell in more detail on the flows to be used in what follows. It is usually assumed that the input flow is of Poisson nature, although sometimes consideration is given to recurrent flows or Markov flows. Systems with more general input flows, in particular, with general stationary flows, are in fact unexplored, and at best only partial results on existence of steady state of system operation and convergence to them are known.

2.1.1. Random flow and its properties

A *random flow* $\{\tau_k,\ k \geq 1\}$ in the time interval $[0, \infty)$ is a nondecreasing sequence of random instants $\tau_1, \tau_2, \dots, \tau_k, \dots,\ \tau_1 \geq 0$, of occurrences of certain events. In what follows, by the random flow is meant the flow of customers arriving at or departing (lost and/or served) from the system and by the flow event, arrival of a customer at the system or departure from it. The random flow in any time interval $[a, b)$, where a and b can take any values, infinite including, is defined in a similar manner.

We note that if we consider the number $\nu(t)$ of customers arrived during the interval $[0, t)$, we obtain a random process which takes nonnegative integer values and is nondecreasing and left-continuous. Conversely, if in the interval $[0, t)$ we consider any nondecreasing left-continuous random process $\nu(t)$ assuming nonnegative integer values, then we see that it defines some random flow. Therefore, there exists a one-to-one correspondence between the random flows and random processes which take nonnegative integer values and are nondecreasing left-continuous.

Usually, the random flows are described in probabilistic terms in either of two ways.

The *first way* lies in defining for any $k \geq 1$ the joint distribution function

$$F_{\xi_1,\xi_2,\dots,\xi_k}(x_1, x_2, \dots, x_k) = \mathbf{P}\{\xi_1 < x_1, \xi_2 < x_2, \dots, \xi_k < x_k\}$$

of the times $\xi_i = \tau_i - \tau_{i-1},\ i \geq 1,\ \tau_0 = 0$, between the subsequent customer arrivals.

The *second way* is based on considering arbitrary families of intervals $[0, t_1), [t_1, t_2), \dots, [t_{k-1}, t_k),\ 0 < t_1 < \dots < t_k, k \geq 1$, and defining the joint distributions

$$G(m_1, \dots, m_k; t_1, \dots, t_k) = \mathbf{P}\{\nu_1 = m_1, \dots, \nu_k = m_k\}$$

of the numbers ν_i, $i = 1, \dots, k$, of customers arrived in the intervals $[t_{i-1}, t_i)$. Here, the following three properties of the random flow can be conveniently highlighted: memoryless, stationarity, and ordinariness.

Memoryless implies independence of the random variables $\nu_1, \dots, \nu_k$. In other words, the number of customers arrived in some interval $[t_{i-1}, t_i)$ does not affect in any way the number of customers arrived in the rest of the intervals $[t_{j-1}, t_j)$, $j \neq i$.

Stationarity means that the probabilistic properties of the flow do not vary with time, that is, the numbers of customers arrived in the intervals $[0, t_1), [t_1, t_2), \dots, [t_{k-1}, t_k)$ are distributed in the same manner as the numbers of customers arrived in the intervals $[T, t_1 + T), [t_1 + T, t_2 + T), \dots, [t_{k-1} + T, t_k + T)$ for any $T > 0$ and $k \geq 1$.

Finally, *ordinariness* of the flow implies that the relation $\mathbf{P}\{\nu(t, t + \Delta) > 1\} = o(\Delta)$, where $\nu(t, t + \Delta)$ is the number of customers arrived in the interval $[t, t + \Delta)$, is true. In actual use, flow ordinariness means that customers can arrive only one at a time.

Let us consider a stationary flow. We denote by $P_1(t)$ the probability that in time t at least one customer arrived and by $\Lambda(t) = \mathbf{E}\nu(t)$ the mean number of customers arrived in time t. By virtue of the stationarity of the flow, $P_1(t)$ and $\Lambda(t)$ are independent of the location of the interval of length t, and additionally, $\Lambda(t)$ satisfies the inequality

$$\Lambda(t + s) = \Lambda(t) + \Lambda(s) \tag{2.1.1}$$

following from the properties of expectation. Since $\Lambda(t)$ is a nondecreasing function of t, it follows from equality (2.1.1) that it must have the form $\Lambda(t) = \lambda t$, $\lambda \geq 0$. The constant

λ which is called the flow *intensity* is the mean number of customers arrived in unit time. Notably, λ can take both finite and infinite values.

It is possible to prove that there exists the limit $\gamma = \lim_{\Delta\to 0} P_1(\Delta)/\Delta$ called the flow *parameter* which can take an infinite value and never exceeds the intensity λ. If $\gamma < \infty$, then the equality $\lambda = \gamma$ holds if and only if the stationary flow is also ordinary.

2.1.2. Poisson flow

The random flow with all three aforementioned properties (memoryless, stationarity, and ordinariness) is called the *Poisson* flow. We determine for it the probabilities $p_i(t)$, $i \geq 0$, that exactly i customers arrive in the interval of length t located anywhere on the time axis $[0, \infty)$. We begin with the probability $p_0(t)$ that no customer arrives in time t. Obviously, $p_0(t)$ is a nonnegative non-increasing function of t. Since the event 'no customer arrives in the interval of length $t + s$' is the intersection of the events 'no customer arrives in the interval of length t' and 'no customer arrives in its right interval of length s', $p_0(t)$ satisfies the equation

$$p_0(t + s) = p_0(t)p_0(s) \tag{2.1.2}$$

by virtue of memoryless and stationarity properties of the Poisson flow.

Any nonnegative monotone non-increasing solution of equation (2.1.2) is known to be of the form $p_0(t) = e^{-\lambda t}$, $0 \leq \lambda < \infty$ or $p_0(t) = 0$. We may ignore the second solution because for it an infinite number of customers arrives with probability one in an arbitrarily small time interval. Also, it is reasonable to exclude the case $\lambda = 0$ because then $p_0(t) = 1$ for any $t > 0$ and customers do not arrive at all. Therefore, we finally obtain

$$p_0(t) = e^{-\lambda t}, \qquad 0 < \lambda < \infty, \tag{2.1.3}$$

which also means that the time before the arrival of the first customer is distributed exponentially with parameter λ.

It follows from (2.1.3) that for $\Delta \to 0$

$$p_0(\Delta) = 1 - \lambda\Delta + o(\Delta). \tag{2.1.4}$$

In turn, the probability $P_1(\Delta) = 1 - p_0(\Delta)$ that at least one customer arrives in time Δ is

$$P_1(\Delta) = \lambda\Delta + o(\Delta). \tag{2.1.5}$$

Hence, we see that λ is a parameter of the Poisson flow.

To determine the remaining $p_i(t)$, $i \geq 1$, we consider the instants t and $t + \Delta$, where Δ is a 'small' time increment. Then,

$$p_i(t + \Delta) = \sum_{j=0}^{i} p_{i-j}(t)p_j(\Delta)$$

in view of stationarity and memoryless properties. Now, $\sum_{j=2}^{\infty} p_j(\Delta) = o(\Delta)$ by virtue of ordinariness of the flow. Therefore, $p_1(\Delta) = P_1(\Delta) + o(\Delta)$; hence, we find from equalities (2.1.4) and (2.1.5) that

$$p_i(t + \Delta) = p_i(t)(1 - \lambda\Delta) + p_{i-1}(t)\lambda\Delta + o(\Delta).$$

Moving $p_i(t)$ to the left-hand side and dividing both sides by Δ we obtain

$$\frac{1}{\Delta}[p_i(t+\Delta) - p_i(t)] = -\lambda p_i(t) + \lambda p_{i-1}(t) + o(1). \tag{2.1.6}$$

We let Δ go to zero and obtain the simultaneous equations

$$\begin{aligned} p_0(t) &= e^{-\lambda t}, \\ p_i'(t) &= -\lambda p_i(t) + \lambda p_{i-1}(t), \qquad i \geq 1, \end{aligned} \tag{2.1.7}$$

with the initial conditions $p_i(0) = 0$, $i \geq 1$. Existence of the derivative $p_i'(t)$ follows from the existence of limit in the right-hand side of (2.1.6). To solve (2.1.7), we substitute $p_i(t) = e^{-\lambda t} q_i(t)$, $i \geq 0$, reducing set of equations (2.1.7) to

$$\begin{aligned} q_0(t) &= 1, \\ q_i'(t) &= \lambda q_{i-1}(t), \qquad i \geq 1, \end{aligned}$$

with the initial conditions $q_i(0) = 0$, $i \geq 1$. By integrating recursively the last system, we obtain

$$q_i(t) = \frac{(\lambda t)^i}{i!}, \qquad i \geq 0,$$

and finally,

$$p_i(t) = \frac{(\lambda t)^i}{i!} e^{-\lambda t}, \qquad i \geq 0.$$

Consequently, the number of customers of the Poisson flow arrived in any interval of length t is distributed by the Poisson law with the parameter λt. The mean number of customers arriving in time t is λt. Therefore, λ is also the intensity, which is obvious in view of ordinariness of the Poisson flow.

In view of properties of the Poisson flow, it is possible to prove that the times between the consecutive arrivals of customers are independent random variables distributed according to the same—exponential with parameter λ—law as the time before the arrival of the first customer.

We note that the number $\nu(t)$ of customers arrived in the interval $[0, t)$ constitutes the Poisson process. We recall that the Poisson process is a pure birth process with $\lambda_i = \lambda$ (see Section 1.4), that is, a homogeneous Markov process with transition intensity matrix $A = (a_{ij})$ whose elements a_{ij} are

$$a_{ij} = \begin{cases} -\lambda, & j = i; \\ \lambda, & j = i+1; \\ 0, & \text{otherwise}. \end{cases}$$

The Poisson flow sometimes can be conveniently determined from its local behaviour in a 'short' time interval $[t, t+\Delta)$. The following lemma is of considerable utility in such cases.

LEMMA 2.1.1. *Let $\lambda\Delta + o(\Delta)$ be, for some ordinary random flow, the probability of customer arrival in the interval $[t, t+\Delta)$, and let λ be independent of the history of flow before t. Then the flow is Poisson.*

We consider now a random process $\{\eta(t),\ t \geq 0\}$, whose value at instant t depends, possibly, on the process of arrival of customers of the Poisson flow, but only before the instant t. Then, since the Poisson flow is memoryless, the random variable $\eta(t)$ and the flow of customers arrived after t are independent. Therefore, the conditional distribution $\eta(t)$ under the condition that the customers arrive in the time interval $[t, t+\Delta)$, $\Delta > 0$, coincides with the unconditional distribution $\eta(t)$. By letting Δ go to 0 and using ordinariness of the Poisson flow, we see that the conditional distribution of the process $\eta(t)$ at time t under the condition that a customer arrives at this instant, coincides with the unconditional distribution of $\eta(t)$. This is the manifestation of the following property which will be invariably used below:

for the queue in a system with the Poisson input flow operating in steady state, at the instant of arrival of a customer at the system the distribution of its characteristics coincides with the unconditional distribution of the same characteristics or, stated differently, the time-stationary distribution of the characteristics of a queue coincides with the stationary distribution at the instants of arrivals of customers at the system.

2.1.3. Non-stationary Poisson flow

Non-stationary Poisson flow appears if in its definition we reject stationarity and leave only the memoryless and ordinariness properties. We assume that the expectation of the number of customers arrived in the interval $[t, t + \Delta)$ is equal to $\lambda(t)\Delta + o(\Delta)$, where $\lambda(t)$ is a nonnegative time function. Then it is reasonable to call $\lambda(t)$ the *instantaneous intensity* of the non-stationary Poisson flow. The instantaneous intensity completely defines the probabilistic properties of the non-stationary Poisson flow.

For the non-stationary Poisson flow, the number of customers arrived in the time interval $[t_1, t_2)$ is distributed by the Poisson law with parameter $\Lambda = \int_{t_1}^{t_2} \lambda(t)\, dt$. For the stationary Poisson flow, $\lambda(t)$ is at the same time the instantaneous flow parameter, that is, $\lambda(t)\Delta$ is the probability to within $o(\Delta)$ that at least one customer arrives in the interval $[t, t + \Delta)$.

2.1.4. Non-ordinary Poisson flow

Now we assume that the random flow is only stationary and has no memory. Such a flow is called the non-ordinary Poisson flow. To clarify its structure, it suffices to analyse the calculations made for the (ordinary) Poisson flow. We note that in fact we did without flow ordinariness when discussing the first customer arrival and times between the consecutive arrivals of customers. Therefore, the non-ordinary Poisson flow can be described as follows: the instants of customer arrivals (call instants) constitute an (ordinary) Poisson flow, but at each call instant precisely k customers may arrive with the probability l_k, $k \geq 1$, $\sum_{k=1}^{\infty} l_k = 1$, which is independent of the flow history.

The parameter γ of the non-ordinary Poisson flow coincides with the parameter (and intensity) γ of the flow of call instants, and the intensity λ obeys the formula $\lambda = \gamma l$, where $l = \sum_{k=1}^{\infty} k l_k$ is the mean number of customers arriving at the call instant.

For the non-ordinary Poisson flow, the following obvious asymptotic representation of the probability $p_k(\Delta)$ of arrival of precisely k customers in time Δ is valid for $\Delta \to 0$:

$$p_k(\Delta) = \lambda l_k \Delta + o(\Delta).$$

Let us determine the probability that an arbitrary customer arrives together with a batch of k customers. To this end we observe that on the average γlt customers arrive in time t with $\gamma k l_k t$ customers on the average arriving in batches of exactly k customers. Therefore, the desired probability is equal to the ratio of $\gamma k l_k t$ to γlt, that is, kl_k/l.

2.1.5. Rarefication and superposition of Poisson flows

In this section we consider two operations which can be applied to a Poisson flow to obtain a Poisson flow again. The first operation called flow *rarefication*, or *sifting* is representable as follows. Let each arrived customer of the Poisson flow of intensity λ be the customer of type k with the probability α_k, $k = 1, \dots, m$, $\sum_{k=1}^{m} \alpha_k = 1$, independent of the flow history. Then, the flow of customers of the type k type is a *rarefied Poisson flow*. The following assertion is valid.

THEOREM 2.1.1 (rarefication of Poisson flow). *Flows of customers of type k, $k = 1, \dots, m$, are Poisson. These flows are independent. The intensity of the kth flow is $\alpha_k \lambda$.*

The second operation, which is inverse to the first one and called the *superposition* or *union* of flows, consists of uniting the arriving m independent Poisson flows of intensities λ_k, $k = 1, \dots, m$, into one flow.

THEOREM 2.1.2 (superposition of Poisson flows). *The superposition of k independent Poisson flows of intensities λ_k, $k = 1, \dots, m$, is again a Poisson flow of intensity $\lambda = \sum_{k=1}^{m} \lambda_k$.*

2.1.6. Markov flow

The needs of applications recently gave rise to new models of flows that are more complex as compared with their Poisson counterpart and in a sense generalise it. These are, in particular, the *Markov arrival processes* (MAP) which allow one to describe conveniently the customer flows lost for some reason in a queueing system or served by it, as well as their superposition.

Let us describe the Markov flow. Again, let $\nu(t)$ be the number of customers arrived in the time interval $[0, t)$ and $\tau_1, \tau_2, \dots$ be the instants of their arrivals. We assume that there also exists a Markov process $\{\xi(t),\ t \geq 0\}$ defined on the finite state set $\mathcal{I} = \{1, 2, \dots, l\}$. We assume that $\eta(t) = (\xi(t), \nu(t))$. The process state set $\{\eta(t),\ t \geq 0\}$ is representable as $\bigcup_{k=0}^{\infty} \mathcal{I}_k$, where $\mathcal{I}_k = \{(i, k), i = 1, \dots, l\}$, $k \geq 0$. Therefore, the process $\{\eta(t),\ t \geq 0\}$ is in the state (i, k), $i = 1, \dots, l$, $k \geq 0$, if k customers arrived at the instant t and the process $\{\xi(t),\ t \geq 0\}$ at t is in the state i.

The customer flow $\{\tau_j,\ j \geq 1\}$ is said to be the *Markov* flow (with respect to the process $\{\xi(t),\ t \geq 0\}$) if the random process $\{\eta(t),\ t \geq 0\}$ is a homogeneous Markov process and its matrix A of transition intensities is of the block form

$$A = \begin{pmatrix} \Lambda & N & 0 & 0 & \cdots \\ 0 & \Lambda & N & 0 & \cdots \\ 0 & 0 & \Lambda & N & \cdots \\ \vdots & \vdots & \vdots & \vdots & \ddots \end{pmatrix},$$

where Λ and N are square matrices of order l. We observe that $\Lambda^* = \Lambda + N$ is the matrix of transition intensities of the Markov process $\{\xi(t),\ t \geq 0\}$.

Obviously, for $j \neq m$ the elements Λ_{jm} of the matrix Λ define the transition intensities of the process $\{\eta(t),\ t \geq 0\}$ which are not related with customer arrivals, and the elements N_{jm} of the matrix N are the transition intensities accompanied by arrivals of customers.

Clearly, if $l = 1$, $\Lambda_{11} = -\lambda$ and $N_{11} = \lambda$, then we get the ordinary Poisson flow. If N is a diagonal matrix, then the Markov flow is called the *Markov modulated Poisson process* (MMPP).

Another important case of the Markov flow, the *interrupted Poisson process* (IPP) which finds wide application for designing the models of telephone networks by the so-called method of equivalent replacements is defined by the matrix N of order $l = 2$ having only one nonzero and strictly positive diagonal element.

Finally, if the matrix N of the order l is representable as $N = \boldsymbol{\nu}\boldsymbol{\alpha}^T$, where $\boldsymbol{\nu}$ and $\boldsymbol{\alpha}$ are some column vectors of dimensionality l and $\boldsymbol{\alpha}$ is a probabilistic vector, then the corresponding Markov flow is called the *phase-type* (PH) flow.

It is well known that if $\{\xi(t),\ t \geq 0\}$ is a stationary Markov process, then the Markov flow generated by the process $\{\eta(t),\ t \geq 0\}$ is stationary.

THEOREM 2.1.3. *Let $\{\xi(t),\ t \geq 0\}$ be a stationary Markov process, $\boldsymbol{p}$ be the vector of stationary probabilities of its states, and $\boldsymbol{p}^T N = \lambda \boldsymbol{p}^T$. Then the corresponding Markov flow is the Poisson flow of intensity λ.*

2.1.7. Recurrent flow

A recurrent flow can be conveniently defined using the first way of describing random flows. By the *recurrent* flow is meant a flow where the inter-arrival times are *independent and identically distributed*, that is,

$$\mathsf{P}\{\xi_1 < x_1, \ldots, \xi_k < x_k\} = \mathsf{P}\{\xi_1 < x_1\} \cdots \mathsf{P}\{\xi_k < x_k\} = A(x_1) \cdots A(x_k),$$

where $A(x)$ is the distribution function of the time between the consecutive arrivals of the customers.

By the *quasi-recurrent flow* is meant the flow where the inter-arrival times are *independent and identically distributed beginning from the second one*, that is,

$$\mathsf{P}\{\xi_1 < x_1, \ldots, \xi_k < x_k\} = \mathsf{P}\{\xi_1 < x_1\} \cdots \mathsf{P}\{\xi_k < x_k\} = A_1(x_1)A(x_2) \cdots A(x_k).$$

Finally, by the *recurrent stationary* or *Palm flow* is meant the flow where the inter-arrival times are *independent and identically distributed beginning from the second one* with the distribution function $A(x)$, and the distribution function $A_1(x)$ is

$$A_1(x) = \frac{1}{a}\int_0^x [1 - A(y)]\,dy, \qquad a = \int_0^\infty x\,dA(x).$$

The simple, general, and stationary renewal processes correspond, respectively, to the recurrent, quasi-recurrent, and recurrent stationary flows.

The Poisson flow of intensity λ is a special case of the recurrent flow distinguished by the fact that the distribution $A(x)$ is exponential with parameter λ. Furthermore, the Poisson flow is also a recurrent stationary flow.

2.1.8. Discussion

As we already mentioned, the results in the queueing theory were mostly obtained for systems with the Poisson or recurrent input arrivals. We conclude this section by a word about correctness of such a description. It is apparent that in the physical queueing system even the memoryless property is satisfied extremely rarely just because in the flow that has it any number of customers can arrive in an arbitrary time interval with a non-zero (although, possibly, very small) probability, which clearly is not the case in actual practice. Nevertheless the practice, which is the criterion for correctness of the queueing theory as an applied theory, shows that in the vast majority of cases the description of the input flow by the Poisson one is quite acceptable. There are also mathematical arguments in favour of such a description. One of them is represented by the Khinchin theorem (Khintchine, 1960) and its generalisations (Grigelionis, 1962a; Grigelionis, 1962b) which state that under rather weak constraints the superposition of many 'rarefied' flows provides a flow close to the Poisson one. The Poisson flow can also be obtained from the recurrent flow as the limit flow if rarefication of the recurrent flow is used and the probability for each customer to remain in the flow tends to zero.

Stationarity, the second property of the Poisson flow, can also be called in question. Indeed, intensity of the input flow depends as a rule on the time of day, year, and so on. A way out of this situation can lie in that if one retains the memoryless and ordinariness properties, then a non-stationary Poisson flow results. Although in some cases one may succeed in developing mathematical methods of systems design with such an input flow, the resulting formulas are too bulky to be used. Therefore, the way out consists of confining oneself when constructing the model to some time interval where intensity of the input flow varies insignificantly.

Finally, if it is known that only ordinariness is not satisfied, then the queueing system input flow can be described by a non-ordinary Poisson flow. Results applicable to the systems with Poisson input flow can be applied virtually without any modification to systems with non-ordinary Poisson flow.

The need for studying systems with recurrent input flow can be easily understood by way of example of the multistage queueing system where the customer successively goes through more than one stage of service by different servers. The output flow of the first stage which is the input into the second stage is, as a rule, a non-Poisson flow. Approximation of the input flow to the second stage by a recurrent flow allows one, on the one hand, to use existing methods to design systems with recurrent input flow and, on the other hand, to improve the design accuracy as compared with the Poisson approximation.

2.2. System structure

In order to define the queueing system structure, one must list all of its servers and define the kinds of customers that can be handled by each server. Here, an individual server can handle customers of several types or, vice versa, the customers of one type can be handled by several servers.

In what follows, we mostly restrict ourselves to the case of queueing system with one or more identical servers which can handle any of the customers. Systems of this kind are called the *single-server* systems (with a single server) or *multi-server full-access* systems (with more than one server). Additionally, we assume that after service the customer departs from the system and never returns, that is, the system is a *single-stage* one.

The queueing networks which will be thoroughly discussed in Chapter 10 give an idea how to design *multistage* queueing systems where the customer must be handled successively by more than one server (go through more than one stage of service). Chapter 10 is devoted to the simplest methods of analysis of the queueing networks.

Buffers of various capacities or *waiting places* can be provided in queueing systems, which provide the customers arrived with the ability to wait for service if at the time of their arrival all servers are occupied. If the buffer capacity is infinite, then the systems are said to be systems of *infinite capacity*; otherwise, the systems are said to be *finite capacity* systems or those with *a finite number of the waiting places*; and if there is no buffer at all (the customer arrived when all servers are busy leaves the system and never returns, as is exemplified by the ordinary telephone systems), then the system is said to be the *loss system*.

2.3. Customer service times

The times of customer handling in servers also issue challenge to formalisation. Generally speaking, along with unpredictable random fluctuations, there can be systematic time variations in the customer service times, the service times can depend on arrival and service of other customers, etc.

However, it is usually assumed that *the service times* of all customers are *identically distributed random variables that are independent of each other and all system processes*. By analogy with the recurrent flow, service here is said to be *recurrent*. Then the time of service of any customer is characterised only by a single distribution function denoted by $B(x)$. If customers of more than one type arrive at the queueing system, then the distribution of the service time can depend on the type of customer.

2.4. Service discipline

As already mentioned, the service discipline is defined by the rules of customer queueing and choosing for service, allocation of the servers between the customers and—in the multistage systems—between the service stages.

We mostly assume—which is reasonable for the first acquaintance with the queueing theory—that the system realizes the natural discipline of *servicing the customers in the order of their arrival* which is known as the first-come-first-served (FCFS) discipline. In the multi-server systems, where the customers form a common queue to all servers, the first queued customer is sent to any released server.

More involved disciplines of service can be used, nevertheless, in the queueing systems. Their simplest examples are as follows: *inverse order of service* known as last-come-first-served (LCFS) discipline where the last arrived customer is served first; *random choice for service* (RANDOM) where any of the n customers is taken with the probability $1/n$; and the discipline of *processor sharing* (PS) where each of the n customers in the system is served with identical rate $1/n$.

A special class of the queueing systems is made by the *priority systems* where the arrived customers have several priorities, the higher-priority customers having advantage over those with lower priorities, that is, being served before them. Priorities can be *non-preemptive* if the higher-priority customers do not interrupt the customers in service and *preemptive* otherwise. In the case of preemptive priorities, various modifications are possible: the under-served interrupted customers leave (are pushed out of) the system, are resumed as soon as the higher-priority customers depart from the system, served anew, etc.

The following factors must be regarded as belonging to the disciplines of service: *'warming up' of the server* prior to handling the next customer or after the arrival of a customer at the empty system, *switching of the server* to customers of another type, service of customers by a *non-reliable server*, *switching off the server* for a random time upon emptying the system and completing service of a predefined number of customers or, stated differently, server vacations, etc.

Finally, the *times of customer sojourn* in the system or *waiting* for service can be *limited*. This finishes the list of service disciplines, although the reader can readily give many other examples of disciplines.

2.5. Performance indices of a queueing system

Now, we describe the queueing system characteristics which are of interest for the user. We call them the *system performance indices*.

The *queue length*, that is, the number of waiting customers and the *waiting time* are the most important indices of service performance. Since both the queue length and the waiting time for the beginning of customer service are random variables, they are characterised by their distributions. Additionally, these distributions depend on the current time. Sometimes, consideration is given to the *total number of customers in the system* along with the queue length and the *total sojourn time of a customer in the system* along with the waiting time.

In the loss systems or in the systems with finite buffers the *customer loss probability* also is regarded as one of the most important characteristics.

In finite-capacity systems, as well as in infinite-capacity systems with quite weak constraints on the initial parameters, the steady state, which is distinguished by the time-constant probabilistic indices of performance of the queueing system, becomes steady. It is precisely determination of the steady characteristics that is considered in the majority of publications on queueing theory, although transient characteristics were discussed in sufficient detail. In this book, we also restrict our consideration mostly to studying the steady characteristics.

Along with the aforementioned indices, some publications consider other performance indices such as the number of served and lost customers in a certain time interval, empty and busy system times, time to the first loss of customer, etc.

2.6. Classification of queueing systems

2.6.1. Kendall classification

As we saw above, the class of physical queueing systems is very diverse. To avoid describing each time the main components of the queueing system, Kendall suggested a classification of the simplest single-stage systems. The *Kendall classification* has four positions of which the first two are occupied by letters or their combinations such as M, E, H (or HM),

HE, *PH*, *D*, or *G* (sometimes, *GI*) and the last two are occupied by letters which take digital values or simply digits.

Letter(s) in the first position characterises the input flow of customers. For the recurrent flow, the letter M is standing for the Poisson input flow; E, for the Erlang flow where the inter-arrival times are distributed by the Erlang law; H (or *HM*) for the hyper-exponential flow where the times between customer arrivals have the hyper-exponential distribution function; *HE*, for the hyper-Erlang flow where the times between customer arrivals have the hyper-Erlang distribution law; *PH*, for the phase-type flow where the times between customer arrivals have the *PH*-distribution; D, for the deterministic, and G, for the general recurrent flow (sometimes, *GI* is written in the case of recurrent flow, the letter G standing for the arbitrary stationary flow). Sometimes, the symbols E and H are accompanied by subscripts indicating values of the integer parameters of the corresponding distributions—for example, E_l or H_l. Attention is given recently to the queueing systems with the Markov input flow which in the Kendall classification is denoted by *MAP* in the first position.

Letter(s) in the second position characterises the distribution function of customer service time. M means that the time of service is distributed by the exponential law; E, by the Erlang law; H (or *HM*), by the hyper-exponential law; *HE*, by the hyper-Erlang law; *PH* means that the time of service has the *PH*-distribution; D means that the time of service is constant; and G means that service is recurrent. It is noteworthy that often in the case of recurrent service the notation *GI* is used, leaving G for the case where the service times can be interdependent or dependent on the input flow.

The character in the third position denotes the number of servers. That in the fourth position, the buffer capacity.

For example, $M/M/1/\infty$ refers to a single-server system with an infinite-capacity buffer (system with waiting), Poisson input flow of customers and exponential distribution of the customer service time; $G/G/n/0$, to the n-server loss system, recurrent input flow, and recurrent service.

2.6.2. Markov models

If in the Kendall classification the first two positions are occupied by the letters M, E, H, *HE*, *PH*, or *MAP* (in the first position), then the queueing system is classified with the so-called *Markov systems*. Their distinguishing feature is that they can be described by a continuous-time Markov process with a discrete state set and, sometimes, even by the birth-and-death process, which enables one to apply to them the well-developed mathematical methods (see Section 1.4). The Markov models will be considered in Chapters 3 and 4.

If arrival of the customers (of one or more types) at the system forms a Poisson process and their times of service are characterised by exponential distributions, then the queueing system is often referred to as *exponential*.

2.6.3. Other classifications

Other classifications can be used for more complicated systems, but none of the existing classifications is as generally accepted as that of Kendall. Their majority represent modifications of the Kendall classification involving additional types of systems such as priority system, multistage systems, etc. We do not dwell on them, although below we sometimes use the modified Kendall notation explained whenever necessary. The readers can find in

(Borovkov, 1976) an example of classification which is somewhat different from the Kendall classification.

2.7. Queueing networks

A multistage queueing system is a simplest example of the *queueing network*. The queueing networks are distinguished by having *more than one queueing system (service nodes)*, and the customers arrived at or permanently circulating in it must go through some or, possibly, all of them and even more than once. We confine our consideration for simplicity to the case where all customers in the queueing network are of the same type or, in other words, are similar. Closed and open queueing networks are distinguished.

In the *closed* networks, the same customers circulate through the nodes according to some probabilistic law. Most frequently, it is assumed that the transitions between the nodes are controlled by some stochastic matrix whose elements are independent of the process history before the customer transition. New customers do not arrive at the closed queueing network, as well as the old ones do not leave it.

In the *open* queueing network, the customers arrive from outside and, having completed certain service cycle, leave the network. As in the case of closed queueing network, the travels of customers between the nodes are controlled by a transition probability matrix which, in distinction to the closed network, is semi-stochastic.

If the distributions of the service times in the network are exponential and, additionally, in the open network the input flows are Poisson, then this network (whether open or closed) is called the *exponential* queueing network.

2.8. Properties of distributions for some types of recurrent input flows and service times

In this section we consider some types of distributions that are used to describe the queueing system in compliance with the Kendall classification. These distributions are characterised by the fact that, on the one hand, they approximate rather well the parameters of the modelled physical objects and, on the other hand, in the majority of cases enable one to simplify substantially the methods of investigating the queueing systems themselves.

2.8.1. Special cases of recurrent flows

According to the definition given in Section 2.1, a recurrent flow is characterised only by one distribution function $A(x)$ of the inter-arrival time between two consecutive customers denoted below by ξ.

Regular or determinate flow (D). Here the customers arrive after a constant time and consequently,

$$A(x) = \begin{cases} 0, & x \le a; \\ 1, & x > a. \end{cases}$$

The Laplace–Stieltjes transform $\alpha(s)$ of the distribution function $A(x)$ is

$$\alpha(s) = \int_0^\infty e^{-sx}\, dA(x) = e^{-sa}.$$

One can easily see that $\mathbf{E}\xi = a$, $\mathbf{E}\xi^k = a^k$, $k \geq 1$, and $\mathbf{Var}\,\xi = 0$. By introducing now the coefficient $C = \sqrt{\mathbf{Var}\,\xi}/\mathbf{E}\xi$ of variation of the random variable ξ, we find that for the determinate flow $C_A^D = 0$.

Poisson flow (M). As mentioned above, for the Poisson flow

$$A(x) = 1 - e^{-\lambda x}, \qquad x > 0,$$

that is, the inter-arrival times are distributed exponentially with the parameter λ, $0 < \lambda < \infty$. Here and below, we determine for brevity the distribution function of a nonnegative random variable only for the positive values of its argument, bearing in mind that for the remaining values the distribution function is zero. For the Laplace–Stieltjes transform of the distribution function $A(x)$, we obtain the formula

$$\alpha(s) = \frac{\lambda}{\lambda + s}.$$

For $|s| < \lambda$, the Laplace–Stieltjes transform $\alpha(s)$ can be expanded into the power series

$$\alpha(s) = \sum_{k=0}^{\infty} (-1)^k \left(\frac{s}{\lambda}\right)^k$$

from which and properties of the Laplace–Stieltjes transform it follows that

$$\mathbf{E}\xi^k = \frac{k!}{\lambda^k}, \qquad k \geq 1,$$

and the mean time between the consecutive arrivals is

$$\mathbf{E}\xi = \frac{1}{\lambda}.$$

Therefore, $\lambda = 1/\mathbf{E}\xi$ has the sense of the mean number of customers arriving in unit time, which, in turn, is nothing other than the intensity of the Poisson flow. The variance of ξ obeys the formula

$$\mathbf{Var}\,\xi = \frac{1}{\lambda^2},$$

and, consequently, the coefficient of variation for the Poisson flow is equal to one, that is, $C_A^M = 1$.

Hyper-exponential flow $(HM$ *or* $H)$. For this flow, the distribution function $A(x)$ is hyper-exponential:

$$A(x) = \sum_{i=1}^{l} \alpha_i (1 - e^{-\lambda_i x}), \qquad x > 0,$$

where $\alpha_i > 0$, $\sum_{i=1}^{l} \alpha_i = 1$ and $0 < \lambda_i < \infty$, $i = 1, \dots, l$, that is, $A(x)$ is a mix of the exponential distributions with parameters λ_i and weights α_i. We rely on the properties of the exponential distribution function to obtain

$$\alpha(s) = \sum_{i=1}^{l} \frac{\alpha_i \lambda_i}{\lambda_i + s}, \quad \mathbf{E}\xi = \sum_{i=1}^{l} \frac{\alpha_i}{\lambda_i}, \quad \mathbf{Var}\,\xi = 2\sum_{i=1}^{l} \frac{\alpha_i}{\lambda_i^2} - (\mathbf{E}\xi)^2,$$

and for the coefficient of variation C_A^H, the inequality $C_A^H \geq 1$, which becomes the strict equality only for $l = 1$, is valid. The proof of this assertion is left to the reader as an exercise.

Erlang flow (E). In this case, $A(x)$ is the Erlang distribution function with the parameters l and λ which is usually defined by its density of distribution

$$a(x) = A'(x) = \frac{\lambda^l x^{l-1}}{(l-1)!} e^{-\lambda x}, \qquad x > 0,$$

where $0 < \lambda < \infty$, and l is a positive integer. Sometimes below we denote the Erlang distribution function by $E_l(x)$. The Laplace–Stieltjes transform of the distribution function $A(x)$ is

$$\alpha(s) = \left(\frac{\lambda}{\lambda + s}\right)^l.$$

For $|s| < \lambda$, the Laplace–Stieltjes transform $\alpha(s)$ is representable as the series

$$\alpha(s) = \sum_{k=0}^{\infty} (-1)^k \frac{s^k}{k!} \frac{(l+k-1)_k}{\lambda^k}.$$

Hence,

$$\mathbf{E}\xi = \frac{l}{\lambda}, \quad \mathbf{E}\xi^k = \frac{(l+k-1)_k}{\lambda^k}, \quad k \geq 1, \quad \mathbf{Var}\,\xi = \frac{l}{\lambda^2}.$$

The coefficient of variation for the Erlang flow is $C_A^E = 1/\sqrt{l}$.

Sometimes, the Erlang distribution is defined by the parameters l and $\lambda' = \lambda/l$. Then for $l \to \infty$ and fixed λ',

$$\alpha(s) = \left(\frac{\lambda}{\lambda + s}\right)^l = \left(1 + \frac{s}{l\lambda'}\right)^{-l} \xrightarrow[l \to \infty]{} e^{-s/\lambda'}.$$

For different $l = 1, 2, \dots$, therefore, the Erlang distribution defines a whole class of flows including the Poisson flow for $l = 1$ and the determinate flow as the limit for $l \to \infty$.

We turn again to the coefficient of variation of the flow which is, in a sense, the indicator of its randomness and see that the following inequalities are true for the considered above flows:

$$0 = C_A^D \leq C_A^E \leq C_A^M = 1 \leq C_A^H.$$

By taking some liberties with speech, we could state that 'the greater the coefficient of flow variation, the 'more random' is the flow itself'. From this standpoint, the hyper-exponential flow is *a priori* the worst of the above flows. Studies and numerical calculations carried out for particular queueing systems are indicative of the fact that as a rule the hyper-exponential flow provides the worst queueing system performance, whereas the deterministic flow for which the coefficient of variation is minimal possesses better characteristics.

Hyper-Erlang flow (HE). The distribution function $A(x)$ of this flow is hyper-Erlang:

$$A(x) = \sum_{i=1}^{l} \alpha_i E_{l_i}(x),$$

where $\alpha_i > 0$, $\sum_{i=1}^{l} \alpha_i = 1$ and $E_{l_i}(x)$, $i = 1, \dots, l$, is the Erlang distribution function with the parameters l_i and λ_i. Similar to the hyper-exponential distribution, the hyper-Erlang one is a mix of the Erlang, rather than exponential, distributions with the weights α_i. From the properties of the Erlang distribution function we obtain

$$\alpha(s) = \sum_{i=1}^{l} \alpha_i \left(\frac{\lambda_i}{\lambda_i + s}\right)^{l_i},$$

$$\mathbf{E}\xi = \sum_{i=1}^{l} \alpha_i \frac{l_i}{\lambda_i}, \quad \mathbf{E}\xi^k = \sum_{i=1}^{l} \alpha_i \frac{(l_i + k - 1)_k}{\lambda_i^k}, \quad k \geq 1.$$

For the hyper-Erlang flow, the coefficient of variation C_A^{HE} can take any value $0 < C_A^{HE} < \infty$.

We note that any distribution $A(x)$ can be approximated (in the sense of weak convergence of the distribution function) with any accuracy by means of the hyper-Erlang distribution. This follows from the fact that, as was seen above, the Erlang distribution can approximate the determinate distribution and, therefore, a mix of the Erlang distribution functions can approximate any step distribution function with a finite number of steps. In turn, the step distribution functions can approximate any distribution $A(x)$.

2.8.2. Some distributions of the service time

Now, according to the above description of the queueing system, the distribution function $B(x)$ is the probabilistic characteristic of the service time. We denote by $\beta(s)$ the corresponding Laplace–Stieltjes transform and by η the random variable—duration of service—itself. The types of the distribution functions of service time that are most frequently used in the queueing theory coincide with the above distributions for the recurrent flow.

For

$$B(x) = \begin{cases} 0, & x \leq b; \\ 1, & x > b, \end{cases}$$

the *service* is *determinate* (D). Then, $\beta(s) = e^{-sb}$, $\mathbf{E}\eta = b$, $\mathbf{Var}\,\eta = 0$ and $C_B^D = 0$.

For the exponential distribution

$$B(x) = 1 - e^{-\mu x}, \quad x > 0, \quad 0 < \mu < \infty,$$

the *service* is said to be *exponential*. By analogy with the Poisson flow, we denote it by M. In this case, $\beta(s) = \mu/(\mu + s)$, $\mathbf{E}\eta = 1/\mu$, $\mathbf{Var}\,\eta = 1/\mu^2$ and $C_B^M = 1$.

If

$$B(x) = \sum_{j=1}^{m} \beta_j (1 - e^{-\mu_j x}), \qquad x > 0, \tag{2.8.1}$$

where $\beta_j > 0$, $\sum_{j=1}^{m} \beta_j = 1$ and $0 < \mu_j < \infty$, $j = 1, \dots, m$, then the *service* is said to be *hyper-exponential* (*HM* or H). In this case,

$$\beta(s) = \sum_{j=1}^{m} \frac{\beta_j \mu_j}{\mu_j + s}, \quad \mathbf{E}\eta = \sum_{j=1}^{m} \frac{\beta_j}{\mu_j}, \quad \mathbf{Var}\,\eta = 2 \sum_{j=1}^{m} \frac{\beta_j}{\mu_j^2} - (\mathbf{E}\eta)^2,$$

and the inequality $C_B^H \geq 1$ is satisfied for the coefficient of variation C_B^H.

If the distribution function $B(x)$ is Erlang, that is,

$$b(x) = B'(x) = \frac{\mu^m x^{m-1}}{(m-1)!} e^{-\mu x}, \quad x > 0, \quad m = 1, 2, \dots, \quad 0 < \mu < \infty, \tag{2.8.2}$$

then the *service* is referred to as *Erlang* (E) as well. Then

$$\beta(s) = \left(\frac{\mu}{\mu + s} \right)^m, \quad \mathbf{E}\eta = \frac{m}{\mu}, \quad \mathbf{Var}\,\eta = \frac{m}{\mu^2}, \quad C_B^E = \frac{1}{\sqrt{m}}. \tag{2.8.3}$$

Finally, if

$$B(x) = \sum_{j=1}^{m} \beta_j E_{m_j}(x),$$

where $\beta_j > 0$, $\sum_{j=1}^{m} \beta_j = 1$ and $E_{m_j}(x)$, $j = 1, \dots, m$, is the Erlang distribution function with parameters m_j and μ_j, then the *service* is referred to as *hyper-Erlang* (*HE*). In this case,

$$\beta(s) = \sum_{j=1}^{m} \beta_j \left(\frac{\mu_j}{\mu_j + s} \right)^{m_j}, \quad \mathbf{E}\eta^k = \sum_{j=1}^{m} \beta_j \frac{(m_j + k - 1)_k}{\mu_j^k}, \quad k \geq 1.$$

2.8.3. Phase-type distributions

We consider below the properties of the phase-type distributions. In fact, we have already dealt with the representatives of this class of distributions such as the Erlang, hyper-exponential, and hyper-Erlang distributions and considered some of their properties. Here, we intend to present another view of these distributions based on their probabilistic interpretation relying on the notion of fictitious phases. The purpose of this section, however, is somewhat wider. It aims at introducing and discussing the properties of the most general representative of the class of phase-type distributions, the *PH*-distribution, including its probabilistic phase interpretation. It is clear that the *PH*-distribution can describe both the recurrent arrival flow and the customer service times. In this case, we speak about the *phase-type flow and/or service time*.

The idea of fictitious phases belongs to A. K. Erlang who used them to markovise the Erlang distribution. This very simple idea is conceptually related with the notion of queueing system or, to put it more precisely, its defining parameters.

We explain this by way of simplistic examples. Let us consider a server and let $B(x)$ be the distribution function of the time of service of the customer arrived. We assume that $B(x)$ is an Erlang distribution function of form (2.8.2) with parameters μ and m. Then by (2.8.3) the Laplace–Stieltjes transform $\beta(s)$ of the distribution function $B(x)$ can be treated as the

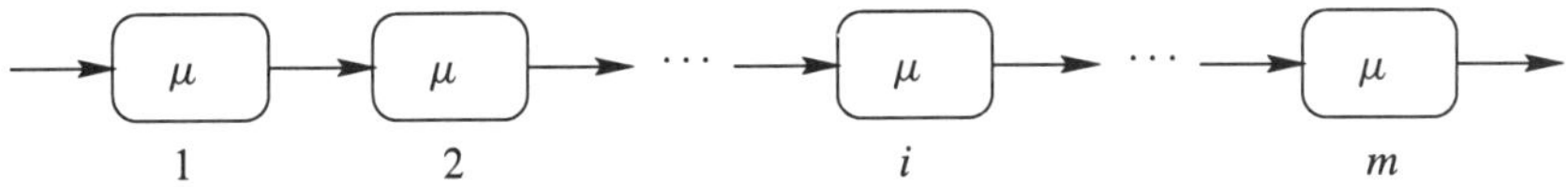

Figure 2.1.

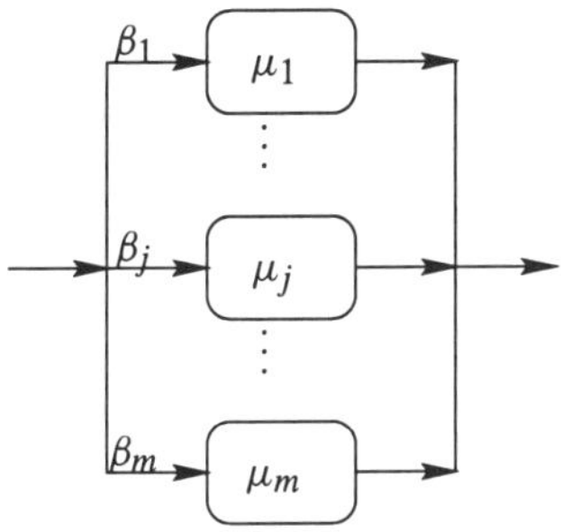

Figure 2.2.

Laplace–Stieltjes transform of the sum of m independent random variables, each distributed exponentially with parameter μ. Consequently, the process of service can be decomposed into m components or, as is the convention, phases which the customer successively goes through one after another. In doing so, the times of going through the phases are mutually independent and exponentially distributed with parameter μ. The process of customer service can be conveniently represented as the diagram in Fig. 2.1, where the rounded rectangles stand for the phase of service to be gone through by the customer in service. The customer arrived at the server must successively go through all the m phases beginning from phase one. Obviously, at any time instant at most one customer can be served, which means that there is no buffer between the phases. Additionally, the customer cannot be simultaneously in two or more phases of service.

Now we consider another case where the service time is distributed hyper-exponentially, that is, by (2.8.1). The expression for the distribution function itself suggests that here one can also extract the phase of service: at the beginning of service, the customer is sent with the probability β_j to the jth phase where it is handled during an exponentially distributed random time with parameter μ_j, and then the process of service is takes as completed. In the diagram of the service process (Fig. 2.2) the phases are arranged in parallel and only one of them is executed. The arrows entering the rounded rectangles (service phases) are tagged with the probabilities of customer arrival at the given phase. Since the server can handle at most one customer, this means that at most one phase can be occupied at each time instant. Therefore, the hyper-exponential distribution, as well as the Erlang one, admits phase interpretation, and in this sense both are phase-type distributions. A strict definition of this type of distributions will be given below.

Now we dwell on some other properties of the hyper-exponential distribution function.

We prove that the distribution function of form (2.8.1) admits the representation

$$B(x) = 1 - \boldsymbol{\beta}^T e^{Mx}\mathbf{1}, \qquad x > 0, \tag{2.8.4}$$

where $\boldsymbol{\beta}^T = (\beta_1, \dots, \beta_m)$ is a row vector and $M = \operatorname{diag}(-\mu_1, \dots, -\mu_m)$ is a diagonal matrix. Indeed,

$$\begin{aligned} e^{Mx} &= \sum_{k=0}^{\infty} \frac{(Mx)^k}{k!} = \operatorname{diag}\left(\sum_{k=0}^{\infty} \frac{(-\mu_1 x)^k}{k!}, \dots, \sum_{k=0}^{\infty} \frac{(-\mu_m x)^k}{k!}\right) \\ &= \operatorname{diag}(e^{-\mu_1 x}, \dots, e^{-\mu_m x}). \end{aligned}$$

Then

$$B(x) = 1 - \boldsymbol{\beta}^T e^{Mx}\mathbf{1} = 1 - \sum_{j=1}^{m} \beta_j e^{-\mu_j x} = \sum_{j=1}^{m} \beta_j (1 - e^{-\mu_j x}).$$

Therefore, (2.8.1) follows from (2.8.4). Obviously, the reverse is true as well, that is, (2.8.4) follows from (2.8.1).

Further, we assume that $\boldsymbol{\mu} = -M\mathbf{1}$ and on the basis of (2.8.4) determine the matrix representation of the Laplace–Stieltjes transform $\beta(s)$ of the hyper-exponential distribution function:

$$\begin{aligned} \beta(s) &= \int_0^{\infty} e^{-sx}\, dB(x) = -\int_0^{\infty} e^{-sx} \boldsymbol{\beta}^T e^{Mx} M\mathbf{1}\, dx \\ &= \boldsymbol{\beta}^T \int_0^{\infty} e^{-(sI-M)x}\, dx\, \boldsymbol{\mu} = \boldsymbol{\beta}^T (sI - M)^{-1} \boldsymbol{\mu}. \end{aligned}$$

Therefore, the Laplace–Stieltjes transform of the hyper-exponential distribution function admits the representation

$$\beta(s) = \boldsymbol{\beta}^T (sI - M)^{-1} \boldsymbol{\mu}. \tag{2.8.5}$$

In order to justify invertibility of the matrix $sI - M$ in the proof of (2.8.5) we use, out of all properties of the hyper-exponent, only the fact that the diagonal elements of the matrix M are strictly negative. Obviously, a certain distribution function of form (2.8.4) with probabilistic vector $\boldsymbol{\beta}$ and matrix M whose eigenvalues have strictly negative real parts also possesses the Laplace–Stieltjes transform of form (2.8.5).

Now we return to the Erlang distribution. It is possible to demonstrate that the Laplace–Stieltjes transform of form (2.8.5) of the distribution function $B(x)$ also admits representation (2.8.3) if one takes the following m-dimensional vector $\boldsymbol{\beta}$ and square $m \times m$ matrix M:

$$\boldsymbol{\beta} = \begin{pmatrix} 1 \\ 0 \\ \vdots \\ 0 \end{pmatrix}, \qquad M = \begin{pmatrix} -\mu & \mu & 0 & 0 & \cdots & 0 & 0 \\ 0 & -\mu & \mu & 0 & \cdots & 0 & 0 \\ 0 & 0 & -\mu & \mu & \cdots & 0 & 0 \\ \vdots & \vdots & \vdots & \vdots & \ddots & \vdots & \vdots \\ 0 & 0 & 0 & 0 & \cdots & -\mu & \mu \\ 0 & 0 & 0 & 0 & \cdots & 0 & -\mu \end{pmatrix}. \tag{2.8.6}$$

The reverse is true as well, that is, the Laplace–Stieltjes transform (2.8.3) admits representation (2.8.5) with the vector $\boldsymbol{\beta}$ and the matrix M defined by (2.8.6). The proof of this fact is left to the reader. Recalling the remark about (2.8.5), we assert that the Erlang distribution function also admits the representation in the matrix form (2.8.4) with the vector $\boldsymbol{\beta}$ and the matrix M of form (2.8.6).

The abovesaid can be summed up as follows:

- the Erlang and hyper-exponential distributions reflect some process of service with fictitious phases;
- their distribution functions are representable in the matrix form (2.8.4); and
- the Laplace–Stieltjes transforms of these distributions admit the representation in the matrix form (2.8.5).

The following question so arises: 'Is it possible to invent a more general distribution function of form (2.8.4) and its corresponding service scheme with fictitious phases encompassing both successive and parallel service?' The answer is positive. This general scheme of service with the fictitious phases is given by the *PH*-distribution proposed by M. F. Neuts. We present a concise description of the main notions and results for the *PH*-distributions.

The distribution function $F(x)$ of a nonnegative random variable is called the *phase-type distribution* or *PH-distribution* if it admits the representation

$$F(x) = 1 - \boldsymbol{f}^T e^{Gx}\mathbf{1}, \qquad x > 0, \tag{2.8.7}$$

where $\boldsymbol{f}$ is an m-dimensional vector such that $\sum_{j=1}^{m} f_j \le 1$, $f_j \ge 0$, $j = 1, \dots, m$, and G is an $m \times m$ matrix such that $\sum_{j=1}^{m} G_{ij} \le 0$; $G_{ij} \ge 0$, $i \ne j$; $G_{ii} < 0$, $i, j = 1, \dots, m$, and $\sum_{j=1}^{m} G_{ij} < 0$ for at least one i. The pair $(\boldsymbol{f}, G)$ is called the *PH-representation of order m* of the distribution function $F(x)$. Below, we often will mention just the *PH*-representation implying by it the pair $(\boldsymbol{f}, G)$, where the vector $\boldsymbol{f}$ and the matrix G have the above properties. We assume further that $\boldsymbol{g} = -G\mathbf{1}$. The *PH*-representation $(\boldsymbol{f}, G)$ is regarded as *irreducible* if $f_0 = 1 - \boldsymbol{f}^T\mathbf{1} \ne 1$ and the matrix

$$A = G + \frac{1}{1 - f_0}\boldsymbol{g}\boldsymbol{f}^T \tag{2.8.8}$$

is indecomposable.

The distribution function of the *PH* type admits probabilistic interpretation based on the concept of phase. Let us discuss this in more detail. Let $\nu_1, \dots, \nu_m$ be some real numbers, $\nu_i \ge -G_{ii}$, $i = 1, \dots, m$, the numbers θ_{ij}, $i, j = 1, \dots, m$, obey the formula

$$\theta_{ij} = \begin{cases} 1 + \dfrac{G_{ii}}{\nu_i}, & i = j; \\ \dfrac{G_{ij}}{\nu_i}, & i \ne j. \end{cases}$$

Then $\sum_{j=1}^{m} \theta_{ij} \le 1$, $\theta_{ij} \ge 0$, $i, j = 1, \dots, m$.

Let us consider now an open queueing network consisting of m nodes (Fig. 2.3) where at most one customer stays at each time, that is, the input flow is blocked if there is a customer inside the network. The arrived customer is sent to some node i, $i = 1, \dots, m$, with

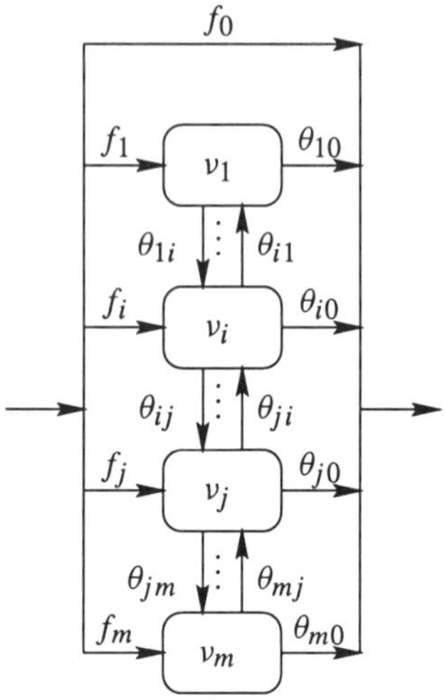

Figure 2.3.

probability f_i and with the complementary probability $f_0 = 1 - \sum_{j=1}^{m} f_j$ he immediately departs from the network bypassing all nodes. The time of customer service in the node i is distributed exponentially with parameter ν_i. After leaving the node i, the customer travels to some node j, $j = 1, \dots, m$, with the probability θ_{ij}, and with the complementary probability $\theta_{i0} = 1 - \sum_{j=1}^{m} \theta_{ij}$ it departs from the network.

Let us consider the process of servicing a customer arrived at an empty network. We denote by τ the customer sojourn time in the network, that is, the time beginning upon his arrival at the network and ending when he leaves the network. Let $\eta(t)$ be the label of the node where the customer stays at time t. The random process $\{\eta(t),\ t \in [0, \tau)\}$ which is defined only for the instants lying within the interval $[0, \tau)$ is a terminating homogeneous Markov process, and G is its matrix of transition intensities. Here

$$G_{ij} = \begin{cases} \nu_i(\theta_{ii} - 1), & i = j; \\ \nu_i \theta_{ij}, & i \neq j. \end{cases}$$

Let $p_{ij}(t)$ be the probability of customer transition from node i to node j in time $[0, t)$ under the condition that the customer arrived at the initial instant 0 is sent to the node i. The matrix of transition probabilities $P(t) = (p_{ij}(t))_{i,j=1,\dots,m}$ satisfies the set of Kolmogorov differential equations

$$\frac{d}{dt} P(t) = P(t)G$$

with the initial condition $P(0) = I$ and solution obeying the formula

$$P(t) = e^{Gt}.$$

Hence

$$\mathsf{P}\{\tau < x\} = 1 - \sum_{i=1}^{m} \sum_{j=1}^{m} f_i p_{ij}(x) = 1 - \boldsymbol{f}^T P(x)\mathbf{1} = 1 - \boldsymbol{f}^T e^{Gx}\mathbf{1}.$$

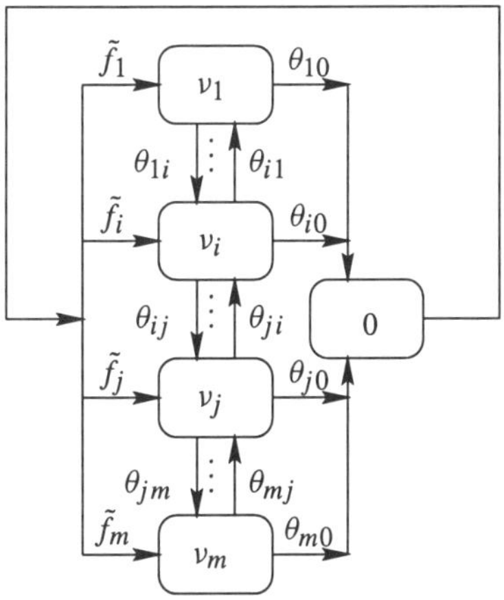

Figure 2.4.

Consequently, $F(x)$ is the distribution function of the instant τ of termination of the process $\{\eta(t),\ t \in [0, \tau)\}$ or, which is the same, of the customer sojourn time in the network. With this probabilistic interpretation of the *PH*-distribution, the hyper-exponential, Erlang, and hyper-Erlang distributions are, obviously, its special cases.

It is also possible to interpret indecomposability of the *PH*-representation in probabilistic terms. To this end, we assume that $\tilde{f}_i = f_i/(1 - f_0)$, $i = 1, \dots, m$, and consider a closed queueing network shown in Fig. 2.4 with exactly one circulating customer that is the same as the above open network, with the only exception that an additional node 0 is introduced from which the customer departing from the network immediately returns to it. In doing so, the customer getting with the probability θ_{i0} from node i, $i = 1, \dots, m$, to node 0 immediately goes through it and is routed with the probability $\{\tilde{f}_j,\ j = 1, \dots, m\}$ to one of the nodes $1, \dots, m$. So, irreducibility of the *PH*-representation means that in the closed network the customer can get from any node i to any other node j, $j \neq i$, $i, j = 1, \dots, m$.

If an irreducible *PH*-representation of the distribution function of the *PH*-type is given, then all key characteristics of this distribution function can be obtained using the vector $\boldsymbol{f}$ and the matrix G. In particular, for the Laplace–Stieltjes transform of the *PH*-distribution of form (2.8.7) with an irreducible *PH*-representation $(\boldsymbol{f}, G)$, we conclude, due to the fact that in this case the matrix G is nondegenerate, that

$$\varphi(s) = f_0 + \boldsymbol{f}^T(sI - G)^{-1}\boldsymbol{g} = 1 - s\boldsymbol{f}^T(sI - G)^{-1}\mathbf{1}, \tag{2.8.9}$$

and for the moments of the distribution function $F(x)$ the following formulas are valid:

$$\mathbf{E}\tau^k = (-1)^k k!\, \boldsymbol{f}^T G^{-k}\mathbf{1}, \qquad k = 1, 2, \dots \tag{2.8.10}$$

We observe that for $f_0 = 0$ the expression of $\varphi(s)$ (2.8.9) was, in fact, validated while deriving (2.8.5) in view of the remark we made there.

Leaving aside some well-known facts about the *PH*-distributions (Neuts, 1981; Bocharov and Pechinkin, 1995; Basharin *et al.*, 1989; Bocharov, 1993), we dwell only on the results that will be required below to prove the assertions of Chapter 4, as well as on the properties important for getting an insight into the *PH*-distribution and acquiring experience of working with them.

Below, we follow (Bellman, 1960) for convenience and refer to G as a *stable* matrix if the condition $\Re\lambda_i < 0$ is satisfied for all its eigenvalues λ_i and as a *semi-stable* matrix if $\Re\lambda_i \leq 0$. It is well known that if the *PH*-representation is irreducible, then all eigenvalues of the matrix G have strictly negative real parts and, consequently, the matrix G is stable. We observe also that *PH*-representations are *matrix-exponential representations* in the sense of (1.6.1) for a distribution function with rational Laplace–Stieltjes transforms.

Prior to formulating other results concerning *PH*-distributions, we present two auxiliary lemmas (Basharin *et al.*, 1989; Simpson, 1970).

LEMMA 2.8.1. *If a matrix G is stable and L is semi-stable, then*

$$\int_0^\infty (e^{Gt} \otimes e^{Lt})\, dt = -(G \oplus L)^{-1}.$$

Here $\otimes$ and $\oplus$ stand, respectively, for the Kronecker matrix product and Kronecker sum (the reader is referred to Section 1.7 for the definitions and base properties of these operations).

LEMMA 2.8.2. *If a matrix N is of the form*

$$N = Q + CB, \tag{2.8.11}$$

where Q is a nondegenerate $m \times m$ matrix, B is an $n \times m$ matrix, C is an $m \times n$ matrix, and the matrix

$$\Phi = I + BQ^{-1}C \tag{2.8.12}$$

is nondegenerate, then the matrix N^{-1} exists and admits the representation

$$N^{-1} = Q^{-1}(I - C\Phi^{-1}BQ^{-1}). \tag{2.8.13}$$

Now, we return to the properties of the *PH*-distributions and in view of their importance present the results below with proofs.

LEMMA 2.8.3. *Let $(\boldsymbol{f}, G)$ be an irreducible PH-representation of a phase-type distribution function $F(x)$, $\boldsymbol{f}^T\mathbf{1} = 1$, $\mathbf{g} = -G\mathbf{1}$, and let a square matrix H be semi-stable. Then*

$$\begin{aligned}\int_0^\infty e^{Ht}\, dF(t) &= I - (I \otimes \boldsymbol{f}^T)(H \oplus G)^{-1}(H \otimes \mathbf{1})\\ &= -(I \otimes \boldsymbol{f}^T)(H \oplus G)^{-1}(I \otimes \mathbf{g})\\ &= I - (\boldsymbol{f}^T \otimes I)(G \oplus H)^{-1}(\mathbf{1} \otimes H) = -(\boldsymbol{f}^T \otimes I)(G \oplus H)^{-1}(\mathbf{g} \otimes I).\end{aligned}$$

PROOF. By virtue of Lemma 2.8.1, we obtain

$$\begin{aligned} I-(I\otimes \mathbf{f}^T)(H\oplus G)^{-1}(H\otimes \mathbf{1}) &= I+(I\otimes \mathbf{f}^T)\int_0^\infty (e^{Ht}\otimes e^{Gt})\,dt(H\otimes \mathbf{1}) \\ &= I+\int_0^\infty f^T e^{Gt}\mathbf{1}\,de^{Ht} \\ &= \int_0^\infty e^{Ht}\,dF(t) = \int_0^\infty e^{Ht}\mathbf{f}^T e^{Gt}\mathbf{g}\,dt \\ &= (I\otimes \mathbf{f}^T)\int_0^\infty (e^{Ht}\otimes e^{Gt})\,dt(I\otimes \mathbf{g}) \\ &= -(I\otimes \mathbf{f}^T)(H\oplus G)^{-1}(I\otimes \mathbf{g}). \end{aligned}$$

The remaining two equalities are proved similarly. □

LEMMA 2.8.4. *Let $(\mathbf{f}, G)$ be an irreducible PH-representation of a phase-type distribution function $F(x)$, $\mathbf{f}^T\mathbf{1} = 1$, $\varphi(s)$ be the Laplace–Stieltjes transform of the distribution function $F(x)$, and an $n\times n$ matrix H be semi-stable with eigenvalues $\gamma_1, \gamma_2, \ldots, \gamma_n$. Then the matrix*

$$\Phi = I-(I\otimes \mathbf{f}^T)(H\oplus G)^{-1}(H\otimes \mathbf{1}) \tag{2.8.14}$$

is nondegenerate if and only if $\varphi(-\gamma_i)\neq 0$ for all γ_i.

PROOF. By virtue of Lemma 2.8.3,

$$\Phi = \int_0^\infty e^{Ht}\,dF(t) = \varphi(-H).$$

Therefore, $\varphi(-\gamma_i)$, $i = 1, \ldots, n$, are the eigenvalues of the matrix Φ (Kato, 1966), which proves the lemma. □

LEMMA 2.8.5. *Let $(\mathbf{f}, G)$ be an irreducible PH-representation of order m of a phase-type distribution function $F(x)$, $\mathbf{f}^T\mathbf{1} = 1$, $\varphi(s)$ be the Laplace–Stieltjes transform of the distribution function $F(x)$, and H be an $n\times n$ semi-stable matrix with eigenvalues $\gamma_1, \gamma_2, \ldots, \gamma_n$. Then the $nm\times nm$ matrix*

$$\tilde{G} = H\otimes(I-\mathbf{1}\mathbf{f}^T)+I\otimes G$$

is nondegenerate if and only if $\varphi(-\gamma_i)\neq 0$ for all γ_i.

PROOF. We reduce the matrix $\tilde{G}$ to

$$\tilde{G} = H\oplus G-(H\otimes \mathbf{1})(I\otimes \mathbf{f}^T), \tag{2.8.15}$$

make use of the equality (Simpson, 1970)

$$\det(Q+CB) = \det Q\,\det(I+BQ^{-1}C),$$

and obtain

$$\det \tilde{G} = \det(H \oplus G) \det[I - (I \otimes \boldsymbol{f}^T)(H \otimes G)^{-1}(H \otimes \mathbf{1})].$$

Since the matrix G is stable and the matrix H is semi-stable, the matrix $H \oplus G$ is stable as well and, in particular, is nondegenerate. Therefore, non-degeneracy of the matrix

$$\Phi = I - (I \otimes \boldsymbol{f}^T)(H \oplus G)^{-1}(H \otimes \mathbf{1})$$

is the necessary and sufficient condition for non-degeneracy of the matrix $\tilde{G}$. Validity of Lemma 2.8.5 now follows in an obvious way from Lemma 2.8.4. □

We observe that the Laplace–Stieltjes transform $\varphi(s)$ of a nonnegative random variable has no zeros in the right-hand half-plane, which, taking into account the last lemma, yields the following result.

LEMMA 2.8.6. *Let $(\boldsymbol{f}, G)$ be an irreducible PH-representation of a phase-type distribution function $F(x)$, $\boldsymbol{f}^T\mathbf{1} = 1$, and let a matrix H have only the real nonpositive eigenvalues. Then the matrix*

$$\tilde{G} = H \otimes (I - \mathbf{1}\boldsymbol{f}^T) + I \otimes G$$

is nondegenerate.

We present another result which is of great importance for calculations involving inversion of matrices of the type considered in Lemma 2.8.5.

LEMMA 2.8.7. *Let $(\boldsymbol{f}, G)$ and $(\boldsymbol{l}, H)$ be irreducible PH-representations of phase-type distribution functions $F(x)$ and $L(x)$, $\boldsymbol{f}^T\mathbf{1} = 1$, and let the matrix $\tilde{G} = H \otimes (I - \mathbf{1}\boldsymbol{f}^T) + I \otimes G$ be nondegenerate. Then the matrix $\tilde{G}^{-1}$ admits the representation*

$$\tilde{G}^{-1} = (H \oplus G)^{-1}[I \oplus (H \otimes \mathbf{1})\Phi^{-1}(I \otimes \boldsymbol{f}^T)(H \oplus G)^{-1}], \tag{2.8.16}$$

where Φ obeys (2.8.14).

Validity of this lemma follows from Lemmas 2.8.2, 2.8.4, and formula (2.8.14).

Now we consider a distribution function $F(x)$ of form (2.8.7) where the constraints $f_j \geq 0$, $j = 1, \dots, m$ are removed. The distribution functions of this kind are denoted by *QPH* (quasi-*PH*), and the pair $(\boldsymbol{f}, G)$ is called the *QPH-representation*. We observe that the condition $f_j \geq 0$ was never used to prove Lemmas 2.8.3–2.8.6, which gives ground to the following result.

LEMMA 2.8.8. *Let $(\boldsymbol{f}, G)$ be a QPH-representation of a QPH-distribution $F(x)$, where $\boldsymbol{f}^T\mathbf{1} = 1$ and G is a stable matrix, $\varphi(s)$ be the Laplace–Stieltjes transform of the distribution function $F(x)$, and H be a square semi-stable matrix with eigenvalues $\{\gamma_i\}$. Then the matrix*

$$\tilde{G} = H \otimes (I - \mathbf{1}\boldsymbol{f}^T) + I \otimes G$$

is nondegenerate if and only if $\varphi(-\gamma_i) \neq 0$ for all γ_i.

3

Elementary Markov models

We proceed to the queueing systems by considering the elementary Markov models. As was already noted, operation of the Markov queueing systems is describable by the continuous-time Markov process with a discrete (finite or enumerable) state set, which permits to use the well-developed and relatively simple mathematical apparatus.

For the queueing system to be Markov, it suffices that the customer arrivals be Poisson and the service time of each customer be exponential. In the four first sections of this chapter, we assume that this requirement is fulfilled. We recall that these models also are referred to as the exponential models. As we will see, this assumption gives rise not only to the continuous-time Markov processes with a discrete state set, but also to the birth-and-death processes for which, as we know from Section 1.5, the stationary probabilities are easily written. The three last sections consider the Markov queueing systems which, if not describable by the birth-and-death process, nevertheless, readily yield to analysis.

We assume throughout this chapter that the customers are taken from the queue to the server in the order of their arrival, that is, in compliance with the FCFS service discipline. The characteristics of queue length obtained using this discipline retain their validity for some other service disciplines such as LCFS and RANDOM.

The main purpose of this chapter is to teach the reader by way of elementary Markov models

(i) to set up equations for the state probabilities of the Markov processes describing the Markov queueing systems, and

(ii) to solve them in both the steady and transient cases.

3.1. $M/M/1/\infty$ system

According to the Kendall classification, the $M/M/1/\infty$ system is a single-server queueing system with an infinite-capacity buffer (system with waiting). A Poisson customer flow of intensity λ arrives at the system. The customer service times are independent random variables distributed exponentially with parameter μ. The $M/M/1/\infty$ system is classified with the exponential queueing systems where the inter-arrival times and the service times are distributed exponentially. Consequently, it is the elementary Markov queueing system, that is, its operation is described, as will be seen below, by a continuous-time Markov process with countable number of states and, moreover, by the birth-and-death process.

3.1.1. Equations for the distribution of the number of customers in the system

We denote by $\nu(t)$ the number of customers in the system at the instant t. On the one hand, the process of customer arrivals after the instant t is independent of system behaviour before this instant because we deal with the Poisson input flow which is independent of the customer service times. On the other hand, according to the description of the queueing system, the service times of the customers that are queued at the instant t or arrive at the system after it, as well as—by virtue of the exponential law (Lemma 1.2.1)—the time the customer remains in service (provided that at the instant t the system is not empty) also are independent of the system behaviour prior to t. Due to stationarity of the Poisson flow and because the customer service times are identically distributed, $\{\nu(t), t \geq 0\}$ is a homogeneous Markov process (see Section 1.5). It is clear that the number of customers in the system can take values $0, 1, 2, \ldots$ Therefore, the state set $\mathscr{X}$ of the process $\{\nu(t), t \geq 0\}$ is $\mathscr{X} = \{0, 1, \ldots\}$.

In what follows, the state of the Markov process describing the queueing system for brevity is sometimes identified with the state of the system itself, and we consider 'system state' instead of 'process state.'

To write the system of (forward) Kolmogorov differential equations, let us see what can happen to the process $\nu(t)$ over the time interval $(t, t + \Delta)$, where Δ is a 'small' time increment, and find expressions for the probabilities $p_{ij}(\Delta)$ of transitions from the state i to the state j in time Δ.

First, over this time no customer arrives with the probability $1 - \lambda\Delta + o(\Delta)$, and if there are customers in the system ($\nu(t) > 0$), then none are served with the probability $1 - \mu\Delta + o(\Delta)$. Consequently, if at the instant t the system was in the state i ($\nu(t) = i$), then it remains in the same state up to the time $t + \Delta$ with the probability

$$p_{00}(\Delta) = 1 - \lambda\Delta + o(\Delta)$$

for $i = 0$ and with the probability

$$p_{ii}(\Delta) = [1 - \lambda\Delta + o(\Delta)][1 - \mu\Delta + o(\Delta)] = 1 - (\lambda + \mu)\Delta + o(\Delta)$$

for $i > 0$.

Second, one customer can arrive with the probability $\lambda\Delta + o(\Delta)$; so, if $\nu(t) = i \geq 0$, then the system goes over the time Δ from the state i to the state $(i + 1)$. Consequently,

$$p_{i,i+1}(\Delta) = \lambda\Delta + o(\Delta), \qquad i \geq 0.$$

Finally, if $\nu(t) = i > 0$, then one customer is served with the probability $\mu\Delta + o(\Delta)$ over the time Δ, and the system goes to the state $(i - 1)$. Therefore,

$$p_{i+1,i}(\Delta) = \mu\Delta + o(\Delta), \qquad i \geq 0.$$

The probabilities of the remaining transitions such as the arrival and/or servicing of two or more customers have the infinitesimality order $o(\Delta)$. For example, the probability that over the time Δ one customer arrives to the system and one customer is served is at most $[\lambda\Delta + o(\Delta)][\mu\Delta + o(\Delta)] = o(\Delta)$. Consequently, $p_{ij}(\Delta) = o(\Delta)$, $|i - j| \geq 2$.

According to (1.5.6) and (1.5.19), $\{\nu(t), t \geq 0\}$ is the birth-and-death process; here, $\lambda_i = \lambda$, $i \geq 0$, and $\mu_i = \mu$, $i \geq 1$. Denoting $p_i(t) = \mathbf{P}\{\nu(t) = i\}$ the probability that there are i customers in the system at the instant t, we obtain by the formula of total probability

$$\begin{aligned} p_0(t+\Delta) &= (1-\lambda\Delta)p_0(t) + \mu\Delta p_1(t) + o(\Delta), \\ p_i(t+\Delta) &= [1-(\lambda+\mu)\Delta]p_i(t) + \lambda\Delta p_{i-1}(t) + \mu\Delta p_{i+1}(t) + o(\Delta), \qquad i \geq 1. \end{aligned}$$

Subtracting $p_i(t)$ from both sides of the equality, dividing by Δ, and passing to the limit as $\Delta \to 0$, we arrive at the Kolmogorov simultaneous differential equations

$$\begin{aligned} p_0'(t) &= -\lambda p_0(t) + \mu p_1(t), \\ p_i'(t) &= -(\lambda+\mu)p_i(t) + \lambda p_{i-1}(t) + \mu p_{i+1}(t), \qquad i \geq 1. \end{aligned} \tag{3.1.1}$$

3.1.2. Stationary queue length distribution

Under a certain condition discussed in what follows, the process $\{\nu(t), t \geq 0\}$ is ergodic, which means that in time the queueing system tends to the steady state, that is, $p_i(t) \xrightarrow[t\to\infty]{} p_i$, the probabilities $p_i > 0$, $i \geq 0$, and are independent of the initial state $\nu(0)$. The stationary probabilities p_i satisfy the simultaneous equilibrium equations

$$\begin{aligned} 0 &= -\lambda p_0 + \mu p_1, \\ 0 &= -(\lambda+\mu)p_i + \lambda p_{i-1} + \mu p_{i+1}, \qquad i \geq 1, \end{aligned} \tag{3.1.2}$$

which is obtained from (3.1.1) by equating the time derivatives to zero.

Accumulating experience in dealing with the Markov models, one can easily write the equilibrium equation systems of type (3.1.2) without reference to the transient case, basing, in the steady state, only on equality of the probability flows entering into each fixed state of the queueing system under consideration and leaving it, and interpreting the equilibrium equations as global balance equations (see Section 1.5). We explain this by way of the example of the ith equation in (3.1.2), $i \geq 1$.

Let us consider the transition diagram in Fig. 3.1, where the rounded rectangles denote the queueing system states and the arrows stand for the immediately possible transitions and indicate their intensities. The total flow of probabilities entering the state i is constituted by the intensity λ of transition from the state $(i-1)$ multiplied by the probability p_{i-1} of this state and the intensity μ of going away from the state $(i+1)$ multiplied by its probability p_{i+1}. The total flow of probabilities emanating from the state i is obtained by summing the intensities λ (due to a new customer arrival) and μ (due to completing customer service) multiplied by the probability p_i of the state i. By equating the total probability flows entering into the state i and emanating from it, we obtain the ith equation of (3.1.2).

In order to solve equations (3.1.2), we make use of another kind of balance, the so-called local balance (Section 1.5), which states here equality of the opposite probability flows between the states i and $(i+1)$. Since the probability flow from the state i to the state $(i+1)$ is λp_i and the opposite flow is μp_{i+1}, we arrive at the equation of local balance

$$\lambda p_i = \mu p_{i+1}, \qquad i \geq 0.$$

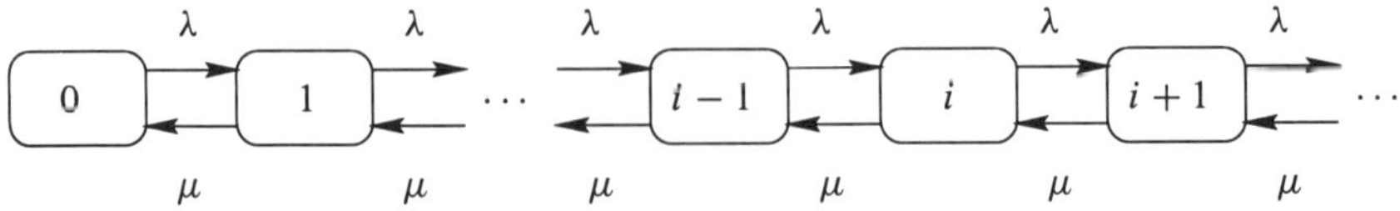

Figure 3.1.

We leave it to the reader to verify that the last equalities can be immediately obtained by successive summation of the equations in (3.1.2) for $i = 0, 1, \ldots$

From the local balance equations we see that the solution of equations (3.1.2) obeys the formula

$$p_i = \frac{\lambda}{\mu} p_{i-1} = \left(\frac{\lambda}{\mu}\right)^2 p_{i-2} = \cdots = p_0 \rho^i, \qquad i \geq 0, \tag{3.1.3}$$

where $\rho = \lambda/\mu$. Of course, the same result is obtained if we make use of the general formula for the stationary birth-and-death process state probabilities (Section 1.5).

In order to find p_0, we make use of the normalisation condition. Then

$$\sum_{i=0}^{\infty} p_i = p_0 \sum_{i=0}^{\infty} \rho^i = 1. \tag{3.1.4}$$

The infinite sum in the right-hand side of (3.1.4) is equal to $1/(1-\rho)$ if and only if $\rho < 1$. We intend to prove that the condition $\rho < 1$ is necessary and sufficient for existence of the limit (stationary) probabilities p_i (that is, for ergodicity of the process $\{\nu(t), t \geq 0\}$). The easiest way to do this is to make use of the findings of Karlin and McGregor (see Section 1.5). Indeed, in our case $\lambda_i = \lambda$ and $\mu_i = \mu$ and, therefore, the series

$$\sum_{k=1}^{\infty} \prod_{i=1}^{k} \frac{\mu_i}{\lambda_i} = \sum_{k=1}^{\infty} \prod_{i=1}^{k} \frac{1}{\rho} = \sum_{k=1}^{\infty} \rho^{-k}$$

diverges, whereas the series

$$\sum_{k=1}^{\infty} \prod_{i=1}^{k} \frac{\lambda_{i-1}}{\mu_i} = \sum_{k=1}^{\infty} \prod_{i=1}^{k} \rho = \sum_{k=1}^{\infty} \rho^k$$

converges if and only if $\rho < 1$, which, by virtue of the results obtained by Karlin and McGregor, is the necessary and sufficient condition for existence of the stationary probabilities.

Therefore, the inequality $\rho < 1$ is necessary and sufficient for existence of the steady state of the $M/M/1/\infty$ system. If it is satisfied, then we see from (3.1.4) that $p_0 = 1 - \rho$, and taking (3.1.3) into account, we arrive at the final expression for p_i

$$p_i = (1-\rho)\rho^i, \qquad i \geq 0, \tag{3.1.5}$$

that is, the stationary distribution of the number of customers in the $M/M/1/\infty$ system is geometrical. The quantity ρ in this distribution is the ratio of the mean number λ of customers arrived at the system in unit time to the mean number μ of customers which the system is able to serve in unit time; it is called the *traffic intensity*. Additionally, the stationary probability u that there is at least one customer in the system, that is, the system is busy, is $u = 1 - p_0 = \rho$. Therefore, in this case ρ also has the sense of the mean fraction u of time used by the system to serve customers in the steady state, which follows from ergodicity of the process $\{\nu(t), t \geq 0\}$. The quantity u is sometimes called the *system utilisation coefficient*.

If $\rho \geq 1$, then the number of customers in the system tends with time to infinity, that is, $p_i(t) \xrightarrow[t\to\infty]{} 0$ for all i. We observe that the queue length tends to infinity even in the case of $\rho = 1$, which is the fundamental difference between this queueing system and the determinate systems.

We give formulas for the stationary mean number N of customers in the system and mean length Q of the queue:

$$N = \sum_{i=0}^{\infty} i p_i = (1-\rho)\sum_{i=0}^{\infty} i\rho^i = \frac{\rho}{1-\rho},$$

$$Q = \sum_{i=1}^{\infty}(i-1)p_i = (1-\rho)\sum_{i=1}^{\infty}(i-1)\rho^i = \frac{\rho^2}{1-\rho}.$$

3.1.3. Stationary distribution of the customer sojourn time

The customer sojourn time is constituted by two independent components: the waiting time and the service time itself. In order to find the distribution function $W(x)$ of the waiting time in the steady state, we first assume that a customer arrived finds i other customers in the system. Since each customer, including that in service, is served in some exponentially distributed with parameter μ time and these times are independent, the total waiting time of a customer which immediately before its arrival finds i other customers in the system has the Erlang distribution $E_i(x)$ with parameters μ and i. If $i = 0$, that is, a customer arrived finds an empty system, then he is served immediately and, consequently, his zero waiting time has the distribution function $E_0(x) = u(x)$, where $u(x)$ is the Heaviside function: $u(x) = 0$ for $x \leq 0$ and $u(x) = 1$ for $x > 0$. Since for the Poisson flow the probability that in the steady state a customer arrived finds exactly i other customers immediately before his arrival is p_i (see Section 2.1), the formula of total probability yields

$$W(x) = \sum_{i=0}^{\infty} E_i(x)p_i = (1-\rho)\sum_{i=0}^{\infty}\rho^i E_i(x).$$

The Laplace–Stieltjes transform can conveniently be used to find $W(x)$ in an explicit form. Since the Erlang distribution $E_i(x)$ has the Laplace–Stieltjes transform $\varepsilon_i(s) = [\mu/(s+\mu)]^i$ and $\varepsilon_0(s) = 1$, we obtain

$$\omega(s) = \int_0^{\infty} e^{-sx}dW(x) = \sum_{i=0}^{\infty}\varepsilon_i(s)p_i$$

$$= (1-\rho)\sum_{i=0}^{\infty}\frac{(\mu\rho)^i}{(s+\mu)^i} = 1-\rho+\rho\frac{\mu(1-\rho)}{s+\mu(1-\rho)}.$$

Passing from the Laplace–Stieltjes transform to the distribution function, we obtain the formula

$$W(x) = 1 - \rho + \rho[1 - e^{-\mu(1-\rho)x}], \qquad x > 0.$$

which has a very simple sense: either the customer arrived is served immediately with the probability $p_0 = 1 - \rho$ or delayed, with the probability $1 - p_0 = \rho$, for a random time distributed exponentially with parameter $\mu(1 - \rho)$.

Since the customer service time is distributed exponentially with parameter μ, the stationary distribution $V(x)$ of the customer sojourn time has the Laplace–Stieltjes transform

$$\varphi(s) = \int_0^\infty e^{-sx} dV(x) = \frac{\mu}{s + \mu}\omega(s) = \frac{\mu(1 - \rho)}{s + \mu(1 - \rho)}.$$

We leave it to the reader to find $V(x)$.

The stationary mean time w of waiting for service beginning and the sojourn time v obey the formulas

$$w = \int_0^\infty x\, dW(x) = -\omega'(0) = \frac{\rho}{\mu(1 - \rho)},$$

$$v = \int_0^\infty x dV(x) = -\varphi'(0) = w + \frac{1}{\mu} = \frac{1}{\mu(1 - \rho)}.$$

3.1.4. Little's formulas

Comparing the stationary mean number N of customers in the system and the stationary mean customer sojourn time v, the system, we see that the relation

$$N = \lambda v$$

holds, which is called the *Little formula*. The intuitive considerations leading to it are obvious. Indeed, let the system be in the steady state over a sufficiently long time interval T. Then, on the average, λT customers arrive at the system during this time. In turn, each customer, on the average, stays in the system for the time v; therefore, the mean time V spent in the system by all λT customers is $\lambda v T$, and at each instant the system, on the average, has $N = V/T = \lambda v$ customers.

The above reasoning can be made more rigorous if one makes use of the ergodic theory of stationary random processes. It is also clear that the area of application of the Little formula is much wider than the $M/M/1/\infty$ system under consideration. This proves useful for analysing systems for which only one characteristic—either N or v—can be calculated.

A similar equality, which is also called the *Little formula*, is valid for Q and w:

$$Q = \lambda w.$$

3.1.5. Transient characteristics

In order to find the transient distribution $\{p_i(t), i \geq 0\}$ of the number of customers in the system, we turn again to differential equations (3.1.1). To solve them, one has to to define the initial distribution $\{p_i(0), i \geq 1\}$ of the number of customers in the system. It is usually assumed that at the initial instant 0 there are no customers in the system and, therefore, $p_0(0) = 1$, $p_i(0) = 0$, $i \geq 1$. We, however, assume that the initial distribution is equal to $p_i(0)$.

To solve (3.1.1), we introduce the generating function

$$P(z,t) = \sum_{i=0}^{\infty} p_i(t) z^i.$$

Multiplying the ith equation in (3.1.1) by z^i and summing all equations, we obtain

$$\frac{\partial}{\partial t} P(z,t) = \left[\lambda(z-1) + \mu\left(\frac{1}{z} - 1\right)\right] P(z,t) + \mu\left(1 - \frac{1}{z}\right) p_0(t). \tag{3.1.6}$$

Now we make use of the Laplace transform

$$\pi(z,s) = \int_0^{\infty} e^{-st} P(z,t)\, dt, \qquad \pi_0(s) = \int_0^{\infty} e^{-st} p_0(t)\, dt,$$

apply it to equation (3.1.6), and obtain

$$s\pi(z,s) - P(z) = \left[\lambda(z-1) + \mu\left(\frac{1}{z} - 1\right)\right] \pi(z,s) + \mu\left(1 - \frac{1}{z}\right) \pi_0(s),$$

where

$$P(z) = \sum_{i=0}^{\infty} p_i(0) z^i$$

is the generating function of the number of customers in the system at the initial instant 0. We solve the last equation and obtain

$$\pi(z,s) = \frac{zP(z) - \mu\pi_0(s)(1-z)}{-\lambda z^2 + (s+\lambda+\mu)z - \mu}, \tag{3.1.7}$$

where $\pi_0(s)$ is unknown. To find it, we consider the denominator of the fraction in (3.1.7). We assume that

$$f(z) = -\lambda z^2 + (s+\lambda+\mu)z - \mu.$$

Since $f(0) = -\mu$ and $f(1) = s$, the quadratic equation

$$-\lambda z^2 + (s+\lambda+\mu)z - \mu = 0 \tag{3.1.8}$$

for all $s > 0$ has two roots

$$z_{1,2} = \frac{s+\lambda+\mu}{2\lambda} \pm \sqrt{\left(\frac{s+\lambda+\mu}{2\lambda}\right)^2 - \frac{\mu}{\lambda}}.$$

Both roots are positive, the least of them being smaller than one and the greater root, greater than one. We denote the minimal root by $z(s)$. The denominator in (3.1.7) vanishes for

$z = z(s)$, $s > 0$. The function $\pi(z, s)$ is defined for all $s > 0$ and z such that $0 \le z \le 1$—in particular, at all points $(s, z = z(s))$. Then, the numerator also must vanish at these points, that is,

$$P(z(s)) - \mu\pi_0(s)\left(\frac{1}{z(s)} - 1\right) = 0,$$

or

$$\pi_0(s) = \frac{z(s)P(z(s))}{\mu - \mu z(s)}.$$

Therefore, $\pi(z, s)$ is determined in terms of double transformation. When considering the transient characteristics, we do not dwell on the issue of transformation inversion and refer the interested reader to special literature or manuals on corresponding transformations.

In the transient case, it is the virtual waiting time $\eta(t)$ at the instant t, that is, the time which a hypothetical customer arrived at time t would wait, that plays the part of the waiting time. If the distribution function of the virtual waiting time $\eta(t)$ is denoted by $W(x, t) = P\{\eta(t) < x\}$, then we see, as in Section 3.1.3, that

$$W(x, t) = \sum_{i=0}^{\infty} E_i(x)p_i(t).$$

Passing to the double transformation

$$\omega(s_1, s_2) = \int_0^\infty \int_0^\infty e^{-s_2 x} W(dx, t)e^{-s_1 t}\, dt$$

—Laplace–Stieltjes transform in x and Laplace transform in t—we obtain

$$\omega(s_1, s_2) = \pi\left(\frac{\mu}{s_2 + \mu}, s_1\right).$$

We leave it to the reader to find in terms of the Laplace transform the mean number of customers in the system and the mean virtual waiting time at the instant t.

3.1.6. Output flow

Let us determine the structure of the output flow, that is, the flow of served customers leaving the queueing system $M/M/1/\infty$, and use to this end the method of time reversal (see Section 1.5.9). First, we assume that the system operates over the interval $[0, \infty)$. As usual, let for definiteness at the initial instant 0 the system be free of customers. We consider the 'reversed' process $\{\tilde{\nu}(t), t \in (-\infty, 0]\}$ which is derived from the process $\{\nu(t), t \ge 0\}$ by substituting the time $-t$ for t, that is, $\tilde{\nu}(t) = \nu(-t)$. As shown in Section 1.5.9, the process $\{\tilde{\nu}(t), t \in (-\infty, 0]\}$ also is Markov one, but non-homogeneous. According to

(1.5.29), its transition intensities $\tilde{a}_{ij}(t)$ are

$$\tilde{a}_{ij}(t) = \begin{cases} \dfrac{\mu p_{i+1}(t)}{p_i(t)}, & j = i+1; \\ \dfrac{\lambda p_{i-1}(t)}{p_i(t)}, & j = i-1; \\ -\left(\dfrac{\mu p_{i+1}(t)}{p_i(t)} + \dfrac{\lambda p_{i-1}(t)}{p_i(t)}\right), & j = i; \\ 0, & |j-i| \geq 2, \end{cases} \qquad i \geq 1,$$

$$\tilde{a}_{0j}(t) = \begin{cases} \dfrac{\mu p_1(t)}{p_0(t)}, & j = 1; \\ -\dfrac{\mu p_1(t)}{p_0(t)}, & j = 0; \\ 0, & j \geq 2. \end{cases} \tag{3.1.9}$$

Let now the system operate infinitely long, that is, $t \in (-\infty, \infty)$. Under the assumption that $\rho = \lambda/\mu < 1$, it must be in the steady state and the probabilities $p_i = p_i(t) = \mathbf{P}\{\nu(t) = i\}$ be time-independent and obey (3.1.5). Therefore, the reversed process $\{\tilde{\nu}(t), -\infty < t < \infty\}$ which is considered now over the interval $t \in (-\infty, \infty)$ is stationary as well. By substituting in (3.1.9) the stationary probabilities p_i for $p_i(t)$, we obtain

$$\tilde{a}_{ij} = \begin{cases} \lambda, & j = i+1; \\ \mu, & j = i-1; \\ -(\lambda+\mu), & j = i; \\ 0, & |j-i| \geq 2, \end{cases} \qquad i \geq 1,$$

$$\tilde{a}_{0j} = \begin{cases} \lambda, & j = 1; \\ -\lambda, & j = 0; \\ 0, & j \geq 2. \end{cases} \tag{3.1.10}$$

Therefore, in the steady state the reversed process $\{\tilde{\nu}(t), -\infty < t < \infty\}$ is the birth-and-death process, the intensities $\tilde{a}_{i,i+1} = \lambda$ being independent of the state i of the process. In terms of the queueing system $M/M/1/\infty$ under consideration, $\tilde{a}_{i,i+1} = \lambda$ is the intensity of the output flow provided that there are i customers in the system. Making use of Lemma 2.1.1, we see that the flow *emanating from $M/M/1/\infty$ operating in the steady state is the Poisson flow of intensity λ.*

This result can easily be interpreted in terms of the reversed queueing system. Indeed, in the transient case the process $\{\tilde{\nu}(t), t \in (-\infty, 0]\}$ describes an abstract queueing system referred to as the reversed queueing system. Let us see what is it and how is it related with the original system. First, the instant of customer arrival at the original system is the instant of customer departure from the reversed system, and vice versa, customer departure from the original system amounts to his arrival at the reversed one. Therefore, for the original and reversed systems, the input and output flows exchange their places. Second, in both systems the customers arrive and depart one at a time, that is, the process $\{\tilde{\nu}(t), t \in (-\infty, 0]\}$ is also of the birth-and-death type. This exhausts the similarity between the systems. Indeed, if a Poisson flow is fed into the original system, then that arrived to the reversed system is not such because, generally speaking, its intensity $\tilde{\lambda} = \tilde{\lambda}_i = \tilde{\lambda}_i(t) = \tilde{a}_{i,i+1}(t)$, depends on the state i in which the process $\tilde{\nu}(t)$ stays at time t and, moreover, on t itself. This fact can be easily understood if one recalls that the reversed process necessarily must get into the zero state at time 0. The same is true for the customer service intensity.

As shown above, in the steady case the customer flow fed into to the reversed system is the Poisson flow with the parameter λ. Furthermore, one can see from formula (3.1.10) that in the reversed system the service time is distributed exponentially with parameter μ. Therefore, in the steady case the reversed system is also the $M/M/1/\infty$ system with the same intensities λ and μ of the input flow and service, respectively.

If $\rho \geq 1$, then the output flow is Poisson at the limit, but with the intensity $\tilde{\lambda} = \mu$, which follows from the fact that for $\rho \geq 1$ the queue length tends with time to infinity and for the infinite queue the inter-departure times are independent and distributed exponentially with parameter μ. As we will see below, some other systems also have Poisson output flows.

3.2. $M/M/n/r$ system

The $M/M/n/r$ system is an n-server queueing system with r waiting places ($r \leq \infty$) which the Poisson flow of intensity λ arrives at, the customer service times being independent and the times of service of each customer by any server being distributed exponentially with parameter μ. If $r < \infty$, then a customer arrived at the full system where all servers and waiting places are occupied is lost and does not return. The $M/M/n/r$ system also is classified with the exponential queueing systems.

3.2.1. Equations for the distribution of the number of customers in the system

If we take $\nu(t)$—the number of customers in the system at instant t—as in Section 3.1.1, then we immediately see that $\{\nu(t), t \geq 0\}$ is a homogeneous Markov process with the state set

$$\mathscr{X} = \{0, 1, \dots\}$$

for $r = \infty$ and

$$\mathscr{X} = \{0, 1, \dots, n + r\}$$

for $r < \infty$. We show below that $\{\nu(t), t \geq 0\}$ is the birth-and-death process.

We write the Kolmogorov differential equations; to this end, we consider the instants of time t and $t + \Delta$, where Δ is 'small.' We assume that at the instant t the process $\nu(t)$ is in the state i, determine where it can be at the instant $t + \Delta$, and find the probabilities $p_{ij}(\Delta)$ of its transitions over the time Δ. Three cases are possible.

In the first case, where $i < n$, all customers in the system are in service (if $i = 0$, then there are no customers at all). The probability that over the time Δ the process $\nu(t)$ does not leave the state i is the product of the probability $1 - \lambda\Delta + o(\Delta)$ that no customer arrives over the time Δ by the probability $(1 - \mu\Delta + o(\Delta))^i$ that during this time none of the i customers are served, that is, it is equal to

$$p_{ii}(\Delta) = 1 - (\lambda + i\mu)\Delta + o(\Delta).$$

The probability of passing over the time Δ to the state $(i + 1)$ is equal to the probability

$$p_{i,i+1}(\Delta) = \lambda\Delta + o(\Delta)$$

of arrival of a customer at the system. Finally, since each server completes service of its customer over the time Δ with the probability $\mu\Delta + o(\Delta)$ and there are i servers, the probability of going to the state $(i-1)$ is $p_{i,i-1}(\Delta) = i\mu\Delta + o(\Delta)$. The probabilities of the rest of transitions are $o(\Delta)$.

The second case, where $n \le i < n+r$, differs from the above in that precisely n customers are served, that is, all servers are busy. Therefore, the probability to remain in the state i after the time Δ is equal to

$$p_{ii}(\Delta) = 1 - (\lambda + n\mu)\Delta + o(\Delta)$$

and that of going at the same time to the state $(i-1)$ is equal to

$$p_{i,i-1}(\Delta) = n\mu\Delta + o(\Delta).$$

In the last case, where $i = n+r$ (this makes sense only for $r < \infty$), the arrived customer is lost. Therefore, over the time Δ with the probability $p_{n+r,n+r}(\Delta) = 1 - n\mu\Delta + o(\Delta)$ one can remain in the state $(n+r)$ and go from it in the same time to the state $(n+r-1)$ with the probability

$$p_{n+r,n+r-1}(\Delta) = n\mu\Delta + o(\Delta).$$

By virtue of (1.5.6) and (1.5.19), we prove in fact that $\{\nu(t), t \ge 0\}$ is the birth-and-death process with intensities $\lambda_i = \lambda$ for $i = 0, \ldots, n+r-1$, $\lambda_{n+r} = 0$, $\mu_i = i\mu$ for $i = 0, \ldots, n-1$, and $\mu_i = n\mu$ for $i = n, \ldots, n+r$.

Letting $p_i(t) = \mathbf{P}\{\nu(t) = i\}$, $i = 0, \ldots, n+r$, denote the distribution of the number of customers in the system at the instant t, we obtain the following expression for $p_i(t+\Delta)$ in the case of $r < \infty$:

$$\begin{aligned}
p_0(t+\Delta) &= (1 - \lambda\Delta)p_0(t) + \mu\Delta p_1(t) + o(\Delta),\\
p_i(t+\Delta) &= [1 - (\lambda + i\mu)\Delta]p_i(t) + \lambda\Delta p_{i-1}(t) + (i+1)\mu\Delta p_{i+1}(t)\\
&\quad + o(\Delta), \qquad i = 0, \ldots, n-1,\\
p_i(t+\Delta) &= [1 - (\lambda + n\mu)\Delta]p_i(t) + \lambda\Delta p_{i-1}(t) + n\mu\Delta p_{i+1}(t)\\
&\quad + o(\Delta), \qquad i = n, \ldots, n+r-1,\\
p_{n+r}(t+\Delta) &= (1 - n\mu\Delta)p_{n+r}(t) + \lambda\Delta p_{n+r-1}(t) + o(\Delta).
\end{aligned}$$

If $r = \infty$, then, obviously, the last relation disappears, and in the next to last relation the subscript i takes the values $i = n, n+1, \ldots$.

Now, subtracting $p_i(t)$ from both sides of the equalities, dividing by Δ, and passing to the limit as $\Delta \to 0$, we obtain the simultaneous differential equations

$$\begin{aligned}
p_0'(t) &= -\lambda p_0(t) + \mu p_1(t),\\
p_i'(t) &= -(\lambda + i\mu)p_i(t) + \lambda p_{i-1}(t) + (i+1)\mu p_{i+1}(t), &\quad i = 1, \ldots, n-1,\\
p_i'(t) &= -(\lambda + n\mu)p_i(t) + \lambda p_{i-1}(t) + n\mu p_{i+1}(t), &\quad i = n, \ldots, n+r-1,\\
p_{n+r}'(t) &= -n\mu p_{n+r}(t) + \lambda p_{n+r-1}(t). &\quad (3.2.1)
\end{aligned}$$

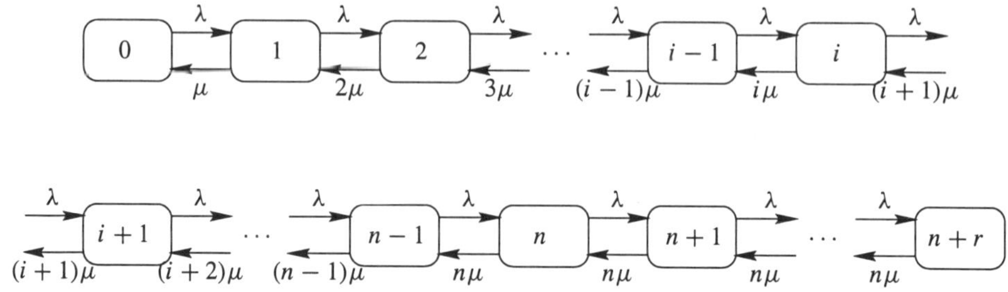

Figure 3.2.

3.2.2. Stationary queue length distribution

For a finite r, the process $\{\nu(t), t \geq 0\}$ is ergodic. It is also ergodic for $r = \infty$ provided that the condition below is fulfilled. Then, as $t \to \infty$, we see from (3.2.1) that the stationary probabilities p_i of the states satisfy the equilibrium equations

$$\begin{aligned} 0 &= -\lambda p_0 + \mu p_1, \\ 0 &= -(\lambda + i\mu)p_i + \lambda p_{i-1} + (i+1)\mu p_{i+1}, i = 1, \dots, n-1, \\ 0 &= -(\lambda + n\mu)p_i + \lambda p_{i-1} + n\mu p_{i+1}, i = n, \dots, n+r-1, \\ 0 &= -n\mu p_{n+r} + \lambda p_{n+r-1} \end{aligned} \tag{3.2.2}$$

derived with the use of the principle of global balance. For example, according to the transition diagram (Fig.. 3.2), for the fixed state i, $i = 1, \dots, n-1$, we see that the total probability flows—entering the state i and leaving it—are equal to $\lambda p_{i-1} + \mu(i+1)p_{i+1}$ and $(\lambda + \mu i)p_i$ respectively. We leave it to the reader to validate equations (3.2.2) for other values of i on the basis of global balance.

With the use of the principle of local balance, we find that the balance of the probability flows between the state i and $(i+1)$ is reflected in the equalities

$$\begin{aligned} \lambda p_i &= (i+1)\mu p_{i+1}, \qquad & i &= 0, \dots, n-1, \\ \lambda p_i &= n\mu p_{i+1}, & i &= n, \dots, n+r-1, \end{aligned} \tag{3.2.3}$$

which are the local-balance equations for the queueing system given. It is left to the reader to verify validity of equalities (3.2.3) by direct summation of the equilibrium equations (3.2.2) in i over $i = 0, 1, \dots, n+r-1$.

Expressing recursively the probabilities p_i in terms of p_0, we obtain from (3.2.3)

$$p_i = \begin{cases} \dfrac{\rho^i}{i!} p_0, & i = 0, \dots, n-1; \\ \dfrac{\rho^i}{n!\, n^{i-n}} p_0, & i = n, \dots, n+r, \end{cases} \tag{3.2.4}$$

where $\rho = \lambda/\mu$ and p_0 is found from the normalisation condition $\sum_{i=0}^{n+r} p_i = 1$, that is,

$$p_0 = \left[\sum_{i=0}^{n-1} \frac{\rho^i}{i!} + \frac{\rho^n}{n!} \sum_{i=0}^{r} \left(\frac{\rho}{n}\right)^i \right]^{-1} = \left[\sum_{i=0}^{n-1} \frac{\rho^i}{i!} + \frac{\rho^n}{n!} \frac{1 - \left(\frac{\rho}{n}\right)^{r+1}}{1 - \frac{\rho}{n}} \right]^{-1}. \tag{3.2.5}$$

It is obvious that formulas (3.2.4) and (3.2.5) can be obtained, as in the case of the $M/M/1/\infty$ system, from the general relations for the stationary probabilities of the birth-and-death process states (see Section 1.5) for the above values of λ_i and μ_i.

In the special case of $n = 1$, that is, for the single-server $M/M/1/r$ system with a buffer of finite capacity r, we obtain

$$p_i = \frac{1-\rho}{1-\rho^{r+2}}\rho^i, \qquad i = 0, \ldots, r+1.$$

If $r = \infty$ (system with waiting), then one easily sees that, as follows from (3.2.5), the probability p_0 is strictly positive, provided that the inequality $\rho/n < 1$ is true. Verification of the Karlin–McGregor conditions shows that this inequality is necessary and sufficient for the steady queueing system state to exist.

If $r < \infty$, then the steady state exists for any ρ, $0 < \rho < \infty$.

Now we give expressions of some characteristics of the queue. The stationary probability $P_{w=0}$ of immediate customer service (service without waiting) coincides with the stationary probability that there are $0, 1, \ldots, n-1$ customers in the system, that is,

$$P_{w=0} = \sum_{i=0}^{n-1} p_i = p_0 \sum_{i=0}^{n-1} \frac{\rho^i}{i!}.$$

If $r < \infty$, then a customer can be lost. This happens if, upon his arrival, all servers are busy and all waiting places occupied. The stationary probability π of losing the customer, because the input flow is the Poisson one, coincides with the stationary probability p_{n+r}, that is,

$$\pi = p_{n+r} = \frac{\rho^{n+r}}{n!n^r} p_0.$$

Also of interest is another special case of $r = 0$ where the system has no waiting places (the $M/M/n/0$ loss system). This system, first studied by A. K. Erlang and often referred to by his name, describes the processes in the elementary telephone networks. For the $M/M/n/0$ system, the stationary probabilities p_i obey the *Erlang formula*

$$p_i = \frac{\rho^i}{i!}\left(\sum_{j=0}^{n} \frac{\rho^j}{j!}\right)^{-1}, \qquad i = 0, \ldots, n.$$

Therefore, the stationary probability of losing a customer obeys the formula

$$\pi = \frac{\rho^n}{n!}\left(\sum_{i=0}^{n} \frac{\rho^i}{i!}\right)^{-1},$$

which is called the *Erlang formula*, too.

Finally, if $n = \infty$, then the $M/M/\infty$ system arises, for which the stationary probabilities exist for any $\rho < \infty$ (we leave the proof of this fact to the reader) and, as follows from the Erlang formulas, and, as $n \to \infty$, are of the form

$$p_i = \frac{\rho^i}{i!} e^{-\rho}, \qquad i \geq 0.$$

The stationary mean length Q of the queue in the $M/M/n/r$ system obeys the formula

$$Q = \sum_{i=1}^{r} i p_{n+i} = p_n \sum_{i=1}^{r} i \left(\frac{\rho}{n}\right)^i$$

or, after summation,

$$Q = \frac{\rho^{n-1}}{(n-1)!} \frac{1 + r\left(\frac{\rho}{n}\right)^{r+1} - (r+1)\left(\frac{\rho}{n}\right)^r}{\left(\frac{n}{\rho} - 1\right)^2} p_0. \tag{3.2.6}$$

Now we turn back to relations (3.2.3), which, after summation over $i = 0, 1, \ldots, n + r - 1$, yield

$$\lambda(1 - \pi) = \mu \left(\sum_{i=1}^{n-1} i p_i + \sum_{i=n}^{n+r} n p_i \right) = \mu \bar{n},$$

where $\bar{n}$ is the mean number of occupied servers. This relation reflects equality of the intensities of the received and served flows in the steady state. Hence we obtain the expression for the system *throughput* λ_D which is defined as the mean number of customers served by the system in a unit time and sometimes is referred to as the *output* or *departure intensity*:

$$\lambda_D = \lambda(1 - \pi) = \mu \bar{n}.$$

It is advisable to use different expressions for the departure intensity to check calculations based on formulas (3.2.4) and (3.2.5).

The expression for the stationary mean number N of customers in the system can be immediately obtained either from the probability distribution (3.2.4) or from the obvious relation $N = Q + \bar{n}$, which the reader can easily do without assistance.

Finally, let us give expressions of the stationary probability p_0 and the stationary mean length of the queue, Q, for $r = \infty$. For $r \to \infty$, they follow from (3.2.5) and (3.2.6), provided that $\rho < n$:

$$p_0 = \left[\sum_{i=0}^{n-1} \frac{\rho^i}{i!} + \frac{\rho^n}{n!} \frac{1}{1 - \frac{\rho}{n}} \right]^{-1}$$

and

$$Q = \frac{\rho^{n-1}}{(n-1)!} \frac{p_0}{\left(\frac{n}{\rho} - 1\right)^2}.$$

3.2.3. Stationary distribution of the customer sojourn time

The stationary distribution $W(x)$ of the time of waiting before a customer accepted by the $M/M/n/r$ system is taken for service is calculated in the same manner as for the $M/M/1/\infty$ system. We observe that a customer which finds i other customers upon his arrival at the system is immediately served if $i < n$ or waits the time required to serve $i - n + 1$ customers by the system with all servers occupied if $n \le i < n + r$. By Lemma 1.2.2, customers depart from the system with all servers occupied in times exponentially distributed with parameter $n\mu$ (not μ). For a customer finding $n+i$, $0 \le i < r$, customers in the system, the waiting time is, therefore, distributed by the Erlang law $E_{i+1}(x)$ with parameters $n\mu$ and $i+1$. Using the total probability formula and keeping in mind that if a customer is got by the system then the distribution function $W(x)$ is conditional, we obtain

$$W(x) = \frac{1}{1-\pi}\left[\sum_{i=0}^{n-1} p_i + \sum_{i=0}^{r-1} p_{n+i} E_{i+1}(x)\right] = \frac{1}{1-\pi}\left[P_{w=0} + \sum_{i=0}^{r-1} p_{n+i} E_{i+1}(x)\right]$$

or, passing to the Laplace–Stieltjes transform,

$$\omega(s) = \int_0^\infty e^{-sx} dW(x) = \frac{1}{1-\pi}\left[P_{w=0} + n\mu p_n \sum_{i=0}^{r-1} \frac{\lambda^i}{(s+n\mu)^{i+1}}\right] = \frac{1}{1-\pi}\left[P_{w=0} + n\mu p_n \frac{1-\left(\frac{\lambda}{s+n\mu}\right)^r}{s+n\mu-\lambda}\right].$$

In view of the fact that the service time is independent of the waiting time, we see that the stationary distribution $V(x)$ of the accepted customer sojourn time has the Laplace–Stieltjes transform

$$\varphi(s) = \int_0^\infty e^{-sx} dV(x) = \frac{1}{1-\pi}\left[P_{w=0} + n\mu p_n \frac{1-\left(\frac{\lambda}{s+n\mu}\right)^r}{s+n\mu-\lambda}\right]\frac{\mu}{s+\mu}.$$

The stationary mean waiting time w and the customer sojourn time v are

$$w = -\omega'(0) = \frac{n\mu - \left(\frac{\rho}{n}\right)^r [(r+1)n\mu - r\lambda]}{(n\mu-\lambda)^2}\frac{p_n}{1-\pi},$$

$$v = -\varphi'(0) = w + \frac{1}{\mu}.$$

They can also be derived from the Little formula.

3.2.4. Transient characteristics

The transient distribution $\{p_i(t), i \ge 0\}$ of the number of customers in the system is obtained by integrating (3.2.1) with the initial distribution $\{p_i(0), i \ge 0\}$.

If $r < \infty$, then (3.2.1) is a homogeneous linear set of ordinary differential equations of the first order with constant coefficients.

If $r = \infty$, then, as in Section 3.1.5, it is convenient to pass to the generating function

$$P(z,t) = \sum_{i=0}^{\infty} p_i(t) z^i.$$

The analogue of equation (3.1.8) has n unknown functions $p_0(t), p_1(t), \dots, p_{n-1}(t)$ whose Laplace transforms are determined by the same reasoning about the zeros of the numerator and denominator as in (3.1.7).

The remaining transient characteristics of the $M/M/n/r$ system—including the virtual waiting time—are determined as above.

3.2.5. Output flow

As in the case of the $M/M/1/\infty$ system, in the steady state the $M/M/n/\infty$ system has a Poisson flow of outgoing customers. The same is true for the output flow of the $M/M/n/r$ system, provided that by this flow is meant the total flow of both served and lost customers. The first fact concerning output flow of the $M/M/n/\infty$ system is proved by the method of time reversal in line with the similar proof for the $M/M/1/\infty$ system, and we leave the proof to the readers.

Now, representing the total flow of served and lost customers which leave the $M/M/n/r$ system as a Markov flow (see Section 2.1), we give another proof of the Poisson nature of such a flow. The advantage of this proof consists of the fact that it permits to do without complicated calculation of the stationary state probabilities.

Let us consider the sequence $\{\tau_k, k \geq 1\}$, $0 < \tau_1 < \tau_2 < \cdots$, of those instants where the customers from the total output flow leave the system. Let $\zeta(t)$ be the number of customers from this flow which leave the queueing system in time t and $\eta(t) = (\nu(t), \zeta(t))$. Obviously, $\{\eta(t), t \geq 0\}$ is a homogeneous Markov process with the transition intensity matrix satisfying the conditions

$$\begin{aligned} \mathsf{P}\{\nu(t+\Delta) = j, \zeta(t+\Delta) \geq k+2 \mid \nu(t) = i, \zeta(t) = k\} &= o(\Delta), \\ \mathsf{P}\{\nu(t+\Delta) = j, \zeta(t+\Delta) = k+1 \mid \nu(t) = i, \zeta(t) = k\} &= N_{ij}\Delta + o(\Delta), \\ \mathsf{P}\{\nu(t+\Delta) = j, \zeta(t+\Delta) = k \mid \nu(t) = i, \zeta(t) = k\} &= \delta_{ij} + \Lambda_{ij}\Delta + o(\Delta) \end{aligned}$$

for $i, j = 0, \dots, n+r$, where N_{ij} and Λ_{ij} are the intensities of transitions of the process $\{\nu(t), t \geq 0\}$ which imply appearance of a customer in the output flow and which do not, respectively, and δ_{ij} is the Kronecker symbol. Therefore, the process $\{\eta(t), t \geq 0\}$ generates the Markov flow $\{\tau_k, k \geq 1\}$. Here the elements of the matrix $N = (N_{ij})_{i,j=0,\dots,n+r}$ that differ from zero are of the form $N_{i,i-1} = \mu_i$, $i = 1, \dots, n+r$, and $N_{n+r,n+r} = \lambda$, while the matrix $\Lambda = (\Lambda_{ij})_{i,j=0,\dots,n+r}$ is found from the equality $\Lambda^* = \Lambda + N$, where Λ^* is the matrix of transition intensities of the process $\{\nu(t), t \geq 0\}$. By virtue of Theorem 2.1.3, the Markov flow $\{\tau_k, k \geq 1\}$ is the Poisson flow of intensity λ if the condition $\boldsymbol{p}^T N = \lambda \boldsymbol{p}^T$, where $\boldsymbol{p}^T = (p_0, p_1, \dots, p_{n+r})$, is fulfilled. Direct verification of this condition for the states $i = 0, \dots, n+r-1$ leads to the local balance equations (3.2.3), and for the state $i = r+n$, identity takes place. Therefore, the total output flow of the served and lost customers is Poisson, like the input flow, and has the same intensity λ.

3.3. $M/M/1/\infty$ system with 'impatient' customers

Let us consider the $M/M/1/\infty$ system with the input Poisson flow of intensity λ and let the service time be exponentially distributed with parameter μ. In actual practice, due to various reasons, a customer may leave the system before the service is completed. Two formulations of the problem are possible here. In the first formulation, a customer may leave the queue only, and if service has been started, then it goes on till completion. In the second formulation, a customer is allowed to leave both the queue and the server.

We consider here by way of example the first formulation where customers are allowed to leave the queue only. We assume that each customer arrived waits for service at most a random time distributed exponentially with parameter γ. By virtue of the above assumptions, this system is exponential as well.

3.3.1. Equations for distribution of the number of customers in the system

As before, we introduce the number $\nu(t)$ of customers in the system at the instant t. One can readily prove relying of the findings of the above sections that $\{\nu(t), t \geq 0\}$ is a homogeneous Markov process with the state set $\mathscr{X} = \{0, 1, \dots\}$. We write the probabilities $p_{ij}(\Delta)$ of transitions from the state i of the process $\{\nu(t), t \geq 0\}$ in the small time Δ. The probability of transition from the state i to the state $(i+1)$ over the time Δ is

$$p_{i,i+1}(\Delta) = \lambda\Delta + o(\Delta).$$

There can be two reasons for the transition from the state $i \geq 1$ to the state $(i-1)$ over the time Δ:

(i) a customer is served with the probability $\mu\Delta + o(\Delta)$, and

(ii) any of the $(i-1)$ queued customers leaves the system with the probability $\gamma\Delta + o(\Delta)$ without being served.

Therefore, the total probability of going to the state $(i-1)$ from a state $i \geq 1$ in Δ is equal to

$$p_{i,i-1}(\Delta) = [\mu + (i-1)\gamma]\Delta + o(\Delta).$$

The probability to remain in the state i for time Δ is

$$p_{ii}(\Delta) = 1 - [\lambda + \mu + (i-1)\gamma]\Delta + o(\Delta).$$

The probability of the rest transitions during the time Δ is equal to $o(\Delta)$. By virtue of (1.5.6) and (1.5.19), the process $\{\nu(t), t \geq 0\}$ is, therefore, the birth-and-death process with $\lambda_i = \lambda$, $i \geq 0$, and $\mu_i = \mu + (i-1)\gamma$, $i \geq 1$.

Without going into detail, we present the Kolmogorov differential equations for the state probabilities $p_i(t) = \mathbf{P}\{\nu(t) = i\}$, $i \geq 0$, which in the case under consideration are of the form

$$\begin{aligned} p_0'(t) &= -\lambda p_0(t) + \mu p_1(t), \\ p_i'(t) &= -[\lambda + \mu + (i-1)\gamma]p_i(t) + \lambda p_{i-1}(t) + (\mu + i\gamma)p_{i+1}(t), \qquad i \geq 1. \end{aligned} \tag{3.3.1}$$

3.3.2. Stationary queue length distribution

As we will see below, the process $\nu(t)$ is ergodic for any values of the parameters λ, μ, and γ. The stationary probabilities p_i that there are i customers in the system satisfy the simultaneous equilibrium equations

$$\begin{aligned} 0 &= -\lambda p_0 + \mu p_1, \\ 0 &= -(\lambda + \mu + (i-1)\gamma)p_i + \lambda p_{i-1} + (\mu + i\gamma)p_{i+1}, \qquad i \geq 1. \end{aligned} \tag{3.3.2}$$

We offer the reader to interpret the equilibrium equations (3.3.2) in probabilistic terms using the principle of the global balance. Without presenting the transition diagram of the system given, we write the local balance equations keeping in mind the fact that only transitions to the neighbouring states are possible:

$$\lambda p_i = (\mu + i\gamma)p_{i+1}, \qquad i \geq 0. \tag{3.3.3}$$

Relations (3.3.3) can be easily proved by the direct summation over $i = 0, 1, \ldots$ of the equilibrium equations (3.3.2). Hence we obtain the expressions of the stationary probabilities

$$p_i = p_0 \frac{\lambda^i}{\mu(\mu+\gamma)\cdots[\mu+(i-1)\gamma]}, \qquad i \geq 1, \tag{3.3.4}$$

which, as above, follow from the general relations for the stationary probabilities of the birth-and-death process states. In order to clear up the existence conditions for the stationary distribution $\{p_i, i \geq 0\}$, we again refer to the Karlin–McGregor conditions. Since for any values of the parameters λ, μ, and $\gamma > 0$ the series

$$\sum_{i=1}^{\infty} \frac{\mu(\mu+\gamma)\cdots[\mu+(i-1)\gamma]}{\lambda^i}$$

diverges and the series

$$\sum_{i=1}^{\infty} \frac{\lambda^i}{\mu(\mu+\gamma)\cdots[\mu+(i-1)\gamma]}$$

converges (we leave it to the readers to verify this), the stationary distribution $\{p_i, i \geq 0\}$ also exists for any values of these parameters.

The probability p_0, which, as usual, is found from the normalisation condition $\sum_{i=0}^{\infty} p_i = 1$, is of the form

$$p_0 = \left[1 + \sum_{i=1}^{\infty} \frac{\lambda^i}{\mu(\mu+\gamma)\cdots[\mu+(i-1)\gamma]}\right]^{-1}.$$

The stationary mean number N of customers in system and the mean length Q of the queue obey the formulas

$$\begin{aligned} N &= \sum_{i=0}^{\infty} i p_i = p_0 \sum_{i=0}^{\infty} \frac{i\lambda^i}{\mu(\mu+\gamma)\cdots[\mu+(i-1)\gamma]}, \\ Q &= \sum_{i=1}^{\infty} (i-1) p_i = p_0 \sum_{i=1}^{\infty} \frac{(i-1)\lambda^i}{\mu(\mu+\gamma)\cdots[\mu+(i-1)\gamma]}. \end{aligned}$$

Let us turn back to the local balance equations. Summing equations (3.3.3) over $i = 0, 1, \dots,$ we obtain the equality

$$\lambda = \mu(1 - p_0) + \gamma \sum_{i=1}^{\infty} (i-1) p_i = \mu(1 - p_0) + \gamma Q,$$

which has an obvious physical sense: in the steady state, the intensity λ of the arriving flow is equal to the sum of intensities $\mu(1 - p_0)$ and γQ of the flows of served and 'impatient' customers, respectively. This equality turns out to be useful in checking whether the stationary probabilities p_i are correctly calculated.

3.3.3. Stationary distribution of the customer sojourn time

We first determine the distribution function $W_i(x)$ of the waiting time for the service beginning of a 'patient' customer, that is, not leaving the queue while finding i other customers upon arrival. The first of these i customers (see Lemma 1.2.2) departs from the system in a time distributed exponentially with parameter $\mu + (i-1)\gamma$ (it leaves either after being served or before the start of service); the next customer leaves the system in a time distributed exponentially with parameter $\mu + (i-2)\gamma$, etc. Therefore, the Laplace–Stieltjes transform of the distribution function $W_i(x)$ is

$$\omega_i(s) = \int_0^{\infty} e^{-sx} dW_i(x) = \frac{\mu + (i-1)\gamma}{s + \mu + (i-1)\gamma} \frac{\mu + (i-2)\gamma}{s + \mu + (i-2)\gamma} \cdots \frac{\mu}{s + \mu}.$$

The distribution function of the total waiting time of a 'patient' customer in the steady state is found from the formula of total probability and has the Laplace–Stieltjes transform

$$\omega(s) = \sum_{i=0}^{\infty} \omega_i(s) p_i = \left[1 + \sum_{i=1}^{\infty} \frac{\lambda}{s + \mu + (i-1)\gamma} \frac{\lambda}{s + \mu + (i-2)\gamma} \cdots \frac{\lambda}{s + \mu} \right] p_0.$$

The stationary probability P_S that a customer is served is one of the main characteristics of the system with 'impatient' customers. As we know, a customer finding i other customers in the system stops waiting over the time interval $(x, x + dx)$ with the probability $dW_i(x)$ if he does not leave the queue before the instant x, the probability of this event being $e^{-\gamma x}$. We make use the formula of total probability to obtain an expression for the probability $P_{S,i}$ that the customer finding i other customers in the system is served:

$$\begin{aligned} P_{S,i} &= \int_0^{\infty} e^{-\gamma x} dW_i(x) = \omega_i(\gamma) \\ &= \frac{\mu + (i-1)\gamma}{\gamma + \mu + (i-1)\gamma} \frac{\mu + (i-2)\gamma}{\gamma + \mu + (i-2)\gamma} \cdots \frac{\mu}{\gamma + \mu} = \frac{\mu}{\mu + i\gamma}. \end{aligned}$$

Making use of the formula of total probability again, we obtain

$$P_S = \sum_{i=0}^{\infty} P_{S,i} p_i = p_0 \left[1 + \sum_{i=1}^{\infty} \frac{\lambda^i}{(\mu + \gamma) \cdots (\mu + i\gamma)} \right] = \frac{\mu}{\lambda}(1 - p_0). \qquad (3.3.5)$$

For comparison purposes, we give another derivation of (3.3.5) based on a purely qualitative reasoning and ergodicity of the process $\{\nu(t), t \geq 0\}$. On the average, λ customers arrive at the system in a unit time. Since the server is, on the average, loaded during the time $1 - p_0$ and under full loading can serve μ customers in unit time, in fact $\mu(1 - p_0)$ customers are served in unit time. Now we can obtain (3.3.5) by dividing the mean number of served customers $\mu(1 - p_0)$ by the mean number of arrived customers. This inference demonstrates that sometimes simple qualitative reasoning is helpful in checking calculations of some system characteristic.

Finally, we determine the stationary distribution $V(x)$ of the customer sojourn time. In the time interval $(x, x + dx)$, the customer finding $i \geq 1$ other customers upon his arrival can with the probability $\gamma e^{-\gamma x}[1 - W_i(x)]\,dx$ depart from the system without waiting for service start or with the probability $e^{-\gamma x} dW_i(x)$ await to be served. In the latter case, before the customer leaves the system, he has been served during an exponentially distributed time with parameter μ. The sojourn time here has the Laplace–Stieltjes transforms e^{-sx} and $e^{-sx}\mu/(s + \mu)$, respectively, for the former and latter cases. Making use of the formula of total probability, we obtain the following formula for the Laplace–Stieltjes transform $\varphi_i(s)$ of the distribution function $V_i(x)$ of the customer sojourn time provided that he finds i other customers in the system:

$$\begin{aligned}\varphi_i(s) &= \int_0^\infty e^{-sx} dV_i(x) \\ &= \int_0^\infty e^{-sx}\gamma e^{-\gamma x}[1 - W_i(x)]\,dx + \int_0^\infty e^{-sx}\frac{\mu}{s+\mu}e^{-\gamma x} dW_i(x).\end{aligned}$$

Taking the first integral by parts, we obtain

$$\begin{aligned}\varphi_i(s) &= \frac{\gamma[1 - \omega_i(s + \gamma)]}{s + \gamma} + \frac{\mu}{s+\mu}\omega_i(s + \gamma) \\ &= \frac{\gamma}{s+\gamma} + \frac{(\mu - \gamma)s}{(s + \mu)(s + \gamma)}\omega_i(s + \gamma), \qquad i \geq 1.\end{aligned}$$

Obviously, for $i = 0$ the equality

$$\varphi_0(s) = \frac{\mu}{\mu + s}$$

is valid. It also follows formally from the preceding relation if one sets $i = 0$.

Using again the formula of total probability, we obtain the Laplace–Stieltjes transform $\varphi(s)$ of the stationary distribution $V(x)$ of the customer sojourn time

$$\varphi(s) = \int_0^\infty e^{-sx} dV(x) = \sum_{i=0}^{\infty} \varphi_i(s) p_i = \frac{\gamma}{s+\gamma} + \frac{(\mu - \gamma)s}{(s + \mu)(s + \gamma)}\omega(s + \gamma).$$

It is left to the readers to calculate the mean stationary waiting time $w = -\omega'(0)$ of a 'patient' customer and the stationary mean customer sojourn time $v = -\varphi'(0)$, as well as to demonstrate validity of the Little formula.

3.3.4. Transient characteristics

In order to find the transient distribution $\{p_i(t), i \geq 0\}$ of the number of customers in the system at the time t, we have to solve, with the initial distribution $\{p_i(0), i \geq 0\}$, the linear homogeneous set of ordinary differential equations of the first order with constant coefficients (3.3.1). This can be done in two ways.

The first, approximate method consists of passing from the infinite set (3.3.1) to a finite one by replacing the queueing system $M/M/1/\infty$ with unlimited buffer by the $M/M/1/r$ system with a limited number of waiting places. If r is large enough, then the characteristics of both systems are practically identical.

The second method consists of introducing the generating function

$$P(z,t) = \sum_{i=0}^{\infty} p_i(t) z^i.$$

Multiplying the ith equation in (3.3.1) by z^i and summing over $i = 0, 1, \ldots$, we obtain

$$\begin{aligned}&\frac{\partial}{\partial t} P(z,t) + \gamma(z-1)\frac{\partial}{\partial z} P(z,t)\\ &\qquad = \left[\lambda(z-1) - \frac{\mu-\gamma}{z}(z-1)\right] P(z,t) + \frac{\mu-\gamma}{z}(z-1)p_0(t).\end{aligned}$$

The last equation is a linear partial differential equation of the first order with the initial condition

$$P(z,0) = P(z) = \sum_{i=0}^{\infty} p_i(0) z^i,$$

which involves the unknown function $p_0(t)$ related to $P(z,t)$ by $p_0(t) = P(0,t)$. Introducing the Laplace transform

$$\pi(z,s) = \int_0^{\infty} e^{-st} P(z,t)\,dt, \qquad \pi_0(s) = \int_0^{\infty} e^{-st} p_0(t)\,dt,$$

we reduce it to the linear differential equation

$$\begin{aligned}&s\pi(z,s) - P(z) + \gamma(z-1)\frac{d}{dz}\pi(z,s)\\ &\qquad = [\lambda(z-1) - \frac{\mu-\gamma}{z}(z-1)]\pi(z,s) + \frac{\mu-\gamma}{z}(z-1)\pi_0(s). \qquad (3.3.6)\end{aligned}$$

For $\mu > \gamma$, taking the initial condition $\pi(0,s) = \pi_0(s)$ into account, we find its solution

$$\begin{aligned}\pi(z,s) &= z^{1-\mu/\gamma}(1-z)^{-s/\gamma} e^{\lambda z/\gamma}\\ &\quad \times \int_0^z \left[\frac{P(u)}{\gamma(u-1)} + \frac{\mu-\gamma}{\gamma u}\pi_0(s)\right] u^{\mu/\gamma-1}(1-u)^{s/\gamma} e^{-\lambda u/\gamma}\,du. \qquad (3.3.7)\end{aligned}$$

In order to find $\pi_0(s)$, we observe that $(1-z)^{-s/\gamma}$ tends to infinity for $z \uparrow 1$ and $s > 0$. Then the integral in the right-hand side of (3.3.7) must tend to zero because $\pi(z, s)$ tends to the finite limit

$$\pi(1, s) = \int_0^\infty e^{-st} dt = 1/s.$$

Hence we obtain

$$\pi_0(s) = \frac{1}{\mu-\gamma} \int_0^1 u^{\mu/\gamma-1}(1-u)^{s/\gamma-1} P(u) e^{-\lambda u/\gamma} du \times \left[\int_0^1 u^{\mu/\gamma-2}(1-u)^{s/\gamma} e^{-\lambda u/\gamma} du \right]^{-1}.$$

If $\mu < \gamma$, then we represent the solution of equation (3.3.6) as

$$\pi(z, s) = z^{1-\mu/\gamma}(1-z)^{-s/\gamma} e^{\lambda z/\gamma} \left\{ \int_0^z \left[\left(\frac{P(u)}{\gamma(u-1)} + \frac{\mu-\gamma}{\gamma u} \pi_0(s) \right) \times u^{\mu/\gamma-1}(1-u)^{s/\gamma} e^{-\lambda u/\gamma} - \frac{\mu-\gamma}{\gamma} \pi_0(s) u^{\mu/\gamma-2} \right] du + z^{\mu/\gamma-1} \pi_0(s) \right\}.$$

The same reasoning as above yields

$$\pi_0(s) = \int_0^1 u^{\mu/\gamma-1}(1-u)^{s/\gamma-1} P(u) e^{-\lambda u/\gamma} du \times \left\{ (\mu-\gamma) \int_0^1 u^{\mu/\gamma-2} \left[(1-u)^{s/\gamma} e^{-\lambda u/\gamma} - 1 \right] du + \gamma \right\}^{-1}.$$

Finally, the case of $\mu = \gamma$ is studied in quite the same way. We do not present here the solution of equation (3.3.1) for this case because for $\mu = \gamma$ the $M/M/1/\infty$ system with 'impatient' customers is equivalent—from the point of view of distribution of the number of customers in the system—to the ordinary infinite-server $M/M/\infty$ system considered in Section 3.2.

Transient characteristics such as the distribution of the virtual waiting time, distribution of the virtual sojourn time (during which the customer that arrived at time t would stay in the system), and virtual probability of service of the customer that arrived at time t are determined precisely as their steady-state analogues.

3.4. System with a finite number of sources

We assume above that the input flow is Poisson, that is, in particular, independent of the number of customers in the system. There exist, however, queueing systems where this assumption is certainly not satisfied. Among them, we point out the system with a finite number of sources, first studied by T. Engset and often referred to by his name.

Let us assume that customers arrive from m identical sources at a single-server queueing system with a buffer of finite capacity m. Each source sends only one customer, and as long as this customer is not served (returns back), the source sends no other customers. The time from the instant when a customer returns to the source to the instant when it again arrives at the system is distributed exponentially with parameter λ. The flow of customers of this kind (with intensity depending on the state of the system) is sometimes called the Poisson flow of the second kind. The service time of each customer is also distributed exponentially with parameter μ. Although the flow fed into the Engset system is not Poisson, it is classified with the exponential systems as well. Servicing m machines by a single engineer is a classical example of the Engset system.

3.4.1. Equations for the distribution of the number of customers in the system

We again consider the number of customers $\nu(t)$ in the system at instant t. The process $\{\nu(t), t \geq 0\}$ has a finite number of states because at most m customers can be in the system. Therefore, it has the state set $\mathcal{X} = \{0, 1, \dots, m\}$. Since that the system under consideration is an exponential queueing system, that is, the inter-arrival times and the service times are distributed exponentially, its Markov nature and homogeneity of $\{\nu(t), t \geq 0\}$ becomes obvious.

In order to find the probabilities $p_{ij}(\Delta)$ of transitions in a 'small' time Δ, we observe that the process in time Δ goes from the state i, $i = 1, \dots, m$, to the state $(i-1)$ with the probability

$$p_{i,i-1}(\Delta) = \mu\Delta + o(\Delta).$$

It goes to the state $(i+1)$ from the state i, $i = 0, \dots, m-1$, with the probability

$$p_{i,i+1}(\Delta) = \lambda(m-i)\Delta + o(\Delta),$$

because if there are i customers in the system, then the number of sources able to send customers is $m-i$, each source sending a customer with the probability $\lambda\Delta + o(\Delta)$. Finally, the probability that the process does not leave the state i is

$$p_{ii}(\Delta) = 1 - [\lambda(m-i) + \mu]\Delta + o(\Delta), \qquad i = 1, \dots, m,$$

and

$$p_{00}(\Delta) = 1 - \lambda m\Delta + o(\Delta).$$

The probabilities of the other transitions are $o(\Delta)$.

We have, thus, demonstrated that—in view of (1.5.6) and (1.5.19)—the process $\{\nu(t), t \geq 0\}$ is the birth-and-death process, $\lambda_i = \lambda(m-i)$, $i = 0, \dots, m-1$, and $\mu_i = \mu$, $i = 1, \dots, m$.

In the usual way, we find that the probabilities $p_i(t) = \mathbf{P}\{\nu(t) = i\}$, $i = 0, \dots, m$, satisfy the simultaneous equations

$$\begin{aligned} p_0'(t) &= -m\lambda p_0(t) + \mu p_1(t), \\ p_i'(t) &= -[(m-i)\lambda + \mu]p_i(t) + (m-i+1)\lambda p_{i-1}(t) + \mu p_{i+1}(t), \quad i = 1, \dots, m-1, \\ p_m'(t) &= -\mu p_m(t) + \lambda p_{m-1}(t). \end{aligned} \tag{3.4.1}$$

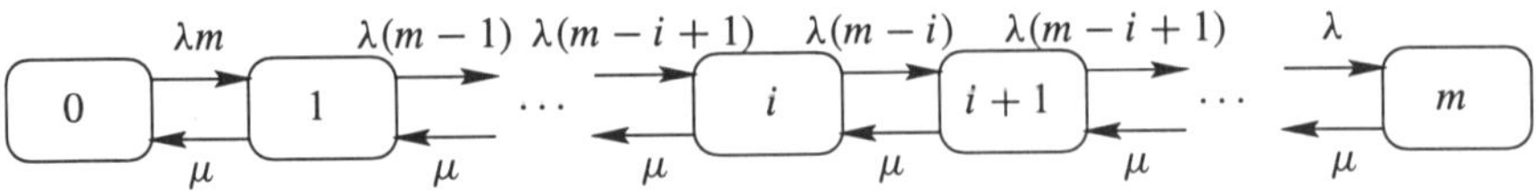

Figure 3.3.

3.4.2. Stationary queue length distribution

Since $\{\nu(t), t \geq 0\}$ is a birth-and-death process with a finite number m of states and all its states are communicating, it is ergodic for any λ and μ and its stationary probabilities p_i satisfy the equilibrium equations

$$\begin{aligned} 0 &= -m\lambda p_0 + \mu p_1, \\ 0 &= -[(m-i)\lambda + \mu]p_i + (m-i+1)\lambda p_{i-1} + \mu p_{i+1}, \qquad i = 1, \dots, m-1, \\ 0 &= -\mu p_m + \lambda p_{m-1}. \end{aligned} \tag{3.4.2}$$

As above, we leave to the readers the exercise to derive the equilibrium equations (3.4.2) starting from the principle of global balance of probability flows (see the diagram in Fig. 3.3).

Making use of the principle of local balance, from the same diagram we arrive at the local balance equations

$$\lambda(m-i)p_i = \mu p_{i+1}, \qquad i = 0, \dots, m-1, \tag{3.4.3}$$

whose solution is

$$\begin{aligned} p_i &= p_0 (m)_i \rho^i, \qquad i = 1, \dots, m, \\ p_0 &= \left[\sum_{i=0}^{m} (m)_i \rho^i \right]^{-1}, \end{aligned} \tag{3.4.4}$$

where $\rho = \lambda/\mu$.

The stationary mean number N of customers in the system obeys the relation

$$N = p_0 \sum_{i=1}^{m} i (m)_i \rho^i.$$

The stationary mean length Q of the queue is

$$Q = p_0 \sum_{i=1}^{m} (i-1)(m)_i \rho^i = N - (1 - p_0).$$

Therefore,

$$N = Q + 1 - p_0. \tag{3.4.5}$$

Another useful relation for the system performance indices can be derived from (3.4.3) by summing these equalities over $i = 0, 1, \dots, m-1$. We obtain

$$\lambda(m - N) = \mu(1 - p_0), \tag{3.4.6}$$

which reflects equality of the intensities of customer flows entering into and served by the system. Hence we obtain an expression for the important characteristic—the throughput λ_D or, which is the same, the intensity of departure from the system:

$$\lambda_D = \mu(1 - p_0) = \lambda(m - N).$$

It is convenient to apply these equalities to checking the calculations.

The stationary probability k_a that an arbitrary source sends a customer is an important characteristic of the Engset system. Recalling the above example of the serviceman, we see that k_a is the stationary probability that the machine is operable. It is called the stationary *availability*. If there are i customers in the system, then any source is among those capable of sending a customer with the probability $(m-i)/m$. By the formula of total probability,

$$k_a = \sum_{i=0}^{m} \frac{m-i}{m} p_i = p_0 \sum_{i=0}^{m} (m-1)_i \rho^i = \frac{m-N}{m} = \frac{\mu}{\lambda m}(1 - p_0).$$

3.4.3. Stationary distribution of the customer sojourn time

For a customer who finds upon his arrival i, $i = 1, \dots, m-1$, other customers, the waiting time is distributed, as in the $M/M/1/\infty$ system, by the Erlang law $E_i(x)$ with parameters μ and i, that is, has the Laplace–Stieltjes transform $\varepsilon_i(s) = (\mu/(s+\mu))^i$. If $i = 0$, then a customer arrived does not have to wait, that is, $\varepsilon_0(s) = 1$.

However, in order to calculate the unconditional stationary distribution $W(x)$ of the waiting time, one must take due regard for the fact that the input flow of customers is no more Poisson and, therefore, the stationary probability that there have been i customers in the system immediately before the arrival of an arbitrary customer does not coincide with the stationary probability p_i. In order to find this probability, we need to introduce additional probabilistic construction. Let us consider the random variable $\nu_n^- = \nu(\tau_n - 0)$, where τ_n is the instant of arrival of the nth customer. We immediately see that the sequence $\{\nu_n^-, n \geq 1\}$ constitutes a homogeneous Markov chain with the state set $\mathscr{X} \setminus \{m\}$. Since the Markov chain is irreducible and aperiodic, there exists its limit (stationary) distribution and

$$\lim_{n\to\infty} \mathbf{P}\{\nu_n^- = i\} = p_i^- > 0.$$

It is exactly p_i^-, which is the stationary probability that an arbitrary customer upon his arrival finds i more customers in the system. Beginning with general results for the Markov (semi-Markov) processes (Theorem 1.6.2), we see that the stationary probabilities p_i and p_i^- are related as follows:

$$p_i^- = \frac{\lambda_i p_i}{\lambda_D}, \qquad i = 0, \dots, m-1,$$

where $\lambda_i = \lambda(m-i)$ and $\lambda_D = \lambda(m-N)$. In view of (3.4.3),

$$p_i^- = p_0 \frac{(m)_{i+1}\rho^i}{m-N}, \qquad i = 0, \ldots, m-1.$$

Let us turn back to the stationary distribution $W(x)$ of the waiting time of an arbitrary customer and make use of the formula of total probability to demonstrate that its Laplace–Stieltjes transform is

$$\omega(s) = \sum_{i=0}^{m-1} \left(\frac{\mu}{\mu+s}\right)^i p_i^- = \frac{p_0}{m-N} \sum_{i=0}^{m-1} \frac{\lambda^i (m)_{i+1}}{(\mu+s)^i}.$$

Since the customer sojourn time is constituted by the waiting time and the exponentially distributed service time, the Laplace–Stieltjes transform $\varphi(s)$ of the stationary distribution $V(x)$ of the sojourn time is

$$\varphi(s) = \frac{\mu\omega(s)}{\mu+s}.$$

The stationary mean queueing time w and sojourn time v obey the formulas

$$\begin{aligned} w = -\omega'(0) &= \frac{p_0}{m-N} \sum_{i=0}^{m-1} \frac{i\lambda^i (m)_{i+1}}{\mu^{i+1}} = \frac{p_0}{\lambda(m-N)} \sum_{i=0}^{m-1} i(m)_{i+1}\rho^{i+1} \\ &= \frac{p_0}{\lambda(m-N)} \sum_{i=1}^{m} (i-1)(m)_i \rho^i = \frac{Q}{\lambda(m-N)} = \frac{Q}{\lambda_D}, \\ v = -\varphi'(0) &= \frac{Q}{\lambda_D} + \frac{1}{\mu}. \end{aligned} \tag{3.4.7}$$

In view of (3.4.5) and (3.4.6), the last formula takes the form

$$v = \frac{1}{\lambda_D}\left(Q + \frac{\lambda(m-N)}{\mu}\right) = \frac{N}{\lambda_D}. \tag{3.4.8}$$

Relations (3.4.7) and (3.4.8) are analogues of the Little formulas for the system under consideration.

3.4.4. Transient characteristics

The transient distribution $\{p_i(t), i \geq 0\}$ is obtained by integrating the finite linear homogeneous set of differential equations of the first order with constant coefficients (3.4.1) with the initial distribution $\{p_i(0), i \geq 0\}$. The transient distributions of the virtual queueing time and sojourn time, as well as their mean values, are found precisely as in the steady case.

3.5. $M^{[X]}/M/1/\infty$ system with batch arrivals

The $M^{[X]}/M/1/\infty$ system differs from the $M/M/1/\infty$ system only in that the customers arrive in batches but not individually. Arrival of customer batches is a Poisson flow with parameter λ, each batch having a random number of customers and the probability that exactly k customers arrive being l_k, $k \geq 1$. The customers are served one at a time, the customer service times being independent and identically distributed by the exponential law with parameter μ. We assume that the customer batches are served in the order of their arrival; inside the batches, serving is random, that is, an arbitrary customer can be with the same probability $1/k$ in any place in a k-customer batch. Obviously, $\sum_{k=1}^{\infty} l_k = 1$ and if $l_1 = 1, l_k = 0, k \geq 2$, then the $M^{[X]}/M/1/\infty$ system reduces to the $M/M/1/\infty$ system.

We again consider the random process $\{\nu(t), t \geq 0\}$, where $\nu(t)$ is the number of customers in the system at the instant t. The queueing system under consideration is exponential; therefore, $\{\nu(t), t \geq 0\}$ is a homogeneous Markov process which can be readily analysed by reasoning similar to that of Section 3.1. Obviously, $\mathscr{X} = \{0, 1, \dots\}$ is the state set of the process $\{\nu(t), t \geq 0\}$.

As the reader has already noticed, up to now (Sections 3.1–3.4) we follow the same basic pattern: having seen that the process $\{\nu(t), t \geq 0\}$ is a Markov one, we prove that it is the birth-and-death process and then, upon finding the transition intensities, go the length of solving the equations for the stationary probabilities of states, although we could have obtained them from the general results on the birth-and-death processes.

However, the list of these 'good' queueing systems that admit the description in the terms of the birth-and-death processes is not lengthy, and we have all but exhausted it. For the queueing system under study, in particular, the process $\{\nu(t), t \geq 0\}$ is not of the birth-and-death kind, which is intuitively clear because of the batch customer arrival. Therefore, we rely on the experience gained with obtaining of the equations describing the behaviour of queueing system to move further along the unbeaten track, each time finding new means to solve the corresponding equations. Although, as we will see below, there are many standard techniques along this track, it will not be as easy to do as before.

3.5.1. Equations for the distribution of the number of customers in the system

Having made the reader ready for the inevitable difficulties, we continue with studying the process $\{\nu(t), t \geq 0\}$. We give a rigorous proof that this process is not the birth-and-death process anymore. Indeed, as for the $M/M/1/\infty$ system, the probability that the process leaves the state i in a 'small' time Δ is $(\lambda + \mu)\Delta + o(\Delta)$ for $i \geq 1$ and $\lambda\Delta + o(\Delta)$ for $i = 0$. Therefore,

$$p_{ii}(\Delta) = 1 - (\lambda + \mu)\Delta + o(\Delta),$$
$$p_{00}(\Delta) = 1 - \lambda\Delta + o(\Delta).$$

The process goes to the state $(i - 1)$ from the state i with the probability

$$p_{i,i-1}(\Delta) = \mu\Delta + o(\Delta).$$

From the state i, the process $\{\nu(t), t \geq 0\}$ may go not only into the state $(i + 1)$, but also to the states $i + 2, i + 3, \dots,$ though. Since the process $\{\nu(t), t \geq 0\}$ goes from the state i to the state $(i + k)$ if there are k customers in the batch, the probability of going to the state $(i + k)$ from the state i in time Δ is

$$p_{i,i+k}(\Delta) = \lambda\Delta l_k + o(\Delta),$$

and therefore, the process $\{\nu(t), t \geq 0\}$ is not of the birth-and-death kind. The probabilities of other possible transitions are $o(\Delta)$.

In view of the aforementioned, we arrive in the usual way at the set of differential Kolmogorov equations for the state probabilities $p_i(t) = \mathbf{P}\{\nu(t) = i\}, i \geq 0$:

$$p_0'(t) = -\lambda p_0(t) + \mu p_1(t),$$
$$p_i'(t) = -(\lambda + \mu) p_i(t) + \lambda \sum_{j=0}^{i-1} p_j(t) l_{i-j} + \mu p_{i+1}(t), \qquad i \geq 1. \tag{3.5.1}$$

3.5.2. Stationary queue length distribution

The stationary state probabilities $p_i, i \geq 0$, are found from the equilibrium equations under the assumption that the process $\{\nu(t), t \geq 0\}$ is ergodic (the ergodicity conditions will be given below):

$$0 = -\lambda p_0 + \mu p_1,$$
$$0 = -(\lambda + \mu) p_i + \lambda \sum_{j=0}^{i-1} p_j l_{i-j} + \mu p_{i+1}, \qquad i \geq 1. \tag{3.5.2}$$

If we proceed from the principle of global balance, we must indicate that the transition diagram for the $M^{[X]}/M/1/\infty$ queueing system is now more complicated than before. We consider individual states of this diagram, where, as before, the arrows (labelled with the appropriate intensities) entering a rounded rectangle show from what states one can reach that state and the outgoing arrows show to what states one can go from this state (Fig. 3.4). Case (a) relates to the state 0, case (b) refers to the rest of the states, $i \geq 1$. Generally speaking, while finding the equilibrium equations, we have no need to show in the diagram to which particular states one can go from the current state but merely specify what event—arrival or service completion—initiates the transition.

In order to write the local balance equations, we introduce the state set $\mathcal{X}_i = \{0, 1, \ldots, i\}$ and proceed as follows (Fig. 3.4c): considering the passages from the set $\mathcal{X}_i$ to the set $\overline{\mathcal{X}}_i = \mathcal{X} \setminus \mathcal{X}_i$ (to the right) and back (to the left), we find that the total probability flow of to the right is $\lambda \sum_{j=0}^{i} p_j L_{i-j}$, where $L_0 = 1$ and $L_i = 1 - \sum_{j=1}^{i} l_j$, $i \geq 1$, is the probability that at most i customers arrive in the batch, and the flow to the left is μp_{i+1}. By equating these flows, we obtain

$$\lambda \sum_{j=0}^{i} p_j L_{i-j} = \mu p_{i+1}, \qquad i \geq 0. \tag{3.5.3}$$

We leave it to the reader to verify equalities (3.5.3) by direct summation of equations (3.5.2) over $i = 0, 1, \ldots$

Since the process $\{\nu(t), t \geq 0\}$ is not of the birth-and-death kind, we cannot write an explicit solution but can obtain recurrence relations to calculate p_i from equalities (3.5.3). Indeed, setting $r_i = p_i/p_0, i \geq 0$, we derive from (3.5.3)

$$r_i = \hat{\rho} \sum_{j=0}^{i-1} r_j L_{i-j-1}, \qquad i \geq 1, \tag{3.5.4}$$

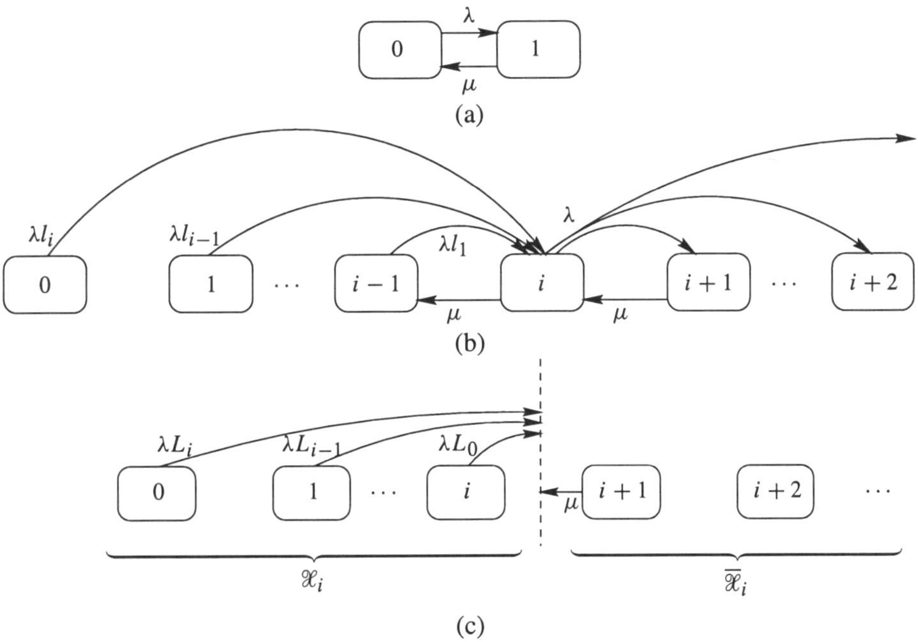

Figure 3.4.

where $\hat{\rho} = \lambda/\mu$. On the other hand, by summing (3.5.3) over $i = 0, 1, \dots$, in view of the normalisation condition $\sum_{i=0}^{\infty} p_i = 1$, we obtain in the left-hand side

$$\lambda \sum_{i=0}^{\infty} \sum_{j=0}^{i} p_j L_{i-j} = \lambda \sum_{i=0}^{\infty} p_i \sum_{j=0}^{\infty} L_j = \lambda \sum_{j=0}^{\infty} \sum_{k=j+1}^{\infty} l_k = \lambda \sum_{k=1}^{\infty} k l_k = l\lambda,$$

where l is the mean number of customers in the batch, and the right-hand side is $\mu(1 - p_0)$, which yields

$$l\lambda = \mu(1 - p_0).$$

Hence,

$$p_0 = 1 - \rho, \tag{3.5.5}$$

where

$$\rho = l\hat{\rho} = l\lambda/\mu.$$

Therefore, for the given distribution $\{l_j, j \geq 1\}$, recurrence relations (3.5.4) together with (3.5.5) allow us to calculate the stationary probabilities $p_i, i \geq 0$.

We observe that necessity of the condition $\rho < 1$ for existence of the steady state follows from (3.5.5). Since all states of the process $\{\nu(t), t \geq 0\}$ are communicating, with the use of the regularity criterion (Theorem 1.5.1) and the Foster theorem 1.5.4 we conclude that this condition is sufficient. The condition $\rho < 1$ is obvious because $l\lambda$ is the mean number of customers arriving in unit time, and μ is the mean number of customers that can be served by the server in unit time.

We have thus demonstrated that $p_0 = 1 - \rho$; then p_i can be recursively calculated from (3.5.4) which, however, gives us no way to find the moments of stationary distribution of the number of customers in the system, because convergence of the series $\sum_{i=0}^{\infty} i p_i$, $\sum_{i=0}^{\infty} i^2 p_i$, etc., does not follow from convergence of the series $\sum_{i=0}^{\infty} p_i$. Therefore, we write the equilibrium equations (3.5.2) also in terms of the generating function.

We introduce

$$P(z) = \sum_{i=0}^{\infty} p_i z^i, \quad L(z) = \sum_{i=1}^{\infty} l_i z^i, \quad |z| \le 1.$$

Multiplying the ith equation in (3.5.2) by z^i and carrying out the summation, we obtain

$$\begin{aligned} 0 &= -(\lambda + \mu) \sum_{i=0}^{\infty} p_i z^i + \lambda \sum_{i=0}^{\infty} p_i z^i \sum_{j=1}^{\infty} l_j z^j + \mu p_0 + \mu \sum_{i=1}^{\infty} p_i z^{i-1} \\ &= -(\lambda + \mu) P(z) + \lambda P(z) L(z) + \frac{\mu}{z} P(z) + \mu p_0 \left(1 - \frac{1}{z}\right). \end{aligned}$$

The solution of this equation is

$$P(z) = \mu(1 - \rho) \frac{1 - z}{\mu(1 - z) - \lambda z[1 - L(z)]}.$$

The moments of the number of customers in the system can be easily obtained by differentiating the appropriate number of times the generating function at the point $z = 1$. For example, the stationary mean number N of customers in the system obeys the formula

$$N = P'(1) = \frac{l^{(2)} + l}{2(1 - \rho)} \hat{\rho},$$

where

$$l^{(2)} = \sum_{k=1}^{\infty} k^2 l_k$$

is the second moment of the number of customers in the batch.

For the stationary mean queue length Q, we obtain, in an obvious way,

$$Q = N - \rho.$$

We see from these formulas that the stationary mean queue is finite if and only if the variance of the number of customers in the batch is finite.

3.5.3. Stationary distribution of the customer sojourn time

As in the case of the $M/M/1/\infty$ system, if the customer batch arrived meets i more customers in the system, then the first customer in the batch awaits either the time distributed by the Erlang law $E_i(x)$ with parameters μ and i, provided that $i \geq 1$, or zero time, that is, the time with the distribution function $E_0(x) = u(x)$ for $i = 0$. Therefore, by virtue of the formula of total probability, the stationary distribution $W^*(x)$ of the waiting time of the first customer in the batch has the Laplace–Stieltjes transform

$$\omega^*(s) = \int_0^\infty e^{-sx} dW^*(x) = \sum_{i=0}^{\infty} \left(\frac{\mu}{\mu+s}\right)^i p_i = P\left(\frac{\mu}{\mu+s}\right).$$

In order to find the stationary distribution of the waiting time of an arbitrary customer, we note that he arrives in the batch of k customers with the probability kl_k/l (see Section 2.1), the probability at any place in the batch being the same, $1/k$. Furthermore, if the customer arrives at the jth place in the batch, then from the beginning of service of the first customer he waits the time distributed by the Erlang law with parameters μ and $j-1$ (obviously, this time is zero for $j = 1$). Therefore, by virtue of the formula of total probability, from the beginning of service of the first customer an arbitrary customer waits the time with the Laplace–Stieltjes transform

$$\sum_{k=1}^{\infty} \frac{kl_k}{l} \sum_{j=1}^{k} \frac{1}{k} \left(\frac{\mu}{\mu+s}\right)^{j-1} = \frac{1}{l} \sum_{k=1}^{\infty} l_k \frac{1-\left(\frac{\mu}{\mu+s}\right)^k}{1-\frac{\mu}{\mu+s}} = \frac{\mu+s}{ls}\left[1 - L\left(\frac{\mu}{\mu+s}\right)\right].$$

Since the total waiting time of an arbitrarily chosen customer is constituted by the waiting time of the first customer in the batch and the time between the beginning of service of the first customer and that of the chosen one, the formula

$$\begin{aligned}\omega(s) &= \int_0^\infty e^{-sx} dW(x) = \omega^*(s) \frac{\mu+s}{ls}\left[1 - L\left(\frac{\mu}{\mu+s}\right)\right] \\ &= \frac{\mu+s}{ls}\left[1 - L\left(\frac{\mu}{\mu+s}\right)\right] P\left(\frac{\mu}{\mu+s}\right)\end{aligned}$$

holds for the Laplace–Stieltjes transform of the stationary distribution $W(x)$ of the waiting time of an arbitrary customer.

The stationary distribution $V(x)$ of the arbitrary customer sojourn time has the Laplace–Stieltjes transform

$$\varphi(s) = \int_0^\infty e^{-sx} dV(x) = \frac{\mu}{\mu+s}\omega(s) = \frac{\mu}{ls}\left[1 - L\left(\frac{\mu}{\mu+s}\right)\right] P\left(\frac{\mu}{\mu+s}\right).$$

The stationary mean waiting time w and the customer sojourn time v obey the relations

$$\begin{aligned} w &= -\omega'(0) = \frac{l^{(2)}+l}{2\mu l(1-\rho)} - \frac{1}{\mu} = \frac{Q}{\bar{\lambda}}, \\ v &= -\varphi'(0) = w + \frac{1}{\mu} = \frac{l^{(2)}+l}{2\mu l(1-\rho)} = \frac{N}{\bar{\lambda}}, \end{aligned}$$

where $\bar{\lambda} = \lambda l$, which are the Little formulas for the queueing system under consideration.

In the system with batch customer arrivals, it is sometimes necessary to find the stationary distribution of the entire batch sojourn time. We leave it to the reader to solve this problem.

3.5.4. Transient characteristics

In order to find the distribution $\{p_i(t), i \geq 0\}$ of the number of customers in the system at the instant t, we introduce the generating function

$$P(z,t) = \sum_{i=0}^{\infty} p_i(t) z^i$$

and then make use of the Laplace transform

$$\pi(z,s) = \int_0^{\infty} e^{-st} P(z,t)\, dt.$$

Then (3.5.1) takes the form

$$s\pi(z,s) - P(z) = \left[\lambda L(z) - \lambda - \mu + \frac{\mu}{z}\right] \pi(z,s) + \mu\pi_0(s)\left(1 - \frac{1}{z}\right),$$

where

$$P(z) = \sum_{i=0}^{\infty} p_i(0) z^i$$

is the generating function of the number of customers in the system at the initial time and

$$\pi_0(s) = \int_0^{\infty} e^{-st} p_0(t)\, dt.$$

The last equation is solved precisely as its analogue for the $M/M/1/\infty$ system, except that the quadratic equation (3.1.8) is replaced by the functional equation

$$s - \lambda L(z) + \lambda + \mu - \frac{\mu}{z} = 0.$$

The transient distributions of the virtual waiting and sojourn times are determined precisely as their stationary analogues.

3.6. $M/E_m/1/\infty$ system

In the queueing systems that in the Kendall classification have either E or H at any (or both) of the two leading places, the number of customers $\nu(t)$ in the system at time t is not a Markov process anymore. However, we can construct for such systems a continuous-time Markov process with a discrete state set which describes their functioning. To this end, the method of 'fictitious phases' is used. It is based on the concept of probabilistic phase interpretation of the *PH*-distributions presented in detail in Section 2.8. Here we recall it briefly as applied both to the arrival flow and the service time of the Erlang type.

For example, if the arrival is an Erlang flow (E_l) with parameters λ and l, then the time between two consecutive arrivals can be decomposed in l subsequent phases (stages), the times of going through the phases being independent of each other and distributed identically by the exponential law with parameter λ. Under such probabilistic interpretation of the Erlang flow, the actual customer arrives at the system only after going through all l phases.

It is assumed for the Erlang distribution of the service time with parameters μ and m that each customer in service goes through m service phases whose durations are independent and exponentially distributed with parameter μ and departs from the system only upon completion of all m phases. As shown in Section 2.8, this approach can also be used if the input customer flows and/or their service times are described by the phase-type distributions (*PH*-distributions) which admit probabilistic interpretation. As we already know, the Erlang, hyper-exponential, and some other distributions are special cases of this distribution. We defer the discussion of the general case until the next chapter. Here we present the method of fictitious phases in terms of the $M/E_m/1/\infty$ queueing system.

Let us consider the $M/E_m/1/\infty$ system with a single server and infinite-capacity buffer which a Poisson flow of customers of intensity λ arrives at. The service times are independent and distributed by the Erlang law with parameters μ and m.

3.6.1. Equations for the Markov process

As usual, $\nu(t)$ stands for the number of customers in the system at the instant t; additionally, let $\xi(t)$ be the number of phases that the current customer has to go through before completing the service. For the system under consideration, the process $\{\nu(t), t \geq 0\}$ is not Markov due to the fact that the Erlang distribution has no memoryless property. Nevertheless, on its basis, taking into account the information about the number $\xi(t)$ of remaining phases, we define a new process $\{\eta(t), t \geq 0\}$ which turns out to be a Markov process. If no customer exists in the system at the instant t, that is, if $\nu(t) = 0$, then we assume that $\eta(t) = \nu(t)$. Since in this case no service takes place, the Markov nature of $\eta(t)$ is determined for these instants by the Poisson nature of the input flow. If at the instant t the system performs service ($\nu(t) > 0$), then we additionally indicate the value of $\xi(t)$ and, therefore, assume that $\eta(t) = (\nu(t), \xi(t))$. By virtue of the exponential nature of the service phases and, again, the Poisson nature of the flow for such time instants, $\eta(t)$ is a Markov process indeed. Thus, $\{\eta(t), t \geq 0\}$ is a Markov homogeneous process. Its state set is

$$\mathscr{X} = \{(0); (i, j), i = 1, 2, \dots, j = 1, \dots, m\},$$

where (0) corresponds to the empty system and i and j in the state (i, j) indicate, respectively, the number of customers in the system and the number of phases until the end of serving the customer.

Now we find the transition probabilities of the process $\{\eta(t), t \geq 0\}$ in the time interval $(t, t + \Delta)$. In the 'small' time Δ, only transition from the state (0) to the state $(1, m)$ with the probability $\lambda\Delta + o(\Delta)$ is possible (a customer arrives and his service begins from the first phase). From the state $(i, j), i \geq 1, j = 2, \dots, m$, transitions are possible to the states $(i + 1, j)$ with the probability $\lambda\Delta + o(\Delta)$ (the service of the same phase of the customer in the server goes on, but another customer arrives) and to $(i, j - 1)$ with the probability

$\mu\Delta+o(\Delta)$ (the current service phase of the customer in the server is complete). Transitions from the state $(i, 1)$, $i \geq 2$, to the state $(i+1, 1)$ with the probability $\lambda\Delta + o(\Delta)$ (a new customer arrives) and the state $(i-1, m)$ with the probability $\mu\Delta + o(\Delta)$ (the final phase of the customer in service is complete, the customer departs from the system, and a newly arrived customer has to go through all m server phases) are possible. Finally, from the state $(1, 1)$ transitions are possible to the state $(2, 1)$ with the probability $\lambda\Delta + o(\Delta)$ (a new customer arrives) and the state (0) with the probability $\mu\Delta + o(\Delta)$ (the service of the last phase of the single customer in system is complete).

The probabilities of all other possible transitions are equal to $o(\Delta)$.

Since, except for (0), any state is directly related with more than two states (for example, from the state $(1, 1)$ one can go directly to the states (0) and $(2, 1)$, and to the state $(1, 1)$ one can go from the state $(1, 2)$), the Markov process $\{\eta(t), t \geq 0\}$, as one could expect, is no more of the birth-and-death kind. Assuming that

$$p_0(t) = \mathbf{P}\{\nu(t) = 0\}, \quad p_{ij}(t) = \mathbf{P}\{\nu(t) = i, \xi(t) = j\}, \quad i \geq 1, \quad j = 1, \dots, m,$$

and leaving aside detailed calculation, we write out the Kolmogorov differential equations

$$\begin{aligned}
p_0'(t) &= -\lambda p_0(t) + \mu p_{11}(t), \\
p_{1j}'(t) &= -(\lambda+\mu) p_{1j}(t) + \mu p_{1,j+1}(t), && j = 1, \dots, m-1, \\
p_{1m}'(t) &= -(\lambda+\mu) p_{1m}(t) + \lambda p_0(t) + \mu p_{21}(t), \\
p_{ij}'(t) &= -(\lambda+\mu) p_{ij}(t) + \lambda p_{i-1,j}(t) + \mu p_{i,j+1}(t), \quad i \geq 2, && j = 1, \dots, m-1, \\
p_{im}'(t) &= -(\lambda+\mu) p_{im}(t) + \lambda p_{i-1,m}(t) + \mu p_{i+1,1}(t), \quad i \geq 2.
\end{aligned} \tag{3.6.1}$$

3.6.2. Stationary distribution of the queue length

We assume that the queueing system operates in the steady state (the condition for the state to exist will be formulated below). Then the stationary probabilities p_0 and p_{ij} satisfy the equilibrium equations

$$\begin{aligned}
0 &= -\lambda p_0 + \mu p_{11}, \\
0 &= -(\lambda+\mu) p_{1j} + \mu p_{1,j+1}, && j = 1, \dots, m-1, \\
0 &= -(\lambda+\mu) p_{1m} + \lambda p_0 + \mu p_{21}, \\
0 &= -(\lambda+\mu) p_{ij} + \lambda p_{i-1,j} + \mu p_{i,j+1}, \quad i \geq 2, && j = 1, \dots, m-1, \\
0 &= -(\lambda+\mu) p_{im} + \lambda p_{i-1,m} + \mu p_{i+1,1}, \quad i \geq 2.
\end{aligned} \tag{3.6.2}$$

As above, we explain the equilibrium equations (3.6.2) by treating it as the global balance equations and using the transition diagram in Fig. 3.5. Then, for example, for the state (i, j), $i \geq 2$, $j = 1, \dots, m-1$ (see Fig. 3.5d), we see that the emanating probability flow is $(\lambda + \mu)p_{ij}$ (we do not indicate to which state the transition from the state (i, j) occurs) and the input flow is $\lambda p_{i-1,j} + \mu p_{i,j+1}$. Equating these flows, we obtain the next to the last equation in (3.6.2), the other equation being derived by the same way.

Let us turn to the state subsets $\mathscr{X}_i$ with at most i customers in the system:

$$\mathscr{X}_0 = \{(0)\}, \quad \mathscr{X}_i = \{(0); (k, j), k = 1, \dots, i, j = 1, \dots, m\}, \quad i \geq 1.$$

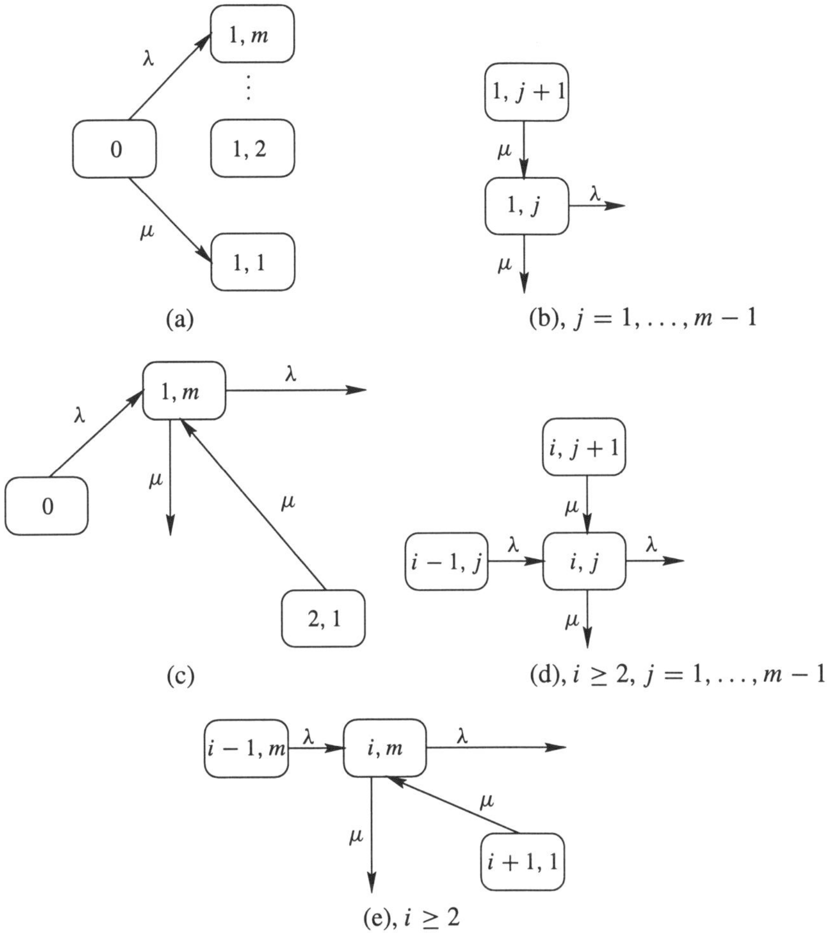

Figure 3.5.

Making use of the local balance principle and equating the opposite probability flows between the sets $\mathcal{X}_i$ and $\mathcal{X}_{i+1}$, $i \geq 0$ (see Fig. 3.6), we obtain the equalities

$$\begin{aligned} \lambda p_0 &= \mu p_{11}, \\ \lambda p_{i,\cdot} &= \mu p_{i+1,1}, \qquad i \geq 1. \end{aligned} \tag{3.6.3}$$

Here and below '$\cdot$' stands for summation over all possible values of the discrete argument of the corresponding function. Equation (3.6.3) can also be obtained by successive summing, first, over $j = 1, 2, \ldots, m$ for fixed i, and then over $i = 1, 2, \ldots$, which is left to the reader.

Let us turn back to equations (3.6.2) and sum them over $i = 0, 1, \ldots$ under fixed $j = 1, \ldots, m-1$. As the result, we obtain $\mu p_{\cdot,j} = \mu p_{\cdot,j+1}$, $j = 1, \ldots, m-1$ and $\mu p_{\cdot,m} =$

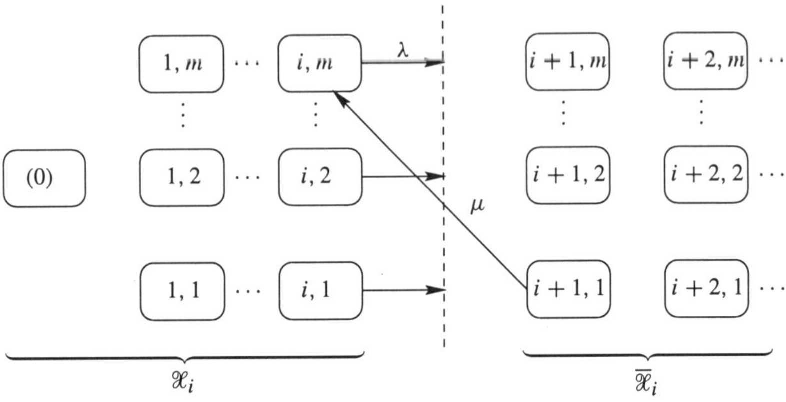

Figure 3.6.

$\mu p_{\cdot,1}$, that is,

$$p_{\cdot,1} = p_{\cdot,2} = \cdots = p_{\cdot,m}. \tag{3.6.4}$$

From (3.6.3) and the normalisation condition $p_0 + p_{\cdot,\cdot} = 1$ we obtain the equality

$$\lambda = \mu p_{\cdot,1}, \tag{3.6.5}$$

which has a transparent physical sense. Indeed, the left-hand side of (3.6.5) is the intensity of the input flow, and the right-hand side is that of the emanating flow; since the system is lossless, in the steady state these intensities must coincide. We find from (3.6.4) and (3.6.5) that

$$p_{\cdot,j} = \hat{\rho}, \qquad j = 1, \ldots, m,$$

where $\hat{\rho} = \lambda/\mu$. Together with the normalisation condition, the last equality in its turn yields

$$p_0 = 1 - \rho, \tag{3.6.6}$$

where $\rho = m\lambda/\mu = \lambda b$ and b is the mean service time. It is reasonable to take ρ as the traffic intensity. Equality (3.6.6) shows that for the given system $\rho < 1$ is also the necessary condition for existence of the steady state. Below we prove that this condition is sufficient.

Knowing p_0, we calculate recursively the probabilities p_{ij} from equations (3.6.2). Indeed, from the first equation in (3.6.2) we calculate p_{11}; from the second equation, p_{1j}, $j = 2, \ldots, m$, from the third equation, p_{21}, from the fourth equation, p_{2j}, $j = 2, \ldots, m$, for $i = 2$, from the fifth equation, p_{31}, then from the fourth and fifth equations p_{3j}, $j = 2, \ldots, m$, and p_{41} for $i = 3$, etc.

Sometimes, these calculations can result in appreciable errors because one has to sum numbers with opposite signs. There are, however, other methods allowing us to calculate p_{ij} more efficiently (in terms of stability). These are the matrix and matrix-geometrical approaches with which we will familiarise ourselves in the next chapter. For the time being, we dwell on the traditional generating function-based solution of the equilibrium equations (3.6.2).

We introduce the generating function

$$P_j(z) = \sum_{i=1}^{\infty} p_{ij} z^i.$$

Multiplying the equation for p_{ij} by z^i and summing, we obtain

$$\begin{aligned} 0 &= -(\lambda + \mu) P_j(z) + \lambda z P_j(z) + \mu P_{j+1}(z), \qquad j = 1, \dots, m-1, \\ 0 &= -\lambda[P_m(z) + p_0] - \mu P_m(z) + \lambda z [P_m(z) + p_0] + \frac{\mu}{z} P_1(z). \end{aligned}$$

We express $P_{j+1}(z)$ in terms of $P_j(z)$ and obtain

$$\begin{aligned} P_j(z) &= [1 + \hat{\rho}(1-z)]^{j-1} P_1(z), \qquad j = 2, \dots, m, \\ P_1(z) &= \frac{\hat{\rho} z (1-z)}{1 - z[1 + \hat{\rho}(1-z)]^m} p_0. \end{aligned}$$

Then the generating function $P(z)$ of the stationary distribution of the number of customers in the system is

$$P(z) = p_0 + P_.(z) = (1-\rho) \frac{1-z}{1 - z[1 + \hat{\rho}(1-z)]^m}. \tag{3.6.7}$$

The stationary mean number N of customers in the system and the stationary mean queue length Q are

$$\begin{aligned} N &= P'(1) = \rho + \frac{\rho^2 \left(1 + \frac{1}{m}\right)}{2(1-\rho)}, \\ Q &= \frac{\rho^2 \left(1 + \frac{1}{m}\right)}{2(1-\rho)} \end{aligned}$$

(the last equality follows from the obvious relation $N = Q + \rho$).

The stationary probabilities $p_i = p_{i,\cdot}$ that there are exactly i customers in the system can be given in the explicit form. To this end, we must determine all zeros $z_0, z_1, \dots, z_m$ in the denominator of the fraction in (3.6.7), that is, solve the algebraic equation of degree $m + 1$

$$1 - z[1 + \hat{\rho}(1-z)]^m = 0. \tag{3.6.8}$$

We note that one zero of the denominator (say, z_0) is quite obvious: $z_0 = 1$ (it is possible to prove that the other zeros are distinct and greater than one in magnitude). Now we represent $P(z)$ as the sum of common fractions

$$P(z) = \sum_{n=1}^{m} \frac{c_n}{z - z_n};$$

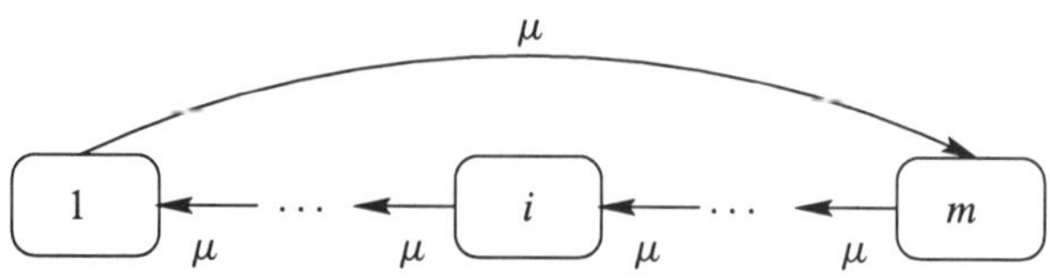

Figure 3.7.

recalling that $p_i = P^{(i)}(0)/i!$, we obtain after differentiation

$$p_i = -\sum_{n=1}^{m} \frac{c_n}{z_n^{i+1}}.$$

The stationary probabilities $p_{ij} = P_j^{(i)}(0)/i!$ are determined precisely in the same manner.

Now we prove that the condition $\rho < 1$ is sufficient for the process $\{\eta(t), t \geq 0\}$ describing queueing system $M/E_m/1/\infty$ be ergodic. To do so, we make use of the findings of Lavenberg (Lavenberg, 1978): for the steady state to exist in the Markov queueing system of a sufficiently general form with an infinite input buffer, it suffices to fulfil the condition $\lambda < \lambda_D^*$, where λ and λ_D^* are, respectively, the intensities of arrivals and departures in the same queueing system where the input flow is replaced by a storage facility with unlimited stock of customers.

We denote by $S/E_m/1$ the newly introduced queueing system with the input storage facility. We easily see that its operation is describable only by the Markov process $\{\xi(t), t \geq 0\}$ (we recall that according to the previous definition $\xi(t)$ is the number of phases that the customer in service has to pass through at the instant of time t) defined on the state set $\mathscr{X} = \{1, 2, \ldots, m\}$. Since all states of the process $\{\xi(t), t \geq 0\}$ are communicating, it is irreducible, and, hence, there are stationary probabilities

$$p_j^* = \lim_{t\to\infty} \mathbf{P}\{\xi(t) = j\}$$

satisfying the equilibrium equations

$$\begin{aligned} 0 &= -\mu p_j^* + \mu p_{j+1}^*, \qquad j = 1, \ldots, m-1, \\ 0 &= -\mu p_m^* + \mu p_1^*. \end{aligned}$$

It is obvious how to derive it with the use of the principle of global balance with due regard for Fig. 3.7.

Solving the last equations with the normalisation condition $\sum_{j=1}^{m} p_j^* = 1$, we find that $p_j^* = 1/m$, $j = 1, \ldots, m$. Since the customer departs from the system only upon the completion of phase one, the departure intensity is $\lambda_D^* = \mu p_1^* = \mu/m$. Therefore, the condition $\lambda < \lambda_D^*$ amounts to the condition $m\lambda/\mu < 1$ or $\rho < 1$.

To use the Lavenberg condition in more general Markov queueing systems, it is required that the modified system with an input storage facility should have no internal losses and be describable by an irreducible Markov process with a finite number of states.

3.6.3. Stationary distribution of the customer waiting time

In the case where the customer in service still has to go through j phases, the waiting time for service of the customer who finds upon his arrival $i \geq 1$ customers in the system has the Erlang distribution $E_l(x)$ with parameters μ and l, where $l = m(i-1)+j$ is the total number of service phases of all customers that are in the system at this instant. If a customer arrives to an empty system, he is immediately served. By virtue of the formula of total probability, we obtain the following formula for the Laplace–Stieltjes transform of the stationary distribution $W(x)$ of the waiting time:

$$\begin{aligned}\omega(s) &= \int_0^\infty e^{-sx}dW(x) = p_0 + \sum_{i=1}^{\infty}\sum_{j=1}^{m}\left(\frac{\mu}{\mu+s}\right)^{(i-1)m+j} p_{ij} \\ &= p_0 + \left(\frac{\mu+s}{\mu}\right)^m \sum_{j=1}^{m}\left(\frac{\mu}{\mu+s}\right)^j P_j\left(\left(\frac{\mu}{\mu+s}\right)^m\right) \\ &= \frac{s(\mu+s)^m(1-\rho)}{(s-\lambda)(\mu+s)^m+\lambda\mu^m}.\end{aligned}$$

The stationary distribution $V(x)$ of the customer sojourn time has the Laplace–Stieltjes transform

$$\varphi(s) = \int_0^\infty e^{-sx}dV(x) = \omega(s)\left(\frac{\mu}{\mu+s}\right)^m = \frac{s\mu^m(1-\rho)}{(s-\lambda)(\mu+s)^m+\lambda\mu^m}.$$

The stationary mean waiting and sojourn times obey the relations

$$\begin{aligned} w &= -\omega'(0) = \frac{\rho(m+1)}{2\mu(1-\rho)} = \frac{Q}{\lambda}, \\ v &= -\varphi'(0) = \frac{m}{\mu}\left[1+\frac{\rho(1+\frac{1}{m})}{2(1-\rho)}\right] = \frac{N}{\lambda}, \end{aligned}$$

that is, we arrive again at the Little formulas relating w and v, respectively, with Q and N.

In complete analogy with what was done while calculating the stationary probabilities p_i, the distribution functions $W(x)$ and $V(x)$ can be written out explicitly. To this end, we have to find $m+1$ roots $s_0, s_1, \ldots, s_m$ of the algebraic equation

$$(s-\lambda)(\mu+s)^m+\lambda\mu^m = 0, \tag{3.6.9}$$

represent $\omega(s)$ and $\varphi(s)$ as a sum of common fractions, and pass to the inverse Laplace transform. We leave it to the reader to carry out these calculations and only say that the roots of equations (3.6.8) and (3.6.9) are related as $s_i = \lambda(1-z_i)$ because the replacement $s = \lambda(1-z)$ reduces equation (3.6.9) to (3.6.8).

3.6.4. Transient characteristics

The transient state probabilities $p_0(t)$ and $p_{ij}(t)$ are found from (3.6.1) by introducing the generating function

$$P_j(z,t) = \sum_{i=1}^{\infty} p_{ij}(t)z^i.$$

Then we obtain

$$\frac{d}{dt}P_j(z,t) = -(\lambda+\mu)P_j(z,t) + \lambda z P_j(z,t) + \mu P_{j+1}(z,t), \qquad j = 1,\dots,m-1,$$

$$\frac{d}{dt}P_m(z,t) + \frac{d}{dt}p_0(t) = -\lambda[P_m(z,t)+p_0(t)] - \mu P_m(z,t) + \lambda z[P_m(z,t)+p_0(t)] + \frac{\mu}{z}P_1(z,t).$$

The last set of differential equations by means of the Laplace transform

$$\pi_j(z,s) = \int_0^\infty e^{-st}P_j(z,t)\,dt, \qquad \pi_0(s) = \int_0^\infty e^{-st}p_0(t)\,dt$$

immediately reduces to the ordinary linear algebraic set

$$s\pi_j(z,s) - P_j(z) = -[\lambda(1-z)+\mu]\pi_j(z,s) + \mu\pi_{j+1}(z,s), \qquad j = 1,\dots,m-1,$$

$$s\pi_m(z,s) - P_m(z) + s\pi_0(s) - p_0 = -[\lambda(1-z)+\mu]\pi_m(z,s) - \lambda(1-z)\pi_0(s) + \frac{\mu}{z}\pi_1(z,s),$$

where

$$P_j(z) = \sum_{i=1}^{\infty} p_{ij}(0)z^i,$$

and $p_0 = p_0(0)$ are the initial conditions. We leave it to the reader to continue the finding of $\pi_j(z,s)$, as well as $\pi_0(s)$.

The transient distributions of the virtual waiting and system times and the mean transient characteristics of the system are determined precisely as in the steady case.

3.7. $M/M/1/0$ system with retrial queue

The majority of physical telephone systems do not yield to description either by the queueing system of the $M/G/n/\infty$ type or queueing system of the $M/G/n/0$ type: on the one hand, the customer finding all n servers busy is rejected and no queue is formed in the system, but on the other hand, the rejected customer retries to establish connection with the subscriber called, that is, a flow of retrial customers arises in the system.

Let us consider this situation by the way of a elementary model, the $M/M/1/0$ system with retrial queue. In this system, as well as in the ordinary $M/M/1/0$ queueing system, the flow of arrivals is Poisson with the parameter λ and the service times are distributed exponentially with the parameter μ. But the rejected customer is put in the retrial queue, the so-called orbit, which differs substantially from the conventional queue.

First, the retrial customer cannot get to the server immediately upon its release but only if a repeated attempt was made and the server was free at this. Each customer makes repeated attempts independently of the other customers in the orbit at random time intervals which are assumed to be distributed exponentially with parameter γ.

Second, after each failed attempt to occupy the server, the customer (both new or retrial) can depart from the system with the probability q.

It goes without saying that even in the presence of retrial customers, the newly arrived customer can get to the server.

3.7.1. Equations for distribution of the number of retrial customers in the system

We consider the number $\nu(t)$ of retrial customers in the system at the instant t and the indicator variable $\chi(t)$ which at time t is equal to one if the server is occupied and to zero otherwise. Then the two-dimensional process $\{\eta(t), t \geq 0\} = \{(\nu(t), \chi(t)), t \geq 0\}$ is a Markov process defined on the state set $\mathscr{X} = \{(i, j), i = 0, 1, \ldots, j = 0, 1\}$, where sojourn in the state (i, j) means that there are i retrial customers in the system (or, which is the same, in the orbit) and the server is empty if $j = 0$ or occupied if $j = 1$.

Since the system under consideration is exponential, there should be no difficulty in proving the Markov nature and homogeneity of the process $\{\eta(t), t \geq 0\}$. Let us present the transition probabilities of the process $\{\eta(t), t \geq 0\}$ in the time interval $(t, t + \Delta)$.

The system goes from the state $(i, 0)$, $i \geq 0$, in a 'small' time Δ to the state $(i, 1)$ with the probability $\lambda\Delta + o(\Delta)$ (a new customer arrives and is immediately taken by the server) and to the state $(i - 1, 1)$ with the probability $i\gamma\Delta + o(\Delta)$ (one of the i customers in the orbit makes a successful repeated attempt).

The system goes from the state $(i, 1)$, $i \geq 0$, in time Δ to the state $(i, 0)$ with the probability $\mu\Delta + o(\Delta)$ (the customer service is complete), to the state $(i + 1, 1)$ with the probability $\lambda(1 - q)\Delta + o(\Delta)$ (a new customer arrives and, finding the server occupied, remains in the system and becomes a retrial customer with the probability $1 - q$), and to the state $(i - 1, 1)$ with the probability $i\gamma q\Delta + o(\Delta)$ (one of the i retrial customers makes an unsuccessful attempt and then leaves the system).

The probabilities of the other transitions are of order $o(\Delta)$.

It is clear that the Markov process $\{\eta(t), t \geq 0\}$ is not of the birth-and-death kind.

We denote by $p_{i0}(t) = \mathbf{P}\{\nu(t) = i, \chi(t) = 0\}$ the probability that at the instant t there are i retrial customers in the orbit and the server is free and by $p_{i1}(t) = \mathbf{P}\{\nu(t) = i, \chi(t) = 1\}$ the probability that there are i retrial customers in the orbit and the server is busy. Then for $p_{ij}(t)$ the following Kolmogorov differential equations are valid:

$$\begin{aligned}
p'_{i0}(t) &= -(\lambda + i\gamma)p_{i0}(t) + \mu p_{i1}(t), && i \geq 0, \\
p'_{01}(t) &= -[\lambda(1 - q) + \mu]p_{01}(t) + \lambda p_{00}(t) + \gamma p_{10}(t) + \gamma q p_{11}, \\
p'_{i1}(t) &= -[\lambda(1 - q) + \mu + i\gamma q]p_{i1}(t) + \lambda(1 - q)p_{i-1,1}(t) + \lambda p_{i0}(t) \\
&\quad + \gamma(i + 1)p_{i+1,0}(t) + \gamma q(i + 1)p_{i+1,1}(t), && i \geq 1.
\end{aligned} \tag{3.7.1}$$

3.7.2. Stationary distribution of the number of retrial customers

As usual, the equilibrium equations for the stationary state probabilities p_{ij}, if any, are obtained from (3.7.1) by replacing the derivatives by zero:

$$\begin{aligned}
0 &= -(\lambda + i\gamma)p_{i0} + \mu p_{i1}, & i \geq 0,\\
0 &= -[\lambda(1-q)+\mu]p_{01} + \lambda p_{00} + \gamma p_{10} + \gamma q p_{11}, &\\
0 &= -[\lambda(1-q)+\mu+i\gamma q]p_{i1} + \lambda(1-q)p_{i-1,1} + \lambda p_{i0} & \\
&\quad + \gamma(i+1)p_{i+1,0} + \gamma q(i+1)p_{i+1,1}, & i \geq 1.
\end{aligned} \tag{3.7.2}$$

We make use of the principle of global balance to explain derivation of (3.7.2). We assume that the reader has gained a sufficient experience to handle the Markov processes describing the exponential queueing systems or even the Markov queueing systems that are reducible to the exponential ones like the $M/E_m/1/\infty$ system. So, when writing the global balance equations we do not draw the transition diagram, but for any state keep in mind all possible arrivals and departures.

For example, the total flow of probabilities $(\lambda + i\gamma)p_{i0}$ of departure from the state $(i, 0)$ is generated by the arrival at the server either of a new customer or one of the i retrial customers, and this state is feasible only when the service is completed (the probability flow is μp_{i1}); equality of these flows provides the first equation in (3.7.2).

The state (0, 1) can be left (the total flow of probabilities is $[\lambda(1-q)+\mu]p_{01}$) either due to the arrival of a new customer which finds the server occupied and (with the probability $1-q$) becomes a retrial customer or upon completing the service. This state can be reached either if a customer arrives to an empty system (the flow of probabilities is λp_{00}), or if an attempt of a single retrial customer is successfully completed (the probability flow is γp_{10}), or finally, if an attempt of a retrial customer fails (the probability flow is $\gamma q p_{11}$); this provides us with the second equation in system (3.7.2). The last equation is interpreted in the same way.

To solve the equilibrium equations (3.7.2), we sum the first relation of this system from $(i+1)$th to ∞ and the third one, from i to ∞. By adding the results, we arrive at the equalities

$$\begin{aligned}
\lambda p_{00} &= \mu p_{01},\\
\lambda p_{i0} + \lambda(1-q)p_{i-1,1} &= (\mu + i\gamma q)p_{i1}, \qquad i \geq 1.
\end{aligned} \tag{3.7.3}$$

We leave it to the readers to derive these equalities with the use of the local balance relations.

Now we express p_{i0} from the first equality in the equilibrium equations (3.7.2) in terms of p_{i1} and substitute it in (3.7.3) in order to obtain the relation

$$\left(\mu + i\gamma q - \frac{\lambda\mu}{\lambda + i\gamma}\right) p_{i1} = \lambda(1-q)p_{i-1,1}, \qquad i \geq 1,$$

which in view of the first equality in (3.7.2) allows us to derive the following expressions for the probabilities p_{ij} up to p_{00}:

$$p_{i0} = p_{00}\sigma_i \prod_{j=0}^{i} \theta_j,$$
$$p_{i1} = p_{00} \prod_{j=0}^{i} \theta_j, \qquad i \geq 0, \tag{3.7.4}$$

where

$$\theta_0 = \rho = \frac{\lambda}{\mu}, \quad \theta_i = \frac{\lambda(1-q)(\lambda + i\gamma)}{i\gamma(\mu + q\lambda + iq\gamma)}, \quad \sigma_i = \frac{\mu}{\lambda + i\gamma}.$$

The probability p_{00} is found from the normalisation condition $p_{\cdot,\cdot} = 1$:

$$p_{00} = \left[\sum_{i=0}^{\infty} (1 + \sigma_i) \prod_{j=0}^{i} \theta_j \right]^{-1}. \tag{3.7.5}$$

It is possible to demonstrate (we leave it to the reader) that if $q > 0$, then series (3.7.5) converges for all λ, μ, and $\gamma > 0$, and if $q = 0$, then it does so only for $\rho = \lambda/\mu < 1$. Moreover, in these cases there exist stationary probabilities p_{ij}, $i \geq 1$, $j = 0, 1$. This fact can be proved following the scheme

(i) prove regularity of the process $\{\eta(t), t \geq 0\}$, beginning with the initial definition and using the property of the arriving flow that over any time interval the number of arriving customers is finite and the number of changes in the states of the $\{\eta(t), t \geq 0\}$ cannot exceed the number of arrivals by a factor greater than three, and

(ii) make use of the Foster theorem 1.5.4.

However, we omit here this proof.

Therefore, formulas (3.7.4) and (3.7.5) allow us to calculate the stationary probabilities p_{ij}. To complete the picture, however, we derive the same formulas using the generating function-based approach. We introduce the generating function

$$P_j(z) = \sum_{i=0}^{\infty} p_{ij} z^i, \qquad j = 0, 1.$$

Multiplying the ith equation by z^i and summing, we obtain

$$0 = -\lambda P_0(z) - \gamma z P_0'(z) + \mu P_1(z),$$
$$0 = \lambda P_0(z) + \gamma P_0'(z) - [\lambda(1-q)(1-z) + \mu] P_1(z) + +\gamma q(1-z) P_1'(z). \tag{3.7.6}$$

The last two simultaneous homogeneous linear differential equations can be reduced to the Bessel equation. For simplicity, we restrict ourselves to the special case of absolutely insistent customers where each customer repeats his attempts until served (the general case is left for the reader who is well familiar with the Bessel functions). Here $q = 0$ and (3.7.6) takes the form

$$0 = -\lambda P_0(z) - \gamma z P_0'(z) + \mu P_1(z),$$
$$0 = \lambda P_0(z) + \gamma P_0'(z) - [\lambda(1-z) + \mu] P_1(z). \tag{3.7.7}$$

Expressing $P_1(z)$ in the former equation in terms of $P_0(z)$ and substituting it in the latter equation, we obtain

$$\alpha(1 - \rho z) P_0'(z) - \rho P_0(z) = 0,$$

where $\rho = \lambda/\mu$ is the traffic intensity and $\alpha = \gamma/\lambda$. This equation has the solution

$$P_0(z) = C(1 - \rho z)^{-1/\alpha}.$$

Hence it follows that

$$P_1(z) = C\rho(1 - \rho z)^{-1-1/\alpha}.$$

The generating function $P(z)$ of the stationary distribution of the number of retrial customers in the system (in the orbit) obeys the relation

$$P(z) = P_0(z) + P_1(z) = C[1 + \rho(1 - z)](1 - \rho z)^{-1-1/\alpha}.$$

The constant C is found from the normalisation condition $P(1) = 1$, that is,

$$C = (1 - \rho)^{1+1/\alpha}.$$

Differentiating the functions $P_0(z)$ and $P_1(z)$ with respect to z and substituting $z = 0$, we obtain the expressions for the probabilities p_{i0} and p_{i1}

$$p_{i0} = \frac{1}{i!} P_0^{(i)}(0) = (1 - \rho)^{1+1/\alpha} \rho^i \binom{1/\alpha + i - 1}{i},$$
$$p_{i1} = \frac{1}{i!} P_1^{(i)}(0) = (1 - \rho)^{1+1/\alpha} \rho^{i+1} \binom{1/\alpha + i}{i}.$$

It is left to the reader to verify that for $q = 0$ the last formulas coincide with (3.7.4).

The stationary mean number N of retrial customers in the system and stationary probability $P_{w=0}$ that the arriving customer is served first (also for $q = 0$) are

$$N = P'(1) = \frac{\rho(1 + \alpha\rho)}{\alpha(1 - \rho)},$$

$$P_{w=0} = \sum_{i=0}^{\infty} p_{i0} = P_0(1) = 1 - \rho.$$

Comparing different techniques to calculate the steady-state characteristics of the queueing systems discussed in this chapter, the reader may conclude that on the whole the simple and visual principles of the global and local balances in the Markov models provide final results much faster, with much smaller diffuculty as compared with the use of generating functions. That is why the generating function method is disregarded in the following chapter which is devoted to studying the steady-state characteristics of more complicated Markov models.

Nevertheless, the situation changes radically if consideration is given to transient characteristics and also non-Markov queueing systems. Here, for the time being, there sometimes exists no good alternative to the generating function, Laplace transform, and Laplace–Stieltjes transform. Practical use of the results obtained brings, of course, to the fore the methods for inversion of the corresponding transforms which, as we already noted, are left out of the scope of this book.

3.7.3. Transient characteristics

Here we restrict our consideration to the case of $q = 0$. In order to find the transient probabilities $p_{ij}(t)$, we introduce the generating function

$$P_j(z,t) = \sum_{i=0}^{\infty} p_{ij}(t) z^i, \qquad j = 0, 1.$$

From (3.7.1) we obtain

$$\begin{aligned} \frac{\partial}{\partial t} P_0(z,t) &= -\lambda P_0(z,t) - \gamma z \frac{\partial}{\partial z} P_0(z,t) + \mu P_1(z,t), \\ \frac{\partial}{\partial t} P_1(z,t) &= \lambda z P_1(z,t) - (\lambda+\mu) P_1(z,t) + \lambda P_0(z,t) + \gamma \frac{\partial}{\partial z} P_0(z,t). \end{aligned}$$

Making use of the Laplace transforms

$$\pi_0(z,s) = \int_0^{\infty} e^{-st} P_0(z,t)\, dt, \qquad \pi_1(z,s) = \int_0^{\infty} e^{-st} P_1(z,t)\, dt,$$

we obtain

$$\begin{aligned} s\pi_0(z,s) - P_0(z) &= -\lambda \pi_0(z,s) - \gamma z \frac{d}{dz} \pi_0(z,s) + \mu \pi_1(z,s), \\ s\pi_1(z,s) - P_1(z) &= \lambda z \pi_1(z,s) - (\lambda+\mu)\pi_1(z,s) + \lambda \pi_0(z,s) + \gamma \frac{d}{dz} \pi_0(z,s), \end{aligned}$$

where

$$P_0(z) = \sum_{i=0}^{\infty} p_{i0}(0) z^i, \qquad P_1(z) = \sum_{i=0}^{\infty} p_{i1}(0) z^i.$$

Expressing $\pi_1(z,s)$ in the former equation in terms of $\pi_0(z,s)$ and substituting it into the latter equation, we obtain

$$\begin{aligned} \frac{d}{dz}\pi_0(z,s) - \frac{(\lambda+s)(\lambda+\mu+s) - \lambda\mu - \lambda(\lambda+s)z}{\gamma[\lambda z^2 - (s+\lambda+\mu)z + \mu]} \pi_0(z,s) \\ = -\frac{(s+\lambda+\mu-\lambda z)P_0(z) + \mu P_1(z)}{\gamma[\lambda z^2 - (s+\lambda+\mu)z + \mu]}. \quad (3.7.8) \end{aligned}$$

The equation

$$\lambda z^2 - (s + \lambda + \mu)z + \mu = 0$$

possesses two positive solutions for all $s > 0$, one being smaller than one (we denote it by $z_1 = z_1(s)$) and the other ($z_2 = z_2(s)$), greater than one. Therefore, resolving the fraction in (3.7.8) into common ones, we write

$$\frac{d}{dz}\pi_0(z, s) - \left(\frac{a}{z - z_1} + \frac{b}{z - z_2}\right)\pi_0(z, s) = \frac{D(z)}{z - z_1},$$

where $a = a(s)$ and $b = b(s)$ are functions of s and $D(z) = D(z, s)$ is a function continuous in z in the domain $|z| < 1$ for all $s > 0$. We can also easily demonstrate that $a < 0$. Solving the last equation, we obtain

$$\pi_0(z, s) = |z - z_1|^a |z - z_2|^b \left[\int D(z)|z - z_1|^{-a}(z - z_1)^{-1}|z - z_2|^{-b} dz + C\right].$$

In order to find C, we observe that $\pi_0(z, s)$ is a function continuous in z for $|z| < 1$ including the point $z = z_1$. Then the bracketed expression must vanish at this point, that is,

$$\pi_0(z, s) = |z - z_1|^a |z - z_2|^b \int_{z_1}^{z} D(u)|u - z_1|^{-a}(u - z_1)^{-1}|u - z_2|^{-b} du.$$

The readers can obtain the formula for $\pi_1(z, s)$ on their own.

Let us find the distribution function of the waiting time (both the stationary and virtual waiting times at the instant t) and begin with the transient case. We assume that the customer under consideration arrives to the system at time t (obviously, at this instant the process $\{\eta(t), t \geq 0\}$ is in the state (i, j) with the probability $p_{ij}(t)$). We consider a new time x counted from the instant t. Our aim is to find the distribution function $W(x, t) = \mathbf{P}\{\tau < x\}$ of the instant τ of arrival of the customer under consideration at the server.

We introduce, with this aim in view, a new process $\{\tilde{\eta}(x), x \geq 0\}$ defined precisely as the process $\{\eta(t), t \geq 0\}$ but with one additional state denoted, say, by (-1). To put it more rigorously, the state (i, j), $i \geq 0$, $j = 0, 1$, corresponds to the case of i retrial customers in the orbit, along with the customer under consideration, and the server is either free ($j = 0$) or occupied ($j = 1$). The state (-1) means that the server begins providing service to the customer under consideration; the subsequent behaviour of the system is of no interest. Therefore, (-1) is the absorbing state.

For the process $\{\tilde{\eta}(x), x \geq 0\}$, the intensities of entering all states (i, j) coincide with the similar intensities for the process $\{\eta(t), t \geq 0\}$. Transition of the process $\{\tilde{\eta}(x), x \geq 0\}$ to the state (-1), however, is possible only from the states $(i, 0)$ and only from them. This transition means that the customer under consideration made a repeated and successful attempt. Obviously, the intensity of transition from any state (i, j) to (-1) is γ. Clearly, since the (-1) is the absorbing state, the intensity of leaving it is zero.

Now we assume that

$$r_{ij}(x) = \mathbf{P}\{\tilde{\eta}(x) = (i, j)\}, \qquad r_{-1}(x) = \mathbf{P}\{\tilde{\eta}(x) = (-1)\},$$

and obtain differential Kolmogorov equations

$$\begin{aligned}
r'_{i0}(x) &= -(\lambda + i\gamma + \gamma)r_{i0}(x) + \mu r_{i1}(x), && i \geq 0, \\
r'_{01}(x) &= -(\lambda + \mu)r_{01}(x) + \lambda r_{00}(x) + \gamma r_{10}(x), \\
r'_{i1}(x) &= -(\lambda + \mu)r_{i1}(x) + \lambda r_{i-1,1}(x) + \lambda r_{i0}(x) + \gamma(i+1)r_{i+1,0}(x), && i \geq 1, \\
r'_{-1}(x) &= \sum_{i=0}^{\infty} \gamma r_{i0}(x).
\end{aligned} \tag{3.7.9}$$

The function $r_{-1}(x)$ is the distribution function $W(x, t)$ of the virtual waiting time at the instant t.

Let us dwell on the initial conditions for (3.7.9). If at the instant t the server is free (the process $\{\eta(t), t \geq 0\}$ is in the state $(i, 0)$), then the customer arrived at this instant is served immediately or, stated differently, $\{\tilde{\eta}(x), x \geq 0\}$ immediately goes to the state (-1) but not to $(i, 0)$. If at the instant t the process $\{\eta(t), t \geq 0\}$ is in the state $(i, 1)$, then at the instant 0 the process $\{\tilde{\eta}(x), x \geq 0\}$ is also in the state $(i, 1)$. To summarise the abovesaid, we obtain the following initial conditions for equations (3.7.9):

$$r_{i0}(0) = 0, \quad r_{-1}(0) = \sum_{i=0}^{\infty} p_{i0}(t), \quad r_{i1}(0) = p_{i1}(t).$$

It is clear that on order to calculate the stationary distribution of the waiting time in the above relations we must substitute the stationary probabilities p_{ij} for $p_{ij}(t)$.

Equations (3.7.9) are solved similarly to (3.7.1). In this case the distribution function $r_{-1}(x) = W(x, t)$ itself is expressed in terms of Laplace transform, rather than Laplace–Stieltjes transform, which can be easily corrected by multiplying by s.

As usual, the customer sojourn time is constituted by the waiting time and the time at server itself.

Other characteristics of the system with retrial queue—for example, the distribution of the number of attempts prior to customer arrival to the server (repeated attempts)—can be determined in a similar manner.

4

Markov systems: algorithmic methods of analysis

In the previous chapter, we familiarised ourselves with some approaches to the analysis of elementary Markov queueing systems. We considered first the systems that admit description in terms of the birth-and-death processes. Here the difficulties involved in finding the stationary queue distribution were quite insignificant: we had just to find the intensities of transitions to the neighbouring states and substitute them into well-known expressions for the birth-and-death processes. Other queueing systems discussed in Chapter 3 are described by the so-called generalised birth-and-death processes. The spectrum of different systems presented there was intended to teach the reader to construct models of Markov processes that are more complicated than the birth-and-death process, as well as to familiarise the reader with methods of their analysis. It is worthwhile to note that the degree of 'generalisation' sufficed to obtain the majority of the queueing system characteristics in explicit form.

The last fact, however, can be taken as an exception, rather than as a rule, because it is mostly impossible to obtain closed formulas while analysing models of queues described by the 'generalised' birth-and-death processes. In particular, if a queueing system has a finite capacity, then the use of the popular queue-theoretical method of generating functions gives rise to considerable difficulties. Although in some cases they are surmountable in principle, the final formulas are not truly closed and require numerical methods.

As for the numerical methods, it would seem reasonable to use them from the very beginning to analyse finite-capacity queueing systems, that is, to solve directly the equilibrium equations. Unfortunately, this simple idea is not so easily realisable because even to store in the computer memory the coefficient matrices of the equilibrium equations one needs—despite their great sparseness—highly efficient algorithms. Moreover, the values of the queueing system parameters such as buffer capacity, number of service phases, and so on, which define the simultaneous equilibrium equations, can make them practically unsolvable because of the high dimensionality of the set of equation. There are, of course, other difficulties encountered in this problem, but these are the key ones. Therefore, other—also computer-aided—approaches and methods are required. Taking into account the state-of-the art of computing and its availability to the researcher, it seems that in some cases it is reasonable to employ the algorithmic approach, which implies solving simultaneous equations and finding a series of queueing system characteristics with the use of recurrence relations and algorithms that are readily programmable. It is the aim of this chapter to familiarise the readers with the algorithmic approach to the analysis of the queueing systems that admits description in terms of Markov models.

4.1. $M/H_m/1/r$ and $H_l/M/1/r$ systems

We begin to present the aforementioned algorithmic approach to calculating steady-state characteristics of queueing systems with relatively simple systems such as $M/H_m/1/r$ and $H_l/M/1/r$. The former, $M/H_m/1/r$, system is a single-server queueing system with a buffer of finite capacity r, $2 \leq r < \infty$, which a Poisson customer flow of intensity λ arrives at. If upon the arrival of a customer there are no empty places on the buffer, the customer leaves the system unserved. The customer service times have a hyper-exponential distribution function

$$B(x) = \sum_{j=1}^{m} \beta_j (1 - e^{-\mu_j x}), \qquad x > 0.$$

We assume, as before, that the customers are served by one of the fixed disciplines such as FCFS, LCFS, or RANDOM.

The second, $H_l/M/1/r$, system may be called dual to the first one. Here the arrival flow is recurrent with the hyper-exponential distribution function

$$A(x) = \sum_{i=1}^{l} \alpha_i (1 - e^{-\lambda_i x}), \qquad x > 0,$$

of the intervals between customer arrivals, and the service times are distributed exponentially with parameter μ.

For the case under consideration, the condition $r \geq 2$, which was introduced for these systems and will be used below for other finite-capacity systems, is not restrictive. It stems only from the desire to avoid considering special cases for $r = 0$ and $r = 1$ which have the nomenclature somewhat different from the general case.

As shown in Section 2.8, the hyper-exponential distribution of the customer service (or generation) time can be treated in the phase terms. It is assumed that at the instant of starting service (generation) of a new customer the customer finds himself at phase i (j) with probability β_j, $j = 1, \ldots, m$ $(\alpha_i, i = 1, \ldots, l)$, and stays here for an exponentially distributed time with parameter μ_j (λ_i). Upon passing phase j (i), the customer serving (generation) is completed. Therefore, we may introduce the corresponding Markov process with the discrete state set by separating the exponentially distributed service phases for the $M/H_m/1/r$ queueing system or the generation phases for the $H_l/M/1/r$ queueing system as in Section 3.6 for the $M/E_k/1/\infty$ queueing system. At the first glance, we should encounter analogous difficulties while analysing these three models because at the descriptive level, that is, at the level of constructing the Markov models, they are strikingly similar. But this is true to only a small extent, because here the queueing systems are studied under a more general assumption of limited buffer capacity than in the case of the $M/E_m/1/\infty$ system.

As we have said, this assumption gives rise to additional difficulties while analysing the corresponding equilibrium equation systems. If we tried to use the generating function in the traditional manner, then we would need to find numerical values of the roots of a polynomial even in order to evaluate the mean characteristics. Inquisitive readers can assure themselves that this is true by doing all calculations required to obtain the generating function.

Another approach presented in this section is related with direct solution of the equilibrium equations on the basis of preliminary analysis which decomposes the original equation system into smaller subsystems related by elementary recurrence relations. In this case, the coefficient matrices of the resulting subsystems sometimes can be explicitly inverted and then—as the result of recursive connectivity of the subsystems—we succeed in getting for the original equilibrium equations recurrence formulas representable in either scalar or matrix form.

4.1.1. Equilibrium equations for the $M/H_m/1/r$ system

Consider, as above, the number of customers, $\nu(t)$, in the system at the instant t. Since the hyper-exponential distribution of the service time is not memoryless, the process $\{\nu(t), t \geq 0\}$, obviously, is not a Markov one. Therefore, we introduce, as in the case of the $M/E_m/1/\infty$ system, the auxiliary variable $\xi(t)$ equal to the phase of customer service at the instant t. We introduce a new process $\{\eta(t), t \geq 0\}$ so that $\eta(t) = \nu(t)$ for the instants t when the server is free, that is, no customer exists in the system, and $\eta(t) = (\nu(t), \xi(t))$ otherwise.

The Poisson nature of the flow and exponentiality of the service phases suggest that the so-defined process $\{\eta(t), t \geq 0\}$ is a homogeneous Markov process. The state set $\mathcal{X}$ of the process $\{\eta(t), t \geq 0\}$ is of the form

$$\mathcal{X} = \{(0); (k, j), k = 1, \ldots, R, j = 1, \ldots, m\},$$

where $R = r + 1$ stands for the total system capacity. The states from the set $\mathcal{X}$ signify that if $\eta(t) = (0)$, then the system is empty, and if $\eta(t) = (k, j)$, then at the instant t there are k customers in the system and the customer in service is at the phase j.

The number of states of the process $\{\eta(t), t \geq 0\}$ is finite and equal to $Rm+1$. We easily see that all states of this process are communicating. Therefore, the process $\{\eta(t), t \geq 0\}$ is ergodic, and there are strictly positive stationary limit probabilities

$$\lim_{t\to\infty} \mathbf{P}\{\eta(t) = (0)\} = p_0, \qquad \lim_{t\to\infty} \mathbf{P}\{\eta(t) = (k, j)\} = p_{kj},$$

which are independent of the initial state of the process $\eta(0)$.

Now we write out the equilibrium equations solved by the stationary probabilities p_0 and p_{kj}

$$0 = -\lambda p_0 + \sum_{s=1}^{m} \mu_s p_{1s}, \tag{4.1.1}$$

$$0 = -(\lambda + \mu_j)p_{1j} + \lambda\beta_j p_0 + \beta_j \sum_{s=1}^{m} \mu_s p_{2s}, \qquad j = 1, \ldots, m, \tag{4.1.2}$$

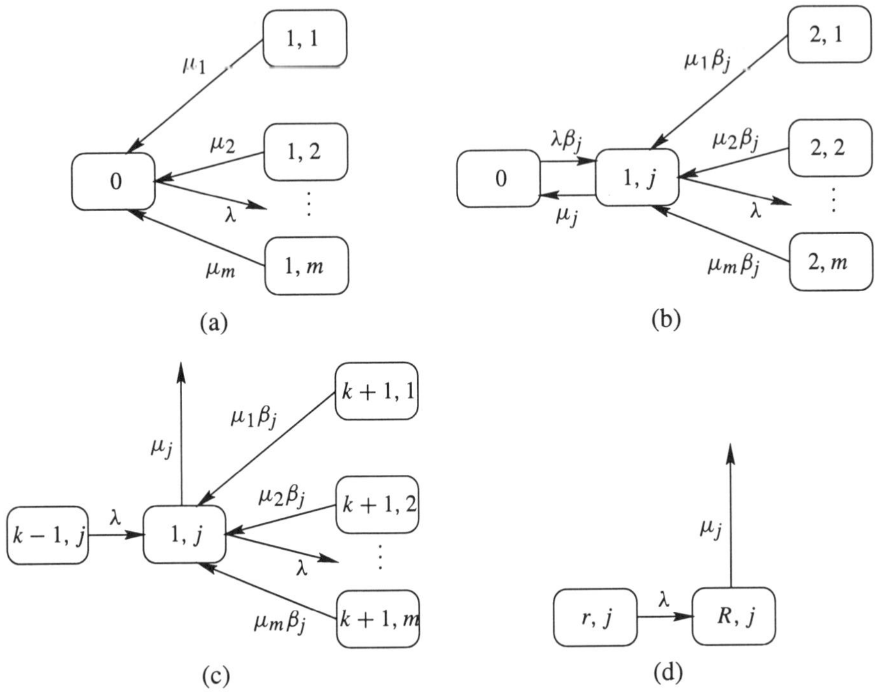

Figure 4.1.

$$0 = -(\lambda + \mu_j)p_{kj} + \lambda p_{k-1,j} + \beta_j \sum_{s=1}^{m} \mu_s p_{k+1,s}, \qquad k = 2, \ldots, r, \quad j = 1, \ldots, m, \tag{4.1.3}$$

$$0 = -\mu_j p_{Rj} + \lambda p_{rj}, \qquad j = 1, \ldots, m. \tag{4.1.4}$$

Taking into account the fact that simultaneous equations (4.1.1)–(4.1.4) is fairly simple, we skip their rigorous derivation based on analysing the process $\{\eta(t), t \geq 0\}$ in the time interval $(t, t + \Delta)$ and just illustrate it relying on the principle of global balance of the probability flow and making use of the queueing system transition diagram (Fig. 4.1). The diagram does not indicate the states which one can go to from a given state, because, as said in Chapter 3, there is no need for it.

As we see in Fig. 4.1a, one goes from the state (0) with intensity λ due to arrivals at the empty system; the probability flow leaving the state (0) is equal to λp_0. One goes to this state from the states $(1, 1), \ldots, (1, m)$ with respective intensities $\mu_1, \ldots, \mu_m$ due to completion of service in one of m phases; the total flow entering the state (0) is $\sum_{j=1}^{m} \mu_j p_{1j}$. We arrive at equation (4.1.1) by equating these flows.

According to Fig. 4.1b, one leaves the state $(1, j)$, both upon arrival of a customer (with the intensity λ) and upon completion of serving the customer which goes through the state j (with the intensity μ_j); the total emanating probability flow is $(\lambda + \mu_j)p_{1j}$. The probability flow entering the state $(1, j)$ consists of the intensity $\lambda\beta_j$ of customer arrival at the empty

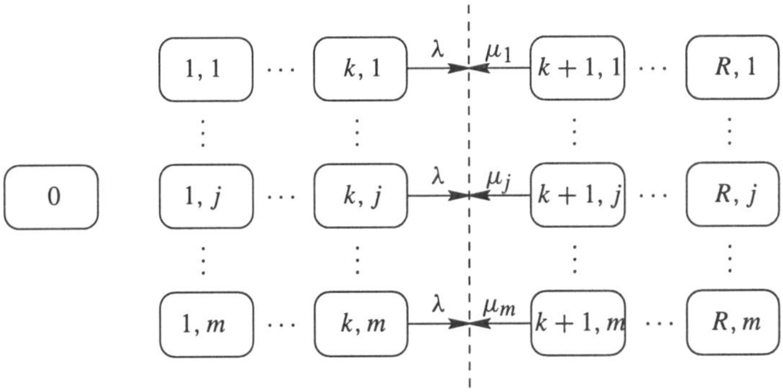

Figure 4.2.

system at phase j multiplied by the probability p_0 of the state (0) and the intensities $\beta_j \mu_s$ of transitions from the states $(2, s)$, $s = 1, \ldots, m$, due to completion of customer service in phase s with account for the probability β_j that serving of a new customer starts at phase j multiplied by the probability p_{2s} that there are two customers in the system and the process of service is at phase s, $s = 1, \ldots, m$; the total input probability flow is $\beta_j \sum_{s=1}^{m} \mu_s p_{2s}$. Hence we obtain equation (4.1.2).

Explanation of equation (4.1.3) differs from the above only in that the state (k, j) is entered due to arrival of a customer not from the empty system state (Fig. 4.1c), but from the state $(k-1, j)$, its intensity is merely λ, and the probability flow is $\lambda p_{k-1,j}$.

Finally, according to Fig. 4.1d, for the last equation we must disregard the intensities of leaving the state (R, j) caused by arrival of a customer (since in this case the system is full and any customer arrived is lost) and entering it from the states with more than R customers due to service completion (there is no such state in the system).

4.1.2. Solution of the equilibrium equations for the $M/H_m/1/r$ system

As usual, we begin solving equilibrium equations (4.1.1)–(4.1.4) with provisional development. We sum equation (4.1.2) over $j = 1, 2, \ldots, m$ and add to (4.1.1). Next, the result obtained is summed with equation (4.1.3) for $k = 2$, which then is summed over $j = 1, 2, \ldots, m$. By reasoning in this manner and taking (4.1.1) into account, we obtain at each step

$$\lambda p_0 = \sum_{j=1}^{m} \mu_j p_{1j},$$

$$\lambda p_{k,\cdot} = \sum_{j=1}^{m} \mu_j p_{k+1,j}, \qquad k = 1, \ldots, r. \tag{4.1.5}$$

Relations (4.1.5) are local balance equations which can be easily obtained from the transition diagram in Fig. 4.2. We leave this as an exercise for the readers.

It is no more possible to obtain from (4.1.5) an expression for the probability p_0 as in Section 3.6 (we come up against some effects of limited buffer), and we find p_0 in a different manner: not at first, but at last. The stationary probabilities p_0 and p_{kj} are known to be determined by (4.1.1)–(4.1.4) up to a constant. Therefore, we can reject one of the equations—say, (4.1.1)—and express all unknowns p_{kj} in terms of the probability p_0 which then can be found from the normalisation condition

$$p_0 + \sum_{k=1}^{R} \sum_{j=1}^{m} p_{kj} = 1. \tag{4.1.6}$$

Relations (4.1.5) prove to be useful here for transforming equations (4.1.2) and (4.1.3) while solving the remaining equations (4.1.2)–(4.1.4). Indeed, replacing the sums in equations (4.1.2) and (4.1.3) by their expressions from (4.1.5), we obtain

$$\begin{aligned} -(\lambda + \mu_j) p_{1j} + \lambda \beta_j p_{1,\cdot} &= -\lambda \beta_j p_0, \qquad j = 1, \ldots, m, \\ -(\lambda + \mu_j) p_{kj} + \lambda \beta_j p_{k,\cdot} &= -\lambda p_{k-1,j}, \quad k = 2, \ldots, r, \quad j = 1, \ldots, m. \end{aligned} \tag{4.1.7}$$

As the result, equations (4.1.2) and (4.1.3), as well as equation (4.1.4) in the form

$$\mu_j p_{Rj} = \lambda p_{rj}, \qquad j = 1, \ldots, m, \tag{4.1.8}$$

are represented as subsets of equations in the unknown vectors $\boldsymbol{p}_k^T = (p_{k1}, p_{k2}, \ldots, p_{km})$ which are recursively related in $k = 1, \ldots, R$. Introducing the matrices $M = \mathrm{diag}(-\mu_1, \ldots, -\mu_m)$, $D = -\lambda I + M$, the vector $\boldsymbol{\lambda} = \lambda \mathbf{1}$ and assuming that $\widetilde{M} = D + \boldsymbol{\lambda}\boldsymbol{\beta}^T$, we represent (4.1.7) and (4.1.8) in the matrix form

$$\begin{aligned} \boldsymbol{p}_1^T \widetilde{M} &= -\lambda \boldsymbol{\beta}^T p_0, \\ \boldsymbol{p}_k^T \widetilde{M} &= -\lambda \boldsymbol{p}_{k-1}^T, \qquad k = 2, \ldots, r, \\ \boldsymbol{p}_R^T M &= -\lambda \boldsymbol{p}_r^T. \end{aligned} \tag{4.1.9}$$

If we prove that $\widetilde{M}$ is a non-degenerate matrix, then, taking $k = 1, 2, \ldots, r$, then we solve successively these subsets and express the unknown vectors $\boldsymbol{p}_k$ in terms of p_0.

Consider the matrix $\widetilde{M}$ which in the general case has an arbitrary non-degenerate matrix D and arbitrary vectors $\boldsymbol{\lambda}$ and $\boldsymbol{\beta}$. For such a matrix, by virtue of Lemma 2.8.2 there exists the inverse matrix

$$\widetilde{M}^{-1} = (D + \boldsymbol{\lambda}\boldsymbol{\beta}^T)^{-1} = D^{-1} - \frac{D^{-1}\boldsymbol{\lambda}\boldsymbol{\beta}^T D^{-1}}{1 + \boldsymbol{\beta}^T D^{-1}\boldsymbol{\lambda}}, \tag{4.1.10}$$

provided that the condition $\boldsymbol{\beta}^T D^{-1}\boldsymbol{\lambda} \neq -1$ is satisfied.

In the case at under consideration, D is a diagonal non-degenerate matrix. Furthermore,

$$1 + \boldsymbol{\beta}^T D^{-1}\boldsymbol{\lambda} = 1 - \lambda \boldsymbol{\beta}^T (\lambda I - M)^{-1}\mathbf{1} = \beta(\lambda),$$

where

$$\beta(\lambda) = \sum_{j=1}^{m} \beta_j \mu_j / (\lambda + \mu_j)$$

is nothing more nor less than the Laplace–Stieltjes transform of the distribution function $B(x)$ at the point λ; hence $\beta(\lambda) > 0$. Therefore, the matrix $\widetilde{M}^{-1}$ exists and is

$$\widetilde{M}^{-1} = D^{-1} - \frac{\lambda D^{-1}\mathbf{1}\boldsymbol{\beta}^T D^{-1}}{\beta(\lambda)}. \tag{4.1.11}$$

We turn back to recurrence relations (4.1.9), introduce the matrices $W = -\lambda\widetilde{M}^{-1}$, $W_R = -\lambda M^{-1}$ and vector $\boldsymbol{w}_0^T = -\lambda\boldsymbol{\beta}^T\widetilde{M}^{-1}$ which, in view of (4.1.11) goes to $\boldsymbol{w}_0^T = -\lambda\boldsymbol{\beta}^T D^{-1}/\beta(\lambda)$, and rewrite (4.1.9) as

$$\begin{aligned} \boldsymbol{p}_1^T &= p_0\boldsymbol{w}_0^T, \\ \boldsymbol{p}_k^T &= \boldsymbol{p}_{k-1}^T W, \qquad k = 2, \dots, r, \\ \boldsymbol{p}_R^T &= \boldsymbol{p}_r^T W_R \end{aligned} \tag{4.1.12}$$

which immediately implies the final result: *the stationary distribution of state probabilities* $\{p_0, \boldsymbol{p}_k, k = 1, \dots, R\}$ *in the* $M/H_m/1/r$ *queueing system admits the representation as the truncated matrix-geometric progression*

$$\boldsymbol{p}_k^T = \begin{cases} p_0\boldsymbol{w}_0^T W^{k-1}, & k = 1, \dots, r; \\ p_0\boldsymbol{w}_0^T W^{r-1} W_R, & k = R. \end{cases} \tag{4.1.13}$$

In order to find p_0, we, as usual, make use of the normalisation condition (4.1.6), substitute (4.1.13) into it, and obtain

$$p_0 = \left[1 + \boldsymbol{w}_0^T\left(\sum_{k=0}^{r-1} W^k + W^{r-1}W_R\right)\mathbf{1}\right]^{-1}. \tag{4.1.14}$$

It is unlikely that anybody will use (4.1.14) in actual practice because it is much simpler to calculate the variables $\tilde{p}_{kj} = p_{kj}/p_0$ and then find p_0 from the condition $p_0 = (1 + \tilde{p}_{\cdot,\cdot})^{-1}$.

In the special case of $B(x) = 1 - e^{-\mu x}$, $x > 0$, that is, for the $M/M/1/r$ queueing system, we obtain $\boldsymbol{\beta} = (1)$, $M = (-\mu)$ and, therefore,

$$W = -\lambda(\lambda - \mu - \lambda)^{-1} = (\lambda/\mu) = (\rho) = \boldsymbol{w}_0, \qquad W_R = -\lambda(-\mu)^{-1} = (\rho).$$

Therefore, $p_k = p_0\rho^k$, $k = 1, \dots, R$.

We have thus demonstrated that upon passing from the $M/M/1/r$ system to the $M/H_m/1/r$ system the geometric distribution of the number of customers in the system becomes the matrix-geometric one!

For calculations of the stationary probabilities p_{kj}, we can use not only explicit matrix formulas (4.1.13) and (4.1.14), but also the simple recurrence formulas below which directly follow from (4.1.11) and (4.1.12):

$$\begin{aligned} p_{1j} &= \frac{\lambda\beta_j p_0}{(\lambda+\mu_j)\beta(\lambda)}, && j = 1, \dots, m, \\ p_{kj} &= \frac{\lambda p_{k-1,j}}{\lambda+\mu_j} + \frac{\lambda^2\beta_j}{(\lambda+\mu_j)\beta(\lambda)}\sum_{i=1}^m \frac{p_{k-1,i}}{(\lambda+\mu_i)}, && j = 1, \dots, m, \quad k = 2, \dots, r, \\ p_{R,j} &= \frac{\lambda}{\mu_j} p_{rj}, && j = 1, \dots, m. \end{aligned} \tag{4.1.15}$$

Matrix formulas (4.1.13) or recurrence formulas (4.1.15) based on them (for sufficiently great buffer capacity r) can be used for approximate calculations (to within the given ε) of the steady-state characteristics of the corresponding infinite-queue queueing systems.

We conclude this section by finding expressions and some useful relations for the performance indices of the system under consideration operating in the steady state. First, we introduce some relations based on qualitative and, at first glance, not too strict reasoning, though it would be possible to give a rigorous foundation on the basis of the ergodic theorems, but we do not intend to dive deeply into this domain.

In the steady state, the mean number of customers arriving at the system in a unit time—intensity of the arrival flow λ_A—must be equal to the mean number of customers served in a unit time—intensity of the served flow λ_D, or the throughput. Now, let π be the customer loss probability; by virtue of the Poisson nature of the flow, $\pi = p_{R,\cdot}$. Then, $1 - \pi$ is the fraction of a long time interval when the system is accessible to the arriving customers, that is, when there are empty places in the buffer, which follows from ergodicity of the process $\{\eta(t), t \geq 0\}$, and therefore,

$$\lambda_A = \lambda(1 - \pi). \tag{4.1.16}$$

Introducing the probability p_0 of system idling and the server utilisation coefficient $u = 1 - p_0$ (again, $1 - p_0$ can be interpreted as the fraction of time over a long interval when the server is busy), we obtain

$$\lambda_D = \mu(1 - p_0), \tag{4.1.17}$$

where μ is the service intensity:

$$\mu = b^{-1} = \left(\sum_{j=1}^{m} \frac{\beta_j}{\mu_j} \right)^{-1}.$$

Then from the balance of intensities of the arrival and output flows we obtain the relation

$$\lambda(1 - \pi) = \mu(1 - p_0) \tag{4.1.18}$$

which may be rewritten as

$$\rho(1 - \pi) = 1 - p_0, \tag{4.1.19}$$

where $\rho = \lambda/\mu$ is the system intensity traffic. The right-hand side of (4.1.18) can be represented in a different manner. Indeed, by summing all equalities in (4.1.5) and taking into account the normalisation conditions, we obtain

$$\lambda(1 - \pi) = \sum_{j=1}^{m} \mu_j p_{\cdot,j} \tag{4.1.20}$$

which, together with (4.1.16) and (4.1.17), yields

$$\lambda_A = \lambda_D = \sum_{j=1}^{m} \mu_j p_{\cdot,j}. \tag{4.1.21}$$

Expressions (4.1.16)–(4.1.21) relate the queueing system characteristics λ_A, λ_D, π, and p_0. Additionally, they can be used to check computations based on (4.1.13) or (4.1.15).

The stationary probabilities p_{kj} can be used to calculate the stationary probabilities $p_k = p_{k,\cdot}$ that there are k customers in the system (disregarding the service phase) which allows us to calculate the stationary mean number $N = \sum_{k=1}^{R} kp_k$ of customers in the system and the mean queue length $Q = \sum_{k=2}^{R}(k-1)p_k$.

4.1.3. Stationary distribution of the waiting time for the $M/H_m/1/r/$FCFS system

Now, for the discipline of servicing in the order of customer arrival, we study the stationary distribution of the waiting time for a customer accepted by the system with the help of the approach used in Chapter 3. Let $W(t)$ be the conditional distribution function of the waiting time for service of a customer under the condition that he is accepted for service, and let $W(t \mid (0))$ and $W(t \mid (k, j))$ be the conditional distribution functions of waiting time of that customer under the condition that upon arrival he finds the system in the states (0) and (k, j) respectively. We denote the Laplace–Stieltjes transforms of these distribution functions by $\omega(s)$, $\omega(s \mid (0))$, and $\omega(s \mid (k, j))$ respectively. Then, if the customer arrives at the empty system, he is immediately sent to the server and, consequently, $\omega(s \mid (0)) = 1$. If the customer finds the system in the state (k, j), $k = 1, \dots, r$, he is not lost but queued until the customer in the service phase j and all $k-1$ customers that have been queued before that customer just arrived are served. In this case, the waiting time is the sum of k random variables of which the first variable distributed exponentially with parameter μ_j and the remaining ones, by the law $B(x)$. Since all these random variables are independent,

$$\omega(s \mid (k, j)) = \frac{\mu_j}{\mu_j + s}\beta^{k-1}(s).$$

Since in the steady state the input flow is Poisson, the probability that the customer arrived finds the system in a certain state coincides with the stationary probability of this state. In view of this fact, we find with the use of the formula of total probability that

$$W(t) = \frac{1}{1-\pi}\left[W(t \mid (0))p_0 + \sum_{j=1}^{m}\sum_{k=1}^{r} W(t \mid (k, j))p_{kj}\right],$$

where π is the customer loss probability. Hence, passing to the Laplace–Stieltjes transform, we obtain

$$\omega(s) = \frac{1}{1-\pi}\left[p_0 + \sum_{j=1}^{m}\sum_{k=1}^{r} p_{kj}\frac{\mu_j}{\mu_j + s}\beta^{k-1}(s)\right], \tag{4.1.22}$$

where

$$\beta(s) = \sum_{j=1}^{m} \beta_j \mu_j/(\mu_j + s)$$

is the Laplace–Stieltjes transform of the distribution function $B(x)$.

For the mean waiting time w, we obtain, by the formula $w = -\omega'(0)$,

$$w = \frac{1}{1-\pi}\left[\sum_{j=1}^{m}\sum_{k=1}^{r}\frac{1}{\mu_j}p_{kj} + \frac{1}{\mu}\sum_{k=1}^{r}(k-1)p_{k,\cdot}\right]. \tag{4.1.23}$$

For the customer sojourn time (provided that he has been accepted), we obtain, in view of independence of the customer service times,

$$\varphi(s) = \omega(s)\beta(s).$$

Hence the mean sojourn time is

$$v = w + \frac{1}{\mu}. \tag{4.1.24}$$

In order to evaluate w and v, we can also make use of the Little formulas, which for the given finite-capacity system are of the form

$$\lambda(1-\pi)w = Q,$$
$$\lambda(1-\pi)v = N.$$

We advise to use them together with (4.1.23) and (4.1.24) to check calculations.

4.1.4. $H_l/M/1/r$ system

Let us briefly discuss key results for the $H_l/M/1/r$ queueing system and demonstrate how it interrelates with the above $M/H_m/1/r$ system. As for the preceding queueing system, introducing the variable $\nu(t)$, which has the same sense, and the phase $\xi(t)$ of the customer generation process at the instant t, assuming that $\eta(t) = (\xi(t), \nu(t))$, and taking into account the probabilistic treatment of the hyper-exponential distribution of the customer generation time, we determine the Markov process $\{\eta(t), t \geq 0\}$ with the state set $\mathcal{X} = \{(i,k), i = 1,\dots,l, k = 0,\dots,R\}$. Here the equality $\eta(t) = (i,k)$ for some instant of time t means that the customer arrived goes through phase i and the system has k customers; the system capacity being again denoted by $R = r+1$.

The number $(R+1)l$ of elements of the set $\mathcal{X}$ is finite, and all states are communicating. Therefore, there exist limit probabilities

$$\lim_{t\to\infty} \mathbf{P}\{\eta(t) = (i,k)\} = p_{ik} > 0$$

which are independent of the initial state of the process $\eta(0)$ and satisfy the equilibrium equations

$$0 = -\lambda_i p_{i0} + \mu p_{i1}, \qquad i = 1,\dots,l, \tag{4.1.25}$$

$$0 = -(\lambda_i + \mu)p_{ik} + \alpha_i \sum_{s=1}^{l} \lambda_s p_{s,k-1} + \mu p_{i,k+1}, \qquad i = 1,\dots,l, \quad k = 1,\dots,r, \tag{4.1.26}$$

$$0 = -(\lambda_i + \mu)p_{iR} + \alpha_i \sum_{s=1}^{l} \lambda_s p_{sr} + \alpha_i \sum_{s=1}^{l} \lambda_s p_{sR}, \qquad i = 1,\dots,l, \tag{4.1.27}$$

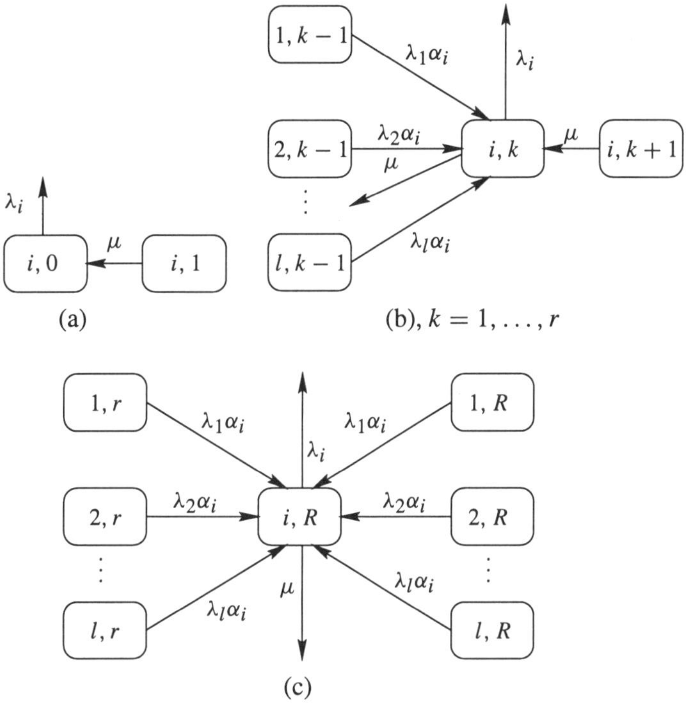

Figure 4.3.

which is explained with the use of the principle of global balance, again, and the queueing system transition diagram shown in Fig. 4.3.

The state $(i, 0)$ (Fig. 4.3a) can be left with intensity λ_i due to completing customer generation in phase i (the probability flow is $\lambda_i p_{i0}$) and entered into from the state $(i, 1)$ with intensity μ due to completing the service of a single customer at the server (the probability flow is μp_{i1}). Equating of the corresponding probability flows yields equation (25).

The state (i, k) (Fig. 3.4b) can also be left upon arrival of a new customer which completed his generation in phase i with intensity λ_i and also due to completion of the customer service with intensity μ; the total flow in this case is $(\lambda_i + \mu)p_{ik}$. This state can be entered into either from the states $(i, k-1)$, $i = 1, \dots, l$, upon arrival of a new customer from one of l phases and in view of the fact that generation of a new customer begins with phase i (the total flow in this case is $\alpha_i \sum_{s=1}^{l} \lambda_s p_{s,k-1}$), or from the state $(i, k+1)$ due to completion of the customer service with intensity μ, the probability flow being $\mu p_{i,k+1}$. Equation (4.1.26) is obtained by summing the flows entering into the state (i, k) and equating them to the emanating flow.

In the last equation, explanation is required only for the last term in the right-hand side, which gives the intensity of entering the state (i, R) from the states $(1, R)$, $(2, R)$, $\dots$, (l, R) —the case of full system—due to completion of generating the customer to be lost, taking into account that the new customer is generated beginning with phase i.

In order to solve the equilibrium equations (4.1.25)–(4.1.27), we may use the same approach as in the case of $M/H_m/1/r$ system, but we proceed differently and consider the single-server queueing system with the Poisson input flow of intensity μ, service time distribution function $A(x)$, and buffer capacity one place more than in the original $H_l/M/1/r$ system. This system is called the system *dual* in traffic to the original system because it is obtained from the original system by replacing $A(x)$ by $B(x)$. We denote it by $\widetilde{M}/\widetilde{H}_l/1/R$. Let $\{\tilde{p}_0, \tilde{p}_{ki}, k = 1, \dots, R+1, i = 1, \dots, l\}$ be the stationary distribution of the state probabilities for the $\widetilde{M}/\widetilde{H}_l/1/R$ queueing system.

Let us prove that for the original $H_l/M/1/r$ queueing system the stationary distribution is

$$p_{ik} = \frac{\tilde{p}_{R-k+1,i}}{1-\tilde{p}_0}, \quad i = 1, \dots, l, \quad k = 0, \dots, R. \tag{4.1.28}$$

To this end, we introduce a constant p_{R+1} defined by the equality

$$\mu p_{R+1} = \sum_{s=1}^{l} \lambda_s p_{s,R}. \tag{4.1.29}$$

Solution of simultaneous equations (4.1.25)–(4.1.27) is known to be determined up to a constant. Therefore, substituting (4.1.29) into (4.1.27), we express all unknowns p_{ik} in terms of the constant p_{R+1}, which is determined by the normalisation condition

$$\sum_{i=1}^{l} \sum_{k=0}^{R} p_{ik} = 1.$$

We assume now that $p_{R+1} = \tilde{p}_0/(1-\tilde{p}_0)$. Substituting (4.1.28) into (4.1.25)–(4.1.27), taking (4.1.29) into account, replacing λ_i by μ_i and μ by λ, and then replacing i by j and $R-k+1$ by k, we obtain simultaneous equations which is nothing but the equilibrium equations for the $M/\widetilde{H}_l/1/R$ queueing system, which proves (4.1.28).

Thus, we find stationary probabilities p_{ik}, which allows us to calculate the corresponding mean characteristics of the queue length, number of customers in the system, and other characteristics. The readers can do this themselves. We cannot still write out an expression for the loss probability π because the flow is no more Poisson; to do this, since the system idling probability p_0 is defined as $p_0 = p_{\cdot,0}$, we use relations such as (4.1.18) or (4.1.19) applicable also to the $H_l/M/1/r$ queueing system. Validity of (4.1.18) and (4.1.19) are easily proved by the same reasoning as for the case of $M/H_m/1/r$ queueing system. Another equivalent expression of the system loss probability for $H_l/M/1/r$ can be obtained by rigorous reasoning. This problem, as well as that of finding the stationary waiting time, will be solved below for a more general queueing system than the $H_l/M/1/r$ system considered here.

4.1.5. Generalisations of the $M/H_m/1/r$ system

It turns out that results obtained for $M/H_m/1/r$ queueing systems may serve as a base for studying a more complicated system. Namely, we assume that the service times of the customers of the ith flow (i-customers) have the distribution function

$$B_i(x) = 1 - e^{-\mu_i x}, \qquad x > 0,$$

and consider the same system, but with m independent Poisson flows of intensities λ_i, $i = 1, \ldots, m$, and total intensity $\lambda = \sum_{i=1}^{m} \lambda_i$. We assume also that the customers are served according to the FCFS discipline and denote such a system as $M_m/M_m/1/r/\text{FCFS}$. Its operation can also be described by the Markov process with a finite state set. Without going into details, we demonstrate the relation between the stationary state probabilities of $M_m/M_m/1/r/\text{FCFS}$ and $M/H_m/1/r$ systems. Let $p_{i_1,i_2,\ldots,i_k}$ be the stationary probability of a state of the $M_m/M_m/1/r/\text{FCFS}$ queueing system such that there is an i_1-customer at the server and $k-1$ customers are queued in the order of their arrival with the flow labels $i_2, i_3, \ldots, i_k$; additionally, let $p_{i,n_1,\ldots,n_m}$ be the stationary probability that the server handles an i-customer and n_1 customers of type 1, n_2 customers of type 2, ..., n_m customers of type m are queued, no matter in what order, $\sum_{s=1}^{m} n_s = k-1$. Relying on properties of the Poisson flow, one easily sees that the following relations are valid (Basharin, 1965):

$$p_{i_1,i_2,\ldots,i_k} = p_{k,i_1} \prod_{j=2}^{k} \alpha_{i_j}, \qquad i_1, \ldots, i_k = 1, \ldots, m, \qquad k = 1, \ldots, R,$$

$$p_{i,n_1,\ldots,n_m} = p_{ki} \frac{(k-1)!\, \alpha_1^{n_1} \ldots \alpha_m^{n_m}}{n_1! \ldots n_m!}, \qquad i = 1, \ldots, m, \quad k-1 = n_1 + \cdots + n_m,$$

where p_{ki} are stationary probabilities of $M/H_m/1/r$ queueing system and $\alpha_i = \lambda_i/\lambda$, $i = 1, \ldots, m$.

The same relations hold for the LCFS discipline (with regard for re-numeration of the order in the queue) and the second relation is valid for the RANDOM discipline. Furthermore, this relation is remains true for these service disciplines in the $M_m/G_m/1/r$ and $M/HG_m/1/r$ systems (Bocharov, 1985), where HG_m stands for the mixture of m arbitrary distributions $B_i(x)$ with weights $\alpha_i = \lambda_i/\lambda$, $i = 1, \ldots, m$.

4.2. $M_2/M/n/r$ system with non-preemptive priority

In the preceding section, we considered an approach providing, as experience shows, fairly efficient recurrence formulas for calculating steady-state queueing system characteristics. In what follows, we use an example of the two-flow non-preemptive fully accessible queueing system to discuss another approach, which also provides recurrence but somewhat different formulas than those for $M/H_m/1/r$ systems.

We consider a system with n identical servers and a buffer of finite capacity r, $2 \leq r < \infty$, which two independent Poisson customer flows of intensities λ_1 and λ_2 and total intensity $\lambda = \lambda_1 + \lambda_2$ arrive at. The customer service times for each flow are independent and have the common distribution function

$$B(x) = 1 - e^{-\mu x}, \qquad x > 0,$$

for all servers. If all servers are busy, then the customer arrived is queued in the buffer, the customers of both flows constituting a common queue. The customers of the first flow (1-customers or priority customers) have non-preemptive priority over the customers of the second flow (2-customers or non-priority customers), which means that if customers from both flows are queued, then a 1-customer is taken for service; a 2-customer is served only if no 1-customer is queued. Within each priority class, the customers are served by any fixed disciplines such as FCFS, LCFS, or RANDOM. If the buffer is full, that is, has no empty places, customers of any type are lost.

Since we consider two types of customers which constitute two queues, though in the common buffer, we strive to obtain characteristics of each queue. But there is no sense in making unnecessary fuss over finding characteristics of the common queue without reference to the customer type, because in this case we can use the results of Section 3.2. Indeed, the sum of two independent Poisson flows of intensities λ_1 and λ_2 is known to yield a Poisson flow of intensity $\lambda = \lambda_1 + \lambda_2$ (see Theorem 2.1.3). Therefore, the characteristics of the common queue can be analysed using the $M/M/n/r$ system with flow of intensity λ.

4.2.1. Equilibrium equations

We proceed to constructing a model of two queues. Let $\nu_i(t)$, $i = 1, 2$, and $\nu(t)$ be, respectively, the number of i-customers in the buffer and the total number of customers in service at time t. We define the process $\{\eta(t), t \geq 0\}$ as follows: if at the instant t the system has less than n customers, then we set $\eta(t) = \nu(t)$; otherwise, we set $\eta(t) = (\nu_1(t), \nu_2(t))$. Since the queueing system under consideration is exponential (Poisson flows and service time distributed exponentially with parameter μ independent of the type of customer and server where he is processed), $\{\eta(t), t \geq 0\}$ will be a homogeneous Markov process. The state set $\mathscr{X}$ of $\{\eta(t), t \geq 0\}$ is

$$\mathscr{X} = \{(i), i = 0, \ldots, n-1; (k, m), 0 \leq k + m \leq r\}.$$

The states from this set are interpreted as follows: if $\eta(t) = (i)$ at some instant t, then i servers are busy; if $\eta(t) = (k, m)$, then all n servers are busy and k 1-customers and m 2-customers are waiting in the buffer.

We set

$$p_i = \lim_{t\to\infty} \mathbf{P}\{\eta(t) = (i)\}, \qquad p_{km} = \lim_{t\to\infty} \mathbf{P}\{\eta(t) = (k, m)\}.$$

Since all states of the process $\{\eta(t), t \geq 0\}$ communicate and there are a finite number $n + (r+1)(r+2)/2$ of them, by virtue of ergodicity of the process $\{\eta(t), t \geq 0\}$ the limit probabilities p_i and p_{km} exist, are strictly positive, independent of the initial conditions, and satisfy the equilibrium equations

$$0 = -(\lambda + i\mu)p_i + u(i)\lambda p_{i-1} + (i+1)\mu p_{i+1}, \qquad i = 0, \ldots, n-1, \tag{4.2.1}$$

$$0 = -(\lambda + n\mu)p_{00} + \lambda p_{n-1} + n\mu p_{10} + n\mu p_{01}, \tag{4.2.2}$$

$$0 = -[\lambda u(r-k) + n\mu]p_{k0} + \lambda_1 p_{k-1,0} + u(r-k)n\mu p_{k+1,0}, \qquad k = 1, \ldots, r, \tag{4.2.3}$$

$$\begin{aligned} 0 = {} & -[\lambda u(r-m) + n\mu]p_{0m} + \lambda_2 p_{0,m-1} \\ & + u(r-m)(n\mu p_{1m} + n\mu p_{0,m+1}), \qquad m = 1, \ldots, r, \end{aligned} \tag{4.2.4}$$

$$\begin{aligned} 0 = {} & -[\lambda u(r-k-m) + n\mu]p_{km} + \lambda_1 p_{k-1,m} + \lambda_2 p_{k,m-1} \\ & + u(r-k-m)n\mu p_{k+1,m}, \qquad m = 1, \ldots, r-1, \\ & \qquad k = 1, \ldots, r-m, \end{aligned} \tag{4.2.5}$$

where some groups of equations are united for brevity using the Heaviside function $u(x)$.

Our experience allows us to write easily the transition probabilities of the process $\{\eta(t), t \geq 0\}$ over the interval $(t, t+\Delta)$ for small Δ and obtain the corresponding transition intensities. We leave the rigorous derivation of equations (4.2.1)–(4.2.5) to the readers, and restrict ourselves only to explaining them on the basis of the principle of global balance.

The first group of these equations is obvious because it coincides with the corresponding equilibrium equations for the $M/M/n/r$ queueing system with Poisson flow of intensity λ. The transition diagram underlying the remaining equations (4.2.2)–(4.2.5) is shown in Fig. 4.4.

If $k+m \neq r$, the state (k, m) can be left either upon arrival of a 1-customer with intensity λ_1 or a 2-customer with intensity λ_2, or upon completing serving a customer at any of n servers with intensity $n\mu$ (the total probability flow is $(\lambda_1+\lambda_2+n\mu)p_{km} = (\lambda+n\mu)p_{km}$). Since for $k+m = r$ a customer arrived is lost, the state $(k, r-k)$ can be left only with intensity $n\mu$ (the probability flow is $n\mu p_{k,r-k}$).

The state $(0, 0)$ (Fig. 4.4a) can be entered into from the state $(n-1)$ upon arrival of a new customer of any type with intensity $\lambda_1+\lambda_2 = \lambda$ (the total flow is λp_{n-1}) or customer service completion at some server in the case of one priority or a non-priority customer in the queue with identical intensities $n\mu$ (the probability flows are equal, respectively, to $n\mu p_{10}$ and $n\mu p_{01}$). The sum of all these flows must be equal to the flow emanating from $(0, 0)$, which leads to equation (4.2.2)).

The state $(k, 0)$, $k = 1, \ldots, r-1$, (Fig. 4.4b) can be entered either from the state $(k-1, 0)$ upon arrival of a new priority customer with intensity λ_1 (the probability flow is $\lambda_1 p_{k-1,0}$) or upon completing the service by a server (with intensity $n\mu$) provided that only $k+1$ priority customers have been queued (the probability flow is $n\mu p_{k+1,0}$). Equating the sum of these flows to the output flow of the state $(k, 0)$, we arrive at equation (4.2.3) for $k \neq r$. The case of $k = r$ (Fig. 4.4c) differs in that the state $(r, 0)$ can be entered only from the state $(r-1, 0)$ upon arrival of a new priority customer with intensity λ_1.

For $k+m \neq r$ (Fig. 4.4f), the state (k, m) can be entered either from the states $(k-1, m)$ and $(k, m-1)$, respectively, upon arrival of a priority customer with intensity λ_1 (the probability flow is $\lambda_1 p_{k-1,m}$) or a non-priority customer with intensity λ_2 (the probability flow is $\lambda_2 p_{k,m-1}$), or from the state $(k+1, m)$ upon completing the service of a customer by one of n servers with intensity $n\mu$, provided that $k+1$ priority and m non-priority customers have been queued (it is the case where priority is taken into consideration upon accepting a customer from the queue, the probability flow being $n\mu p_{k+1,m}$); equating the resulting probability flow to the output probability flow of the state (k, m) (equal to $(\lambda+n\mu)p_{km}$)), we obtain equation (4.2.5) for $k+m \neq r$. The case of $k+m = r$ differs from the last case only in that the state $(k, r-k)$ cannot be entered upon customer service completion. Equation (4.2.4) is explained in a similar manner, and we leave this to the readers.

4.2.2. Solution of equilibrium equations

Prior to proceeding to solving simultaneous equations (4.2.1)–(4.2.5), we introduce the stationary probability p_{n+i} that there are i customers in the queue (or $n+i$ customers in the

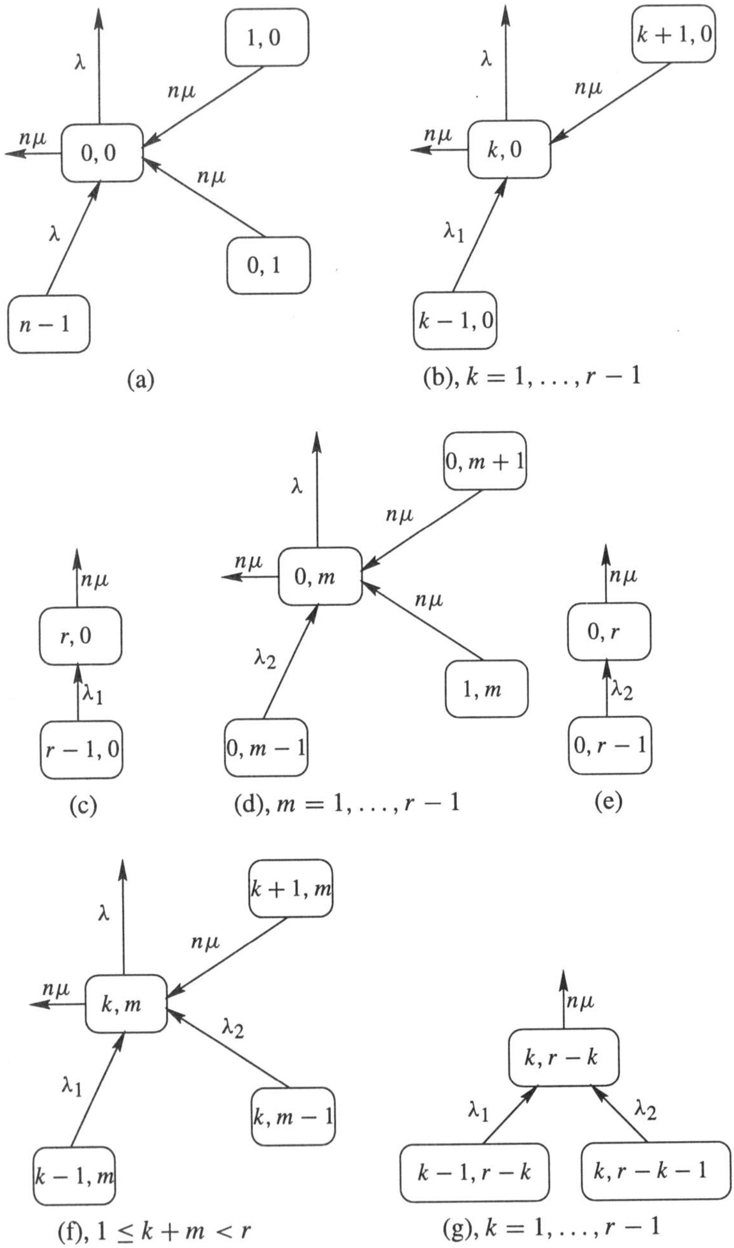
λ
1, 0
nμ
nμ
0, 0
nμ
λ
0, 1
n − 1
(a)
λ
k + 1, 0
nμ
nμ
k, 0
λ1
k − 1, 0
(b), k = 1, . . . , r − 1
λ
0, m + 1
nμ
nμ
0, m
nμ
λ2
1, m
0, m − 1
(d), m = 1, . . . , r − 1
nμ
r, 0
λ1
r − 1, 0
(c)
nμ
0, r
λ2
0, r − 1
(e)
λ
k + 1, m
nμ
nμ
k, m
λ2
λ1
k, m − 1
k − 1, m
(f), 1 ≤ k + m < r
nμ
k, r − k
λ1
λ2
k − 1, r − k
k, r − k − 1
(g), k = 1, . . . , r − 1

Figure 4.4.

system), irrespective of their type. Obviously,

$$p_{n+i} = \sum_{k+m=i} p_{km}. \tag{4.2.6}$$

It is not difficult to see that for a fixed $i = 0, \dots, r$, combining equations (4.2.2)–(4.2.5) so that the sum of subscripts $k + m$ of the probabilities p_{km} in the leading terms in the right-hand sides of these equations equals i, we arrive at equations which are satisfied by the probabilities p_{n+i} and which together with (4.1.1) constitute a set of equilibrium equations for the $M/M/n/r$ queueing system.

We consider now equations (4.2.3) and (4.2.5) and represent them in the matrix form. To this end, we divide the equations by $n\mu$ and introduce the matrices $D_0 = (1)$, $D_1 = (-1)$, and D_j, $j \geq 2$, of order j

$$D_j = \begin{pmatrix} -1-\rho & 1 & 0 & \cdots & 0 & 0 & 0 \\ \rho_1 & -1-\rho & 1 & \cdots & 0 & 0 & 0 \\ 0 & \rho_1 & -1-\rho & \cdots & 0 & 0 & 0 \\ \vdots & \vdots & \vdots & \ddots & \vdots & \vdots & \vdots \\ 0 & 0 & 0 & \cdots & \rho_1 & -1-\rho & 1 \\ 0 & 0 & 0 & \cdots & 0 & \rho_1 & -1 \end{pmatrix},$$

where $\rho = \rho_1 + \rho_2$, $\rho_i = \lambda_i/(n\mu)$, $i = 1, 2$, and the vectors $\boldsymbol{p}_m^T = (p_{1m}, \dots, p_{r-m,m})$, $\boldsymbol{e}_1^T = (1, 0, \dots, 0)$. Then in the matrix notation equations (4.2.3) and (4.2.5) take the form

$$\begin{aligned} D_r \boldsymbol{p}_0 &= -\rho_1 p_n \boldsymbol{e}_1, \\ D_{r-m} \boldsymbol{p}_m &= -\rho_1 p_{0m} \boldsymbol{e}_1 - \rho_2 \boldsymbol{p}_{m-1}, \qquad m = 1, \dots, r-1. \end{aligned} \tag{4.2.7}$$

Therefore, the set of equations (4.2.3) and (4.2.5) is decomposed into r subsets of lower order in unknown vectors $\boldsymbol{p}_0, \boldsymbol{p}_1, \dots, \boldsymbol{p}_{r-1}$ (the mth subset is of order $r - m$) which can be solved successively. Indeed, since the probability p_n is known (it satisfies (3.2.4)), we can find $\boldsymbol{p}_0$ from the first subset in (4.2.7) having the Jacobian—hence non-degenerate—matrix of coefficients D_r. Further, we set $m = 1$ in (4.2.7) and arrive at an equation in $\boldsymbol{p}_1$ with a Jacobian matrix D_{r-1}. In the right-hand side of this equation, however, there is an undefined probability p_{01}. To find it, we represent (4.2.6) as

$$p_{0m} = p_{n+m} - \sum_{k=1}^{m} p_{k,m-k}, \qquad m = 1, \dots, r. \tag{4.2.8}$$

Hence p_{01} is expressed in terms of the known variables p_{n+1} and p_{10}. Now we are able to obtain the solution $\boldsymbol{p}_1$ of (4.2.7) for $m = 1$. Then for $m = 2$ we find p_{02} from (4.2.8) and solve (4.2.2) again, but for $m = 2$, etc. In principle, taking into account that there exist standard routines for solving simultaneous linear algebraic equations with Jacobian matrix, one could put a full stop here. However, solution of these equations admits an analytic representation with the use of the Cramér rule. To this end, we consider the determinant of the matrix D_j denoted by d_j and expand it in the elements of the first column. As the result, we arrive at the recurrence relations

$$d_0 = 1, \quad d_1 = -1, \quad d_j = -(1+\rho)d_{j-1} - \rho_1 d_{j-2}, \quad j \geq 2. \tag{4.2.9}$$

Additionally, we need the matrices $\tilde{D}_j$, $j \geq 2$, derived from D_j by replacing the element -1 in the rightmost lower angle by $-1-\rho$ and matrices $\tilde{D}_0 = (1)$ and $\tilde{D}_1 = (-1-\rho)$. Then their determinants $\tilde{d}_j$ also are computable by means of recurrence formulas of form (4.2.9), but with the initial conditions $\tilde{d}_0 = 1$ and $\tilde{d}_1 = -1-\rho$. Further, we consider the determinant $d_{r-m,k}$ which is obtained from d_{r-m} by replacing the kth column by $\boldsymbol{p}_{m-1}$. Expanding the determinant $d_{r-m,k}$ in the elements of the kth column, we obtain

$$d_{r-m,k} = d_{r-m-k}\sum_{s=1}^{k-1}(-1)^{k+s}p_{s,m-1}\rho_1^{k-s}\tilde{d}_{s-1} + \tilde{d}_{k-1}\sum_{s=k}^{r-m}(-1)^{k+s}p_{s,m-1}d_{r-m-s}, \qquad k=1,\ldots,r-m, \tag{4.2.10}$$

where it is assumed that $\sum_{s=1}^{0} = 0$.

We apply the Cramér rule to equations (4.2.7) and obtain

$$\begin{aligned} p_{k0} &= \frac{(-\rho_1)^k p_n d_{r-k}}{d_r}, && k=1,\ldots,r, \\ p_{km} &= \frac{-\rho_2 d_{r-m,k} + (-\rho_1)^k d_{r-m-k} p_{0m}}{d_{r-m}}, && k=1,\ldots,r-m, \quad m=1,\ldots,r-1. \end{aligned} \tag{4.2.11}$$

In order to calculate the variables d_j that enter into (4.2.10) and (4.2.11) and obey recurrence formulas (4.2.9), explicit expressions can also be obtained. Indeed, (4.2.9) can be taken as a homogeneous difference equation of the second order with constant coefficients. Then its characteristic equation

$$z^2 + (1+\rho)z + \rho_1 = 0$$

has the roots

$$z_{1,2} = -\frac{1+\rho}{2}\left[1 \mp \left(1 - \frac{4\rho_1}{(1+\rho)^2}\right)^{1/2}\right].$$

Defining the arbitrary constants in the solution $d_j = C_1 z_1^j + C_2 z_2^j$ by means of the initial conditions, we obtain

$$d_j = \frac{1+z_2}{z_2-z_1}z_1^j + \frac{1+z_1}{z_1-z_2}z_2^j, \qquad j \geq 0. \tag{4.2.12}$$

Analogously we obtain

$$\tilde{d}_j = \frac{1+\rho+z_2}{z_2-z_1}z_1^j + \frac{1+\rho+z_1}{z_1-z_2}z_2^j, \qquad j \geq 0. \tag{4.2.13}$$

Depending on the queueing system parameters and the desired accuracy of calculations, d_j and $\tilde{d}_j$ can be found using recurrence relations (4.2.9) or explicit formulas (4.2.12)–(4.2.13).

We obtain recurrence relations (4.2.8), (4.2.10), and (4.2.11) which allow us to calculate the joint stationary distribution of two queues which underlie the marginal distributions of the lengths of each of them. Let q_{ij} be the stationary probability that n servers are busy and j customers of the ith type wait in the buffer. Then $q_{1j} = p_{j,\cdot}$, $j = 0, \dots, r$, where $p_{j,\cdot} = \sum_{m=0}^{r-j} p_{jm}$. Similarly, $q_{2j} = p_{\cdot,j}$, $j = 0, \dots, r$. Then the stationary mean length Q_i of the queue of i-customers obeys the formula

$$Q_i = \sum_{j=1}^{r} j q_{ij}, \qquad i = 1, 2. \tag{4.2.14}$$

Obviously, irrespective of the customer type, the stationary mean queue length Q is

$$Q = Q_1 + Q_2. \tag{4.2.15}$$

Relation (4.2.15) can be used to check calculations, because the left-hand side can be calculated using (3.2.6) and the right-hand one, (4.2.14). Finally, since the customers of both types accumulate in the common buffer, the customer of any type arrived to the full system is lost with the probability $\pi = p_{n+r}$.

4.2.3. Stationary distribution of the waiting times of priority customers

We consider now the distribution of the waiting times of a priority customer accepted by the system in the steady state. We also assume that the customers of the same priority class are served in the order of their arrival, that is, according to the FCFS discipline. Since the non-priority customers, both queued and arrived, do not affect the waiting time of a priority customer, the problem of determining the waiting time distribution of the 1-customer is close to the similar problem for the $M/M/n/r$ queueing system, which is used below.

We denote by $W_1(t)$ the distribution function of the waiting time of a 1-customer accepted by the system operating in the steady state. Now, let $W_1(t \mid x)$ be the conditional distribution function of the waiting time of some (marked) 1-customer which arrived to the system when it was in the state $x \in \mathscr{X}'$, $\mathscr{X}' = \mathscr{X} \backslash \mathscr{X}_{n+r}$, and $\omega_1(s \mid x)$ be its Laplace–Stieltjes transform, where $\mathscr{X}_{n+r}$ denotes, as before, the state with $n + r$ customers in the system, that is, $\mathscr{X}_{n+r} = \{(k, r-k), k = 0, \dots, r\}$. Since the arriving flows are Poisson, and therefore, the stationary probability that the arriving customer finds the system in a state x coincides with the time-stationary probability p_x of this state, we obtain from the formula of total probability

$$W_1(t) = \frac{1}{1-\pi} \sum_{x \in \mathscr{X}'} p_x W_1(t \mid x)$$

or, passing to the Laplace–Stieltjes transform,

$$\omega_1(s) = \frac{1}{1-\pi} \sum_{x \in \mathscr{X}'} p_x \omega_1(s \mid x). \tag{4.2.16}$$

It is clear that if the customer under consideration finds the system in the state (i), $i = 0, \dots, n-1$, he is immediately taken by the server, and in this case $\omega(s \mid (i)) = 1$. If the tagged customer finds k queued 1-customers, that is, finds the system in any of the states

$(k, 0), (k, 1), \ldots, (k, r-k-1)$, then he is not rejected, a queue of k 1-customers precedes him, and his service starts as soon as, beginning from the instant of his arrival, serving of $k+1$ customers becomes completed. For $k = 0, \ldots, r-1$, therefore,

$$\omega_1(s \mid (k, m)) = \left(\frac{n\mu}{s + n\mu}\right)^{k+1}, \qquad m = 0, \ldots, r-k-1.$$

With this reasoning, we obtain from (16) the final expression

$$\omega_1(s) = \frac{1}{1-\pi}\left[p_{w=0} + \sum_{k=0}^{r-1}\sum_{m=0}^{r-k-1} p_{km}\left(\frac{n\mu}{s+n\mu}\right)^{k+1}\right], \tag{4.2.17}$$

where $p_{w=0} = \sum_{i=0}^{n-1} p_i$ is the probability of serving a customer without waiting.

Differentiating (4.2.17) with respect to s, we find the lth moment $w_1^{(l)}$ of the waiting time of a 1-customer by the formula $w_1^{(l)} = (-1)^l \omega_1^{(l)}(0)$:

$$w_1^{(l)} = \sum_{k=0}^{r-1} (k+l)_l (p_{k,\cdot} - p_{k,r-k}).$$

For $l = 1$, in particular, the expression for the mean waiting time w_1 of a 1-customer follows immediately:

$$w_1 = \sum_{k=0}^{r-1} (k+1)(p_{k,\cdot} - p_{k,r-k}). \tag{4.2.18}$$

We note that w_1 can also be calculated using the Little formula, which takes the form

$$\lambda_1(1-\pi)w_1 = Q_1. \tag{4.2.19}$$

To check the calculations, it is advisable to bear this fact in mind. We suggest that by way of exercise the readers prove formula (4.2.19) on the basis of (4.2.18) and the equilibrium equations (4.2.3)–(4.2.5).

4.3. $M/PH/1/r$ and $PH/M/1/r$ systems

In the preceding chapter, we applied the method of fictitious phases to the $M/E_m/1/\infty$ queueing system and in Section 3.1, to the $M/H_m/1/r$ queueing system, which allowed us to markovise the model. As shown in Chapter 2, however, service can be decomposed into 'exponential' phases in the much more general case of PH-distributions. Let us discuss this in more detail on the example of the $M/PH/1/r$ system and extend the results to the $PH/M/1/r$ system.

First, we consider the $M/PH/1/r$ queueing system by constructing its Markov model and presenting, in order to analyse it, a matrix technique which leads to a matrix-geometric solution for the stationary queue distribution. The material of this section presents no difficulties to the reader who mastered Section 4.1 well, because their philosophies are extremely close.

According to the Kendall classification, $M/PH/1/r$ is a single-server queueing system with a buffer of finite capacity r, $2 \leq r \leq \infty$, which a Poisson flow of customers arrives at with intensity λ, the customer service times having the phase-type distribution function

$$B(x) = 1 - \boldsymbol{\beta}^T e^{Mx}\mathbf{1}, \quad x > 0, \quad \boldsymbol{\beta}^T\mathbf{1} = 1.$$

It is assumed that $B(x)$ admits an irreducible PH-representation $(\boldsymbol{\beta}^T, M)$ of order m (see Section 2.8). One of the fixed service disciplines FCFS, LCFS, or RANDOM is used to take customers to be served. As usual, we specify below, if necessary, a certain type of discipline. If the buffer is full, the customer arrived is lost.

As concerns the $PH/M/1/r$ system, it can be regarded as dual to the above system, which means that the flow arriving to the $PH/M/1/r$ system has phase structure, that is, is recurrent with the distribution function

$$A(x) = 1 - \boldsymbol{\alpha}^T e^{\Lambda x}\mathbf{1}, \quad x > 0, \quad \boldsymbol{\alpha}^T\mathbf{1} = 1,$$

admitting an irreducible PH-representation $(\boldsymbol{\alpha}^T, \Lambda)$ of order l, the service times being distributed exponentially with parameter μ.

4.3.1. Equilibrium equations for the $M/PH/1/r$ queueing system

We proceed now to constructing a mathematical model of the $M/PH/1/r$ system. We rely on the probabilistic interpretation of the PH-representation (Section 2.8) to describe behaviour of the system by means of a homogeneous Markov process $\{\eta(t), t \geq 0\}$ with the state set $\mathscr{X} = \{(0); (k, j), k = 1, \dots, R, j = 1, \dots, m\}$, where $R = r + 1$, as before, stands for the total capacity of the system. The states of this set have the following meanings: if $\eta(t) = (0)$ at some instant t, then the system is empty, and if $\eta(t) = (k, j)$, then the system has k customers and the customer in service undergoes phase j.

If the state of the process $\eta(t)$ at the instant t is known, then—by virtue of the memoryless property of the Poisson flow and exponential distribution of the durations of the service phases—its behaviour after t is independent of its history, that is, of the states preceding the instant t, and also of the choice of this instant on the time axis. Therefore, $\{\eta(t), t \geq 0\}$ is a homogeneous Markov process.

The number $Rm + 1$ of states of the process $\{\eta(t), t \geq 0\}$ is finite, and all states are communicating. Therefore, the process is ergodic and there exist limit probabilities

$$p_0 = \lim_{t\to\infty} p_0(t) > 0, \qquad p_{kj} = \lim_{t\to\infty} p_{kj}(t) > 0,$$

which are independent of the initial state of $\eta(0)$ and coincide with the stationary state probabilities of the process.

Let us write out the equilibrium equations for the probabilities p_0 and p_{kj} and introduce to this end some additional notation. Let $\boldsymbol{\beta}^T = (\beta_1, \dots, \beta_m)$, $M = (M_{ij})_{i,j=1,\dots,m}$ and $\boldsymbol{\mu} = -M\mathbf{1}$. We recall that, according to the probabilistic interpretation of the PH-distribution, the component β_j of the vector $\boldsymbol{\beta}$ is the probability of beginning service in the corresponding open exponential network (server, in this case) from the node (phase) j, the element M_{ij} of the matrix M is the intensity of customer transition in the network (server) from the node (phase) i to the node (phase) j, $i \neq j$, and the component μ_j of the vector $\boldsymbol{\mu}$ is the intensity of customer departure from the network (server) from the node (phase) j. Then the queueing system under consideration obeys the equilibrium equations

$$0 = -\lambda p_0 + \sum_{i=1}^{m} \mu_i p_{1i}, \tag{4.3.1}$$

$$0 = -\lambda p_{1j} + \sum_{i=1}^{m} M_{ij} p_{1i} + \lambda \beta_j p_0 + \beta_j \sum_{i=1}^{m} \mu_i p_{2i}, \qquad j = 1, \ldots, m, \tag{4.3.2}$$

$$0 = -\lambda p_{kj} + \sum_{i=1}^{m} M_{ij} p_{ki} + \lambda p_{k-1,j} + \beta_j \sum_{i=1}^{m} \mu_i p_{k+1,i}, \quad k = 2, \ldots, r, \quad j = 1, \ldots, m, \tag{4.3.3}$$

$$0 = \sum_{i=1}^{m} M_{ij} p_{Ri} + \lambda p_{Rj}, \qquad j = 1, \ldots, m. \tag{4.3.4}$$

Additionally, the normalisation condition

$$p_0 + \sum_{k=1}^{R} \sum_{j=1}^{m} p_{kj} = 1 \tag{4.3.5}$$

must be satisfied.

We explain now the derivation of the equilibrium equations and introduce to this end the sets $\mathcal{X}_0 = \{(0)\}$ and $\mathcal{X}_k = \{(k, j), j = 1, \ldots, m\}, k = 1, \ldots, R$; $\mathcal{X}_k$ is the set of states in which there are k customers in the system. Let us consider the possible changes in the process $\{\eta(t), t \geq 0\}$ in the time interval $(t, t + \Delta)$, where Δ is a small increment. Let $\eta(t + \Delta) = (0)$. This event can occur in two cases:

- at the instant t $\eta(t) = (0)$ and no new customer arrives in the time interval Δ, the probability of this event being $p_0(t)[1 - \lambda\Delta + o(\Delta)]$; or
- at the instant t $\eta(t) \in \mathcal{X}_1$, that is, it is in one of the states $(1, j)$, $j = 1, \ldots, m$, and in the time interval Δ the customer service is completed (with regard for the fact that the probability of completing the customer service in phase j is $\mu_j \Delta + o(\Delta)$, the total probability of this event being $\sum_{j=1}^{m} p_{1j}(t)\mu_j\Delta + o(\Delta)$).

The probabilities of other transitions in the time Δ are $o(\Delta)$.

After summation, we see that the probability $p_0(t + \Delta)$ of the event $\{\eta(t + \Delta) = (0)\}$ is

$$p_0(t + \Delta) = (1 - \lambda\Delta) p_0(t) + \sum_{j=1}^{m} p_{1j}(t)\mu_j \Delta + o(\Delta).$$

Usual conversions—rearranging of $p_0(t)$ into the left-hand side, dividing by Δ, and passing to the limit for as $\Delta \to 0$—yield

$$p_0'(t) = -\lambda p_0(t) + \sum_{j=1}^{m} \mu_j p_{1j}(t)$$

which implies equation (4.1.1) as $t \to \infty$.

Let now $\eta(t+\Delta) = (1, j)$ for some fixed $j = 1, 2, \ldots, m$. Then, in the time Δ the following transitions are possible in the system that lead to this event:

- $\eta(t) = (1, j)$ and in the time Δ no customer arrives and also service in phase j is not completed, the probability of this event being
$$p_{1j}(t)[1 - \lambda\Delta + o(\Delta)][1 + M_{jj}\Delta + o(\Delta)] = p_{1j}(t)[1 - (\lambda - M_{jj})\Delta + o(\Delta)];$$
- $\eta(t) \in \mathscr{X}_1$, that is, the process $\eta(t)$ at the instant t is in any of the states $(1, i)$, $i = 1, \ldots, m$, $i \neq j$, and in the time Δ the service phase changes from i to j (taking into account that the passage from phase i to j occurs with the probability $M_{ij}\Delta + o(\Delta)$, we see that the probability of this event is equal to $\sum_{\substack{i=1 \\ i \neq j}}^{m} p_{1i}(t) M_{ij}\Delta + o(\Delta)$);
- $\eta(t) = (0)$ and in the time Δ a customer arrives (transition 'from below') which is served beginning from phase j, the probability of this event being $p_0(t)\lambda\Delta\beta_j + o(\Delta)$; and
- $\eta(t) \in \mathscr{X}_2$, that is, at the instant t the process $\eta(t)$ is in any of the states $(2, i)$, $i = 1, \ldots, m$, the customer service completes in the time Δ (transition 'from above'), and the service of a new customer begins from phase j, the probability of this transition being $\beta_j \sum_{i=1}^{m} p_{2i}(t)\mu_i\Delta + o(\Delta)$.

The probabilities of other possible transitions is $o(\Delta)$.

The probability of the event $\eta(t+\Delta) = (1, j)$ is hence
$$\begin{aligned} p_{1j}(t+\Delta) = p_{1j}(t)[1 - (\lambda - M_{jj})\Delta] + \sum_{\substack{i=1 \\ i \neq j}}^{m} p_{1i}(t) M_{ij}\Delta \\ + p_0(t)\lambda\Delta\beta_j + \beta_j \sum_{i=1}^{m} p_{2i}(t)\mu_i\Delta + o(\Delta), \end{aligned}$$
which yields
$$p'_{1j}(t) = -\lambda p_{1j}(t) + \sum_{i=1}^{m} M_{ij} p_{1i}(t) + \lambda\beta_j p_0(t) + \beta_j \sum_{i=1}^{m} \mu_i p_{2i}(t), \qquad j = 1, \ldots, m,$$
and leads, as $t \to \infty$, to equation (4.3.2).

In the general case of $k = 2, 3, \ldots, r$, the transitions in the time Δ which result in the event $\eta(t+\Delta) = (k, j)$ for some fixed $j = 1, 2, \ldots, m$ are similar to the case of $k = 1$, with the exception for the transition 'from below;' if $\eta(t) = (k-1, j)$, then in time Δ a customer can arrive (and be queued in the buffer), the probability of this event being $\lambda\Delta + o(\Delta)$. Therefore, we obtain by analogy with the case $k = 1$
$$\begin{aligned} p_{kj}(t+\Delta) = [1 - (\lambda - M_{jj})\Delta] p_{kj}(t) + \sum_{\substack{i=1 \\ i \neq j}}^{m} p_{ki}(t) M_{ij}\Delta \\ + \lambda\Delta p_{k-1,j}(t) + \beta_j \sum_{i=1}^{m} p_{k+1,i}(t)\mu_i\Delta + o(\Delta); \end{aligned}$$

hence we obtain the equations

$$p'_{kj}(t) = -\lambda p_{kj}(t) + \sum_{i=1}^{m} M_{ij}\, p_{ki}(t) + \lambda p_{k-1,j}(t)$$
$$+ \beta_j \sum_{i=1}^{m} \mu_i\, p_{k+1,i}(t), \quad j = 1, \dots, m, \quad k = 2, \dots, r,$$

which leads, as $t \to \infty$, to equation (4.3.3).

Finally, we consider the case $\eta(t + \Delta) = (R, j)$ for fixed $j = 1, 2, \dots, m$. Here the following transition leading to this event can occur in the time Δ:

- $\eta(t) = (R, j)$ and in the time Δ this state does not change, the probability of this event being $p_{Rj}(t)[1 + M_{jj}\Delta + o(\Delta)]$, (the arrival of a new customer does not alter the state because the customer is lost; therefore,—in contrast to the above cases for $k = 0, \dots, r$—the probability of non-arrival of a customer in the time Δ here is disregarded);
- $\eta(t) \in \mathscr{X}_R$ and in the time Δ the transition from phase i to j, $i \neq j$, occurs with the probability $\sum_{i=1,\, i\neq j}^{m} p_{Ri}(t) M_{ij}\,\Delta + o(\Delta)$; and
- $\eta(t) = (r, j)$ and in the time Δ one customer arrives at the system (the probability of this event is $p_{rj}(t)\lambda\Delta + o(\Delta)$).

The probability of the remaining transitions is $o(\Delta)$.

So, we see that the probability of the event $\eta(t + \Delta) = (R, j)$ is

$$p_{Rj}(t + \Delta) = p_{Rj}(t)[1 + M_{jj}\Delta] + \sum_{\substack{i=1\\ i\neq j}}^{m} p_{Ri}(t) M_{ij}\,\Delta + p_{rj}(t)\lambda\Delta + o(\Delta),$$

which leads to the equation

$$p'_{Rj}(t) = \sum_{i=1}^{m} M_{ij}\, p_{Ri}(t) + \lambda p_{rj}(t), \qquad j = 1, \dots, m,$$

and to equation (4.3.4) as $t \to \infty$.

We suggest the readers to derive, as an exercise, equations (4.3.1)–(4.3.4) on the base of the principle of global balance by writing first the diagram of transitions in the queueing system.

4.3.2. Matrix-geometric solution of equilibrium equations for $M/PH/1/r$ system

The experience gained in the above sections with the equilibrium equations suggests that the preliminary analysis can substantially simplify the problem. Here we follow the lines of Section 4.1, but make use of the matrix apparatus. It may happen that the matrix notation and operations over equations—especially their probabilistic interpretation—will not seem as straightforward as the scalar ones, but, as certain habits are acquired, the advantages of the matrix approach in writing and analysing the equations become evident, the probabilistic interpretation of the results, if any, also becoming quite transparent. And again, we repeat that all the readers need is experience, and we try to help them to acquire it by using the matrices above and in full measure below.

We turn back to the equilibrium equations (4.3.1)–(4.3.4), introduce the vector $\boldsymbol{p}_k^T = (p_{k1}, \dots, p_{km})$, and represent the system in the matrix form

$$-\lambda p_0 + \boldsymbol{p}_1^T \boldsymbol{\mu} = 0, \tag{4.3.6}$$

$$\boldsymbol{p}_1^T(-\lambda I + M) + \lambda \boldsymbol{\beta}^T p_0 + \boldsymbol{p}_2^T \boldsymbol{\mu} \boldsymbol{\beta}^T = \mathbf{0}^T, \tag{4.3.7}$$

$$\boldsymbol{p}_k^T(-\lambda I + M) + \lambda \boldsymbol{p}_{k-1}^T + \boldsymbol{p}_{k+1}^T \boldsymbol{\mu} \boldsymbol{\beta}^T = \mathbf{0}^T, \qquad k = 2, \dots, r, \tag{4.3.8}$$

$$\boldsymbol{p}_R^T M + \lambda \boldsymbol{p}_r^T = \mathbf{0}^T, \tag{4.3.9}$$

where the vector $\mathbf{0}$ consists only of zeros. The normalisation condition (4.3.5) admits the representation

$$\sum_{k=0}^{R} p_k = 1, \tag{4.3.10}$$

where $p_k = \boldsymbol{p}_k^T \mathbf{1}$ is the stationary probability that there are k customers in the system. By multiplying now each of equations (4.3.7)–(4.3.9) from the right by the vector $\mathbf{1}$ and then summing successively the resulting equalities with account for (6), we arrive at the relations

$$\lambda p_k = \boldsymbol{p}_{k+1}^T \boldsymbol{\mu}, \qquad k = 0, \dots, r, \tag{4.3.11}$$

which are the equations of local balance between the groups of states with k and less customers (from the left) and $k+1$ and more customers (from the right) in the system, respectively. Indeed, we can go to the right only from the states (k, j), $j = 1, \dots, m$, due to arrival of a new customer (the total probability flow being λp_k) and to the left only from the states $(k+1, j)$, $j = 1, \dots, m$, due to service completion (the total probability flow being $\sum_{j=1}^m \mu_j p_{k+1,j}$). Equality of these flows implies (4.3.11).

Now we turn back to equations (4.3.6)–(4.3.9). We easily see that the recurrence matrix equations of the second order in $\boldsymbol{p}_k$ (4.3.7) and (4.3.8) reduce by means of (4.3.11) to the recurrence relations of the first order

$$\boldsymbol{p}_1^T(-\lambda I + M) + \lambda p_1 \boldsymbol{\beta}^T + \lambda p_0 \boldsymbol{\beta}^T = \mathbf{0}^T,$$
$$\boldsymbol{p}_k^T(-\lambda I + M) + \lambda p_k \boldsymbol{\beta}^T + \lambda \boldsymbol{p}_{k-1}^T = \mathbf{0}^T, \qquad k = 2, \dots, r.$$

We introduce the vector $\boldsymbol{\lambda} = \lambda \mathbf{1}$ and matrices $D = -\lambda I + M$ and $\widetilde{M} = D + \boldsymbol{\lambda} \boldsymbol{\beta}^T$, and finally obtain in view of (4.3.9)

$$\begin{aligned} \boldsymbol{p}_1^T \widetilde{M} &= -\lambda \boldsymbol{\beta}^T p_0, \\ \boldsymbol{p}_k^T \widetilde{M} &= -\lambda \boldsymbol{p}_{k-1}^T, \qquad k = 2, \dots, r, \\ \boldsymbol{p}_R^T M &= -\lambda \boldsymbol{p}_r^T. \end{aligned} \tag{4.3.12}$$

Thus, we have just obtained quite simple relations for the vectors $\boldsymbol{p}_k$ which allow for their recursive calculation, provided that the matrix $\widetilde{M}$ is non-degenerate. Hopefully, the attentive readers already noticed that relations (4.3.12) completely coincide in notation with relations (4.1.9), the only difference consisting in that the matrix M used there to determine the matrix $\widetilde{M}$ was diagonal, but here it is arbitrary, of course, satisfying the properties that are necessary for $(\boldsymbol{\beta}, M)$ to be a PH-representation of the distribution function $B(x)$. We can also use (4.1.10) to invert the matrix $\widetilde{M}$, but in doing so we have to check whether the matrix D is invertible under the initial assumptions and when the condition $\boldsymbol{\beta}^T D^{-1}\boldsymbol{\lambda} \neq -1$ for existence of the matrix $\widetilde{M}^{-1}$ is satisfied. By virtue of irreducibility of the PH-representation $(\boldsymbol{\beta}^T, M)$, all the eigenvalues of the matrix M have negative real parts, that is, the matrix M is stable (see Section 2.8), therefore, the matrix $D = -\lambda I + M$ also is stable and hence, non-degenerate. It remains to check under what conditions $\boldsymbol{\beta}^T D^{-1}\boldsymbol{\lambda} = \lambda\boldsymbol{\beta}^T D^{-1}\mathbf{1} \neq -1$. Let $\beta(s)$ denote the Laplace–Stieltjes transform of the distribution function $B(t)$. Then, as follows from (2.8.9), $1 + \lambda\boldsymbol{\beta}^T D^{-1}\mathbf{1} = 1 - \lambda\boldsymbol{\beta}^T(\lambda I - M)^{-1}\mathbf{1} = \beta(\lambda) > 0$. Therefore, under the above assumptions the matrix $\widetilde{M}^{-1}$ always exists and is

$$\widetilde{M}^{-1} = D^{-1} - \frac{\lambda D^{-1}\mathbf{1}\boldsymbol{\beta}^T D^{-1}}{\beta(\lambda)}. \tag{4.3.13}$$

Invertibility of $\widetilde{M}$ could be obtained using the general Lemma 2.8.5, assuming that $(\boldsymbol{f}, G) = (\boldsymbol{\beta}, M)$ and $H = (-\lambda)$. We preferred, however, to obtain this result directly in order to help the readers to acquire skill in handling matrices of this kind.

Turning again back to recurrence relations (4.3.12), we can obtain now explicit expressions for $\boldsymbol{p}_k$ as in Section 4.1 for the $M/H_m/1/r$ queueing system. We introduce the same matrices $W = -\lambda\widetilde{M}^{-1}$, $W_R = -\lambda M^{-1}$ and the vector $\boldsymbol{w}_0 = -\lambda\boldsymbol{\beta}D^{-1}/\beta(\lambda)$, and reduce (4.3.12) to

$$\begin{aligned} \boldsymbol{p}_1^T &= p_0\boldsymbol{w}_0^T, \\ \boldsymbol{p}_k^T &= \boldsymbol{p}_{k-1}W, \qquad k = 2, \ldots, r, \\ \boldsymbol{p}_R^T &= \boldsymbol{p}_r W_R, \end{aligned} \tag{4.3.14}$$

which formally coincides with (4.1.12). As the result, we can assert that the stationary distribution of the probabilities $\{p_0, \boldsymbol{p}_k, k = 1, \ldots, R\}$ of states of the $M/PH/1/r$ queueing system is expressed as a truncated matrix progression

$$\boldsymbol{p}_k^T = \begin{cases} p_0\boldsymbol{w}_0^T W^{k-1}, & k = 1, \ldots, r; \\ p_0\boldsymbol{w}_0^T W^{r-1} W_R, & k = R. \end{cases} \tag{4.3.15}$$

The probability p_0 is found from normalisation conditions (4.3.10), and the resulting explicit expression for p_0 coincides with (4.1.14). However, this expression is rarely used in actual practice: first the variables $\tilde{p}_{kj} = p_{kj}/p_0$ are found and then $p_0 = (1 + \tilde{p}_{\cdot,\cdot})^{-1}$ is calculated from the normalisation condition.

We leave it to the readers as an exercise to derive explicit expressions for the matrices W, W_R and $\boldsymbol{w}_0$ in the case of the Erlang service time distribution with parameters μ and m, that is, for the $M/E_m/1/r$ queueing system. Given explicit expressions for the matrix W or, which is the same, for $\widetilde{M}^{-1}$, the stationary state probabilities for the $M/E_m/1/r$

system can be calculated not only from the matrix progression (4.3.15), but also from the recurrence relations that immediately follow from (4.3.14). As mentioned in Section 4.1, the matrix formulas (4.3.15) or the recurrence relations obtained from them as in case of the $M/H_m/1/r$ system can be used for approximate calculation of the steady-state characteristics in the similar infinite-queue systems. As practice shows, this provides more stable results than, say, calculation by he recurrence formulas directly from the equilibrium equations for the $M/E_m/1/\infty$ queueing system in Section 3.6.2.

Now we briefly outline the basic relations for the performance indices of the $M/PH/1/r$ queueing system in the steady state of operation. The same qualitative reasoning as in Section 4.1 allows us to demonstrate that for the intensities of the accepted, λ_A, and served, λ_D, flows the following relations are valid:

$$\begin{aligned} \lambda_A &= \lambda(1-\pi), \\ \lambda_D &= \mu(1-p_0), \end{aligned} \tag{4.3.16}$$

where, due to Poisson nature of the arriving flow, the loss probability π obeys the relation

$$\pi = \boldsymbol{p}_R^T \mathbf{1} = p_R$$

and μ is the service intensity obeying, in view of (2.8.10), the relation

$$\mu = -(\boldsymbol{\beta}^T M^{-1} \mathbf{1})^{-1}.$$

Then from equality of the intensities λ_A and λ_D we see that in the steady state

$$\lambda_D = \lambda(1-\pi) = \mu(1-p_0). \tag{4.3.17}$$

The last equality in (4.3.17) is equivalent to

$$\rho(1-\pi) = 1 - p_0, \tag{4.3.18}$$

where $\rho = \lambda/\mu$.

We derive now another, equivalent, expression for the right-hand side of (4.3.18). To this end, we sum equations (4.3.8) over $k = 2, 3, \ldots, r$, add the result to equations (4.3.7) and (4.3.9), and obtain

$$\sum_{k=1}^{R} \boldsymbol{p}_k^T M + \lambda \boldsymbol{\beta}_0^T p_0 + \sum_{k=2}^{R} \boldsymbol{p}_k^T \mu \boldsymbol{\beta}^T = \mathbf{0}^T.$$

Multiplying this relation by the vector $\boldsymbol{\beta}^T$ and adding the result to (4.3.6), we obtain

$$\sum_{k=1}^{R} \boldsymbol{p}_k^T (M + \boldsymbol{\mu}\boldsymbol{\beta}^T) = \mathbf{0}^T.$$

Multiplying this equality from the right by $M^{-1}\mathbf{1}$, in view of the normalisation condition $\sum_{k=0}^{R} p_k = 1$ we obtain

$$1 - p_0 = \frac{1}{\mu} \sum_{k=1}^{R} \boldsymbol{p}_k^T \boldsymbol{\mu} \tag{4.3.19}$$

which together with (4.3.16) yields

$$\lambda_A = \lambda_D = \sum_{k=1}^{R} \boldsymbol{p}_k^T \boldsymbol{\mu}. \tag{4.3.20}$$

Formulas (4.3.16)–(4.3.20) relate the characteristics of the queueing system λ_A, λ_D, π and p_0. They can also be used for checking calculations by formulas (4.3.15) or recurrence relations obtained from (4.3.14).

The just calculated vectors $\boldsymbol{p}_k$ are used to find the stationary probabilities $p_k = \boldsymbol{p}_k^T \mathbf{1} = p_{k,\cdot}$ that there are k customers the system, the stationary mean number $N = \sum_{k=1}^{R} k p_k$ of customers in the system, the stationary mean queue length $Q = \sum_{k=2}^{R} (k-1) p_k$, and also moments of higher order.

4.3.3. Stationary distribution of waiting time for $M/PH/1/r/$FCFS system

In order to find the stationary distribution of the time for waiting a service for a queued customer in the case where the customers are served in the order of their arrival, we make use of a somewhat formalised approach which was used above in Chapter 3 and Section 4.1 to evaluate this characteristic. Let $W(t \mid x)$ denote the conditional distribution function of the waiting time of a tagged customer under the condition that he finds the system in the state $x \in \mathcal{X}' = \mathcal{X} \setminus \mathcal{X}_R$ (in this case, the customer is not lost and is accepted for service). For the Poisson flow, as we have said, the stationary probabilities for an arbitrary instant coincide with the stationary state probabilities of the queueing system considered at the instants $t - 0$ of customer arrivals. In view of this fact and with understanding that the customer was accepted by the system, for the distribution function $W(t)$ of the customer waiting time we see that

$$W(t) = \frac{1}{1-\pi} \sum_{x \in \mathcal{X}'} p_x W(t \mid x), \tag{4.3.21}$$

where p_x is the stationary probability of the state x and $\pi = p_R$ is the customer loss probability. Passing on to the Laplace–Stieltjes transform in (4.3.21), we obtain

$$\omega(s) = \frac{1}{1-\pi} \sum_{x \in \mathcal{X}'} p_x \omega(s \mid x). \tag{4.3.22}$$

Let us find the functions $\omega(s \mid x)$. Let $x = (0)$, that is, let the customer under consideration find an empty system. In this case, he is immediately sent to the server, his waiting time is zero, and therefore, $\omega(s \mid (0)) = 1$. If $x = (k, j)$ for some fixed $k = 1, 2, \ldots, r$ and $j = 1, 2, \ldots, m$, then the waiting time of the customer consists of the time required to complete the service of a customer which, upon arrival of the customer under consideration, is in phase j and the service times of $k-1$ customers queued. Since the customer service times are independent, $\omega(s \mid (k, j)) = \beta^{k-1}(s)\beta_j(s)$, where $\beta_j(s)$ is the Laplace–Stieltjes transform of the distribution of the residual service time which started from phase j. Since the phases have exponentially distributed durations and relying on Section 2.8, we see that $(\beta_1(s), \ldots, \beta_m(s))^T = (sI - M)^{-1}\boldsymbol{\mu}$. Then, introducing the vector

$\boldsymbol{\omega}_k^T(s) = (\omega(s \mid (k,1)), \ldots, \omega(s \mid (k,m)))$ and using the matrix notation, we obtain

$$\boldsymbol{\omega}_k(s) = (sI - M)^{-1}\boldsymbol{\mu}\beta^{k-1}(s).$$

By virtue of (4.3.22), we arrive at the final expression for $\omega(s)$:

$$\omega(s) = \frac{1}{1-\pi}\left[p_0 + \sum_{k=1}^{r} \boldsymbol{p}_k^T(sI - M)^{-1}\boldsymbol{\mu}\beta^{k-1}(s)\right]. \tag{4.3.23}$$

Differentiating $\omega(s)$, we easily obtain from (23) the moments of the waiting time. In particular, for the mean waiting time $w = -\omega'(0)$, we obtain

$$w = \frac{1}{1-\pi}\left[-\sum_{k=1}^{r} \boldsymbol{p}_k^T M^{-1}\mathbf{1} + b(Q - r\pi)\right], \tag{4.3.24}$$

where b is the mean service time and Q is the mean queue length. Another equivalent expression for w can be obtained from the Little formula, which has been already mentioned more than once and which, taking into account the possible losses in the system, takes the form

$$\lambda(1-\pi)w = Q. \tag{4.3.25}$$

Therefore, it is advisable to use either (4.3.24) or (4.3.25) to check calculations.

4.3.4. $PH/M/1/r$ system

Having met an analogy between the $M/H_m/1/r$ and $M/PH/1/r$ systems, one might expect that the same holds true for the $H_l/M/1/r$ and $PH/M/1/r$ systems. We introduce, as in Section 4.1.4, the random process $\{\eta(t) = (\xi(t), \nu(t)), t \geq 0\}$, where $\xi(t)$ is the number of the phase undergone at the instant t by the process of customer generation and $\nu(t)$ again is the number of customers in the system at this instant. Then $\{\eta(t), t \geq 0\}$, in view of the probabilistic interpretation of the PH-distribution $A(x)$, is a homogeneous Markov process. Its state set $\mathscr{X} = \{(i,k), i = 1, \ldots, l, k = 0, \ldots, R\}$, where $R = r+1$, coincides with the state set for the $H_l/M/1/r$ system, the states (i,k) themselves having the same physical sense as for the $H_l/M/1/r$ system: i is the phase of the generated customer and k is the number of customers in the system.

Since the number of states of $\mathscr{X}$ is finite (equal to $(R+1)l$) and all states are communicating, the process $\{\eta(t), t \geq 0\}$ is ergodic and there exist stationary probabilities

$$p_{ik} = \lim_{t\to\infty} \mathbf{P}\{\eta(t) = (i,k)\}.$$

Introducing the vectors $\boldsymbol{p}_k^T = (p_{1k}, \ldots, p_{lk})$ and $\boldsymbol{\lambda} = -\Lambda\mathbf{1}$, we write immediately in the matrix form the equilibrium equations which the distribution $\{\boldsymbol{p}_k, k = 0, \ldots, R\}$ satisfies:

$$\mathbf{0}^T = \boldsymbol{p}_0^T\Lambda + \mu\boldsymbol{p}_1^T, \tag{4.3.26}$$

$$\mathbf{0}^T = \boldsymbol{p}_k^T(\Lambda - \mu I) + \boldsymbol{p}_{k-1}^T\boldsymbol{\lambda}\boldsymbol{\alpha}^T + \mu\boldsymbol{p}_{k+1}^T, \qquad k = 1, \ldots, r, \tag{4.3.27}$$

$$\mathbf{0}^T = \boldsymbol{p}_R^T(\Lambda - \mu I) + \boldsymbol{p}_r^T\boldsymbol{\lambda}\boldsymbol{\alpha}^T + \boldsymbol{p}_R^T\boldsymbol{\lambda}\boldsymbol{\alpha}^T. \tag{4.3.28}$$

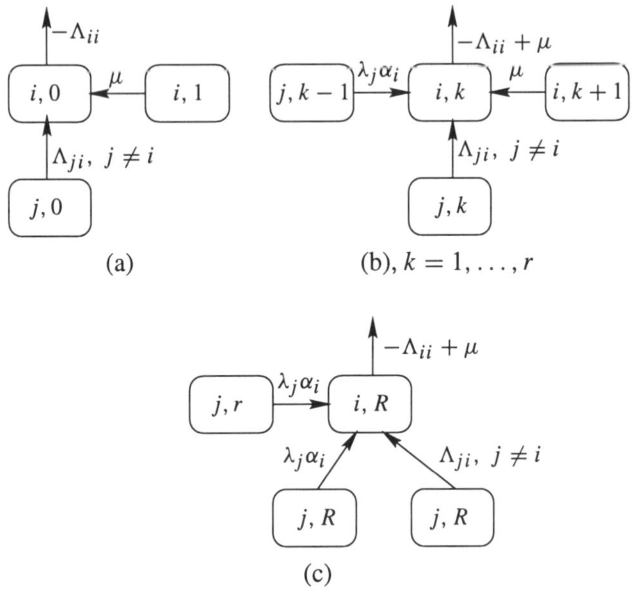

Figure 4.5.

We omit the detailed derivation of equations (4.3.26)–(4.3.28) and only present the queueing system transition diagram (Fig. 4.5) which can be used by the readers to obtain these equilibrium equations on the basis of the principle of global balance. The following notation is used in Fig. 4.5: Λ_{ij} is the (i, j)th element of the matrix Λ (intensity of transition from phase i to phase jth), λ_i is the ith component of the vector $\boldsymbol{\lambda}$ (intensity of leaving phase i by the generated customer), and α_i is the ith component of the vector $\boldsymbol{\alpha}$ (probability that customer generation starts from phase ith).

In order to solve the equilibrium equations (4.3.26)–(4.3.28), we use the same approach as for the $H_l/M/1/r$ system, that is, consider a single-server queueing system with a buffer of capacity one place more than in the original $PH/M/1/r$ system where the change $A(x) \leftrightarrow B(x)$ is carried out or, stated differently, the system dual to the original one. We denote this system by $\widetilde{M}/\widetilde{PH}/1/R$. The stationary probabilities of its states $\tilde{p}_0$ and $\tilde{p}_{ki}$, $k = 1,\ldots, R+1$, $i = 1,\ldots,l$, obey formulas similar to (4.3.14) up to the corresponding changes in notation ($\boldsymbol{\beta} \leftrightarrow \boldsymbol{\alpha}$, $M \leftrightarrow \Lambda$, $m \leftrightarrow l$, $\mu \leftrightarrow \lambda$, and $r \leftrightarrow R$). Now we introduce the vector $\tilde{\boldsymbol{p}}_k^T = (\tilde{p}_{k1},\ldots,\tilde{p}_{kl})$. The readers should try to prove the assertion that

$$\boldsymbol{p}_k = \frac{\tilde{\boldsymbol{p}}_k}{1-\tilde{p}_0} \qquad k = 0,\ldots,R,$$

which is proved similarly to (4.1.28) for the $H_l/M/1/r$ queueing system.

Given the distribution $\{p_{ik}, i = 1,\ldots,l, k = 0,\ldots,R\}$, we can obtain a number of indices of the $PH/M/1/r$ queueing system. In particular, the system idling probability, p_0, is defined as $p_0 = p_{\cdot,0}$, and that of losses, π, can be found from (4.3.17) or (4.3.18) which are proved for the $PH/M/1/r$ queueing system by the same reasoning as for the

$M/PH/1/r$ system. Another, equivalent expression for the loss probability can also be obtained by rigorous reasoning, but we set this aside for Section 4.5, where a more general $PH/PH/1/r$ system with non-Poisson flow will be considered along with the problem of customer waiting time. The stationary mean number of customers in the system and the stationary queue length obey the formulas

$$N = \sum_{k=1}^{R} k p_{\cdot,k}, \qquad Q = \sum_{k=2}^{R} (k-1) p_{\cdot,k}$$

respectively.

4.4. $M/PH/1/r$ system with server vacations and flow dependent on the queue state

We demonstrate now how the approach used in the last section to analyse the $M/PH/1/r$ queueing system can be extended to the case where the server is allowed to turn off for a random time or 'go on vacation,' and the arrival flow is dependent on the queue length. Solution here is no more matrix-geometric, but its notation, nevertheless, is multiplicative. We again assume that the arrival flow is Poisson and has intensity λ, but let additionally the customer arrived which finds k other queued customers, $k = 0, \ldots, r-1$, be accepted by the queueing system with the probability f_k, $0 < f_k \leq 1$, or be rejected and lost with the probability $1 - f_k$. It is reasonable that the customer is lost also if at the instant of its arrival the system is full, that is, the buffer has r customers. Customer service times are distributed according to distribution function $B_1(x)$. Upon emptying the system, the server may either turn off or stay empty waiting for a new customer with the probabilities θ_1 and $\bar{\theta}_1 = 1 - \theta_1$ respectively. The duration of server vacation has the distribution function $B_0(x)$. At the end of the vacation, the server again turns off with the probability θ_0 for a random time with the distribution function $B_0(x)$ or remains empty and waits for a new customer with the complementary probability $\bar{\theta}_0 = 1 - \theta_0$. If at the end of the vacation there is at least one customer in the system, then the server resumes operation. This discipline of server vacation is referred to as *exhaustive*.

We consider the case where both distribution functions $B_i(x)$, $i = 0, 1$, are *PH*-distributions

$$B_i(x) = 1 - \boldsymbol{\beta}_i^T e^{M_i x} \mathbf{1}, \quad x > 0, \quad \boldsymbol{\beta}_i^T \mathbf{1} = 1,$$

admitting the irreducible *PH*-representations $(\boldsymbol{\beta}_i^T, M_i)$ of order m_i, $i = 0, 1$.

4.4.1. Equilibrium equations

We proceed to constructing the mathematical model of the above queueing system. Its stochastic behaviour can be analysed with the use of the homogeneous Markov process $\{\eta(t), t \geq 0\}$ with the state set

$$\mathcal{X} = \{(0); (i, k, j), i = 0, 1, j = 1, \ldots, m_i, \qquad k = 0, \ldots, r\},$$

where for some instant of time t the state (0) corresponds to the case of empty system and the server waiting for a customer, the state (i, k, j) means that there are k queued customers and the process of server vacation (for $i = 0$) or operation (for $i = 1$) undergoes phase j. There is no difficulty in proving that the process $\{\eta(t), t \geq 0\}$ is Markov and homogeneous, because, upon considering such states, the system is exponential and the probabilities f_k, $k = 0, \ldots, r-1$, with which the customer is queued after t, depend only on the system state at the arrival time and are independent of the system history before t. Obviously, all states of the process $\{\eta(t), t \geq 0\}$ are communicating, and their number $R(m_0 + m_1) + 1$ is finite. Therefore, the limit probabilities

$$p_0 = \lim_{t\to\infty} \mathsf{P}\{\eta(t) = (0)\}, \qquad p_{ikj} = \lim_{t\to\infty} \mathsf{P}\{\eta(t) = (i, k, j)\}$$

exist, are strictly positive, independent of the initial distribution, and coincide with the stationary probabilities.

Now we write the equilibrium equations which the stationary probabilities p_0 and p_{ikj} satisfy. Taking into account the previous experience gained in constructing the equilibrium equations for various systems—$M/PH/1/r$ (Section 4.3), in particular, where this procedure was considered in detail—we restrict ourselves only to the matrix notation of the equilibrium equations and explain them at the level of the scheme of transitions in small time Δ. We introduce to this end the vectors $\boldsymbol{p}_{ik}^T = (p_{ik1}, \ldots, p_{ikm})$. The stationary distribution $\{p_0, \boldsymbol{p}_{ik}, i = 0, 1, k = 0, \ldots, r\}$ satisfies the equilibrium equations

$$0 = -\lambda_0 p_0 + \sum_{i=0}^{1} \bar{\theta}_i \boldsymbol{p}_{i0}^T \boldsymbol{\mu}_i, \tag{4.4.1}$$

$$\boldsymbol{0}^T = \boldsymbol{p}_{00}^T(-\lambda_0 I + M_0) + \boldsymbol{\beta}_0^T \sum_{i=0}^{1} \theta_i \boldsymbol{p}_{i0}^T \boldsymbol{\mu}_i, \tag{4.4.2}$$

$$\boldsymbol{0}^T = \boldsymbol{p}_{10}^T(-\lambda_0 I + M_1) + \lambda_0 \boldsymbol{\beta}_1^T p_0 + \boldsymbol{\beta}_1^T \sum_{i=0}^{1} \boldsymbol{p}_{i1}^T \boldsymbol{\mu}_i, \tag{4.4.3}$$

$$\boldsymbol{0}^T = \boldsymbol{p}_{0k}^T(-\lambda_k I + M_0) + \lambda_{k-1} \boldsymbol{p}_{0,k-1}^T, \qquad k = 1, \ldots, r-1, \tag{4.4.4}$$

$$\boldsymbol{0}^T = \boldsymbol{p}_{1k}^T(-\lambda_k I + M_1) + \lambda_{k-1} \boldsymbol{p}_{1,k-1}^T + \boldsymbol{\beta}_1^T \sum_{i=0}^{1} \boldsymbol{p}_{i,k+1}^T \boldsymbol{\mu}_i, \qquad k = 1, \ldots, r-1, \tag{4.4.5}$$

$$\boldsymbol{0}^T = \boldsymbol{p}_{ir}^T M_i + \lambda_{r-1} \boldsymbol{p}_{i,r-1}^T, \qquad i = 0, 1, \tag{4.4.6}$$

where $\boldsymbol{\mu}_i = -M_i \boldsymbol{1}$ and $\lambda_k = \lambda f_k$, $k = 0, \ldots, r-1$. Furthermore, the normalisation condition

$$p_0 + \sum_{i=0}^{1} \sum_{k=0}^{r} \boldsymbol{p}_{ik}^T \boldsymbol{1} = 1 \tag{4.4.7}$$

must be satisfied.

We explain the derivation of the equilibrium equations. Fig. 4.6 gives the scheme of transitions between the state subsets of the process $\{\eta(t), t \geq 0\}$ in a small time Δ; of course, the transitions do not occur just into this or that subset, but to the appropriate state defined according to the order defined on the subsets and the matrix notation corresponding to this order. The following notation is used in the figure: the diagonal of the square matrix

$A = (a_{ij})_{i,j=1,\ldots,n}$ is denoted by $A_{\text{diag}} = \text{diag}(a_{11}, a_{22}, \ldots, a_{nn})$ and A^0_{diag} is the matrix A with zeros instead of all its diagonal elements; $\mathcal{X}_0 = \{0\}$ and $\mathcal{X}_{ik} = \{(i, k, j), j = 1, \ldots, m_i\}$, $i = 0, 1$, $k = 0, \ldots, r$. Additionally, the terms $o(\Delta)$ are omitted everywhere.

This scheme helps to explain, for example, the derivation of equation (5) under fixed $k = 1, \ldots, r-1$ (Fig. 4.6e). One goes in the time Δ from the subset $\mathcal{X}_{i,k-1}$ to the subset $\mathcal{X}_{ik}$ due to arrival of a new customer, which occurs with intensity $\lambda_{k-1} = \lambda f_{k-1}$. One also goes to $\mathcal{X}_{1k}$ in the time Δ from the subsets $\mathcal{X}_{0,k+1}$ and $\mathcal{X}_{1,k+1}$, respectively, through server vacation or service of the customer, which is characterised by the vectors of intensities $\boldsymbol{\mu}_0$ and $\boldsymbol{\mu}_1$. In this case, serving of a new customer starts immediately, the initial phase of his service being dependent on the probability vector $\boldsymbol{\beta}_1$. Additionally, there are two cases of not leaving the subset $\mathcal{X}_{1k}$ in the time Δ. In the first case, the current service phase is not completed in the time Δ and no new customer arrives at the system. This case is characterised by the intensities equal to the elements of the principal diagonal of the matrix M_1 but of the opposite sign and intensities $\lambda_k I$ of arrival of new customers. In the second case, the current service phase can change in the time Δ. This situation is characterised by the intensities equal to the off-diagonal elements of the matrix M_1. The rest of the cases shown in Fig. 4.6 a–d, f are explained along the same lines.

The readers can themselves interpret equilibrium equations (4.4.1)–(4.4.5) using the scalar notation of the principle of global balance. Given the probabilities of transition in the time Δ shown in Fig. 4.6, one can immediately cope with this problem.

4.4.2. Matrix-multiplicative solution of the equilibrium equations

As we have seen in Section 4.3, apart from simplifying the equilibrium equations notation, matrices offer great possibilities for solving them analytically also in the matrix form. Below, we extend the approach of Section 4.3 and demonstrate that it is also possible to obtain an explicit matrix solution of equilibrium equations (4.4.1)–(4.4.6) in the form of a product of certain matrices constructed on the basis of the initial (scalar and matrix) parameters of the queueing system which define its traffic.

Prior to solving equilibrium equations (4.4.1)–(4.4.6), we represent them in another form and introduce the matrices

$$N = \begin{pmatrix} M_0 & 0 \\ 0 & M_1 \end{pmatrix}, \qquad N_0 = \begin{pmatrix} M_0 + \boldsymbol{\mu}_0\boldsymbol{\beta}_0^T\theta_0 & 0 \\ \boldsymbol{\mu}_1\boldsymbol{\beta}_0^T\theta_1 & M_1 \end{pmatrix}$$

and the vectors

$$\begin{aligned} \boldsymbol{p}_k^T &= (\boldsymbol{p}_{0k}^T, \boldsymbol{p}_{1k}^T), \quad k = 0, \ldots, r, \qquad & \boldsymbol{\nu} &= -N\mathbf{1}, \\ \boldsymbol{\nu}_0 &= -N_0\mathbf{1}, & \boldsymbol{\beta}^T &= (\mathbf{0}^T, \boldsymbol{\beta}_1^T). \end{aligned}$$

We immediately see that in this notation equilibrium equations (4.4.1)–(4.4.6) take the form

$$0 = -\lambda_0 p_0 + \boldsymbol{p}_0^T\boldsymbol{\nu}_0, \tag{4.4.8}$$

$$\mathbf{0}^T = \boldsymbol{p}_0^T(-\lambda_0 I + N_0) + \lambda_0 p_0 \boldsymbol{\beta}^T + \boldsymbol{p}_1^T\boldsymbol{\nu}\boldsymbol{\beta}^T, \tag{4.4.9}$$

$$\mathbf{0}^T = \boldsymbol{p}_k^T(-\lambda_k I + N) + \lambda_{k-1}\boldsymbol{p}_{k-1}^T + \boldsymbol{p}_{k+1}^T\boldsymbol{\nu}\boldsymbol{\beta}^T, \qquad k = 1, \ldots, r-1, \tag{4.4.10}$$

$$\mathbf{0}^T = \boldsymbol{p}_r^T N + \lambda_{r-1}\boldsymbol{p}_{r-1}^T. \tag{4.4.11}$$

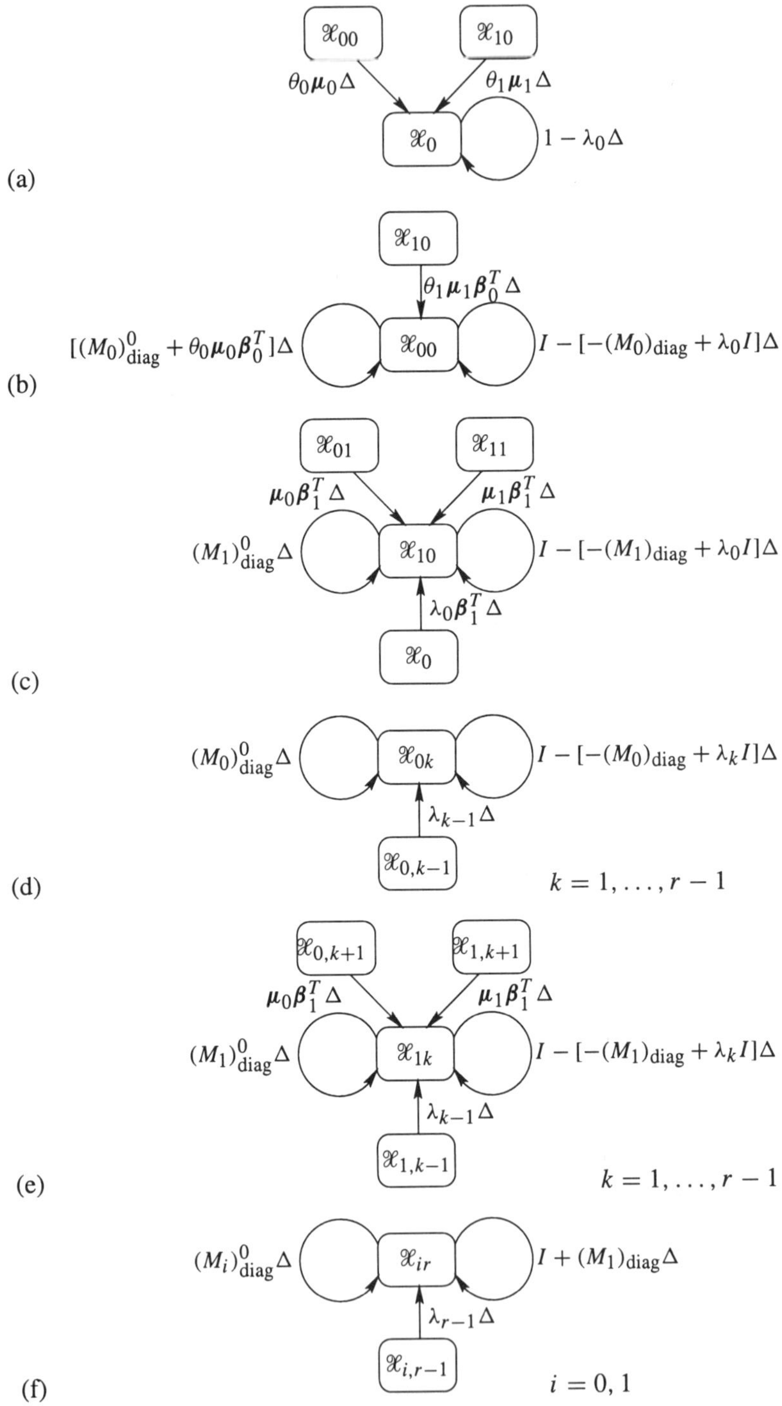
(a)
$\mathscr{X}_{00}$
$\mathscr{X}_{10}$
$\theta_0\boldsymbol{\mu}_0\Delta$
$\theta_1\boldsymbol{\mu}_1\Delta$
$\mathscr{X}_0$
$1-\lambda_0\Delta$
(b)
$\mathscr{X}_{10}$
$\theta_1\boldsymbol{\mu}_1\boldsymbol{\beta}_0^T\Delta$
$[(M_0)^0_{\text{diag}}+\theta_0\boldsymbol{\mu}_0\boldsymbol{\beta}_0^T]\Delta$
$\mathscr{X}_{00}$
$I-[-(M_0)_{\text{diag}}+\lambda_0 I]\Delta$
(c)
$\mathscr{X}_{01}$
$\mathscr{X}_{11}$
$\boldsymbol{\mu}_0\boldsymbol{\beta}_1^T\Delta$
$\boldsymbol{\mu}_1\boldsymbol{\beta}_1^T\Delta$
$(M_1)^0_{\text{diag}}\Delta$
$\mathscr{X}_{10}$
$I-[-(M_1)_{\text{diag}}+\lambda_0 I]\Delta$
$\lambda_0\boldsymbol{\beta}_1^T\Delta$
$\mathscr{X}_0$
(d)
$(M_0)^0_{\text{diag}}\Delta$
$\mathscr{X}_{0k}$
$I-[-(M_0)_{\text{diag}}+\lambda_k I]\Delta$
$\lambda_{k-1}\Delta$
$\mathscr{X}_{0,k-1}$
$k=1,\ldots,r-1$
(e)
$\mathscr{X}_{0,k+1}$
$\mathscr{X}_{1,k+1}$
$\boldsymbol{\mu}_0\boldsymbol{\beta}_1^T\Delta$
$\boldsymbol{\mu}_1\boldsymbol{\beta}_1^T\Delta$
$(M_1)^0_{\text{diag}}\Delta$
$\mathscr{X}_{1k}$
$I-[-(M_1)_{\text{diag}}+\lambda_k I]\Delta$
$\lambda_{k-1}\Delta$
$\mathscr{X}_{1,k-1}$
$k=1,\ldots,r-1$
(f)
$(M_i)^0_{\text{diag}}\Delta$
$\mathscr{X}_{ir}$
$I+(M_1)_{\text{diag}}\Delta$
$\lambda_{r-1}\Delta$
$\mathscr{X}_{i,r-1}$
$i=0,1$

Figure 4.6.

We proceed to solving simultaneous equations (4.4.8)–(4.4.11). First, we observe that the notation in these equations resembles equilibrium equations (4.3.6)–(4.3.11)) for the $M/PH/1/r$ queueing system where the flow intensity depends on the number of customers in the system and the PH-distributions of service time of the first and subsequent customers after the idle period have the PH-representations $(\boldsymbol{\beta}, N_0)$ and $(\boldsymbol{\beta}, N)$ respectively. This similarity, however, is formal, which is due to the fact that, as seen from the probabilistic interpretation of PH-distributions, the PH-representation $(\boldsymbol{\beta}, N)$ is not irreducible. Therefore, service 'lumping' does not make the states of the 'new' service phases communicating. Despite this, generally speaking, fundamental discrepancy between the aforementioned equations, we analyse equilibrium equations (4.4.8)–(4.4.11) using the same approach as in Section 4.3, although we have to prove anew that the matrices arising under this approach are invertible.

So, following the lines of matrix analysis (Section 4.3) of the equilibrium equations, we first obtain the local balance equations. Equation (4.4.8) immediately yields the relation

$$\lambda_0 p_0 = \boldsymbol{p}_0^T \boldsymbol{\nu}_0. \tag{4.4.12}$$

Multiplying equations (4.4.9) and (4.4.10) by the vector $\mathbf{1}$ over $k = 1, \ldots, r$, summing successively the resulting equalities, and using (4.4.12), we arrive at the relations

$$\lambda_k \boldsymbol{p}_k^T \mathbf{1} = \boldsymbol{p}_{k+1}^T \boldsymbol{\nu}, \qquad k = 0, \ldots, r-1. \tag{4.4.13}$$

From the physical viewpoint, (4.4.12) and (4.4.13) could be constructed at once without summing equilibrium equations (4.4.8)–(4.4.11). Indeed, we easily see that (4.4.12) and (4.4.13) account for the balance of probability flows between the state subsets $\mathscr{X}_0 = \{(0)\}$ and $\mathscr{X}_k$, $k = 1, \ldots, r$, and subsets $\mathscr{X}_l$, $l = 0, \ldots, k$, and $\mathscr{X}_n$, $n = k+1, \ldots, r$, $k = 1, \ldots, r$, where $\mathscr{X}_k = \bigcup_{i=0}^{1} \mathscr{X}_{ik}$ and hence are local balance equations.

Substituting (4.4.12) and (4.4.13) into (4.4.9) and (4.4.10) and introducing the matrices $\tilde{N}_0 = \lambda_0 I - N_0 - \lambda_0 \mathbf{1}\boldsymbol{\beta}^T$ and $\tilde{N}_k = \lambda_k I - N - \lambda_k \mathbf{1}\boldsymbol{\beta}^T$, we obtain

$$\begin{aligned} \boldsymbol{p}_0^T \tilde{N}_0 &= \lambda_0 p_0 \boldsymbol{\beta}^T, \\ \boldsymbol{p}_k^T \tilde{N}_k &= \lambda_{k-1} \boldsymbol{p}_{k-1}^T, \qquad k = 1, \ldots, r-1. \end{aligned} \tag{4.4.14}$$

It follows immediately from (4.4.11) that

$$\boldsymbol{p}_r^T N = -\lambda_{r-1} \boldsymbol{p}_{r-1}^T. \tag{4.4.15}$$

We see from (4.4.14) and (4.4.15) that if the matrices $\tilde{N}_0$, $\tilde{N}_k$, $k = 1, \ldots, r-1$, and N are non-degenerate, then the vectors $\boldsymbol{p}_k$, $k = 0, \ldots, r$, admit a recursive representation in terms of the probability p_0 found from the normalisation condition.

Non-degeneracy of the matrix N follows from irreducibility of the PH-representations $(\boldsymbol{\beta}_i, M_i)$ since there exist M_i^{-1}, $i = 0, 1$; then the matrix inverse to N is

$$N^{-1} = \operatorname{diag}(M_0^{-1}, M_1^{-1}).$$

Irreducibility of the initial PH-representations is, however, insufficient for $\tilde{N}_0$ and $\tilde{N}_k$ to be non-degenerate, and one has to impose on the parameters θ_i, $i = 0, 1$, some additional conditions formulated in the lemma below.

LEMMA 4.4.1. *Let the PH-representations* $(\boldsymbol{\beta}_i, M_i)$, $i = 0, 1$, *be irreducible,* $\theta_1 \neq 1$, *or* $\theta_1 = 1$, $\theta_0 \neq 1$. *Then* $\tilde{N}_k$, $k = 0, \ldots, r-1$, *are non-degenerate.*

PROOF. Assume that $D_0 = \lambda_0 I - N_0$, $D_k = \lambda_k I - N$, $k = 1, \ldots, r-1$, and $\boldsymbol{u}_k = \lambda_k \mathbf{1}$, $k = 0, \ldots, r-1$, and, as usual, let $\beta_i(s)$ stand for the Laplace–Stieltjes transform of $B_i(x)$, $i = 0, 1$. Then the matrices $\tilde{N}_k$, $k = 0, \ldots, r-1$, admit the representation $\tilde{N}_k = D_k - \boldsymbol{u}_k \boldsymbol{\beta}^T$. By (4.1.10) where $D = D_k$ and $\boldsymbol{\lambda} = -\boldsymbol{u}_k$, we conclude that if there exists D_k^{-1} and

$$\boldsymbol{\beta}^T D_k^{-1} \boldsymbol{u}_k \neq 1, \tag{4.4.16}$$

then the inverse matrix $\tilde{N}_k^{-1}$ exists and is equal to

$$\tilde{N}_k^{-1} = D_k^{-1} + \frac{D_k^{-1} \boldsymbol{u}_k \boldsymbol{\beta}^T D_k^{-1}}{1 - \boldsymbol{\beta}^T D_k^{-1} \boldsymbol{u}_k}. \tag{4.4.17}$$

The matrix D_k, $k = 0, \ldots, r-1$, must be non-degenerate (see the proof of (4.1.10)). Let us consider the matrix $\tilde{N}_k$ for a fixed $k = 1, 2, \ldots, r-1$ and prove that (4.4.16) holds for these k. Indeed,

$$\boldsymbol{\beta}^T D_k^{-1} \boldsymbol{u}_k = \lambda_k (\mathbf{0}^T, \boldsymbol{\beta}_1^T (\lambda_k I - M_1)^{-1}) \mathbf{1} = \lambda_k \boldsymbol{\beta}_1^T (\lambda_k I - M_1)^{-1} \mathbf{1}.$$

Therefore,

$$1 - \boldsymbol{\beta}^T D_k^{-1} \boldsymbol{u}_k = 1 - \lambda_k \boldsymbol{\beta}_1^T (\lambda_k I - M_1)^{-1} \mathbf{1} = \beta_1(\lambda_k) > 0.$$

Hence, (4.4.16) is true for $k = 1, \ldots, r-1$. Therefore, the matrices $\tilde{N}_k^{-1}$, $k = 1, \ldots, r-1$, exist and are of the form (4.4.17).

Let us establish invertibility of the matrix $\tilde{N}_0$. To do this, we have to show, in view of non-degeneracy of the matrix D_0, that inequality (4.4.16) is valid for $k = 0$ as well. With this aim in view, we first find the explicit form of D_0^{-1} using the Frobenius formula (Gantmacher, 1959):

$$D_0^{-1} = (\lambda_0 I - N_0)^{-1} = \begin{pmatrix} (Q - \boldsymbol{\mu}_0 \boldsymbol{\beta}_0^T \theta_0)^{-1} & 0 \\ C & (\lambda_0 I - M_1)^{-1} \end{pmatrix},$$

where

$$Q = \lambda_0 I - M_0, \qquad C = (\lambda_0 I - M_1)^{-1} \boldsymbol{\mu}_1 \boldsymbol{\beta}_0^T \theta_1 (Q - \boldsymbol{\mu}_0 \boldsymbol{\beta}_0^T \theta_0)^{-1}.$$

Then

$$\begin{aligned} \boldsymbol{\beta}^T D_0^{-1} \boldsymbol{u}_0 &= (\boldsymbol{\beta}_1^T C, \boldsymbol{\beta}_1^T (\lambda_0 I - M_1)^{-1}) \boldsymbol{u}_0^T \\ &= \boldsymbol{\beta}_1^T (\lambda_0 I - M_1)^{-1} \boldsymbol{\mu}_1 \boldsymbol{\beta}_0^T \theta_1 (Q - \boldsymbol{\mu}_0 \boldsymbol{\beta}_0^T \theta_0)^{-1} \lambda_0 \mathbf{1} + \boldsymbol{\beta}_1^T (\lambda_0 I - M_1)^{-1} \lambda_0 \mathbf{1}. \end{aligned}$$

Since the matrix Q is non-degenerate and, moreover,

$$\theta_0 \boldsymbol{\beta}_0^T Q^{-1} \boldsymbol{\mu}_0 = \theta_0 \beta_0(\lambda_0) \neq 1,$$

the matrix $Q - \mu_0 \beta_0^T \theta_0$ is non-degenerate and its inverse matrix is

$$(Q - \mu_0 \beta_0^T \theta_0)^{-1} = Q^{-1} + \frac{Q^{-1} \mu_0 \beta_0^T Q^{-1}}{1 - \theta_0 \beta_0(\lambda_0)} \theta_0.$$

Let us prove that (4.4.16) holds true for $k = 0$. Indeed,

$$\begin{aligned} 1 - \beta^T D_0^{-1} u_0 &= \beta_1(\lambda_0) - \beta_1(\lambda_0)\theta_1 \left[1 - \beta_0(\lambda_0) + \theta_0 \frac{\beta_0(\lambda_0)(1 - \beta_0(\lambda_0))}{1 - \theta_0 \beta_0(\lambda_0)} \right] \\ &= \beta_1(\lambda_0) - \theta_1 \beta_1(\lambda_0)(1 - \beta_0(\lambda_0))\beta_1(\lambda_0) \left[1 + \frac{\theta_0 \beta_0(\lambda_0}{1 - \beta_0(\lambda_0)} \right] \\ &= \beta_1(\lambda_0) \left[1 - \theta_1 \beta_0(\lambda_0)) \frac{1 - \beta_0(\lambda_0)}{1 - \theta_0 \beta_0(\lambda_0)} \right]. \end{aligned}$$

Hence (4.4.16) is valid under the hypotheses of the lemma, which completes the proof. □

Setting

$$w_0^T = \lambda_0 \beta^T \tilde{N}_0^{-1}, \quad W_k = \lambda_{k-1} \tilde{N}_k^{-1}, \quad k = 1, \ldots, r-1, \quad W_r = -\lambda_{r-1} N^{-1},$$

we finally find from (4.4.14) and (4.4.15) that if the hypotheses of Lemma 4.4.1 are fulfilled, then the stationary distribution $\{p_0, p_k, k = 0, \ldots, r\}$ admits the representation

$$p_k^T = p_0 w_0^T \prod_{i=1}^{k} W_i, \qquad k = 0, \ldots, r, \tag{4.4.18}$$

where $\prod_{i=1}^{0} \equiv 1$. In order to find p_0, we make use of the normalisation condition

$$p_0 + \sum_{k=0}^{r} p_k^T \mathbf{1} = 1$$

which, together with (4.4.18), allows us to obtain an explicit matrix expression for the probability p_0. This expression, however, can serve only as an ornamentation—as the explicit expression (4.1.14) for p_0 in the case of the $M/H_m/1/r$ system or a similar expression for the $M/PH/1/r$ queueing system—because, as said in Sections 4.1 and 4.3, in practical calculations these formulas are not used.

Consider now the special case of $f_k = f$, $k = 0, \ldots, r-1$. Then $\lambda_k = \lambda f$, $k = 0, \ldots, r-1$. We assume that $\tilde{N} = \tilde{N}_k$, $k = 1, \ldots, r-1$. Under the hypotheses of Lemma 4.4.1, we conclude for $f_k = f, k = 0, \ldots, r-1$, that

$$p_k^T = \begin{cases} p_0 w_0^T W^k, & k = 0, \ldots, r-1; \\ p_0 w_0^T W^{r-1} W_r, & k = r, \end{cases} \tag{4.4.19}$$

where $W = \lambda f \tilde{N}^{-1}$.

It goes without saying that the explicit matrix expressions have—mostly analytical—advantages. In practical calculations, however, the recurrence formulas are preferred. After inversion of the matrices N, $\tilde{N}_0$ and $\tilde{N}_k$, formulas (4.4.14) and (4.4.15) can be taken as such. We offer to the readers to derive from (4.4.14) and (4.4.15) recurrence relations for the vectors $\boldsymbol{p}_{0k}$ and $\boldsymbol{p}_{1k}$ using the explicit form (4.4.17) of the inverse matrix $\tilde{N}_k^{-1}$ and the Frobenius formula.

We conclude this section by traditional writing out of expressions for the key performance indices of the queueing system under consideration. Assume that

$$p_{ik} = \boldsymbol{p}_{ik}^T \mathbf{1}, \quad p_k = p_{0k} + p_{1,k-1}, \quad k = 1, \dots, r, \quad p_R = p_{1r},$$

and also introduce

$$q_0 = p_{\cdot,0} + p_0, \quad q_k = p_{\cdot,k}, \quad k = 1, \dots, r.$$

The quantities p_k, $k = 0, \dots, r+1$, define the stationary distribution of the number of customers in the system, and q_k, $k = 0, \dots, r$, define the stationary distribution of the queue length. We observe that p_0 is the probability that the system is empty and the server waits for a new customer. We denote by p_S and p_W, respectively, the stationary probabilities that the server is busy or deactivated. Then

$$p_S = p_{1,\cdot}, \qquad p_W = p_{0,\cdot}.$$

Let μ_1 be the intensity of customer serving at the server,

$$\mu_1^{-1} = -\boldsymbol{\beta}^T{}_1 M_1^{-1} \mathbf{1},$$

and let π denote, as before, the customer loss probability. Using the balance of the intensities of the accepted and served flows, we see that the intensity of the departing flow of served customers is

$$\lambda_D = \lambda(1 - \pi) = \mu_1 p_{1,\cdot}. \tag{4.4.20}$$

Another expression for the right-hand side of the equality obtained can be found by summing (4.4.5) over $k = 1, 2, \dots, r-1$ and then adding the result to (4.4.3) and (4.4.6) for $i = 1$. Multiplying the obtained equalities from the right by $M_1^{-1}\mathbf{1}$, after straightforward simplifying transformation we obtain

$$\mu_1 p_{1,\cdot} = \boldsymbol{p}_{1,\cdot}^T \boldsymbol{\mu}_1.$$

Thus, (4.4.2) admits the representation

$$\lambda_D = \lambda(1 - \pi) = \boldsymbol{p}_{1,\cdot}^T \boldsymbol{\mu}_1 \tag{4.4.21}$$

which gives rise to the following expression for the probability:

$$\pi = 1 - \frac{p_S}{\rho_1}, \tag{4.4.22}$$

where $\rho_1 = \lambda/\mu_1$. Relation (4.4.21) may be used to check calculations. The probability π can also be obtained from the following reasoning. By virtue of the initial assumptions, the probability that the system loses a customer which finds k other customers in the buffer is equal to $1 - f_k$, $k = 0, \dots, r-1$, and $1 - f_r = 1$. Then, because the input customer flow is Poisson, we obtain with the use of the formula of total probability

$$\pi = \sum_{k=0}^{r} (1 - f_k) q_k.$$

This formula also proves useful for checking calculations.

4.4.3. Stationary distribution of the waiting time for the FCFS discipline

We assume here that the customers are served in the order of their arrival and that the same approach as in Section 4.3.3 can be used for analysing the waiting time. We denote by $W(t)$ the stationary distribution of the time of waiting for service of an arbitrary accepted customer, and by $W(t \mid x)$, the conditional distribution function of the waiting time of a customer who arrives at the instant $t - 0$ when the system is in the state $x \in \mathscr{X}$. Let $\omega(s)$ and $\omega(s \mid x)$ denote, respectively, the Laplace–Stieltjes transforms $W(t)$ and $W(t \mid x)$. We introduce the vector $\mathbf{W}_{ik}^T(t) = (W(t \mid (i,k,1)), W(t \mid (i,k,2)), \dots, W(t \mid (i,k,m_i)))$, and denote its Laplace–Stieltjes transform by $\boldsymbol{\omega}_{ik}(s)$.

The distribution function of the waiting time of an arbitrary accepted customer is $(1-\pi)W(t)$. Then, using the formula of total probability and taking into account that the arriving flow is Poisson, we obtain

$$(1-\pi)W(t) = f_0 p_0 W(t \mid (0)) + \sum_{i=0}^{1} \sum_{k=0}^{r-1} f_k \boldsymbol{p}_{ik}^T \mathbf{W}_{ik}(t).$$

Now, applying to both sides of this equality the Laplace–Stieltjes transformation and dividing by $1 - \pi$, we obtain

$$\omega(s) = \frac{1}{1-\pi} \left[f_0 p_0 \omega(s \mid (0)) + \sum_{i=0}^{1} \sum_{k=0}^{r-1} f_k \boldsymbol{p}_{ik}^T \boldsymbol{\omega}_{ik}(s) \right]. \tag{4.4.23}$$

Now we observe that if at the instant $t - 0$ of arrival of a customer the system is in the state (0), then the customer is immediately served. Therefore, $\omega(s \mid (0)) = 1$. If the customer finds the system in the state (i,k,j), then he is served as soon as the other customer completes the service (for $i = 1$) or the server is enabled (for $i = 0$) and k customers from the buffer have been served. Since the customer service time and that of server vacation are independent, we obtain

$$\boldsymbol{\omega}_{ik}(s) = (sI - M_i)^{-1} \boldsymbol{\mu}_i \beta_1^k(s), \qquad i = 0, 1, k = 0, \dots, r-1.$$

Together with (4.4.23), this reasoning yields the relation

$$\omega(s) = \frac{1}{1-\pi} \left[f_0 p_0 + \sum_{i=0}^{1} \sum_{k=0}^{r-1} f_k \boldsymbol{p}_{ik}^T (sI - M_i)^{-1} \boldsymbol{\mu}_i \beta_1^k(s) \right], \tag{4.4.24}$$

where

$$\beta_1(s) = \boldsymbol{\beta}_1^T (sI - M_1)^{-1} \boldsymbol{\mu}_1.$$

Recalling that $w = -\omega'(0)$, we obtain the expression of the mean waiting time

$$w = \frac{1}{1-\pi}\left(\frac{1}{\mu_1}\sum_{k=1}^{r-1} k f_k q_k - \sum_{i=0}^{1}\sum_{k=0}^{r-1} f_k \boldsymbol{p}_{ik}^T M_i^{-1}\mathbf{1}\right). \tag{4.4.25}$$

For the Laplace–Stieltjes transform $\varphi(s)$ of the customer sojourn time in the system, we obtain the obvious expression

$$\varphi(s) = \omega(s)\beta_1(s).$$

Then the mean customer sojourn time v is

$$v = w + \frac{1}{\mu_1}.$$

We also observe that w and v can be obtained as usual from the Little formulas:

$$\lambda(1-\pi)w = Q,$$
$$\lambda(1-\pi)v = N,$$

where

$$Q = \sum_{k=1}^{r} k q_k$$

is the stationary mean queue length and

$$N = \sum_{k=1}^{R} k p_k$$

is the stationary mean number of customers in the system.

4.5. $PH/PH/1/r$ system

After reading the preceding sections of this chapter, the reader may ask whether the requirement that the arrival flow must be Poisson is a key factor for representing the stationary queue distribution in the matrix-geometric form. The answer is negative. In what follows, we extend the results of Sections 4.1 and 4.3 to the case of a recurrent arrival flow defined by a phase-type distribution function under rather weak constraints on parameters of the corresponding PH-representation.

Like in Sections 4.1 and 4.3, we consider here a single-server queueing system with a buffer of capacity r, $2 \le r < \infty$, and assume that the customer flow is recurrent and has the phase-type distribution function $A(x)$:

$$A(x) = 1 - \boldsymbol{\alpha}^T e^{\Lambda x}\mathbf{1}, \quad x > 0, \quad \boldsymbol{\alpha}^T\mathbf{1} = 1,$$

which admits an irreducible PH-representation $(\boldsymbol{\alpha}, \Lambda)$ of order l. The customer service times have the phase-type distribution function

$$B(x) = 1 - \boldsymbol{\beta}^T e^{Mx}\mathbf{1}, \quad x \ge 0, \quad \boldsymbol{\beta}^T\mathbf{1} = 1,$$

which also admits an irreducible PH-representation of order m.

For the distribution functions $A(x)$ and $B(x)$, we require, additionally, that for all eigenvalues $\{\gamma_s\}$ of the matrix Λ the condition $\beta(-\gamma_s) \neq 0$ be satisfied for the Laplace–Stieltjes transform $\beta(s)$ of the distribution function $B(x)$.

The customer arrived at the full buffer is lost and does not return to the system. The customers are taken from the queue using any of the fixed disciplines FCFS, LCFS, and RANDOM.

4.5.1. Equilibrium equations

We rely again on the probabilistic interpretation of customer arrivals and service processes in order to describe the queueing system under consideration by some homogeneous Markov process $\{\eta(t), t \geq 0\}$ with the state set

$$\mathcal{X} = \bigcup_{k=0}^{R} \mathcal{X}_k,$$

where, as before, $R = r + 1$ is the system capacity and

$$\mathcal{X}_0 = \{(i, 0), i = 1, \ldots, l\},$$
$$\mathcal{X}_k = \{(i, k, j), i = 1, \ldots, l, j = 1, \ldots, m\}, \quad k = 1, \ldots, R.$$

Here, for an arbitrary instant of time t the state $(i, 0)$ corresponds to the case where the system is empty and the arrival process is in phase i (in the ith node of the network interpreting this process) and the state (i, k, j) reflects the situation where there are k customers in the system and the arrival and service processes are in the phases i and j, respectively (in the ith and jth nodes of the corresponding networks).

As follows from irreducibility of the initial PH-representations, all states of the process $\{\eta(t), t \geq 0\}$ are communicating. Therefore, the process $\{\eta(t), t \geq 0\}$ with a finite number $(Rm + 1)l$ of states is ergodic; then the limit probabilities

$$p_{i0} = \lim_{t\to\infty} \mathbf{P}\{\eta(t) = (i, 0)\}, \qquad p_{ikj} = \lim_{t\to\infty} \mathbf{P}\{\eta(t) = (i, k, j)\}$$

are strictly positive, independent of the initial distribution, and coincide with the stationary probabilities.

We assume that, having mastered the preceding sections of this chapter, the reader gained a sufficient experience in writing the equilibrium equations for the queueing systems described by the phase-type distributions and also acquired some skill in writing these equations in the matrix form. Therefore, we immediately write the equilibrium equations in the matrix form and leave it to the readers to write its scalar analogue. With this aim in mind, we introduce the vectors

$$\boldsymbol{p}_0^T = (p_{10}, \ldots, p_{l0}), \qquad \boldsymbol{p}_k^T = (p_{1k1}, \ldots, p_{1km}, \ldots, p_{lk1}, p_{lk2}, \ldots, p_{lkm}).$$

The stationary probabilities $\{\boldsymbol{p}_k, k = 0, \ldots, R\}$ form a unique solution of the simultaneous equilibrium equations

$$\mathbf{0}^T = \boldsymbol{p}_0^T \Lambda + \boldsymbol{p}_1^T (I \otimes \boldsymbol{\mu}), \tag{4.5.1}$$

$$\mathbf{0}^T = \boldsymbol{p}_0^T (\boldsymbol{\lambda\alpha}^T \otimes \boldsymbol{\beta}^T) + \boldsymbol{p}_1^T (\Lambda \oplus M) + \boldsymbol{p}_2^T (I \otimes \boldsymbol{\mu\beta}^T), \tag{4.5.2}$$

$$\mathbf{0}^T = \boldsymbol{p}_{k-1}^T (\boldsymbol{\lambda\alpha}^T \otimes I) + \boldsymbol{p}_k^T (\Lambda \oplus M) + \boldsymbol{p}_{k+1}^T (I \otimes \boldsymbol{\mu\beta}^T), \qquad k = 2, \ldots, r, \tag{4.5.3}$$

$$\mathbf{0}^T = \boldsymbol{p}_r^T (\boldsymbol{\lambda\alpha}^T \otimes I) + \boldsymbol{p}_R^T (\Lambda \oplus M + \boldsymbol{\lambda\alpha}^T \otimes I) \tag{4.5.4}$$

with the normalisation condition

$$\sum_{k=0}^{R} \boldsymbol{p}_k^T \mathbf{1} = 1, \tag{4.5.5}$$

where $\boldsymbol{\lambda} = -\Lambda\mathbf{1}$, $\boldsymbol{\mu} = -M\mathbf{1}$, $A \otimes B$ is the Kronecker product of matrices A and B, and $A \oplus B = A \otimes I + I \otimes B$ is the Kronecker sum of matrices A and B.

As mentioned, we do not go in particulars of the derivation of equilibrium equations and just explain it by means of the scheme of transitions of the process $\{\eta(t), t \geq 0\}$ in the interval $(t, t + \Delta)$, where Δ is a small time interval (Fig. 4.7). As in Section 4.4, we agree to denote by A_{diag} the matrix $A_{\text{diag}} = \text{diag}(a_{11}, a_{22}, \ldots, a_{nn})$ for the square matrix $A = (a_{ij})_{i,j=1,\ldots,n}$ and by A_{diag}^0, the matrix A where all elements on the principal diagonal are replaced by zeros. We also agree to simplify the notation by omitting the symbol $o(\Delta)$ in Fig. 4.7.

We begin with explaining the derivation of equation (4.5.3) for fixed $k = 2, 3, \ldots, r$. In the system, one goes in the time Δ from the subset $\mathscr{X}_{k-1}$ to the state subset $\mathscr{X}_k$ with k customers due to arrival of a new customer which occurs with intensity characterised by the vector $\boldsymbol{\lambda}$. In doing so, generation of the new customer begins immediately and the initial phase (node of the interpreting network) of his generation is defined by the probability vector $\boldsymbol{\alpha}$. Furthermore, one goes to the subset $\mathscr{X}_k$ in the time Δ from the subset $\mathscr{X}_{k+1}$ due to customer service completion with intensity defined by the vector $\boldsymbol{\mu}$. We must also keep in mind that the initial service phase of the next customer is chosen according to the probability vector $\boldsymbol{\beta}$. Finally, it is possible not to leave the set $\mathscr{X}_k$ in the time Δ. This can be done in two ways. In the first case, the current phases of customer generation and service are not completed in the time Δ. This possibility is reflected by the intensities equal to the elements of the principal diagonal of the matrix $\Lambda \oplus M$ with the opposite sign. In the second case, the current phases of customer generation and service can change in this time. This situation is reflected by the intensities equal to the off-diagonal elements of the matrix $\Lambda \oplus M$.

Equation (4.5.2) is derived with account for the fact that a customer whose generation is complete is immediately sent to the server and chooses the initial phase according to the vector $\boldsymbol{\beta}$. The situations which are shown in Fig. 4.7a and 4.7d and underlie derivation of equations (4.5.1) and (4.5.4) are explained in a similar way, and we leave it to the readers. Moreover, we recommend them to obtain equations (4.5.1)–(4.5.4) on the basis of the principle of global balance.

4.5.2. Matrix-geometric solution of the equilibrium equations

Prior to solving directly the equilibrium equations, we carry out their analysis, which, in fact, provides the desired result, the matrix-geometric form for the stationary queue distribution. We first explain the idea of this analysis and recall that we managed to reduce the

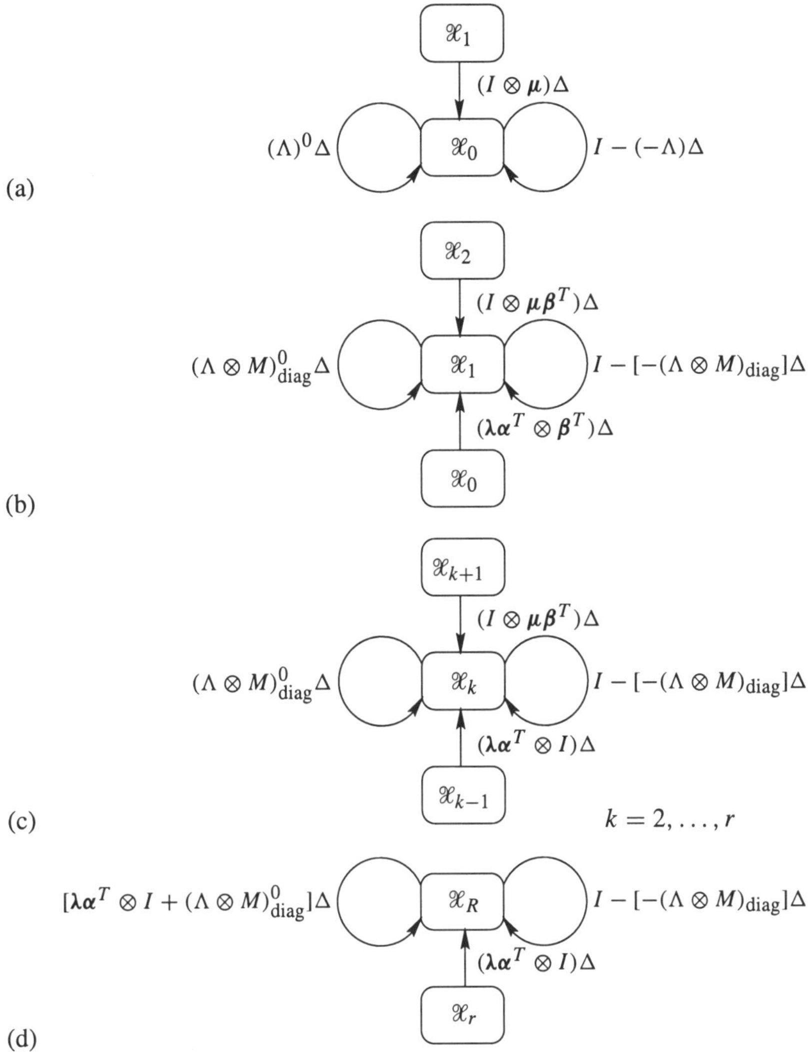

Figure 4.7.

order of recurrence of the matrix equation from two to one for the $M/PH/1/r$ queueing system by removing the highest term with the use of equations of local balance only; then it was not hard to 'bootstrap' the resulting recurrence relation. In its pure form, this does not work for the given queueing system, but the common sense suggests that one could apply this technique to the recurrence matrix relation (4.5.3) of the second order to get rid, successively, of the highest and minor terms, and then use the equations of local balance to obtain the recurrence relation of the first order. We will realise this idea in what follows.

First, we obtain the equations of the local balance for which end we multiply both sides of equality (4.5.1) from the right by the vector $\mathbf{1}$ and equalities (4.5.2)–(4.5.4) by the vector

$\mathbf{1} \otimes \mathbf{1}$ (in the scalar notation, this is equivalent to summing each equation in p_{ikj} over all values of the components i and j), and then we successively sum the resulting equalities to obtain the relations

$$\begin{aligned} \boldsymbol{p}_0^T \boldsymbol{\lambda} &= \boldsymbol{p}_1^T (\mathbf{1} \otimes \boldsymbol{\mu}), \\ \boldsymbol{p}_k^T (\boldsymbol{\lambda} \otimes \mathbf{1}) &= \boldsymbol{p}_{k+1}^T (\mathbf{1} \otimes \boldsymbol{\mu}), \qquad k = 1, \dots, r, \end{aligned} \tag{4.5.6}$$

which are the local balance equations for the queueing system under consideration. Naturally, these equalities can also be obtained for each fixed $k = 0, 1, \dots, r$ from the balance of the flows of probabilities circulating between the groups of the subsets $\mathscr{X}_0, \mathscr{X}_1, \dots, \mathscr{X}_k$ and $\mathscr{X}_{k+1}, \mathscr{X}_{k+2}, \dots, \mathscr{X}_R$.

Multiplying both sides of equalities (4.5.2) and (4.5.3) from the right by the matrix $I \otimes (\mathbf{1}\boldsymbol{\beta}^T - I)$ and taking (4.5.1) into account, we obtain

$$\boldsymbol{p}_1^T \widetilde{M} = \boldsymbol{p}_1^T (I \otimes \boldsymbol{\mu}\boldsymbol{\beta}^T) = -\boldsymbol{p}_0^T (\Lambda \otimes \boldsymbol{\beta}^T), \tag{4.5.7}$$

$$\boldsymbol{p}_k^T \widetilde{M} = \boldsymbol{p}_{k-1}^T (\boldsymbol{\lambda}\boldsymbol{\alpha}^T \otimes I) + \boldsymbol{p}_k^T [(I - \mathbf{1}\boldsymbol{\alpha}^T) \otimes \boldsymbol{\mu}\boldsymbol{\beta}^T], \qquad k = 2, \dots, r, \tag{4.5.8}$$

where

$$\widetilde{M} = \Lambda \otimes (\mathbf{1}\boldsymbol{\beta}^T - I) - I \otimes M.$$

Finally, we multiply both sides of (4.5.2) and (4.5.3) from the right by the matrix $(\mathbf{1}\boldsymbol{\alpha}^T - I) \otimes I$ and obtain

$$\boldsymbol{p}_k^T \tilde{\Lambda} = \boldsymbol{p}_k^T (\boldsymbol{\lambda}\boldsymbol{\alpha}^T \otimes I) + \boldsymbol{p}_{k+1}^T [(I - \mathbf{1}\boldsymbol{\alpha}^T) \otimes \boldsymbol{\mu}\boldsymbol{\beta}^T], \qquad k = 1, \dots, r, \tag{4.5.9}$$

where

$$\tilde{\Lambda} = (\mathbf{1}\boldsymbol{\alpha}^T - I) \otimes M - \Lambda \otimes I.$$

With the use of (4.5.6), we easily see from (4.5.7)–(4.5.9) and (4.5.4) that the recurrence relations

$$\boldsymbol{p}_1^T \widetilde{M} = -\boldsymbol{p}_0^T (\Lambda \otimes \boldsymbol{\beta}^T), \tag{4.5.10}$$

$$\boldsymbol{p}_k^T \widetilde{M} = \boldsymbol{p}_{k-1}^T \tilde{\Lambda}, \qquad k = 2, \dots, r, \tag{4.5.11}$$

$$\boldsymbol{p}_R^T \tilde{\widetilde{M}} = -\boldsymbol{p}_r^T (\boldsymbol{\lambda}\boldsymbol{\alpha}^T \otimes I), \tag{4.5.12}$$

where

$$\tilde{\widetilde{M}} = \Lambda \oplus M + \boldsymbol{\lambda}\boldsymbol{\alpha}^T \otimes I,$$

are valid.

Lemma 2.8.5 allows us to assert that under the initial assumptions about the *PH*-distributions $A(x)$ and $B(x)$ the matrix $\widetilde{M}$ is non-degenerate. It follows from the definition of the *PH*-distribution that the matrix $\tilde{\widetilde{M}}$ is non-degenerate as well. Therefore, equations (4.5.10)–(4.5.12) can be solved recursively, and all unknown vectors $\boldsymbol{p}_k$, $k = 1, \dots, R$ can be expressed in terms of the vector $\boldsymbol{p}_0$. To formulate the final result, we introduce the matrices $W_0 = -(\Lambda \otimes \boldsymbol{\beta}^T)\widetilde{M}^{-1}$, $W = \tilde{\Lambda}\widetilde{M}^{-1}$, $W_r = -(\boldsymbol{\lambda}\boldsymbol{\alpha}^T \otimes I)\tilde{\widetilde{M}}^{-1}$, and $Z = W_0 W^{r-1}(W_r \widetilde{M} - \tilde{\Lambda})(I \otimes \mathbf{1})$. Now the final result stated in the title of this section can be formulated, in view of its importance and bulky proof, as a theorem.

THEOREM 4.5.1. *If the PH-representations* $(\boldsymbol{\alpha}, \Lambda)$ *and* $(\boldsymbol{\beta}, M)$ *are irreducible and the condition* $\beta(-\gamma_s) \neq 0$ *is satisfied for all eigenvalues* $\{\gamma_s\}$ *of the matrix* Λ, *then the stationary distribution* $\{\boldsymbol{p}_k,\ k = 0, \dots, R\}$ *of the state probabilities of the* $PH/PH/1/r$ *queueing system is admits the representation as the truncated matrix progression*

$$\boldsymbol{p}_k^T = \begin{cases} \boldsymbol{p}_0^T W_0 W^{k-1}, & k = 1, \dots, r; \\ \boldsymbol{p}_0^T W_0 W^{r-1} W_r, & k = R, \end{cases} \tag{4.5.13}$$

where the vector $\boldsymbol{p}_0$ *is uniquely determined by the simultaneous equations*

$$\boldsymbol{p}_0^T Z = \mathbf{0}^T \tag{4.5.14}$$

and the normalisation condition (4.5.5).

PROOF. It follows from the definition of the matrices W, W_0, W_r, and (4.5.10)–(4.5.12) that the stationary distribution of the process $\{\eta(t), t \geq 0\}$ is of the form (4.5.13). However, transformations reducing the equilibrium equations (4.5.1)–(4.5.4) to (4.5.10)–(4.5.12) are not equivalent. Therefore, in order to make sure that expressions (4.5.13) define a solution of the equilibrium equations, we have to substitute them into equations (4.5.1)–(4.5.4). Obviously, equation (4.5.4) is identically satisfied on substitution of (4.5.13) into it. Before turning to the rest of equations, we present some relations which follow from the definition of the matrices $\tilde{\Lambda}$ and $\widetilde{M}$ and will be used below in the proof. First,

$$\boldsymbol{\lambda\alpha}^T \otimes I = \tilde{\Lambda}(\mathbf{1}\boldsymbol{\alpha}^T + I),\ I \otimes \boldsymbol{\mu\beta}^T = \widetilde{M}(I \otimes \mathbf{1}\boldsymbol{\beta}^T). \tag{4.5.15}$$

We introduce the matrices

$$H = I \otimes I - \mathbf{1}\boldsymbol{\alpha}^T \otimes I - I \otimes \mathbf{1}\boldsymbol{\beta}^T + \mathbf{1}\boldsymbol{\alpha}^T \otimes \mathbf{1}\boldsymbol{\beta}^T, \qquad F = (\Lambda \oplus M)H.$$

With the use of (4.5.15), we easily see that

$$F = -\widetilde{M}H = -\tilde{\Lambda}H, \tag{4.5.16}$$

which yields

$$\widetilde{M}H = \tilde{\Lambda}H. \tag{4.5.17}$$

Consider equation (4.5.3) for fixed $k = 2, 3, \dots, r-1$, which we have to transform to identity by means of (4.5.13) or their equivalents (4.5.10)–(4.5.12). To this end, we use (4.5.11) and (4.5.16) to transform (4.5.3) to

$$\mathbf{0}^T = \boldsymbol{p}_k^T F, k = 2, \dots, r-1.$$

From the last equality we obtain, using equalities (4.5.11) and (4.5.16) over and over again,

$$\mathbf{0}^T = \boldsymbol{p}_{k-1}^T \tilde{\Lambda} H = -\boldsymbol{p}_{k-1}^T F.$$

We make use of induction to prove that

$$\mathbf{0}^T = \boldsymbol{p}_k^T F = \boldsymbol{p}_{k-1}^T F = \dots = \boldsymbol{p}_1^T F.$$

Let us turn back to equation (4.5.2). In view of the relation

$$\boldsymbol{\lambda}\boldsymbol{\alpha}^T \otimes \boldsymbol{\beta}^T = -(\Lambda \otimes \boldsymbol{\beta}^T)(\mathbf{1}\boldsymbol{\alpha}^T \otimes I)$$

and (4.5.10) and (4.5.11), we transform (4.5.2) to

$$\mathbf{0}^T = \boldsymbol{p}_1^T F.$$

Therefore, successive substitution of (4.5.10) and (4.5.11), which are equivalent to (4.5.13), into equations (4.5.2) and (4.5.3) for $k = 2, \dots, r-1$ yields

$$\boldsymbol{p}_k^T F = \boldsymbol{p}_{k-1}^T F = \cdots = \boldsymbol{p}_1^T F.$$

In turn, from (4.5.10) and the definition of the matrices F and H, because

$$(I \otimes \boldsymbol{\beta}^T)H \equiv \mathbf{0}^T,$$

we obtain

$$\begin{aligned}\boldsymbol{p}_1^T F &= -\boldsymbol{p}_0^T(\Lambda \otimes \boldsymbol{\beta}^T)\widetilde{M}^{-1}F = \boldsymbol{p}_0^T(\Lambda \otimes \boldsymbol{\beta}^T)\widetilde{M}^{-1}\widetilde{M}H \\ &= \boldsymbol{p}_0^T(\Lambda \otimes I)(I \otimes \boldsymbol{\beta}^T)H \equiv \mathbf{0}^T.\end{aligned}$$

Consider equation (4.5.1). The second term in the right-hand side of this is transformed using (4.5.10) and (4.5.15) to

$$\boldsymbol{p}_1^T(I \otimes \boldsymbol{\mu}) = \boldsymbol{p}_1^T \widetilde{M}(I \otimes \mathbf{1}) = -\boldsymbol{p}_0^T(\Lambda \otimes \boldsymbol{\beta}^T)(I \otimes \mathbf{1}) = -\boldsymbol{p}_0^T \Lambda.$$

Substituting this into (4.5.1), we recast it into identity. Finally, substitution of expression (4.5.13) of $\boldsymbol{p}_k$ into (4.5.3) for $k = r$ leads to equations (4.5.14).

Thus, equations (4.5.14) of order l in the unknown vector $\boldsymbol{p}_0$ are obtained as the result of substituting expressions (4.5.13) of the stationary distribution into equations (4.5.1)–(4.5.4). Since under the normalisation condition (4.5.5) equations (4.5.1)–(4.5.4) have a unique positive solution, the unknowns p_{i0}, $i = 1, \dots, l$, are found from equations (4.5.14) up to a constant defined uniquely from (4.5.5), which proves the theorem. □

Another equation system can be obtained to find the vector $\boldsymbol{p}_0$. To do this, we multiply (4.5.1)–(4.5.4) from the right by the matrix $I \otimes \mathbf{1}$ and sum the results over $k = 0, 1, \dots, R$. We thus obtain

$$\mathbf{0}^T = \left[\boldsymbol{p}_0^T + \sum_{k=1}^{R} \boldsymbol{p}_k^T(I \otimes \mathbf{1})\right](\Lambda + \boldsymbol{\lambda}\boldsymbol{\alpha}^T). \tag{4.5.18}$$

It is easy to demonstrate, and we leave this to the readers, that by virtue of (4.5.13) this equality is equivalent to (4.5.14). We assume that

$$V = I + \left(\sum_{k=0}^{r-1} W_0 W^k + W_0 W^{r-1} W_r\right)(I \otimes \mathbf{1}).$$

Furthermore, let $\boldsymbol{y}$ be a solution of the equilibrium equations

$$\boldsymbol{y}^T(\Lambda + \boldsymbol{\lambda}\boldsymbol{\alpha}^T) = \boldsymbol{0}^T,$$
$$\boldsymbol{y}^T\boldsymbol{1} = 1. \tag{4.5.19}$$

Since by irreducibility of the PH-representation $(\boldsymbol{\alpha}, \Lambda)$ the matrix $\Lambda + \boldsymbol{\lambda}\boldsymbol{\alpha}^T$ is irreducible, these equations have a unique solution. By substituting (4.5.13) for $\boldsymbol{p}_k$ into (4.5.18), we see that the vector $\boldsymbol{p}_0 V$ solves the equations

$$\boldsymbol{p}_0^T V(\Lambda + \boldsymbol{\lambda}\boldsymbol{\alpha}^T) = \boldsymbol{0}^T.$$

Therefore, $\boldsymbol{p}_0^T V = c\boldsymbol{y}^T$, where c is a constant. It follows from the normalisation condition (4.5.5) and (4.5.13) that $c = 1$. Therefore, the vector $\boldsymbol{p}_0$ solves the simultaneous linear algebraic equations

$$\boldsymbol{p}_0^T V = \boldsymbol{y}^T, \tag{4.5.20}$$

where the vector $\boldsymbol{y}$ is uniquely determined by equations (4.5.19). In turn, since, as mentioned above, (4.5.18) and (4.5.14) are equivalent, we conclude that the solution for $\boldsymbol{p}_0$ found from (4.5.20) with account for (4.5.19) is unique as well.

4.5.3. Generalisations of the system

Let us consider a more general case where the constraints $\alpha_i > 0$, $i = 1, \ldots, l$, or $\beta_j > 0$, $j = 1, \ldots, m$, on the distribution functions $A(x)$ or $B(x)$ are removed. Of course, in this case $A(x)$ and $B(x)$ cannot be PH-distributions any more, and we agree to denote them by QPH (see Section 2.8).

$QPH/PH/1/r$ system. It admits description by a Markov chain with a semi-Markov control (see Section 1.6) $\{\eta(t) = (\nu(t), \xi(t)), t \geq 0\}$, where $\{\nu(t), t \geq 0\}$ is the Markov process defined on the state set $\mathcal{X} = \{(0), (k, j), k = 1, \ldots, R, j = 1, \ldots, m\}$, and $\{\xi(t), t \geq 0\}$ is a semi-Markov process with state set $\mathcal{Y} = \{0\}$. The states of the process $\{\nu(t), t \geq 0\}$ are interpreted as follows: $\nu(t) = (0)$ if the system is empty; $\nu(t) = (k, j)$ if there are k customers and service is in phase j. The process $\{\xi(t), t \geq 0\}$ has only one state 0, and the distribution function $F_0(x)$ of the sojourn time of $\xi(t)$ in the state 0 coincides with $A(x)$. Thus, in the notation of Section 1.6, $G_0 = -\Lambda$, $\boldsymbol{\gamma}_0 = \boldsymbol{\alpha}$, and the non-zero probabilities $p^{v,w} = p_{0,0}^{v,w}$ are

$$p^{(0),(1,j)} = \beta_j, \quad p^{(k,j),(k+1,j)} = p^{(R,j),(R,j)} = 1, \quad j = 1, \ldots, m, \quad k = 1, \ldots, r.$$

We assume now that $x_{i0} = x(i, (0), 0)$ and $x_{ikj} = x(i, (k, j), 0)$, and introduce the vectors $\boldsymbol{x}_0$ and $\boldsymbol{x}_k$ which are similar to the vectors $\boldsymbol{p}_0$ and $\boldsymbol{p}_k$, $k = 1, \ldots, R$, in structure and dimensionality. With regard for the changed notation $p \leftrightarrow x$, equations (1.6.2), (1.6.3) then reduce to (4.5.1)–(4.5.5). For brevity, we refer to the resulting simultaneous equations as quasi-equilibrium equations. Let us assume that $p_0 = p((0), 0)$ and $p_{kj} = p((k, j), 0)$. Now we see that if we follow the proof of Theorem 4.5.1, then the condition $\alpha_i > 0$ is used implicitly only to establish non-degeneracy of the matrices $\widetilde{M}$ and $\tilde{\widetilde{M}}$. It follows, however, from Lemmas 2.8.2, 2.8.4, and 2.8.8 that these matrices are non-degenerate also in the case where the condition $\alpha_i > 0$ is not fulfilled. Therefore, Theorem 1 can be used to find the vectors $\boldsymbol{x}_k$, $k = 0, \ldots, R$, with regard for changed notation $p \leftrightarrow x$; and if the variables x_{i0} and x_{ikj} are known, we find by means of (1.6.5) the probabilities p_0 and p_{kj}

$$p_0 = x_{\cdot,0}, \quad p_{kj} = x_{\cdot,k,j}, \quad k = 1, \ldots, R, \quad j = 1, \ldots, m. \tag{4.5.21}$$

PH/QPH/1/r system. This queueing system also admits description by means of a Markov chain with semi-Markov control $\{\eta(t) = (\nu(t), \xi(t)), t \geq 0\}$, where the Markov process $\{\nu(t), t \geq 0\}$ is defined on the set $\mathscr{X} = \{(i,k), i = 1, \ldots, l, k = 0, \ldots, R\}$, and the semi-Markov process $\{\xi(t), t \geq 0\}$, on $\mathscr{Y} = \{0, 1, \ldots l\}$. Here $\nu(t) = (i,k)$ if the arriving customer is in phase i and k customers are in the system. The distribution function $F_0(x)$ of the state 0 sojourn time $\xi(t)$ coincides with $B(x)$. Recall (see Section 2.8) that in the probabilistic interpretation of the phase-type distribution function $A(x)$ as a process of random walk in some open queueing network it is assumed that the customer sojourn time in phase i is distributed exponentially with some parameter ν_i; the transitions from phase i to phase j occur with the probability θ_{ij}; and the customer departs from the network with the probability $\theta_{i0} = 1 - \sum_{j=1}^{l} \theta_{ij}$. So, we assume that the distribution function $F_i(x)$ of the sojourn time $\xi(t)$ in the ith state, $i = 1, \ldots, l$, is exponential with parameter ν_i. Therefore, $\boldsymbol{\gamma}_0 = \boldsymbol{\beta}$, $G_0 = -M$, $\boldsymbol{\gamma}_i = (1)$, $G_i = (\nu_i)$, $i = 1, \ldots, l$. The non-zero probabilities $p_{n,m}^{v,w}$ are

$$\begin{aligned}
p_{0,0}^{(i,k+1),(i,k)} &= p_{0,i}^{(i,1),(i,0)} = 1, && i = 1, \ldots, l, \quad k = 0, \ldots, r, \\
p_{i,s}^{(i,0),(s,0)} &= \theta_{is}, && i, s = 1, \ldots, l, \\
p_{s,0}^{(s,0),(i,1)} &= \alpha_i \theta_{s0}, && s, i = 1, \ldots, l.
\end{aligned}$$

We assume that $x_{i0} = x((i,0), i)$ and $x_{ikj} = x(j, (i,k), 0)$, and introduce the same vectors $\boldsymbol{x}_k$ as for the *QPH/PH/1/r* queueing system. Additionally, let $p_{i0} = p((i,0), i)$ and $p_{ik} = p((i,k), 0)$.

Reasoning by analogy with the *QPH/PH/1/r* system and using again (2.8.5), we conclude that

$$\begin{aligned}
p_{i0} &= x_{i0}, && i = 1, \ldots, l, \\
p_{ik} &= x_{i,k,\cdot}, && i = 1, \ldots, l, \quad k = 1, \ldots, R,
\end{aligned} \tag{4.5.22}$$

where, as above, x_{i0} and x_{ikj} are found with the use of Theorem 4.5.1 (with account for p changed for x).

The attentive reader, possibly, has noticed that we arrive here at a somewhat unexpected result. Indeed, for the *QPH/PH/1/r* and *PH/QPH/1/r* systems, which, in the general case, are not Markov, we construct equation systems (quasi-equilibrium equations) which are indistinguishable from the equilibrium equations for similar queueing systems where *QPH* is replaced by *PH*, but—since their transition intensities cannot be negative—are not such! Yet, we disregard this fact and solve the quasi-equilibrium equations as the conventional equilibrium equations for the *PH/PH/1/r* queueing system, that is, roughly speaking, use the ready-made solution where x is substituted for p. In doing so, some variables x_{i0} and x_{ikj} can be negative, although their sum is equal to one. Summing the variables

x over the subscripts i or j which have no physical meaning, that is, do not already denote the generation of service phases (we had no right to separate these phases because the distribution functions $A(x)$ or $B(x)$ of QPH type now do not admit the phase probabilistic interpretation), we finally obtained the desired and, naturally, correct result, the stationary state probabilities of the corresponding (non-Markov) systems. Therefore, we succeeded in extending the algorithmic approach based on the matrix apparatus to analysing the non-Markov queueing systems.

4.5.4. Main system performance indices

Denote by λ and μ the intensities of customer arrival and service respectively. By virtue of (2.8.10),

$$\lambda^{-1} = -\boldsymbol{\alpha}^T \Lambda^{-1}\mathbf{1}, \qquad \mu^{-1} = -\boldsymbol{\beta}^T M^{-1}\mathbf{1}.$$

As before, we denote the system traffic intensity by $\rho = \lambda/\mu$. Assume that

$$\boldsymbol{p} = \sum_{k=1}^{R} \boldsymbol{p}_k.$$

Then $u = \boldsymbol{p}^T\mathbf{1}$ is the server utilisation coefficient and

$$p_0 = \boldsymbol{p}_0^T\mathbf{1} = 1 - u$$

is the probability of system idling. The system throughput, or departure intensity, is denoted by λ_D as before, and the intensity of the arriving flow is denoted by λ_A. In addition, we introduce the stationary customer loss probability π.

The same reasoning as in the preceding sections proves that in the steady state of queueing system operation

$$\begin{aligned} \lambda_D &= \mu(1 - p_0), \\ \lambda_A &= \lambda(1 - \pi). \end{aligned} \tag{4.5.23}$$

From the last equality we obtain

$$\pi = 1 - \frac{\lambda_A}{\lambda}. \tag{4.5.24}$$

Since in the steady state $\lambda_A = \lambda_D$, from (4.5.23) the physically obvious relation follows:

$$\rho(1 - \pi) = 1 - p_0. \tag{4.5.25}$$

The left-hand side of (4.5.25) defines the intensity traffic accepted by the system; since the system has one server, this value coincides with the utilisation coefficient.

Now, we again turn to the equilibrium equations (4.5.1)–(4.5.4) and, multiplying both sides of (4.5.2)–(4.5.4) by the matrix $I \otimes (\mathbf{1}\boldsymbol{\beta}^T - I)$ and then summing all the resulting equalities, obtain

$$\mathbf{0}^T = \boldsymbol{p}^T[\lambda\boldsymbol{\alpha}^T \otimes (\mathbf{1}\boldsymbol{\beta}^T - I) + (\Lambda \oplus M)(I \otimes (\mathbf{1}\boldsymbol{\beta}^T - I))]. \tag{4.5.26}$$

Multiplying, in a similar manner, both sides of (4.5.2)–(4.5.4) from the right by the matrix $(\mathbf{1}\boldsymbol{\alpha}^T - I) \otimes I$ and then summing the resulting equalities, taking (4.5.1) and (4.5.6) into account, we obtain

$$\mathbf{0}^T = \boldsymbol{p}^T[(\Lambda \oplus M)((\mathbf{1}\boldsymbol{\alpha}^T - I) \otimes I) - I \otimes \boldsymbol{\mu}\boldsymbol{\beta}^T + \boldsymbol{\lambda}\boldsymbol{\alpha}^T \otimes \mathbf{1}\boldsymbol{\beta}^T] \\ - \boldsymbol{p}_0^T(\Lambda \otimes \boldsymbol{\beta}^T) - \boldsymbol{p}_R^T(\boldsymbol{\lambda}\boldsymbol{\alpha}^T \otimes \mathbf{1}\boldsymbol{\beta}^T). \quad (4.5.27)$$

Subtracting (4.5.27) from (4.5.26), we obtain

$$\boldsymbol{p}^T(\Lambda \otimes \mathbf{1}\boldsymbol{\beta}^T) + \boldsymbol{p}_0^T(\Lambda \otimes \boldsymbol{\beta}^T) = \boldsymbol{p}^T(\mathbf{1}\boldsymbol{\alpha}^T \otimes M) - \boldsymbol{p}_R^T(\boldsymbol{\lambda}\boldsymbol{\alpha}^T \otimes \mathbf{1}\boldsymbol{\beta}^T). \quad (4.5.28)$$

Multiplying both sides of this equality by the matrix $I \otimes M^{-1}\mathbf{1}$ and taking (4.5.1)into account, we obtain

$$-\frac{1}{\mu}[\boldsymbol{p}^T(\Lambda \otimes \mathbf{1}) + \boldsymbol{p}_0^T\Lambda] = \boldsymbol{p}^T(\mathbf{1}\boldsymbol{\alpha}^T \otimes \mathbf{1}) + \frac{1}{\mu}\boldsymbol{p}_R^T(\boldsymbol{\lambda}\boldsymbol{\alpha}^T \otimes \mathbf{1}).$$

In view of the relation

$$\boldsymbol{p}^T(\mathbf{1}\boldsymbol{\alpha}^T \otimes \mathbf{1}) = (1 - p_0)\boldsymbol{\alpha}^T,$$

which follows from the normalisation condition (4.5.5), the last equality reduces to

$$-\frac{1}{\mu}[\boldsymbol{p}^T(\Lambda \otimes \mathbf{1}) + \boldsymbol{p}_0^T\Lambda] = (1 - p_0)\boldsymbol{\alpha}^T + \frac{1}{\mu}\boldsymbol{p}_R^T(\boldsymbol{\lambda}\boldsymbol{\alpha}^T \otimes \mathbf{1}). \quad (4.5.29)$$

Multiplying both sides of this equality from the right by the matrix $-\Lambda^{-1}\mathbf{1}$ and again taking the normalising condition (4.5.5) into account, we obtain

$$1 - p_0 = \rho\left[1 - \frac{1}{\lambda}\boldsymbol{p}_R^T(\boldsymbol{\lambda} \otimes \mathbf{1})\right]. \quad (4.5.30)$$

It follows from (4.5.29) and (4.5.30) that

$$-[\boldsymbol{p}^T(\Lambda \otimes \mathbf{1}) + \boldsymbol{p}_0^T\Lambda] = \lambda\boldsymbol{\alpha}^T. \quad (4.5.31)$$

Multiplying both sides of (4.5.31) from the right by the matrix $-\Lambda^{-1}$, we obtain

$$\boldsymbol{p}_0^T + \boldsymbol{p}^T(I \otimes \mathbf{1}) = -\lambda\boldsymbol{\alpha}^T\Lambda^{-1}, \quad (4.5.32)$$

where the ith component of the vector in the left-hand side of the equality is the probability $p_{i0} + p_{i,\cdot,\cdot}$ that in the steady state at an arbitrary instant of time the customer generated at the system input is in phase i.

Multiplying both sides of (4.5.31) by the vector $\mathbf{1}$, we also obtain

$$\boldsymbol{p}_0^T\boldsymbol{\lambda} + \boldsymbol{p}^T(\boldsymbol{\lambda} \otimes \mathbf{1}) = \lambda. \quad (4.5.33)$$

Relations (4.5.25) and (4.5.30) provide us with another equivalent expression for the loss probability:

$$\pi = \frac{1}{\lambda}\boldsymbol{p}_R^T(\boldsymbol{\lambda} \otimes \mathbf{1}). \quad (4.5.34)$$

In view of (4.5.32), equality (4.5.28) now admits the representation

$$\lambda(1-\pi)(\boldsymbol{\alpha}^T \otimes \boldsymbol{\beta}^T) = -\boldsymbol{p}^T(\mathbf{1}\boldsymbol{\alpha}^T \otimes M). \tag{4.5.35}$$

Multiplying both its sides from the right by the matrix $\mathbf{1} \otimes M^{-1}$, we obtain

$$-\lambda(1-\pi)\boldsymbol{\beta}^T M^{-1} = \boldsymbol{p}^T(\mathbf{1} \otimes I),$$

where the jth component $p_{\cdot,\cdot,j}$ of the vector in the right-hand side of the equality is the probability that in the steady state at an arbitrary instant of time the served customer is in phase j.

Finally, multiplying both sides of (4.5.35) from the right by the vector $\mathbf{1} \otimes \mathbf{1}$, we obtain another simple relation

$$\lambda(1-\pi) = \boldsymbol{p}^T(\mathbf{1} \otimes \boldsymbol{\mu}) \tag{4.5.36}$$

which, in view of (4.5.25) and the normalisation condition, yields

$$\boldsymbol{p}^T(\mathbf{1} \otimes \boldsymbol{\mu}) = \mu \boldsymbol{p}^T \mathbf{1}. \tag{4.5.37}$$

The above relations allow us to obtain further insight into relation between the queueing system performance indices. Furthermore, these relations allow us to perform a rather full check of computations using the algorithm of Theorem 4.5.1. We conclude the section by noting that by means of similar calculations one can demonstrate that relations (4.5.23)–(4.5.37) are valid (with p changed for x) for the $QPH/PH/1/r$ and $PH/QPH/1/r$ queueing systems.

4.5.5. Stationary state probabilities for the embedded Markov chains

Let us consider the $QPH/PH/1/r$ queueing system and continue to study the Markov chain with the semi-Markov control defined for this system in Section 4.5.3. In the notation of Section 1.6, we set $p_{A,0}^- = p^-((0),0)$, $p_A^{\pm}(k,j) = p^{\pm}((k,j),0)$, and introduce the vectors $\boldsymbol{p}_{A,k}^{T\pm} = (p_A^{\pm}(k,1),\dots,p_A^{\pm}(k,m))$.

The stationary distribution $\{\boldsymbol{p}_{A,k}^{T+}, k=1,\dots,R\}$ of the embedded Markov chain generated by the process $\{\eta(t), t \geq 0\}$ immediately after the instants τ_n of customer arrival at the $QPH/PH/1/r$ system is obtained from formulas (1.6.4) and equality (4.5.33), which—with regard for the substitution of x for p—is valid, as stated before, for the queueing system under consideration as well. The distribution is

$$\boldsymbol{p}_{A,k}^{T+} = \begin{cases} \dfrac{1}{\lambda}\boldsymbol{x}_0^T(\boldsymbol{\lambda} \otimes \boldsymbol{\beta}^T), & k=1; \\ \dfrac{1}{\lambda}(\boldsymbol{x}_{k-1}^T + \delta_{k,R}\boldsymbol{x}_R^T)(\boldsymbol{\lambda} \otimes I), & k=2,\dots,R, \end{cases} \tag{4.5.38}$$

where $\delta_{k,n}$ is the Kronecker symbol.

Let us prove that the stationary distribution $\{p_{A,0}^-, \boldsymbol{p}_{A,k}^-, k=1,\dots,R\}$ of the Markov chain embedded at the instants $\tau_n - 0$ of customer arrivals at the $QPH/PH/1/r$ queueing system is determined by the relations

$$p_{A,0}^{-} = \frac{1}{\lambda} x_0^T \boldsymbol{\lambda}, \tag{4.5.39}$$

$$p_{A,k}^{T-} = \frac{1}{\lambda} x_k^T (\boldsymbol{\lambda} \otimes I), \qquad k = 1, \dots, R. \tag{4.5.40}$$

Relations (4.5.39) and (4.5.40) for $k = 1, \dots, r-1$ follow immediately from (1.6.6). From (1.6.6) with $k = R$ it also follows that

$$p_{A,R}^{+} = p_{A,r}^{-} + p_{A,R}^{-}. \tag{4.5.41}$$

We cannot derive from (1.6.6) any other equations to find $p_{A,r}^{-}$ and $p_{A,R}^{-}$; in order to obtain additional information, we need to consider the embedded Markov chain $\{\eta_n^{-}, n \geq 0\}$.

We easily see that for this Markov chain the equilibrium equations imply that

$$p_{A,R}^{T-} = (p_{A,r}^{T-} + p_{A,R}^{T-}) Q, \tag{4.5.42}$$

where

$$Q = \int_0^\infty e^{Mt} dA(t).$$

We observe that the element Q_{sj} of the matrix Q is the conditional probability that in the time between arrivals of two subsequent customers the current customer does not complete his service and the process of service at the instant of arrival of a new customer is in phase j under the condition that at the instant of arrival of the preceding customer the service process is in phase s. By virtue of Lemma 2.8.3, we obtain

$$Q = -(\boldsymbol{\alpha}^T \otimes I)(\Lambda \oplus M)^{-1}(\boldsymbol{\lambda} \otimes I). \tag{4.5.43}$$

The relation

$$p_{A,R}^{T-} = p_{A,R}^{T+} Q \tag{4.5.44}$$

follows from (4.5.41) and (4.5.42).

Consider equation (4.5.4) of the simultaneous quasi-equilibrium equations. Multiplying both its sides from the right by the matrix $(\Lambda \oplus M)^{-1}(\boldsymbol{\lambda} \otimes I)/\lambda$, we obtain, in view of (4.5.43), the equality

$$\frac{1}{\lambda} x_R^T (\boldsymbol{\lambda} \otimes I) = \frac{1}{\lambda} (x_r^T + x_R^T)(\boldsymbol{\lambda} \otimes I) Q,$$

which, due to (4.5.38), reduces for $k = R$ to

$$\frac{1}{\lambda} x_R^T (\boldsymbol{\lambda} \otimes I) = p_{A,R}^{T+} Q,$$

which together with (4.5.44) implies validity of (4.5.40) for $k = R$. Finally, it follows from (4.5.38) and (4.5.40) for $k = R$ and from (4.5.41) that (4.5.40) is also true for $k = r$.

The above results allow us to find the loss probability π in the $QPH/PH/1/r$ queueing system. To do this, we first observe that the customer loss probability is the probability that at the instant $\tau_n - 0$ of customer arrival the buffer is full. Hence, $\pi = \boldsymbol{p}_{A,R}^{T-}\mathbf{1}$, and therefore, in view of (4.5.40), we obtain

$$\pi = \frac{1}{\lambda}\boldsymbol{x}_R^T(\boldsymbol{\lambda} \otimes \mathbf{1}). \tag{4.5.45}$$

Consider the $PH/QPH/1/r$ queueing system. In the notation used in Section 4.5.3 and 1.6, we define the set

$$\tilde{Z} = \{(((i, k+1), 0), ((i, k), 0)), \quad i = 1, \dots, l, \quad k = 1, \dots, r,$$
$$(((i, 1), 0), ((i, 0), i)), \quad i = 1, \dots, l\}.$$

Then, by virtue of Theorem 1.6.4, the sequences $\{\tilde{\eta}_n^{\pm}, n \geq 1\}$ defined for the Markov chain with the semi-Markov control (see Section 4.5.3) are the Markov chains embedded at the instants $s_n \pm 0$ of completion of customer service. Assume that $p_D^+(i, 0) = \tilde{p}^+((i, 0), i)$, $p_D^{\pm}(i, k) = \tilde{p}^{\pm}((i, k), 0)$, and introduce the vectors $\boldsymbol{p}_{D,k}^{T\pm} = (p_D^{\pm}(1, k), \dots, p_D^{\pm}(l, k))$. In view of equality (4.5.36)) which holds true also for the $PH/QPH/1/r$ system, from (1.6.7) and (1.6.8) it follows that the stationary distributions $\{\boldsymbol{p}_{D,k}^+, k = 0, \dots, r\}$ and $\{\boldsymbol{p}_{D,k}^-, k = 1, \dots, R\}$ of the Markov chains embedded at the instants $s_n \pm 0$ of completion of customer service in the $PH/QPH/1/r$ queueing system obey the relations

$$\boldsymbol{p}_{D,k}^{T+} = \frac{1}{\lambda(1-\pi)}\boldsymbol{x}_{k+1}^T(I \otimes \boldsymbol{\mu}), \qquad k = 0, \dots, r, \tag{4.5.46}$$

$$\boldsymbol{p}_{D,k}^{T-} = \frac{1}{\lambda(1-\pi)}\boldsymbol{x}_k^T(I \otimes \boldsymbol{\mu}), \qquad k = 1, \dots, R. \tag{4.5.47}$$

where the loss probability π is defined by (4.5.34) with x substituted for p.

We assume now that $p_k = \boldsymbol{p}_k^T(\mathbf{1} \otimes \mathbf{1})$, $p_{A,k}^{\pm} = \boldsymbol{p}_{A,k}^{T\pm}\mathbf{1}$ and $p_{D,k}^{\pm} = \boldsymbol{p}_{D,k}^{T\pm}\mathbf{1}$, where p_k, $p_{A,k}^{\pm}$, and $p_{D,k}^{\pm}$ are the stationary probabilities that k customers are in the system, respectively, at arbitrary time instants, instants $\tau_n \pm 0$ of customer arrival, and instants $s_n \pm 0$ of completion of customer service. Let us consider some examples.

EXAMPLE 4.5.1 ($M/PH/1/r$ QUEUEING SYSTEM). As we might expect, for the Poisson flow we see from (4.5.39) and (4.5.40) that the distributions p_A^- and p coincide. The distributions p_A^+ and p in this case are related as follows:

$$p_{A,k}^+ = \begin{cases} p_0, & k = 1; \\ p_{k-1} + \delta_{k,R}\, p_R, & k = 2, \dots, R. \end{cases}$$

For the distribution p_D^+, from (4.5.46) it follows that

$$p_{D,k}^+ = \frac{1}{1-\pi}p_k, \qquad k = 0, \dots, r.$$

EXAMPLE 4.5.2 ($E_l/E_m/1/r$ QUEUEING SYSTEM). For the distributions $\pi_A^{\pm}$, expressions (4.5.38)–(4.5.40) take in the scalar notation the form

$$p_A^+(k,j) = \begin{cases} lp_{10}, & j = m, k = 1; \\ l(p_{1,k-1,j} + \delta_{k,R}p_{1Rj}), & j = 1,\dots,m, k = 2,\dots,R, \end{cases}$$

$$p_{A,0}^- = lp_{10},$$

$$p_A^-(k,j) = lp_{1kj}, \quad j = 1,\dots,m, \quad k = 1,\dots,R.$$

In view of (4.5.46) and (4.5.47), the distributions $p_D^{\pm}$ in this case are

$$p_D^+(i,k) = \frac{\nu p_{i,k+1,1}}{\lambda(1-\pi)}, \quad i = 1,\dots,l, \quad k = 0,\dots,r;$$

$$p_D^-(i,k) = \frac{\nu p_{ik1}}{\lambda(1-\pi)}, \quad i = 1,\dots,l, \quad k = 1,\dots,R.$$

4.5.6. Stationary distribution of waiting time for the FCFS discipline

We consider the steady state of the $QPH/PH/1/r$ queueing system and assume that the customers are served in the order of their arrival, that is, according to the FCFS discipline. For this system, we consider the waiting time for service of a customer accepted by the system. Let $W(t)$ denote the corresponding conditional distribution function.

Let us consider, as before, a tagged customer, and denote by $W(t \mid x)$ the conditional distribution function of his waiting time under the condition that, immediately before his arrival, he finds the system in a state x, $x \in \mathscr{X}' = \bigcup_{k=0}^{r} \mathscr{X}_{A,k}^-$, where

$$\mathscr{X}_{A,k}^- = \begin{cases} \{0\}, & k = 0; \\ \{(k,j), j = 1,\dots,m\}, & k = 1,\dots,r. \end{cases}$$

Let $\omega(s \mid x)$ stand for the Laplace–Stieltjes transform of the distribution function $W(t \mid x)$ and introduce the vector $\boldsymbol{\omega}_k^T(s) = (\omega(s \mid (k,1)),\dots,\omega(s \mid (k,m)))$. Then with the use of the total probability formula we obtain the expression for $W(t)$ which admits the representation in terms of Laplace–Stieltjes transform as

$$\omega(s) = \frac{1}{1-\pi}\left[p_{A,0}^-\omega(s \mid (0)) + \sum_{k=1}^{r} \mathbf{p}_{A,k}^{T-}\boldsymbol{\omega}_k(s)\right], \tag{4.5.48}$$

where π is the customer loss probability.

As above, we have to find an expression of $\omega(s \mid x)$. If $x = (0)$, then the waiting time is, of course, zero. Therefore, $\omega(s \mid (0)) = 1$. If just before the customer arrival the system is in any of the states of the subset $\mathscr{X}_{A,k}^-$, $k = 1,\dots,r$, then the customer we chose is served only after the customer in service and all of $k-1$ customers preceding him in the queue. It follows from independence of the service time and properties of the PH-distribution that

$$\boldsymbol{\omega}_k(s) = (sI - M)^{-1}\boldsymbol{\mu}\beta^{k-1}(s),$$

where, in view of (2.8.9), the Laplace–Stieltjes transform $\beta(s)$ of the distribution function $B(t)$ is

$$\beta(s) = \boldsymbol{\beta}^T(sI - M)^{-1}\boldsymbol{\mu}.$$

Substituting the expressions of $\omega(s \mid (0))$ and $\boldsymbol{\omega}_k(s)$ into (4.5.48), we arrive at the final expression

$$\omega(s) = \frac{1}{1-\pi}\left[p^{-}_{A,0} + \sum_{k=1}^{r} \boldsymbol{p}^{T-}_{A,k}(sI - M)^{-1}\boldsymbol{\mu}\beta^{k-1}(s)\right]. \tag{4.5.49}$$

For the first two moments w and $w^{(2)}$ of the waiting time, the following expressions can be obtained from (4.5.49) with the use of the formula $w^{(i)} = (-1)^i\omega^{(i)}(0)$ and, generally speaking, not so simple calculations, which we recommend to the readers to carry out themselves:

$$\begin{aligned}\lambda(1-\pi)w &= q_1,\\ \lambda(1-\pi)w^{(2)} &= 2[bq_2 - \boldsymbol{q}_1^T(\mathbf{1}\otimes M^{-1}\mathbf{1})].\end{aligned}$$

Here Q_1 is the mean queue length, the component $q_n(i,j)$ of the vector $\boldsymbol{q}_n$ is

$$q_n(i,j) = \sum_{k=n+1}^{R}\binom{k-1}{n}x_{ikj},$$

$q_n = q_n(\cdot,\cdot)$ is the binomial moment of order n of queue length, and b is the mean service time.

We note that the second of the above relations is an extension of well-known Little formula relating the second moments of queue length and waiting time in matrix form.

4.5.7. Output flow

We consider now the distribution of the interval between departures of neighbouring served customers from the $PH/QPH/1/r$ queueing system operating in the steady state. We make use of the same approach which has been repeatedly used to find the distribution of the waiting time of various queueing systems.

Let $D(t \mid x)$ be the conditional distribution function of the interval between departures of neighbouring served customers under the condition that immediately after completion of service of the first customer the system is in the state x, $x \in \mathscr{X}^+_D = \bigcup_{k=0}^r \mathscr{X}^+_{D,k}$, where $\mathscr{X}^+_{D,k} = \{(i,k), i = 1,\dots,l\}$, $k = 0,\dots,r$. Let $\delta(s \mid x)$ stand for the Laplace–Stieltjes transform of $D(t \mid x)$, and introduce the vector $\boldsymbol{\delta}^T_k(s) = (\delta(s \mid (1,k)), \dots \delta(s \mid (l,k)))$.

From the formula of total probability for the Laplace–Stieltjes transform $\delta(s)$ of the unconditional distribution function $D(x)$ of the intervals between departures of neighbouring served customers, we obtain

$$\delta(s) = \sum_{k=0}^{r}\boldsymbol{p}^{T+}_{D,k}\boldsymbol{\delta}_k(s). \tag{4.5.50}$$

Since the expressions for the probabilities $\boldsymbol{p}^+_D(i,k)$ have been obtained, it remains to find $\boldsymbol{\delta}_k(s)$. If, upon departure of a customer, the system is busy, that is, is in one of the states from the subset $\mathscr{X}^+_{D,k}$, $k = 1, \dots, r$, then the time before departure of the next customer is equal to the time of his service; hence,

$$\boldsymbol{\delta}_k(s) = \beta(s)\mathbf{1}, \qquad k = 1, \dots, r. \tag{4.5.51}$$

If the customer served leaves an empty system, that is, in any state from the subset $\mathscr{X}^+_{D,0}$, then the time before the departure of the next customer is equal to the sum of the time remained before the arrival of a customer generated at the system and his service time. Since these variables are independent, we obtain

$$\boldsymbol{\delta}_0(s) = (sI - \Lambda)^{-1}\boldsymbol{\lambda}\beta(s). \tag{4.5.52}$$

Substituting (4.5.51) and (4.5.52) into (4.5.50), we obtain the final expression for the Laplace–Stieltjes transform of the distribution function $D(t)$

$$\delta(s) = [1 - p^+_{D,0} + \boldsymbol{p}^{T+}_{D,0}(sI - \Lambda)^{-1}\boldsymbol{\lambda}]\beta(s). \tag{4.5.53}$$

This allows us to determine the initial instants $\delta^{(n)} = (-1)^n\delta^{(n)}(0)$ of the inter-departure times. We omit the intermediate calculations and give the final expression

$$\delta^{(n)} = b^{(n)} + \boldsymbol{p}^{T+}_{D,0}\sum_{j=1}^{n}(-1)^j(n)_j\Lambda^{-j}\mathbf{1}b^{(n-j)}, \qquad n \geq 1, \tag{4.5.54}$$

where

$$b^{(n)} = \int_0^\infty t^n dB(t).$$

In particular, another equivalent expression of $\delta^{(1)} = \delta$ can be derived from (4.5.54) with the use of (4.5.23) and (4.5.32):

$$\delta = \frac{1}{\lambda(1-\pi)},$$

where the loss probability π obeys (4.5.34) with p changed for x.

Further, let σ_B^2, σ_0^2, and σ_D^2 be the variances of the service time, server idling time, and inter-departure time respectively. Then

$$\sigma_D^2 = \sigma_B^2 + \sigma_0^2,$$

where

$$\sigma_0^2 = 2\boldsymbol{p}^{T+}_{D,0}\Lambda^{-2}\mathbf{1} - (\boldsymbol{p}^{T+}_{D,0}\Lambda^{-1}\mathbf{1})^2.$$

The last relation follows immediately from (4.5.54).

4.6. Markov systems described by generalised birth-and-death process

Let the functioning of a queueing system be described by some homogeneous Markov process $\{\eta(t), t \geq 0\}$ with the three-diagonal block matrix of transition intensities

$$Q = \begin{pmatrix} N_0 & \Lambda_1 & 0 & 0 & \cdots & 0 & 0 & 0 \\ M_0 & N_1 & \Lambda_2 & 0 & \cdots & 0 & 0 & 0 \\ 0 & M_1 & N_2 & \Lambda_3 & \cdots & 0 & 0 & 0 \\ \vdots & \vdots & \vdots & \vdots & \ddots & \vdots & \vdots & \vdots \\ 0 & 0 & 0 & 0 & \cdots & M_{r-2} & N_{r-1} & \Lambda_r \\ 0 & 0 & 0 & 0 & \cdots & 0 & M_{r-1} & N_r \end{pmatrix}.$$

Here N_k are square matrices of order n_k and M_k, Λ_k are rectangular $n_{k+1} \times n_k$ and $n_{k-1} \times n_k$ matrices respectively. The state set $\mathscr{X}$ of this process is the union $\mathscr{X} = \bigcup_{k=0}^{r} \mathscr{X}_k$ of non-overlapping subsets $\mathscr{X}_k$, which we agree to refer to as the *strata*. The number of states in the kth stratum $\mathscr{X}_k$ is n_k. Transitions are possible only between states of the same stratum or between states of neighbouring strata.

It is reasonable to classify the Markov process $\{\eta(t), t \geq 0\}$ of this kind, as well as processes considered in the preceding sections of this chapter, with the *generalised birth-and-death processes*. Furthermore, by virtue of sufficient generality of the matrix Q, the so-defined process can be used, in fact, as the definition of the generalised birth-and-death process.

The process under consideration is said to have homogeneous strata if for some integers a and b, $0 \leq a < b \leq r$, the equalities $\Lambda_k = \Lambda$, $N_k = N$, and $M_k = M$ are true for $a < k < b$. The integers a and b are called, respectively, the lower and upper homogeneity boundaries.

We prove below that under somewhat weak constraints the components $\boldsymbol{p}_k$ of the stationary distribution $\boldsymbol{p}^T = (\boldsymbol{p}_0^T, \boldsymbol{p}_1^T, \ldots, \boldsymbol{p}_r^T)$ of a Markov process with homogeneous strata admit the representation

$$\boldsymbol{p}_k^T = \boldsymbol{f}^T R^{k-a} + \boldsymbol{g}^T S^{b-k}, \qquad a \leq k \leq b,$$

where $\boldsymbol{f}$ and $\boldsymbol{g}$ are some vectors, and R and S are the roots of certain quadratic matrix equations.

We conclude this section by considering two examples of particular queueing systems that are described by generalised birth-and-death processes and write out for them the state set $\mathscr{X}$ and the matrix Q.

4.6.1. Auxiliary results

Let us consider the simultaneous equations

$$z_{k-1}^T A + z_k^T B + z_{k+1}^T C = \mathbf{0}^T, \qquad k = 1, \ldots, l-1, \tag{4.6.1}$$

where A, B, and C are some square matrices of the same size and $z_0, z_1, \ldots, z_l$ are the desired vectors. In order to solve it, we need to know the roots R and S of the quadratic matrix equations

$$A + RB + R^2C = 0, \qquad S^2A + SB + C = 0 \tag{4.6.2}$$

for which $SA + B + RC$ is a non-degenerate matrix.

On the base of (4.6.2), we immediately see that the equalities

$$(I - SR)(B + RC) = (I - RS)(B + SA) = SA + B + RC \tag{4.6.3}$$

are true. Therefore, the matrices $I - SR$, $I - RS$, $B + RC$, and $B + SA$ are non-degenerate as well.

LEMMA 4.6.1. *Let $A+RB+R^2C = 0$, $S^2A+SB+C = 0$, and the matrix $SA+B+RC$ be non-degenerate. Then each of the following conditions is equivalent to the fact that the vectors $z_0, \dots, z_l$ solve simultaneous equations* (4.6.1)*:*

(1) *for* $\mathbf{g}^T = (z_l^T - z_{l-1}^T R)(I - SR)^{-1}$,

$$z_k^T = z_{k-1}^T R + \mathbf{g}^T S^{l-k}(I - SR), \qquad k = 1, \dots, l; \tag{4.6.4}$$

(2) *for* $f^T = (z_0^T - z_1^T S)(I - RS)^{-1}$,

$$z_k^T = z_{k+1}^T S + f^T R^k (I - RS), \qquad k = 0, \dots, l - 1. \tag{4.6.5}$$

PROOF. We first obtain some useful relations required to prove the lemma. We set

$$U = -(B + RC)^{-1}, \qquad V = -(B + SA)^{-1}.$$

Then

$$R = AU, \qquad S = CV, \tag{4.6.6}$$

hence the matrices U and V satisfy the equations

$$U = -(B + AUC)^{-1}, \qquad V = -(B + CVA)^{-1}. \tag{4.6.7}$$

In turn, equations (4.6.7) are equivalent to

$$\begin{aligned} I + BU + AUCU = 0, &\qquad I + BV + CVAV = 0, \\ I + UB + UAUC = 0, &\qquad I + VB + VCVA = 0. \end{aligned} \tag{4.6.8}$$

Hence the identities

$$\begin{aligned} AU(I - AUCV) &= (I - AUCV)AV, \\ CV(I - CVAU) &= (I - CVAU)CU \end{aligned} \tag{4.6.9}$$

follow, which yield

$$\begin{aligned} AV &= (I - RS)^{-1}R(I - RS), \\ CU &= (I - SR)^{-1}S(I - SR). \end{aligned} \tag{4.6.10}$$

We prove (4.6.4) only, because (4.6.5) are proved similarly. Since

$$z_{k-1}^T A + z_k^T B + z_{k+1}^T C = (z_k^T - z_{k-1}^T R)(B + RC) + (z_{k+1}^T - z_k^T R)C,$$

equalities (4.6.1) are true if and only if

$$z_k^T - z_{k-1}^T R = (z_{k+1}^T - z_k^T R)CU, \qquad k = 1, \dots, l-1, \tag{4.6.11}$$

which, in turn, is equivalent to the equalities

$$z_k^T = z_{k-1}^T R + (z_l^T - z_{l-1}^T R)(CU)^{l-k}, \qquad k = 1, \dots, l,$$

coinciding, in view of (4.6.10), with (4.6.4), which is what we set out to prove. □

THEOREM 4.6.1. *Let $A + RB + R^2C = 0$, $S^2A + SB + C = 0$, and the matrix $SA + B + RC$ be non-degenerate. Then the general solution of equations* (4.6.1) *is*

$$z_k^T = \boldsymbol{f}^T R^k + \boldsymbol{g}^T S^{l-k}, \qquad k = 0, \dots, l, \tag{4.6.12}$$

where $\boldsymbol{f}$ and $\boldsymbol{g}$ are arbitrary vectors.

PROOF. As follows from Lemma 4.6.1, for all $k = 1, 2, \dots, l$

$$\begin{aligned} z_k^T &= [z_k^T S + \boldsymbol{f}^T R^{k-1}(I - RS)]R + \boldsymbol{g}^T S^{l-k}(I - SR) \\ &= z_k^T SR + (\boldsymbol{f}^T R^k + \boldsymbol{g}^T S^{l-k})(I - SR). \end{aligned} \tag{4.6.13}$$

Furthermore, from (4.6.4) and (4.6.5) it follows that for all $k = 0, 1, \dots, l-1$

$$\begin{aligned} z_k^T &= [z_k^T R + \boldsymbol{g}^T S^{l-k-1}(I - SR)]S + \boldsymbol{f}^T R^k(I - RS) \\ &= z_k^T RS + (\boldsymbol{f}^T R^k + \boldsymbol{g}^T S^{l-k})(I - RS). \end{aligned} \tag{4.6.14}$$

Formula (4.6.13) implies that the components z_k, $k = 1, \dots, l$, of the solution of equations (4.6.1) admit representation (4.6.12). Validity of (4.6.12) for the component z_0 follows from (4.6.14).

Conversely, the equalities for the vectors defined by (4.6.12)

$$z_k^T - z_{k-1}^T R = \boldsymbol{g}^T S^{l-k}(I - SR), \qquad k = 1, \dots, l. \tag{4.6.15}$$

are true. Hence we obtain for $k = l$

$$\boldsymbol{g}^T = (z_l^T - z_{l-1}^T R)(I - SR)^{-1}. \tag{4.6.16}$$

Therefore, the vectors defined by (4.6.12) satisfy condition 1 of Lemma 4.6.1 and hence solve equations (4.6.1). □

The above result is applicable to equations (4.6.1) with any matrices A, B, and C. Consider its application to the equilibrium equations.

4.6.2. Stationary distributions

Let us turn back to the generalised birth-and-death process $\{\eta(t), t \geq 0\}$ under consideration, which is defined by the three-diagonal block matrix Q of transition intensities. We assume that the state set $\mathscr{X}$ of the process $\{\eta(t), t \geq 0\}$ has homogeneous strata, the numbers a and b defining the lower and upper homogeneity boundaries. Under the assumption that all states of the process $\{\eta(t), t \geq 0\}$ are communicating, the process is ergodic.

Let $\boldsymbol{p}^T = (\boldsymbol{p}_0^T, \boldsymbol{p}_1^T, \dots, \boldsymbol{p}_r^T)$ be the vector of stationary probabilities of its states. Then it follows from the equilibrium equations $\boldsymbol{p}^T Q = \boldsymbol{0}^T$ that the components $\boldsymbol{p}_a, \boldsymbol{p}_{a+1}, \dots, \boldsymbol{p}_b$ of the vector $\boldsymbol{p}$ satisfy the equations

$$\boldsymbol{p}_{k-1}^T \Lambda + \boldsymbol{p}_k^T N + \boldsymbol{p}_{k+1}^T M = \boldsymbol{0}^T, \qquad a < k < b, \tag{4.6.17}$$

and can be found with the use of Theorem 4.6.1.

Indeed, the vectors $z_k = \boldsymbol{p}_{k+a}$, $k = 0, \dots, b-a$, satisfy equations (4.6.1) with $l = b-a$. Therefore, if $\Lambda + RN + R^2 M = 0$, $S^2\Lambda + SN + M = 0$, and the matrix $S\Lambda + N + RM$ is non-degenerate, then the components $\boldsymbol{p}_a, \boldsymbol{p}_{a+1}, \dots, \boldsymbol{p}_b$ of the stationary distribution admit the representation

$$\boldsymbol{p}_k^T = \boldsymbol{f}^T R^{k-a} + \boldsymbol{g}^T S^{b-k}, \qquad k = a, \dots, b. \tag{4.6.18}$$

We consider two most frequent cases of $a = 1, b = r - 1$ and $a = 0, b = r$. They will be discussed in more detail, and the results will be stated as theorems below.

THEOREM 4.6.2. *Let*

$$\Lambda + RN + R^2 M = 0, \qquad S^2\Lambda + SN + M = 0, \tag{4.6.19}$$

and the matrix $S\Lambda + N + RM$ be non-degenerate. The vector $\boldsymbol{p}^T = (\boldsymbol{p}_0^T, \dots, \boldsymbol{p}_r^T)$ satisfies the equations $\boldsymbol{p}^T Q = \boldsymbol{0}^T$ with the matrix

$$Q = \begin{pmatrix} N_0 & \Lambda_1 & 0 & 0 & \cdots & 0 & 0 & 0 & 0 \\ M_0 & N & \Lambda & 0 & \cdots & 0 & 0 & 0 & 0 \\ 0 & M & N & \Lambda & \cdots & 0 & 0 & 0 & 0 \\ \vdots & \vdots & \vdots & \vdots & \ddots & \vdots & \vdots & \vdots & \vdots \\ 0 & 0 & 0 & 0 & \cdots & M & N & \Lambda & 0 \\ 0 & 0 & 0 & 0 & \cdots & 0 & M & N & \Lambda_r \\ 0 & 0 & 0 & 0 & \cdots & 0 & 0 & M_{r-1} & N_r \end{pmatrix} \tag{4.6.20}$$

if and only if the equalities

$$\boldsymbol{p}_k^T = \boldsymbol{f}^T R^{k-1} + \boldsymbol{g}^T S^{r-1-k}, \qquad k = 1, \dots, r-1, \tag{4.6.21}$$

$$\boldsymbol{p}_0^T N_0 + (\boldsymbol{f}^T + \boldsymbol{g}^T S^{r-2}) M_0 = \boldsymbol{0}^T, \tag{4.6.22}$$

$$\boldsymbol{p}_0^T \Lambda_1 + \boldsymbol{f}^T (N + RM) - \boldsymbol{g}^T S^{r-1} \Lambda = \boldsymbol{0}^T, \tag{4.6.23}$$

$$\boldsymbol{p}_r^T M_{r-1} + \boldsymbol{g}^T (N + S\Lambda) - \boldsymbol{f}^T R^{r-1} M = \boldsymbol{0}^T, \tag{4.6.24}$$

$$\boldsymbol{p}_r^T N_r + (\boldsymbol{f}^T R^{r-2} + \boldsymbol{g}^T)\Lambda_r = \boldsymbol{0}^T \tag{4.6.25}$$

are valid for some vectors $\boldsymbol{f}$ and $\boldsymbol{g}$.

PROOF. We rewrite the simultaneous equations $\boldsymbol{p}^T Q = \mathbf{0}^T$ with matrix (4.6.20) as

$$\boldsymbol{p}_0^T N_0 + \boldsymbol{p}_1^T M_0 = \mathbf{0}^T, \tag{4.6.26}$$
$$\boldsymbol{p}_0^T \Lambda_1 + \boldsymbol{p}_1^T N + \boldsymbol{p}_2^T M = \mathbf{0}^T, \tag{4.6.27}$$
$$\boldsymbol{p}_{k-1}^T \Lambda + \boldsymbol{p}_k^T N + \boldsymbol{p}_{k+1}^T M = \mathbf{0}^T, \qquad k = 2, \dots, r-2, \tag{4.6.28}$$
$$\boldsymbol{p}_{r-2}^T \Lambda + \boldsymbol{p}_{r-1}^T N + \boldsymbol{p}_r^T M_{r-1} = \mathbf{0}^T, \tag{4.6.29}$$
$$\boldsymbol{p}_{r-1}^T \Lambda_r + \boldsymbol{p}_r^T N_r = \mathbf{0}^T. \tag{4.6.30}$$

By virtue of Theorem 4.6.1, (4.6.28) hold if and only if (4.6.21) are true for some vectors $\boldsymbol{f}$ and $\boldsymbol{g}$. Substituting these expressions into (4.6.26) and (4.6.30), we obtain (4.6.22) and (4.6.25). If we substitute them into (4.6.27) and (4.6.29) and make use of the fact that R and S are the roots of the equations $\Lambda + RN + R^2M = 0$ and $S^2\Lambda + SN + M = 0$, we obtain (4.6.23) and (4.6.24). □

The theorem below is proved similarly.

THEOREM 4.6.3. *Let* $\Lambda + RN + R^2M = 0$, $S^2\Lambda + SN + M = 0$ *and the matrix* $S\Lambda + N + RM$ *be non-degenerate. In this case, the vector* $\boldsymbol{p}^T = (\boldsymbol{p}_0^T, \boldsymbol{p}_1^T, \dots, \boldsymbol{p}_r^T)$ *satisfies the equations* $\boldsymbol{p}^T Q = \mathbf{0}^T$ *with the matrix*

$$Q = \begin{pmatrix} N_0 & \Lambda & 0 & 0 & \cdots & 0 & 0 & 0 & 0 \\ M_0 & N & \Lambda & 0 & \cdots & 0 & 0 & 0 & 0 \\ 0 & M & N & \Lambda & \cdots & 0 & 0 & 0 & 0 \\ \vdots & \vdots & \vdots & \vdots & \ddots & \vdots & \vdots & \vdots & \vdots \\ 0 & 0 & 0 & 0 & \cdots & M & N & \Lambda & 0 \\ 0 & 0 & 0 & 0 & \cdots & 0 & M & N & \Lambda_r \\ 0 & 0 & 0 & 0 & \cdots & 0 & 0 & M & N_r \end{pmatrix} \tag{4.6.31}$$

if and only if the equalities

$$\boldsymbol{p}_k^T = \boldsymbol{f}^T R^k + \boldsymbol{g}^T S^{r-k}, \qquad k = 0, \dots, r, \tag{4.6.32}$$
$$\boldsymbol{f}^T (N_0 + RM_0) + \boldsymbol{g}^T S^{r-1}(SN_0 + M_0) = \mathbf{0}^T, \tag{4.6.33}$$
$$\boldsymbol{f}^T R^{r-1}(RN_r + \Lambda_r) + \boldsymbol{g}^T (N_r + S\Lambda_r) = \mathbf{0}^T \tag{4.6.34}$$

are true for some vectors $\boldsymbol{f}$ *and* $\boldsymbol{g}$.

Thus, solution of the simultaneous equations $\boldsymbol{p}^T Q = \mathbf{0}^T$ with the matrices Q of the form (4.6.20) or (4.6.31) reduces to finding the roots of certain quadratic matrix equations, checking the matrix $S\Lambda + N + RM$ for non-degeneracy, and solving set of equations of low order. Since the process $\{\eta(t), t \geq 0\}$ is ergodic, solution of the equilibrium equations is unique within a constant factor. Therefore, solutions of equations (4.6.22)–(4.6.25) and (4.6.33)–(4.6.34) are also unique within a constant factor, which can be established from the normalisation condition.

4.6.3. Roots of the square matrix equations

Let us see whether or not it is possible to find the roots of equations (4.6.19). We observe that the matrices Λ and M are non-negative. The matrix N has non-negative off-diagonal elements, and $\Lambda + N + M$ is the matrix of transition intensities. We assume below that the matrix $\Lambda + N + M$ is indecomposable. Then there exists a unique positive solution $\boldsymbol{\pi}$ of the equilibrium equations

$$\boldsymbol{\pi}^T(\Lambda + N + M) = \mathbf{0}^T \tag{4.6.35}$$

with the normalisation condition

$$\boldsymbol{\pi}^T\mathbf{1} = 1. \tag{4.6.36}$$

Let ω denote the absolute value of the maximal diagonal element of the matrix N. Then the matrix $N + \omega I$ is non-negative. We consider the sequence of matrices

$$\begin{aligned} R_0 &= 0, \quad R_k = \frac{1}{\omega}[\Lambda + R_{k-1}(N+\omega I) + R_{k-1}^2 M], \quad k \geq 1, \\ S_0 &= 0, \quad S_k = \frac{1}{\omega}[S_{k-1}^2\Lambda + S_{k-1}(N+\omega I) + M], \quad k \geq 1. \end{aligned} \tag{4.6.37}$$

Obviously, $R_1 \geq R_0$, and it follows from the inequality $R_k \geq R_{k-1}$ that

$$R_{k+1} \geq \frac{1}{\omega}[\Lambda + R_{k-1}(N+\omega I) + R_{k-1}^2 M] = R_k.$$

Therefore, $R_0 \leq R_1 \leq R_2 \leq \dots$

It is also obvious that $\boldsymbol{\pi}^T R_0 \leq \boldsymbol{\pi}^T$, and it follows from the inequality $\boldsymbol{\pi}^T R_k \leq \boldsymbol{\pi}^T$ that

$$\boldsymbol{\pi}^T R_{k+1} = \frac{1}{\omega}\boldsymbol{\pi}^T[\Lambda + R_k(N+\omega I) + R_k^2 M] \leq \frac{1}{\omega}\boldsymbol{\pi}^T(\Lambda + N + M + \omega I) = \boldsymbol{\pi}^T.$$

Therefore, $\boldsymbol{\pi}^T R_k \leq \boldsymbol{\pi}^T$ for all $k \geq 0$.

Let us demonstrate that the elements of the non-negative matrices X satisfying the inequality $\boldsymbol{\pi}^T X \leq \boldsymbol{\pi}^T$ are uniformly bounded. To do this, we observe that the inequalities

$$X_{ij} \leq C = \frac{1}{\min_{s=1,\dots,n} \pi_s}, \qquad i,j = 1,\dots,n,$$

hold for each such $X = (X_{ij})_{i,j=1,\dots,n}$. Indeed, it follows from the inequality $X_{i^*j^*} > C$ that

$$\sum_{i=1}^{n} \pi_i X_{ij^*} \geq \pi_{i^*} X_{i^*j^*} > \pi_{i^*} C \geq 1 \geq \pi_{j^*}.$$

But the resulting inequality contradicts the condition $\boldsymbol{\pi}^T X \leq \boldsymbol{\pi}^T$ and hence, X_{ij} are bounded uniformly by the constant C.

Since the sequence of matrices R_k, $k \geq 0$, is monotonically non-decreasing and bounded, there exists a limit $R = \lim_{k\to\infty} R_k$ and furthermore, the inequality $\boldsymbol{\pi}^T R \leq \boldsymbol{\pi}^T$ is true.

Similarly, the sequence of matrices S_k converges monotonically to $S = \lim_{k\to\infty} S_k$ and $\pi^T S \le \pi^T$. Passing to the limit in recurrence relations (4.6.37), we conclude that the non-negative matrices R and S solve equations (4.6.19).

Let X be some non-negative matrix such that $\Lambda + XN + X^2 M = 0$. Obviously, $X \ge R_0 = 0$. It follows from the inequality $X \ge R_k$ that

$$X = \frac{1}{\omega}[\Lambda + X(N + \omega I) + X^2 M] \ge \frac{1}{\omega}[\Lambda + R_k(N + \omega I) + R_k^2 M] = R_{k+1}.$$

Therefore, $X \ge R_k$ for all $k = 0, 1, \ldots$, and $X \ge R = \lim_{k\to\infty} R_k$. Similarly, if Y is a non-negative root of the equation $Y^2\Lambda + YN + M = 0$, then $Y \ge S$. The matrices R and S are referred to as the *minimum non-negative roots* of equations (4.6.19). There exist other methods of calculating the minimum nonnegative roots R and S (Latouche and Ramaswami, 1993).

Any non-negative matrix X is known to have a real eigenvalue $\rho(X)$ which possesses the following property: the absolute value of any other eigenvalue does not exceed $\rho(X)$. This eigenvalue is called the *spectral radius of the matrix* X which the non-negative eigenvector corresponds to. If the matrix X is indecomposable, then its spectral radius is a prime eigenvalue (Berman and Plemmons, 1979).

Let $\boldsymbol{x}$ and $\boldsymbol{y}$ be non-negative eigenvectors of the matrices R and S corresponding to $\rho(R)$ and $\rho(S)$. Then it follows from the above inequalities $\boldsymbol{\pi}^T R \le \boldsymbol{\pi}^T$ and $\boldsymbol{\pi}^T S \le \boldsymbol{\pi}^T$ that

$$\boldsymbol{\pi}^T \boldsymbol{x} \ge \boldsymbol{\pi}^T R\boldsymbol{x} = \rho(R)\boldsymbol{\pi}^T \boldsymbol{x},$$
$$\boldsymbol{\pi}^T \boldsymbol{y} \ge \boldsymbol{\pi}^T S\boldsymbol{y} = \rho(S)\boldsymbol{\pi}^T \boldsymbol{y}.$$

Since $\boldsymbol{\pi}^T \boldsymbol{x} > 0$ and $\boldsymbol{\pi}^T \boldsymbol{y} > 0$, we obtain

$$\rho(R) \le 1, \qquad \rho(S) \le 1.$$

The following assertion whose proof is rather sophisticated and follows from more general results given in (Naumov, 1995) is true.

THEOREM 4.6.4. *Let the matrix* $\Lambda + N + M$ *be indecomposable, the determinant of the matrix* $\Lambda + sN + s^2 M$ *be not identically zero, and* R *and* S *be the minimum non-negative roots of equations* (4.6.19). *In this case,*

(1) $\rho(R) = 1$ *if and only if* $\boldsymbol{\pi}^T \Lambda \mathbf{1} \ge \boldsymbol{\pi}^T M\mathbf{1}$*; here* $\boldsymbol{\pi}^T R = \boldsymbol{\pi}^T$*; and*

(2) $\rho(S) = 1$ *if and only if* $\boldsymbol{\pi}^T \Lambda \mathbf{1} \le \boldsymbol{\pi}^T M\mathbf{1}$*; here* $\boldsymbol{\pi}^T S = \boldsymbol{\pi}^T$.

In particular, it follows from Theorem 4.6.4 that for the matrix $S\Lambda + N + RM$ to be non-degenerate, the condition $\boldsymbol{\pi}^T \Lambda \mathbf{1} \ne \boldsymbol{\pi}^T M\mathbf{1}$ must be satisfied.

Indeed, if $\boldsymbol{\pi}^T \Lambda \mathbf{1} = \boldsymbol{\pi}^T M\mathbf{1}$, then $\boldsymbol{\pi}^T RS = \boldsymbol{\pi}^T$, and therefore, the matrix $I - RS$ is degenerate. By (4.6.3), the matrix $S\Lambda + N + RM$ in this case is degenerate as well. Unfortunately, simple sufficient conditions for its non-degeneracy are not known yet.

4.6.4. Infinite-capacity Markov system

Let the stochastic behaviour of an infinite-capacity queueing system be described by a homogeneous irreducible Markov process $\{\eta(t), t \geq 0\}$ with transition intensity matrix

$$Q = \begin{pmatrix} N_0 & \Lambda_1 & 0 & 0 & 0 & \cdots \\ M_0 & N & \Lambda & 0 & 0 & \cdots \\ 0 & M & N & \Lambda & 0 & \cdots \\ 0 & 0 & M & N & \Lambda & \cdots \\ \vdots & \vdots & \vdots & \vdots & \vdots & \ddots \end{pmatrix}. \tag{4.6.38}$$

We do not discuss this case in detail and, referring the interested readers to the monograph (Neuts, 1981), just formulate the main result. In this case, it suffices to calculate the minimum non-negative root of the matrix R of the matrix equation $\Lambda + RN + R^2M = 0$ in order to determine the stationary process distribution.

THEOREM 4.6.5. *The irreducible Markov process* $\{\eta(t), t \geq 0\}$ *with the transition intensity matrix* (4.6.38) *is ergodic if and only if* $\rho(R) < 1$ *and the simultaneous equations*

$$\begin{aligned} \boldsymbol{p}_0^T N_0 + \boldsymbol{p}_1^T M_0 &= \boldsymbol{0}^T, \\ \boldsymbol{p}_0^T \Lambda_1 + \boldsymbol{p}_1^T (N + RM) &= \boldsymbol{0}^T, \\ \boldsymbol{p}_0^T \boldsymbol{1} + \boldsymbol{p}_1^T (I - R)^{-1} \boldsymbol{1} &= 1 \end{aligned} \tag{4.6.39}$$

have a unique positive solution $(\boldsymbol{p}_0^T, \boldsymbol{p}_1^T)$. *In this case, the stationary distribution* $\boldsymbol{p}^T = (\boldsymbol{p}_0^T, \boldsymbol{p}_1^T, \ldots)$ *of the process* $\{\eta(t), t \geq 0\}$ *is*

$$\boldsymbol{p}_k^T = \boldsymbol{p}_1^T R^{k-1}, \qquad k = 1, 2, \ldots \tag{4.6.40}$$

This theorem follows from more general results given in (Neuts, 1981).

The probability vector $\boldsymbol{p}^T = (\boldsymbol{p}_0^T, \boldsymbol{p}_1^T, \ldots)$ whose components $\boldsymbol{p}_1, \boldsymbol{p}_2, \ldots$ are of the same length and are related by (4.6.40) is referred to as the *modified geometric distribution*. If all components of the vector $\boldsymbol{p}_k$, $k \geq 0$, are of the same length and are related by

$$\boldsymbol{p}_k^T = \boldsymbol{p}_0^T R^k, \qquad k = 0, 1, \ldots, \tag{4.6.41}$$

then the process $\{\eta(t), t \geq 0\}$ is said to have the matrix-geometric distribution which is intrinsic to the processes with matrices (4.6.38) if $\Lambda_1 = \Lambda$. Indeed, it follows from (4.6.39) that in this case $\boldsymbol{p}_1^T = -\boldsymbol{p}_0^T \Lambda(N+RM)^{-1} = \boldsymbol{p}_0^T R$. Non-degeneracy of the matrix $N+RM$ is proved in (Neuts, 1981). In order to check calculations of the matrix R by recurrence relations (4.6.37) one may use the relation $RM\boldsymbol{1} = \Lambda\boldsymbol{1}$, which follows from the results in (Neuts, 1981).

4.6.5. $PH/PH/2/r$ system with non-homogeneous servers

Application of the results obtained in Sections 4.6.1–4.6.3 is illustrated by the queueing system with two non-homogeneous servers and common buffer of finite capacity r, $1 < r < \infty$. The arrival flow is recurrent and has the phase-type distribution function $A(x)$ with PH-representation $(\boldsymbol{\alpha}, \Lambda)$ of order l, $\boldsymbol{\alpha}^T\boldsymbol{1} = 1$. The customer service times at the server s

are independent of each other, independent of the service at the other server, and also have the phase-type distribution function $B_s(x)$ with *PH*-representation $(\boldsymbol{\beta}_s, M_s)$ of order m_s, $\boldsymbol{\beta}_s^T \mathbf{1} = 1$, $s = 1, 2$. The *PH*-representations $(\boldsymbol{\alpha}, \Lambda)$ and $(\boldsymbol{\beta}_s, M_s)$, $s = 1, 2$, are assumed to be irreducible. The customer can be taken from the queue using any of fixed disciplines FCFS, LCFS, and RANDOM. The customer arrived at an empty system is sent to server 1 with the probability q or to server 2 with the complementary probability $\bar{q} = 1 - q, 0 < q < 1$.

In view of the probabilistic interpretation of the *PH*-distributions (see Section 2.8), the stochastic behaviour of the queueing system under consideration is described by the homogeneous Markov process $\{\eta(t), t \geq 0\}$ with the state set

$$\mathscr{X} = \bigcup_{k=0}^{R} \mathscr{X}_k,$$

where

$$\begin{aligned}
\mathscr{X}_0 &= \{(i, 0),\ i = 1, \dots, l\}, \\
\mathscr{X}_1 &= \mathscr{X}_{11} \cup \mathscr{X}_{12}, \\
\mathscr{X}_{1s} &= \{(i, j, s),\ i = 1, \dots, l, j = 1, \dots, m_s\}, && s = 1, \dots, 2, \\
\mathscr{X}_k &= \{(i, k, j_1, j_2), i = 1, \dots, l, j_s = 1, \dots, m_s, s = 1, 2\}, && k = 2, \dots, R,
\end{aligned}$$

and $R = r + 2$ is the total system capacity.

The system states are interpreted as follows: $\eta(t) = (i, 0)$ if the system is empty and the customer generation process undergoes phase i; $\eta(t) = (i, j, s)$ if there is one customer in the system and the customer at server s undergoes phase j, i having the same sense as before; $\eta(t) = (i, k, j_1, j_2)$ if the system has k customers and the customer at server s undergoes phase j_s, i having the same sense as before.

Under the above assumptions, all states of the process $\{\eta(t), t \geq 0\}$ are communicating and hence, the process is ergodic. Therefore, there exists the limit probabilities

$$\begin{aligned}
p_{i0} &= \lim_{t\to\infty} \mathbf{P}\{\eta(t) = (i, 0)\}, \\
p_{ijs} &= \lim_{t\to\infty} \mathbf{P}\{\eta(t) = (i, j, s)\}, \\
p_{ikj_1j_2} &= \lim_{t\to\infty} \mathbf{P}\{\eta(t) = (i, k, j_1, j_2)\}
\end{aligned}$$

which are strictly positive and coincide with the stationary probabilities.

We do not dwell on the derivation of the equilibrium equations but present them immediately in the matrix form reduced to (4.6.20), leaving it to the readers to try their skills in dealing with *PH*-distributions and matrix equations using the Kronecker matrix product. To avoid confusion, we agree to add the superscript * to the matrices Λ_i, N_i, M_i, and index r that are used to describe the transition intensity matrix Q.

We introduce the vectors

$$\begin{aligned}
\boldsymbol{p}_{00}^T &= (p_{10}, p_{20}, \dots, p_{l0}),\\
\boldsymbol{p}_{1s}^T &= (p_{11s}, \dots, p_{1m_s s}, p_{21s}, \dots, p_{2m_s s}, \dots, p_{lm_s s}),\\
\boldsymbol{p}_0^T &= (\boldsymbol{p}_{00}^T, \boldsymbol{p}_{11}^T, \boldsymbol{p}_{12}^T),\\
\boldsymbol{p}_k^T &= (p_{1,k+1,11}, \dots, p_{1,k+1,1m_2}, \dots, p_{1,k+1,m_1m_2}, \dots, p_{l,k+1,11},\\
&\qquad \dots, p_{l,k+1,1m_2}, \dots, p_{l,k+1,m_1m_2}), \qquad k = 1, \dots, r+1.
\end{aligned}$$

Setting $r^* = r + 2$, we see that the transition intensity matrix of the process $\{\eta(t), t \geq 0\}$ is (4.6.20), and its elements are

$$\begin{aligned}
N_0^* &= \begin{pmatrix} \Lambda & q\boldsymbol{\lambda}\boldsymbol{\alpha}^T \otimes \boldsymbol{\beta}_1^T & \bar{q}\boldsymbol{\lambda}\boldsymbol{\alpha}^T \otimes \boldsymbol{\beta}_2^T \\ I \otimes \boldsymbol{\mu}_1 & \Lambda \oplus M_1 & 0 \\ I \otimes \boldsymbol{\mu}_2 & 0 & \Lambda \oplus M_2 \end{pmatrix},\\
M_0^* &= (0, I \otimes I \otimes \boldsymbol{\mu}_2, I \otimes \boldsymbol{\mu}_1 \otimes I),\\
\Lambda_1^* &= \begin{pmatrix} 0 \\ \boldsymbol{\lambda}\boldsymbol{\alpha}^T \otimes I \otimes \boldsymbol{\beta}_2^T \\ \boldsymbol{\lambda}\boldsymbol{\alpha}^T \otimes \boldsymbol{\beta}_1^T \otimes I \end{pmatrix},\\
N^* &= \Lambda \otimes I \otimes I + I \otimes M_1 \otimes I + I \otimes I \otimes M_2,\\
M^* &= I \otimes \boldsymbol{\mu}_1\boldsymbol{\beta}_1^T \otimes I + I \otimes I \otimes \boldsymbol{\mu}_2\boldsymbol{\beta}_2^T,\\
\Lambda^* &= \boldsymbol{\lambda}\boldsymbol{\alpha}^T \otimes I \otimes I,\\
N_{r^*}^* &= (\Lambda + \boldsymbol{\lambda}\boldsymbol{\alpha}^T) \otimes I \otimes I + I \otimes M_1 \otimes I + I \otimes I \otimes M_2,\\
M_{r^*-1} &= M^*,\\
\Lambda_{r^*}^* &= \Lambda^*.
\end{aligned}$$

Here we use the notation $\boldsymbol{\lambda} = -\Lambda\mathbf{1}$ and $\boldsymbol{\mu}_s = -M_s\mathbf{1}$, which is common for the *PH*-distributions.

Therefore, if we find—in terms of Λ^*, N^*, and M^*—the minimum non-negative roots of the matrix equations (4.6.19) and find that the matrix $S\Lambda^* + N^* + RM^*$ is non-degenerate, then by virtue of Theorem 4.6.2 the stationary probability distribution of the states of the queueing system under consideration should be sought in the form (4.6.21), the vectors $\boldsymbol{p}_0$, $\boldsymbol{p}_{r^*}$, $\boldsymbol{f}$ and $\boldsymbol{g}$ being found from the equations (4.6.22)–(4.6.24) and the normalisation condition $\boldsymbol{p}_k^T\mathbf{1} = 1$.

4.6.6. *MAP/PH/2/r* system with non-homogeneous servers

The system considered in this section differs from the above *PH/PH/2/r* queueing system only in that the customer flow is Markov. According to the description of the Markov flow of customers (see Section 2.1), we assume that it is characterised by matrices Λ and N of order l. We recall that the matrix Λ includes only those transition intensities in the phase set $\{1, 2 \dots, l\}$ under which no new customer appears and the matrix N includes the transition intensities in the same phase set which lead to the arrival of a new customer. We assume that the matrix $\Lambda + N$ is indecomposable.

We for the first time encounter the case of non-recurrent flow of customers. This fact, nevertheless, does not lead to an intractable mathematical model of the queueing system. Moreover, the model is almost the same as in the last case. Indeed, we can describe the system under consideration by a homogeneous Markov process with the same—both in notation and physical sense—state set $\mathscr{X}$ as for the $PH/PH/2/r$ queueing system, which is quite evident if one recalls that the durations of the customer generation phases are distributed exponentially. But this fact does not exhaust the similarities between these models. It turns out that the transition matrices Q_{MAP} and Q_{PH} of the Markov processes describing, respectively, the $MAP/PH/2/r$ and $PH/PH/2/r$ queueing systems coincide, with the exception that in Q_{PH} the matrix N must be substituted for the matrix $\boldsymbol{\lambda}\boldsymbol{\alpha}^T$. Indeed, the (i,s)th element $\lambda_i\alpha_s$ of the matrix $\boldsymbol{\lambda}\boldsymbol{\alpha}^T$ is the intensity of transition of the customer generation process from phase i to phase s accompanied by arrival of a new customer. As mentioned before, the N_{is}th element of the matrix N has the same sense. Therefore, Theorem 4.6.2 can be used again to find the stationary probability distribution of the states of the $MAP/PH/2/r$ queueing system, with the comments being the same as in the last section.

4.6.7. $PH/PH/1/\infty$ system

To illustrate the results of Section 4.6.4, we consider the $PH/PH/1/\infty$ system which differs from that considered in Section 4.5 queueing system only in an infinite-capacity buffer. We again assume that the PH-representations $(\boldsymbol{\alpha}, \Lambda)$ and $(\boldsymbol{\beta}, M)$ of the corresponding orders l and m are irreducible. Other constraints on the system parameters will be discussed below.

In view of the probabilistic phase interpretation of the PH-distributions, the $PH/PH/1/\infty$ queueing system is described by a homogeneous Markov process $\{\eta(t), t \geq 0\}$ with state set $\mathscr{X} = \bigcup_{k=0}^{\infty} \mathscr{X}_k$, where $\mathscr{X}_0 = \{(i,0),\ i = 1,\dots,l\}$, $\mathscr{X}_k = \{(i,k,j),\ i = 1,\dots,l,\ j = 1,\dots,m\}$, $k \geq 1$. The states of the process $\{\eta(t), t \geq 0\}$ are interpreted here as for the case of finite buffer: $\eta(t) = (i,0)$ if the system is empty and the customer generated at the input is in phase i; $\eta(t) = (i,k,j)$ if k customers are in the system and that in service undergoes phase j, the index i having the same sense as before. Since the PH-representations $(\boldsymbol{\alpha}, \Lambda)$ and $(\boldsymbol{\beta}, M)$ are irreducible, we immediately see that the Markov process $\{\eta(t), t \geq 0\}$ is irreducible as well.

Labelling the matrices N_i, M_i, and Λ_i involved in the description of the transition matrix Q by the superscript * again, we see that it is of the form (4.6.38), and its elements are

$$\begin{aligned} N_0^* &= \Lambda, & M_0^* &= I \otimes \boldsymbol{\mu}, & \Lambda_1^* &= \boldsymbol{\lambda}\boldsymbol{\alpha}^T \otimes \boldsymbol{\beta}^T, \\ N^* &= \Lambda \oplus M, & M^* &= I \otimes \boldsymbol{\mu}\boldsymbol{\beta}^T, & \Lambda^* &= \boldsymbol{\lambda}\boldsymbol{\alpha}^T \otimes I, \end{aligned}$$

where, as usual, $\boldsymbol{\lambda} = -\Lambda\mathbf{1}$ and $\boldsymbol{\mu} = -M\mathbf{1}$.

We consider the matrix $\Lambda^* + N^* + M^*$ and prove that under the assumption that the PH-representations $(\boldsymbol{\alpha}, \Lambda)$ and $(\boldsymbol{\beta}, M)$ are irreducible, this matrix is indecomposable. We denote it by A and represent as

$$\begin{aligned} A &= \Lambda^* + N^* + M^* = (\Lambda + \boldsymbol{\lambda}\boldsymbol{\alpha}^T) \otimes I + I \otimes (M + \boldsymbol{\mu}\boldsymbol{\beta}^T) \\ &= \hat{\Lambda} \otimes I + I \otimes \widehat{M} = \hat{\Lambda} \oplus \widehat{M}. \end{aligned}$$

Using the probabilistic phase interpretation of irreducibility of the PH-representations (see Section 2.8), we easily see that A is the transition intensity matrix of the Markov process describing the random walk of two customers in two independent closed queueing networks (one customer in each network) of which the first network interprets irreducibility of the PH-representation $(\boldsymbol{\alpha}, \Lambda)$, and the second one, of $(\boldsymbol{\beta}, M)$. Hence,

$$P(t) = e^{\hat{\Lambda}t} \otimes e^{\widehat{M}t} = e^{(\hat{\Lambda}\oplus\widehat{M})t}$$

is the transition matrix of this process, all its states are communicating, and the matrix $A = \hat{\Lambda} \oplus \widehat{M}$ is indecomposable.

LEMMA 4.6.2. *Let the PH-representations $(\boldsymbol{\alpha}, \Lambda)$ and $(\boldsymbol{\beta}, M)$ be irreducible. Then the solution of the equilibrium equations*

$$\boldsymbol{\pi}^T A = \mathbf{0}^T, \qquad \boldsymbol{\pi}^T \mathbf{1} = 1 \tag{4.6.42}$$

admits the representation

$$\boldsymbol{\pi} = \boldsymbol{\pi}_1 \otimes \boldsymbol{\pi}_2,$$

where $\boldsymbol{\pi}_1$ and $\boldsymbol{\pi}_2$ solve the equilibrium equations

$$\begin{aligned} \boldsymbol{\pi}_1^T(\Lambda + \boldsymbol{\lambda}\boldsymbol{\alpha}^T) &= \mathbf{0}^T, \qquad \boldsymbol{\pi}_1^T\mathbf{1} = 1; \\ \boldsymbol{\pi}_2^T(M + \boldsymbol{\mu}\boldsymbol{\beta}^T) &= \mathbf{0}^T, \qquad \boldsymbol{\pi}_2^T\mathbf{1} = 1. \end{aligned} \tag{4.6.43}$$

PROOF. We demonstrate that the expression $\boldsymbol{\pi} = \boldsymbol{\pi}_1 \otimes \boldsymbol{\pi}_2$, where $\boldsymbol{\pi}_1$ and $\boldsymbol{\pi}_2$ solve the equilibrium equations (4.6.43), solves the equilibrium equations (4.6.42). Indeed, it follows from the properties of the Kronecker product that

$$\begin{aligned} \boldsymbol{\pi}^T A &= (\boldsymbol{\pi}_1^T \otimes \boldsymbol{\pi}_2^T)[(\Lambda + \boldsymbol{\lambda}\boldsymbol{\alpha}^T) \otimes I + I \otimes (M + \boldsymbol{\mu}\boldsymbol{\beta}^T)] \\ &= \boldsymbol{\pi}_1^T(\Lambda + \boldsymbol{\lambda}\boldsymbol{\alpha}^T) \otimes \boldsymbol{\pi}_2^T + \boldsymbol{\pi}_1^T \otimes \boldsymbol{\pi}_2^T(M + \boldsymbol{\mu}\boldsymbol{\beta}^T) = \mathbf{0}^T \otimes \boldsymbol{\pi}_2^T + \boldsymbol{\pi}_1^T \otimes \mathbf{0}^T = \mathbf{0}^T. \end{aligned}$$

Similarly,

$$\begin{aligned} \boldsymbol{\pi}^T\mathbf{1} &= (\boldsymbol{\pi}_1^T \otimes \boldsymbol{\pi}_2^T)\mathbf{1} = (\boldsymbol{\pi}_1^T \otimes \boldsymbol{\pi}_2^T)(\mathbf{1} \otimes \mathbf{1}) \\ &= \boldsymbol{\pi}_1^T\mathbf{1} \otimes \boldsymbol{\pi}_2^T\mathbf{1} = 1 \otimes 1 = 1. \end{aligned}$$

Therefore, $\boldsymbol{\pi}_1 \otimes \boldsymbol{\pi}_2$ solves (4.6.42). By virtue of indecomposability of the matrix A, the equilibrium equations (4.6.42) have a unique solution $\boldsymbol{\pi} > \mathbf{0}$; hence, $\boldsymbol{\pi} = \boldsymbol{\pi}_1 \otimes \boldsymbol{\pi}_2$, which proves the lemma. □

In order to obtain the modified matrix-geometric solution of the form (4.6.40) for the stationary distribution $\boldsymbol{p}^T = (\boldsymbol{p}_0^T, \boldsymbol{p}_1^T, \dots)$ of the process $\{\eta(t), t \geq 0\}$, if any, we have to find the minimum non-negative root of the equation $\Lambda^* + RN^* + R^2M^* = 0$ and establish the conditions under which the spectral radius $\rho(R)$ of the solution R obtained is smaller than one. By virtue of Theorem 4.6.4, the condition $\rho(R) < 1$ amounts to

$$\boldsymbol{\pi}^T(I \otimes \boldsymbol{\mu}\boldsymbol{\beta}^T)\mathbf{1} > \boldsymbol{\pi}^T(\boldsymbol{\lambda}\boldsymbol{\alpha}^T \otimes I)\mathbf{1}, \tag{4.6.44}$$

where $\boldsymbol{\pi}$ solves (4.6.42). This inequality is equivalent to

$$\boldsymbol{\pi}_2^T\boldsymbol{\mu} > \boldsymbol{\pi}_1^T\boldsymbol{\lambda}, \tag{4.6.45}$$

where $\boldsymbol{\pi}_1$ and $\boldsymbol{\pi}_2$ solve (4.6.43). Indeed, (4.6.44) can be rewritten as

$$(\boldsymbol{\pi}_1^T \otimes \boldsymbol{\pi}_2^T)(I \otimes \boldsymbol{\mu}\boldsymbol{\beta}^T - \boldsymbol{\lambda}\boldsymbol{\alpha}^T \otimes I)(\mathbf{1} \otimes \mathbf{1}) > 0$$

or, which is the same,

$$(\boldsymbol{\pi}_1^T \otimes \boldsymbol{\pi}_2^T)(\mathbf{1} \otimes \boldsymbol{\mu}\boldsymbol{\beta}^T\mathbf{1} - \boldsymbol{\lambda}\boldsymbol{\alpha}^T\mathbf{1} \otimes \mathbf{1}) > 0.$$

Since $\boldsymbol{\alpha}^T\mathbf{1} = 1$ and $\boldsymbol{\beta}^T\mathbf{1} = 1$, we obtain

$$\boldsymbol{\pi}_1^T\mathbf{1} \otimes \boldsymbol{\pi}_2^T\boldsymbol{\mu} - \boldsymbol{\pi}_1^T\boldsymbol{\lambda} \otimes \boldsymbol{\pi}_2^T\mathbf{1} > 0.$$

Hence, by virtue of the equalities $\boldsymbol{\pi}_1^T\mathbf{1} = 1$ and $\boldsymbol{\pi}_2^T\mathbf{1} = 1$, we obtain (4.6.45).

Let us turn back to the equilibrium equations (4.6.43) and consider the first equation. In view of the fact that the matrix Λ is non-degenerate, we multiply both sides of the equality $\boldsymbol{\pi}_1^T(\Lambda + \boldsymbol{\lambda}\boldsymbol{\alpha}^T) = \mathbf{0}^T$ from the right by the matrix $\Lambda^{-1}\mathbf{1}$. Recalling that $-\boldsymbol{\alpha}^T\Lambda^{-1}\mathbf{1} = \lambda^{-1}$, where λ is the arriving flow intensity, we obtain

$$\boldsymbol{\pi}_1^T\mathbf{1} + \boldsymbol{\pi}_1^T\boldsymbol{\lambda}\boldsymbol{\alpha}^T\Lambda^{-1}\mathbf{1} = 1 - \frac{1}{\lambda}\boldsymbol{\pi}_1^T\boldsymbol{\lambda} = 0,$$

which yields

$$\boldsymbol{\pi}_1^T\boldsymbol{\lambda} = \lambda. \tag{4.6.46}$$

Similarly we prove that

$$\boldsymbol{\pi}_2^T\boldsymbol{\mu} = \mu, \tag{4.6.47}$$

where μ is the service intensity, $\mu^{-1} = -\boldsymbol{\beta}^T M^{-1}\mathbf{1}$.

It follows from (4.6.45)–(4.6.47) that condition (4.6.45) is equivalent to the inequality $\lambda < \mu$ or, introducing the system intensity traffic $\rho = \lambda/\mu$, to the well-known condition for other infinite-capacity queueing systems:

$$\rho < 1. \tag{4.6.48}$$

By virtue of Theorem 4.6.5, condition (4.6.48) is, therefore, the necessary condition for ergodicity of the process $\{\eta(t), t \geq 0\}$.

Let us prove that (4.6.48) is also a sufficient condition. To this end, we make use of the result of Lavenberg (see Section 3.6) and consider the $S/PH/1$ system with an input buffer that corresponds to the $PH/PH/1/\infty$ system. It is described by the Markov process $\{\xi(t), t \geq 0\}$ with the state set $\mathscr{X} = \{(i, j),\ i = 1, \ldots, l,\ j = 1, \ldots, m\}$, where i and j in the notation of the state (i, j) refer, as before, to the customer generation phase at the input and customer service phase respectively. Since the PH-representations $(\boldsymbol{\alpha}, \Lambda)$ and $(\boldsymbol{\beta}, M)$ are irreducible, all states of the process $\{\xi(t), t \geq 0\}$ are communicating, and therefore, there exist the limit probabilities

$$\pi_{ij} = \lim_{t\to\infty} \mathbf{P}\{\xi(t) = (i, j)\}.$$

Let $\boldsymbol{\pi}^T = (\pi_{11}, \ldots, \pi_{1m}, \pi_{2m}, \ldots, \pi_{lm})$. We easily see that the equilibrium equations for the process $\{\xi(t), t \geq 0\}$ are of the form (4.6.42), where $A = \widehat{\Lambda} \oplus \widehat{M}$; by virtue of Lemma 4.6.2, $\boldsymbol{\pi} = \boldsymbol{\pi}_1 \otimes \boldsymbol{\pi}_2$, where $\boldsymbol{\pi}_1$ and $\boldsymbol{\pi}_2$ solve (4.6.43).

According to the Lavenberg condition, for the steady state of the $PH/PH/1/\infty$ queueing system to exist, it suffices that the inequality $\lambda < \lambda_D^*$ holds, where λ_D^* is the intensity of departure from the $S/PH/1$ queueing system. Obviously,

$$\lambda_D^* = \boldsymbol{\pi}^T(\mathbf{1} \otimes \boldsymbol{\mu}) = (\boldsymbol{\pi}_1^T \otimes \boldsymbol{\pi}_2^T)(\mathbf{1} \otimes \boldsymbol{\mu}) = \boldsymbol{\pi}_2^T \boldsymbol{\mu} = \mu.$$

Therefore, $\lambda < \mu$ or its equivalent $\rho < 1$ is a sufficient and, as shown above, necessary condition for the process $\{\xi(t), t \geq 0\}$ to be ergodic. Therefore, by virtue of Theorem 4.6.5, the stationary distribution $\boldsymbol{p}^T = (\boldsymbol{p}_0^T, \boldsymbol{p}_1^T, \ldots)$ admits representation (4.6.40), and the vectors $\boldsymbol{p}_0$ and $\boldsymbol{p}_1$ are uniquely determined by (4.6.39).

Imposing on the distribution functions $A(x)$ and $B(x)$ some additional constraints discussed below, we obtain in the matrix-geometric form a solution for the vector $\boldsymbol{p}$ up to the vector $\boldsymbol{p}_0$. To this end, we consider two equations of (4.6.39) which in this case are

$$\boldsymbol{p}_0^T \Lambda + \boldsymbol{p}_1^T (I \otimes \boldsymbol{\mu}) = \mathbf{0}^T, \tag{4.6.49}$$

$$\boldsymbol{p}_0^T (\boldsymbol{\lambda\alpha}^T \otimes \boldsymbol{\beta}^T) + \boldsymbol{p}_1^T [\Lambda \oplus M + R(I \otimes \boldsymbol{\mu\beta}^T)] = \mathbf{0}^T. \tag{4.6.50}$$

Multiplying both sides of (4.6.50) from the right by the matrix $I \otimes (\mathbf{1}\boldsymbol{\beta}^T - I)$, we obtain

$$\boldsymbol{p}_1^T \widetilde{M} = \boldsymbol{p}_1^T (I \otimes \boldsymbol{\mu\beta}^T),$$

where $\widetilde{M} = \Lambda \otimes (\mathbf{1}\boldsymbol{\beta}^T - I) - I \otimes M$; this together with (4.6.49) yields

$$\boldsymbol{p}_1^T \widetilde{M} = -\boldsymbol{p}_0^T (\Lambda \otimes \boldsymbol{\beta}^T). \tag{4.6.51}$$

We have obtained relations of this kind for the $PH/PH/1/r$ queueing system in Section 4.5, where we seen that if $\beta(-\gamma_n) \neq 0$, where $\beta(s)$ is the Laplace–Stieltjes transform of the distribution function $B(x)$, for all eigenvalues $\{\gamma_n\}$ of the matrix Λ, then the matrix $\widetilde{M}$ is invertible. Introducing, as in Section 4.5, the matrix $W_0 = -(\Lambda \otimes \boldsymbol{\beta}^T)\widetilde{M}^{-1}$, we obtain from (4.6.51)

$$\boldsymbol{p}_1^T = \boldsymbol{p}_0^T W_0. \tag{4.6.52}$$

Substituting (4.6.52) into (4.6.50), we obtain

$$\boldsymbol{p}_0^T [\boldsymbol{\lambda\alpha}^T \otimes \boldsymbol{\beta}^T + W_0(\Lambda \oplus M + R(I \otimes \boldsymbol{\mu\beta}^T))] = \mathbf{0}^T, \tag{4.6.53}$$

which combined with

$$\boldsymbol{p}_0^T [\mathbf{1} + W_0(I - R)^{-1}(\mathbf{1} \otimes \mathbf{1})] = 1 \tag{4.6.54}$$

derived from the last equation of (4.6.39) and (4.6.52), defines uniquely the vector $\boldsymbol{p}_0$, which proves the theorem below.

THEOREM 4.6.6. *If the condition $\rho < 1$ is fulfilled and $\beta(-\gamma_n) \neq 0$ for all eigenvalues $\{\gamma_n\}$ of the matrix Λ, then the stationary distribution of the probabilities of states of the $PH/PH/1/\infty$ queueing system admits the matrix-geometric representation*

$$\boldsymbol{p}_k^T = \boldsymbol{p}_0^T W_0 R^{k-1}, \qquad k = 1, 2, \ldots,$$

where the vector $\boldsymbol{p}_0$ is uniquely determined by (4.6.53) *and* (4.6.54).

The matrix R is calculated from the recurrence relations (4.6.37), and takes the form

$$R_0 = 0,$$

$$R_k = \frac{1}{\omega}[\boldsymbol{\lambda}\boldsymbol{\alpha}^T \otimes I + R_{k-1}(\Lambda \oplus M) + R_{k-1}^2(I \otimes \boldsymbol{\mu}\boldsymbol{\beta}^T)], \qquad k \geq 1.$$

Calculation of the matrix R can be substantially simplified if one takes into account the block structure of the matrices involved in these relations. For more detailed discussion of the procedure of calculation of the matrix R for the $PH/PH/1/\infty$ queueing system, the readers are referred to (Neuts, 1981). To check the calculations, one can use the equality $RM^*\mathbf{1} = \Lambda^*\mathbf{1}$ which takes the form $R(\mathbf{1} \otimes \boldsymbol{\mu}) = \boldsymbol{\lambda} \otimes \mathbf{1}$.

4.6.8. $MAP/PH/1/\infty$ system

It is obvious that, with account for Section 4.6.6, the results of this section can be easily extended to the $MAP/PH/1/\infty$ system with the Markov arrival flow characterised by matrices Λ and N of order l. The matrix Q_{MAP} of the transition intensities of the Markov process describing the $MAP/PH/1/\infty$ queueing system differs from the matrix Q_{PH} of transition intensities for the $PH/PH/1/\infty$ system only in that in the matrices Q_{PH} one has to replace $\boldsymbol{\lambda}\boldsymbol{\alpha}^T$ by N. In doing so, the condition for irreducibility of the PH-representation $(\boldsymbol{\alpha}, \Lambda)$ is replaced by that for the matrix $\Lambda + N$. Under these assumptions, Lemma 4.6.2 of is true also for the given queueing system—of course, with $\boldsymbol{\lambda}\boldsymbol{\alpha}^T$ replaced by N—because all arguments used to prove it are repeated word for word.

Let us consider the vectors $\boldsymbol{p}_k$, $k = 0, 1, \ldots$ (structured as those for the $PH/PH/1/\infty$ queueing system) which define the stationary probability distribution, if any, of the states of the $MAP/PH/1/\infty$ queueing system. Then the equilibrium equations admits the representation (4.5.1)–(4.5.3) for $\boldsymbol{\lambda}\boldsymbol{\alpha}^T = N$ and $k = 2, 3 \ldots$ in (4.5.3).

By analogy with (4.5.18), we see that the relation

$$\left[\boldsymbol{p}_0^T + \sum_{k=1}^{\infty} \boldsymbol{p}_k^T (I \otimes \mathbf{1})\right](\Lambda + N) = \mathbf{0}^T \tag{4.6.55}$$

is true. Hence, taking into account the property of the Markov flow

$$(\Lambda + N)\mathbf{1} = \mathbf{0} \tag{4.6.56}$$

and the normalisation condition

$$\boldsymbol{p}_0^T \mathbf{1} + \sum_{k=1}^{\infty} \boldsymbol{p}_k^T (\mathbf{1} \otimes \mathbf{1}) = 1, \tag{4.6.57}$$

we conclude that equations (4.6.55) and (4.6.57) are equivalent to the first set of equations in (4.6.43), that is,

$$\boldsymbol{\pi}_1^T (\Lambda + N) = \mathbf{0}^T, \qquad \boldsymbol{\pi}_1^T \mathbf{1} = 1. \tag{4.6.58}$$

In view of uniqueness of the solution of equations (4.6.58) or, which is the same, of (4.6.55) and (4.6.57), we obtain

$$\boldsymbol{\pi}_1^T = \boldsymbol{p}_0^T + \sum_{k=1}^{\infty} \boldsymbol{p}_k^T (I \otimes \mathbf{1}).$$

Now we assume, as before, that $\boldsymbol{\lambda} = -\Lambda\mathbf{1}$. Then it follows from (4.6.56) that $\boldsymbol{\lambda} = N\mathbf{1}$, that is, the ith component of the vector $\boldsymbol{\lambda}$ is the arrival intensity of the customers completing their generation in phase i. Taking this fact into account, we see that for the intensity λ of the flow arriving at the $MAP/PH/1/\infty$ queueing system the equality $\lambda = \boldsymbol{\pi}_1^T\boldsymbol{\lambda}$ is true. Therefore, the formulas (4.6.46) and (4.6.47) (the proof of (4.6.47) is similar) for the $MAP/PH/1/\infty$ queueing system are true, as before. Therefore, condition $\rho < 1$ (4.6.48) is necessary for existence of the limit distribution $\{\boldsymbol{p}_k, k \geq 0\}$ for the $MAP/PH/1/\infty$ system as well. Here Lavenberg's result, which was obtained for the recurrent arrival flow of customers, is inadequate for proving sufficiency of this condition. Sufficiency of the condition $\rho < 1$ can be shown using more sharp results for general queueing systems with stationary arrival flow of customers (Borovkov, 1976). Taking into account this remark and repeating word for word the reasoning for the $PH/PH/1/\infty$ queueing system, we conclude that Theorem 4.6.6 is true for the $MAP/PH/1/\infty$ system as well, with the only difference that the matrix $\boldsymbol{\lambda}\boldsymbol{\alpha}^T$ in (4.6.53) is replaced by the matrix N.

5

$M/G/1/\infty$ system: investigation methods

The $M/G/1/\infty$ system is a single-server queueing system with an infinite buffer. Its input is a Poisson flow of intensity λ. The symbol G in the notation means that customer service times are independent of one another, do not depend on the arrival instants, and are identically distributed by some law $B(t)$. We assume that a customer for service is chosen from the queue in the order of arrival, i.e., according to the FCFS discipline; though, as mentioned in Chapter 3, the results concerning queue length characteristics belong to the LCFS and RANDOM disciplines as well.

If $B(t)$ is not a phase-type distribution function, it is not possible to construct a process $\eta(t)$ that might describe the system and be a continuous-time Markov process with a discrete state set. In particular, the number $\nu(t)$ of customers in the system at an instant t is not such a process, because the distribution of the residual service time of the customer at the server, in distinction to the exponential case, depends on the time during which the customer has already been served.

Various methods are applied to studying the $M/G/1/\infty$ system. In this chapter, we discuss a few of them. Our aim is not so much as to analyse the $M/G/1/\infty$ system in full depth, but to describe main approaches used in the modern queueing theory; this is done best with the use of the $M/G/1/\infty$ system as an illustrative example. As will become clear in the succeeding chapters, not all but only a few of the methods described here are applicable to studying complicated systems, and that too with the sophisticated tools of the theory of random processes.

5.1. Embedded Markov chain

Historically, the method of embedded Markov chains was first applied to investigate the $M/G/1/\infty$ system. We also start the study of the $M/G/1/\infty$ system with precisely this method, because its application is based on the use of the simplest class of random processes—discrete Markov chains.

It has a serious disadvantage: it is helpful only in computing certain characteristics of the number of customers in the system at specific instants. Soon it will be clear that even determination of the (time-) stationary distribution of the number of customers in the system from the characteristics thus obtained requires the use of probabilistic constructions that are

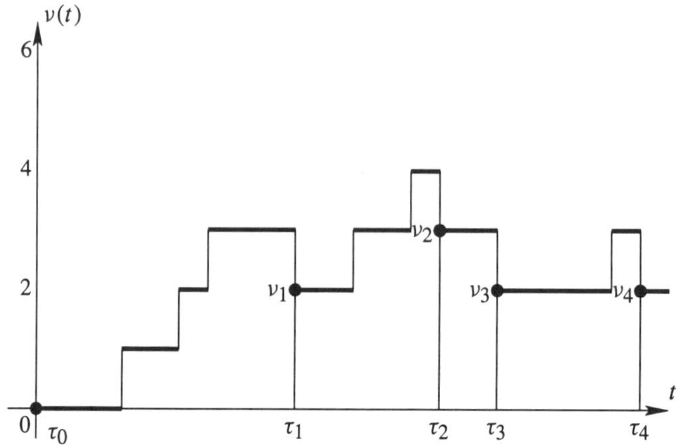

Figure 5.1.

more complicated than Markov chains.

5.1.1. Embedded Markov chain

The method of embedded Markov chains consists of the following: in many queueing systems, there exist, in general, certain random instants at which the non-Markovian process describing the system becomes a Markov process. In the $M/G/1/\infty$ system, such are the instants $\tau_1, \tau_2, \ldots, \tau_n, \ldots$ of termination of service of the first, second, ... , nth customer. Indeed, let $\nu_n = \nu(\tau_n + 0)$ be the number of customers in the system immediately after the instant τ_n (Fig.5.1). For the sake of definiteness, we assume that τ_0 is the initial instant 0 of operation of the system. At this instant, the system is empty, i.e., $\nu_0 = \nu(\tau_0 + 0) = 0$. Let us show that the sequence $\{\nu_n,\ n \geq 0\}$ is a (homogeneous) Markov chain.

Turning back to the description of the $M/G/1/\infty$ system, let us recall that the arrival and service processes for this system have the following properties:

(A) The input flow is Poisson, i.e., the conditions for memoryless and stationarity properties are satisfied.

(B) Service times are independent random variables, do not depend on the input flow, and are identically distributed (with distribution function $B(t)$).

If $\nu_n \geq 1$, then the time interval $(\tau_{n+1} - \tau_n)$ coincides with the service time of the $(n+1)$th customer and does not depend, by virtue of properties A and B, on the functioning of the system before the instant τ_n, and its distribution function $B(t)$ does not depend on the number n. But if $\nu_n = 0$, then the interval $(\tau_{n+1} - \tau_n)$ is the sum of the times from τ_n to the instant of arrival of the $(n+1)$th customer and his service time, i.e., does not depend on the 'past' due to properties A and B. In this case, the distribution function of the difference $(\tau_{n+1} - \tau_n)$ does not depend on the number n (but is the convolution of an exponential distribution function with parameter λ and function $B(t)$). Thus, we have found one more property of the arrival and service processes of the $M/G/1/\infty$ system:

(C) For a fixed ν_n, the random variable $(\tau_{n+1} - \tau_n)$ does not depend on the behaviour of the system up to the instant τ_n, and its distribution does not depend on the number n.

The following property is derived from the properties A and C along similar lines:

(D) For a fixed ν_n, the number of customers arriving in the interval $(\tau_n, \tau_{n+1}]$ also does not depend on the operation of the system up to the instant τ_n and, in particular, on $\nu_0, \nu_1, \dots, \nu_{n-1}$. Obviously, its distribution also does not depend on the number n (however, this must be emphasised, depends on ν_n).

Now, since the number ν_{n+1} of customers at the instant τ_{n+1} in the system is the sum of the number ν_n of customers at the instant τ_n and the number of customers that have arrived in the interval $(\tau_n, \tau_{n+1}]$ minus one served customer, from property D we find that the sequence $\{\nu_n = \nu(\tau_n + 0),\ n \geq 0\}$ is a (homogeneous) Markov chain.

Obviously, the state set $\mathscr{X}$ of the embedded Markov chain $\{\nu_n, n \geq 0\}$ is a set of non-negative integers $\mathscr{X} = \{0, 1, 2, \dots\}$.

An attentive reader would have noticed that in demonstrating the Markov nature of the sequence $\nu_n = \nu(\tau_n + 0)$ we have used much stronger conditions than those stipulated in describing the arrival and service processes, because the number of arrived customers and their service times were assumed to be independent of the random instants τ_n, but not of deterministic instants, as required by the description. This intentional simplification can be eliminated through a strict Markov property, which we have avoided as it is rather complicated for an unprepared reader.

Also note that in the proof the input flow is not assumed to be ordinary. This must be so, because the results of this section, like the results of this chapter, with slight modification can be extended to an $M/G/1/\infty$ system with batch arrivals (a non-ordinary Poisson input flow) and, additionally, many other systems with batch arrivals can be investigated virtually along the same lines as analogous systems with ordinary input flows.

We now find the transition probabilities p_{ij} for the embedded Markov chain $\{\nu_n, n \geq 0\}$.

First, let us find the probability β_k that k new customers arrive during the time of service of a customer. Since every customer is served with probability $dB(t)$ in the interval $(t, t + dt)$ and exactly k customers arrive in time t with probability $(\lambda t)^k \exp\{-\lambda t\}/k!$, applying the total probability formula, we obtain

$$\beta_k = \int_0^\infty \frac{(\lambda t)^k}{k!} e^{-\lambda t} dB(t).$$

The number β_k is called the kth exponential moment of the distribution function $B(t)$. Differentiating k times the Laplace–Stieltjes transform

$$\beta(s) = \int_0^\infty e^{-st} dB(t),$$

we express the exponential moment β_k as

$$\beta_k = \frac{(-\lambda)^k}{k!} \beta^{(k)}(\lambda).$$

It is easy to compute β_k if the Laplace–Stieltjes transform $\beta(s)$ admits a simple representation; in particular, if $B(t)$ is a hyper-Erlang distribution.

Now let $\nu_n = i > 0$. For j customers to exist at the instant τ_{n+1} in the system, $(j-i+1)$ customers must arrive at the system with regard for the departure of the customer that has been served. Therefore, $p_{ij} = \beta_{j-i+1}, i = 1, 2, \dots, j = i-1, i, i+1, \dots$ But if $\nu_n = 0$, then the service of the $(n+1)$th customer begins immediately upon his arrival, i.e., p_{0j} is equal to p_{1j}. Thus, the matrix of transition probabilities p_{ij} of the embedded Markov chain ν_n is of the form

$$P = (p_{ij}) = \begin{pmatrix} \beta_0 & \beta_1 & \beta_2 & \cdots \\ \beta_0 & \beta_1 & \beta_2 & \cdots \\ 0 & \beta_0 & \beta_1 & \cdots \\ \vdots & \vdots & \vdots & \ddots \end{pmatrix}.$$

5.1.2. Stationary state probabilities of the embedded Markov chain

Under certain constraints, which will be described below, stationary probabilities

$$p_i^* = \lim_{n\to\infty} \mathbf{P}\{\nu_n = i\}$$

exist for the embedded Markov chain $\{\nu_n,\ n \geq 0\}$. The simultaneous equilibrium equations for p_i^*, $i \geq 0$, take the form

$$p_i^* = p_0^*\beta_i + \sum_{k=1}^{i+1} p_k^*\beta_{i-k+1}, \qquad i \geq 0. \tag{5.1.1}$$

Equations (5.1.1) can be numerically solved recursively. Indeed, the equation for p_0^* yields

$$p_1^* = p_0^*\frac{1-\beta_0}{\beta_0} = p_0^* r_1,$$

the equation for p_1^* yields

$$p_2^* = \frac{p_1^*(1-\beta_1) - p_0^*\beta_1}{\beta_0} = p_0^*\frac{r_1(1-\beta_1)-\beta_1}{\beta_0} = p_0^* r_2$$

etc. Thus,

$$p_i^* = p_0^* r_i, i \geq 0,$$

where r_i is defined by the recurrence relation

$$r_0 = 1, \qquad r_1 = \frac{1-\beta_0}{\beta_0},$$

$$r_i = \frac{1}{\beta_0}\left[r_{i-1} - \beta_{i-1} - \sum_{k=1}^{i-1} r_k\beta_{i-k}\right], \qquad i \geq 2.$$

In Sections 5.3 and 5.4, we will derive convenient formulas for application in the algorithm to find p_i^*, more exactly, time-stationary probabilities

$$p_i = \lim_{t\to\infty} \mathbf{P}\{\nu(t) = i\}$$

of states. But in Section 5.1.3 we will show that p_i^* and p_i coincide.

The probability p_0^* is computed using the normalisation condition $\sum_{i=0}^{\infty} p_i^* = 1$, i.e., the formula

$$p_0^* = \left(\sum_{i=0}^{\infty} r_i\right)^{-1}.$$

But the vital point here is the accuracy of computation of p_0^*. Although the sum $\sum_{i=0}^{\infty} r_i$, by the assumption that the stationary probabilities p_i^* exist, must converge, we cannot predict when the computation must be terminated so that a given accuracy of computation p_0^* is guaranteed. Moreover, the recurrence relation thus obtained is not helpful in computing the moments of the distribution of the number of customers in the system, because even if convergence of the series $\sum_{i=0}^{\infty} r_i$ is established, it does not imply convergence of the series $\sum_{i=0}^{\infty} i r_i$, $\sum_{i=0}^{\infty} i^2 r_i$, etc. Therefore, to solve (5.1.1) we use the generating function

$$P^*(z) = \sum_{i=0}^{\infty} p_i^* z^i.$$

Another method of computing p_0^*, indeed under an additional assumption, will be described in Section 5.1.5.

Multiplying the ith equation by z^i and summing over $i = 0, 1, \ldots$, we obtain

$$P^*(z) = p_0^* \sum_{i=0}^{\infty} \beta_i z^i + \sum_{i=0}^{\infty} \sum_{k=1}^{i+1} p_k^* \beta_{i-k+1} z^i.$$

Since

$$\sum_{i=0}^{\infty} \beta_i z^i = \sum_{i=0}^{\infty} \int_0^{\infty} \frac{(\lambda t)^i}{i!} e^{-\lambda t} dB(t) z^i = \int_0^{\infty} e^{-\lambda t(1-z)} dB(t) = \beta(\lambda - \lambda z),$$

we obtain

$$P^*(z) = p_0^* \sum_{i=0}^{\infty} \beta_i z^i + \frac{1}{z} \sum_{i=0}^{\infty} z^i \left(\sum_{k=0}^{i} p_k^* \beta_{i-k} - p_0^* \beta_i\right).$$

Using the convolution property of the generating function, we find

$$P^*(z) = p_0^* \beta(\lambda - \lambda z) + \frac{1}{z}[P^*(z) - p_0^*]\beta(\lambda - \lambda z).$$

Hence

$$P^*(z) = \frac{(1-z)\beta(\lambda - \lambda z)}{\beta(\lambda - \lambda z) - z} p_0^*.$$

The probability p_0^* is determined from the normalisation condition

$$\sum_{i=0}^{\infty} p_i^* = P^*(1) = 1.$$

Applying the L'Hospital rule, we obtain

$$P^*(1) = \frac{p_0^*}{1 - \lambda b} = 1.$$

Hence

$$p_0^* = 1 - \rho,$$

where

$$b = \int_0^\infty t \, dB(t)$$

is the mean service time and $\rho = \lambda b$ is the traffic intensity of the system. Thus, the final expression for $P^*(z)$ is

$$P^*(z) = (1-\rho)\frac{(1-z)\beta(\lambda - \lambda z)}{\beta(\lambda - \lambda z) - z}, \tag{5.1.2}$$

which is known as the *Pollaczek–Khinchin formula*.

Since the probability p_0^* is necessarily positive, the condition $\rho < 1$ is necessary for existence of stationary probabilities of the states of the embedded Markov chain. Below we will demonstrate that the condition $\rho < 1$ is not only necessary, but also sufficient for the embedded Markov chain $\{\nu_n, \ n \geq 0\}$ to be ergodic.

Differentiating formula (5.1.2) with respect to z at the point $z = 1$, we obtain the mean number N of customers in the system in terms of the embedded Markov chain in the steady state:

$$N = P^{*\prime}(1) = \rho + \frac{\lambda^2 b^{(2)}}{2(1-\rho)}. \tag{5.1.3}$$

Here

$$b^{(2)} = \int_0^\infty t^2 dB(t)$$

is the second moment of the service time. According to formula (5.1.3), for the stationary mean queue length to be finite, it is necessary and sufficient that $b^{(2)}$ exist. Similar formulas can be easily derived also for higher moments of the stationary queue length.

Introducing the variation coefficient for the service time

$$C_B = (b^{(2)} - b^2)^{1/2}/b,$$

we can rewrite formula (5.1.3) as

$$N = \rho + \frac{\rho^2(1 + C_B^2)}{2(1-\rho)},$$

which is called the *Pollaczek–Khinchin formula for the mean number of customers* in the system in the steady state. Since $C_B = 1$, in particular, for exponential distributions, we obtain the well-known result

$$N = \rho/(1-\rho)$$

for the $M/M/1/\infty$ system.

Finally, we prove that the condition $\rho < 1$ is also sufficient for the existence of a stationary distribution of the embedded Markov chain $\{\nu_n, n \geq 0\}$. Applying the Moustafa criterion (Theorem 1.4.4), in which setting $x_i = i$, $i \geq 0$, and $i_0 = 1$, let us write explicit expressions of the sums in inequalities (1.4.3) and (1.4.4). As shown above,

$$p_{ij} = \begin{cases} \beta_j, & i = 0; \\ \beta_{j-i+1}, & j \geq i-1,\ i \geq 1; \\ 0 & \text{otherwise.} \end{cases}$$

Then the left-hand side of inequality (1.4.3) for $i \geq i_0 = 1$ takes the form

$$\sum_{j=0}^{\infty} j p_{ij} = \sum_{j=i-1}^{\infty} j\beta_{j-i+1} = \sum_{j=0}^{\infty}(j+i-1)\beta_j = \sum_{j=0}^{\infty} j\beta_j + i - 1 = \rho + i - 1.$$

Similarly, the left-hand side of inequality (1.4.4) for $i = 0 < i_0 = 1$ takes the form

$$\sum_{j=0}^{\infty} j p_{0j} = \sum_{j=0}^{\infty} j\beta_j = \rho.$$

Hence, if inequalities (1.4.3) and (1.4.4) are satisfied, then two conditions hold: $\rho + i - 1 \leq i - \varepsilon$ and $\rho < \infty$ or, which is the same, $\rho \leq 1 - \varepsilon$. In turn, if $\rho < 1$, then the last inequality is obviously true for a suitable ε.

Finally, we find that the condition $\rho < 1$ is sufficient and, as shown before, necessary for the existence of the stationary probabilities p_i^*, $i \geq 0$.

5.1.3. Time-stationary queue distribution

In the previous section, we found stationary probabilities of the states of an embedded Markov chain generated by the instants of departure of customers from the system. But, for practical purposes, of great interest are the time-stationary probabilities

$$p_i = \lim_{t\to\infty} \mathbf{P}\{\nu(t) = i\},$$

because, since the input flow is Poisson, precisely these probabilities characterise the queue distribution at the instant of arrival of a customer at the system in the steady state (see Section 2.1). We now show that the $M/G/1/\infty$ system obeys the *Khinchin stationary queue law*: the time-stationary distribution of the number of customers in the system coincides with the stationary distribution of the number of customers in the system for the embedded Markov chain generated by the instants of departure of customers from the system, i.e.,

$$\lim_{t\to\infty} \mathbf{P}\{\nu(t) = i\} = \lim_{n\to\infty} \mathbf{P}\{\nu_n = i\}.$$

Therefore, $P^*(z)$ is also the generating function

$$P(z) = \sum_{i=0}^{\infty} p_i z^i$$

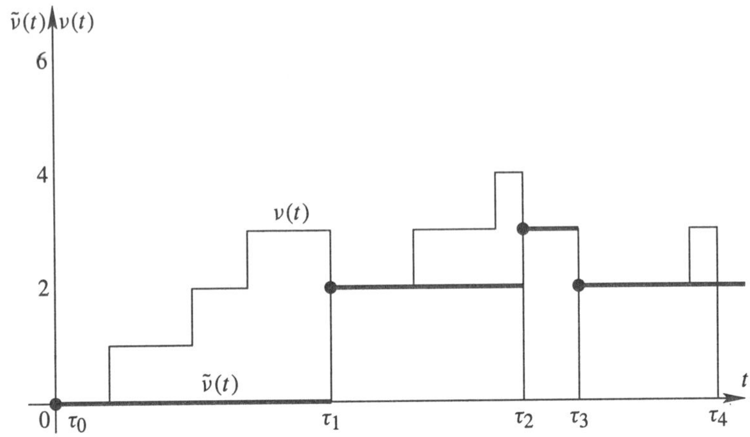

Figure 5.2.

of the (time-) stationary distribution of the number of customers in the system.

To derive the Khinchin stationary queue law, let us consider a process $\{\tilde{\nu}(t),\ t \geq 0\}$ that takes a constant value $\tilde{\nu}(t) = \nu_n$ on the interval $\tau_n \leq t < \tau_{n+1}$. This value is interpreted as the state of the process on this time interval (in Fig. 5.2, the process $\{\tilde{\nu}(t), t \geq 0\}$ is shown by a thick line). In other words, $\tilde{\nu}(t)$ is the number of customers that exist in the system immediately after the instant τ_n of departure of the nth customer from the system. The process $\{\tilde{\nu}(t), t \geq 0\}$ is semi-Markov (this is demonstrated exactly along the same lines as the Markov nature of the embedded chain $\{\nu_n, n \geq 0\}$). Furthermore, if $\nu_n > 0$, then the time $(\tau_{n+1} - \tau_n)$ between the the nth and $(n+1)$th instants of change of state of this process is distributed by the law $B(t)$. But if $\nu_n = 0$ (the system at the instant τ_n is empty), then the time $(\tau_{n+1} - \tau_n)$ is the sum of two independent times: an exponentially distributed time with parameter λ up to instant of arrival of a customer at an empty system and the service time of the arrived customer, which is distributed by the law $B(t)$. It is also obvious that the embedded Markov chains $\{\nu_n, n \geq 0\}$ and $\{\tilde{\nu}_n = \tilde{\nu}(\tau_n + 0),\ n \geq 0\}$ for the initial process $\{\nu(t), t \geq 0\}$ describing the number of customers at instant t and the semi-Markov process $\{\tilde{\nu}(t), t \geq 0\}$ coincide.

By virtue of Theorem 1.6.2, the semi-Markov process $\{\tilde{\nu}(t), t \geq 0\}$ converges to a stationary process as $t \to \infty$ if and only if the embedded Markov process $\{\nu_n,\ n \geq 0\}$ also converges to a stationary process as time grows. But this theorem cannot be applied to finding the stationary probabilities of states p_i, $i \geq 0$, because the processes $\{\tilde{\nu}(t), t \geq 0\}$ and $\{\nu(t), t \geq 0\}$ are different. Therefore, we will use Theorem 1.6.1, according to which the probability that a stationary semi-Markov process passes to the state j in an arbitrary time interval $(t, t + dt)$ is $p_j^* dt / \left(\sum_k f_k p_k^*\right)$, where f_k is the mean sojourn time of the process in the state k and p_j^* is the stationary probability of the state j of the embedded Markov chain.

We now apply this result to the process $\tilde{\nu}(t)$. Obviously, $f_k = b$ for $k > 0$ and $f_0 =$

$b + 1/\lambda$. Hence,

$$\sum_{k=0}^{\infty} f_k p_k^* = b + p_0^*/\lambda = b(1 + p_0^*/\rho) = 1/\lambda,$$

and, therefore, the probability that the semi-Markov process $\{\tilde{\nu}(t), -\infty < t < \infty\}$ constructed for the system $M/G/1/\infty$ in the steady state passes to the state j in any time interval of length dt is $\lambda p_j^*\, dt$. In terms of the system, this is exactly the probability that the service of a customer ends in an interval of length dt and j customers remain in the system after the departure of this customer.

Now we are in a position to find the time-stationary probabilities p_i of the states of the $M/G/1/\infty$ system. Let $A(i)$, $i \geq 0$, denote the event that 'at an arbitrary instant (which, without loss of generality, can be taken to be zero in the steady state) i customers are in the system.' Let us represent this event as the union of 'infinitesimal' events $A(i, j, x)$, $i \geq 0$, $j = 0, \ldots, i$, $x > 0$: 'a customer quits the system in the time interval $(-x - dx, -x)$ and j customers remain in the system after his departure, and no customer is served in the interval $[-x, 0)$ and $(i - j)$ customers arrive in this interval.' Obviously, the events $A(i, j, x)$ are mutually non-overlapping for different pairs (j, x) and their union over j and x is the event $A(i)$.

Let us find the probabilities of the events $A(i, j, x)$, considering the cases $j \geq 1$ and $j = 0$ separately.

For $j \geq 1$, the probability of the event $A(i, j, x)$ is equal to the probability that the service of a customer ends in the interval $(-x - dx, -x)$ and j customers remain in the system after his departure (this probability, as shown above, is $\lambda p_j^*\, dx$), the service of the customer that was taken up for service at the instant $-x$ (with the probability $1 - B(x)$) is not completed in the residual time x, and $(i - j)$ customers arrive in this time interval (since the input flow is Poisson, the probability of the latter event is $(\lambda x)^{i-j} e^{-\lambda x}/(i - j)!$). Since the events described above are independent, we have

$$\mathbf{P}(A(i, j, x)) = \lambda p_j^*\, dx[1 - B(x)](\lambda x)^{i-j} e^{-\lambda x}/(i - j)!.$$

For $j = 0$, the probability that the service of a customer is completed in the time interval $(-x - dx, -x)$ and the system is empty after his departure, as before, is $\lambda p_0^*\, dx$. But here, too, we must consider two cases. If $i = 0$, then no customer must arrive at the system in time x (the probability of this event is $e^{-\lambda x}$). Therefore,

$$\mathbf{P}(A(0, 0, x)) = \lambda p_0^*\, dx\, e^{-\lambda x}.$$

If $i \geq 1$, a customer must arrive (with the probability $\lambda e^{-\lambda y}\, dy$) at the system in some intermediate time interval $(-x+y, -x+y+dy), 0 < y < x$, after the instant $-x$. Its service is not completed in time $(x-y)$ (with the probability $1-B(x-y)$) and $(i-1)$ customers must arrive at the system in this time $(x-y)$ (with the probability $[\lambda(x-y)]^{i-1} e^{-\lambda(x-y)}/(i-1)!$). Therefore, by the total probability formula,

$$\mathbf{P}(A(i, 0, x)) = \lambda p_0^*\, dx \int_0^x \lambda e^{-\lambda y}[1 - B(x - y)] \frac{[\lambda(x - y)]^{i-1}}{(i - 1)!} e^{-\lambda(x-y)} dy.$$

Now using the probability addition formula, we obtain

$$p_0 = \lambda p_0^* \int_0^\infty e^{-\lambda x} dx,$$

$$p_i = \sum_{j=1}^{i} \lambda p_j^* \int_0^\infty \frac{(\lambda x)^{i-j}}{(i-j)!} e^{-\lambda x}[1 - B(x)]\,dx$$

$$+ \lambda p_0^* \int_0^\infty dx \int_0^x e^{-\lambda(x-y)} \lambda e^{-\lambda y} \frac{[\lambda(x-y)]^{i-1}}{(i-1)!}[1 - B(x-y)]\,dy, \qquad i \geq 1.$$

Hence, introducing the generating function

$$P(z) = \sum_{i=0}^{\infty} p_i z^i,$$

we obtain

$$P(z) = \lambda[P^*(z) - p_0^*] \int_0^\infty e^{-\lambda x(1-z)}[1 - B(x)]\,dx$$

$$+ \lambda p_0^* z \int_0^\infty dx \int_0^x [1 - B(x-y)] e^{-\lambda(x-y)(1-z)} \lambda e^{-\lambda y} dy + \lambda p_0^* \int_0^\infty e^{-\lambda x} dx.$$

Integrating the equality

$$\beta(s) = \int_0^\infty e^{-sx} dB(x)$$

by parts, we obtain

$$\int_0^\infty e^{-sx}[1 - B(x)]\,dx = \frac{1 - \beta(s)}{s},$$

and thus we arrive at the formula

$$\int_0^\infty e^{-\lambda x(1-z)}[1 - B(x)]\,dx = \frac{1 - \beta(\lambda - \lambda z)}{\lambda - \lambda z}.$$

After simple transformations, with the use of the convolution formula, we obtain

$$P(z) = \lambda[P^*(z) - p_0^*] \frac{1 - \beta(\lambda - \lambda z)}{\lambda - \lambda z} + \lambda p_0^* z \frac{1 - \beta(\lambda - \lambda z)}{\lambda - \lambda z} + p_0^*$$

$$= P^*(z) \frac{1 - \beta(\lambda - \lambda z)}{1 - z} + p_0^* \beta(\lambda - \lambda z).$$

Replacing $P^*(z)$ by its expression (5.1.2), we finally obtain

$$P(z) = P^*(z).$$

This completes the proof of the Khinchin stationary queue law.

The reader is advised to carry out the above computations independently, because we apply similar steps in the sequel, omitting the details.

The Khinchin stationary queue law implies, in particular, that the mean number N of customers in the system in steady state is also given by formula (5.1.3).

Note that the Khinchin stationary queue law also holds for certain other systems.

5.1.4. Non-stationary state probabilities for the embedded Markov chain

Let $p_i(n)$ denote the probability that the embedded Markov chain $\{\nu_n,\ n \geq 0\}$ exists in state i after the nth step. In this section, we find the non-stationary probabilities $p_i(n)$ in terms of the double transformation

$$P(z_1, z_2) = \sum_{n=0}^{\infty}\sum_{i=0}^{\infty} p_i(n) z_1^n z_2^i.$$

According to the theory of Markov chains (Section 1.4), in order to find $p_i(n)$ it suffices to calculate the matrix $(p_{ij}^{(n)})$ of transition probabilities for n steps, which is nth power of the matrix (p_{ij}), and apply it to the vector of initial probabilities $\boldsymbol{p}_0^T(0) = (p_0(0), p_1(0), \dots)$. Since it is not easy to calculate an nth-degree matrix, particularly, an infinite-degree matrix, we use a different method from the renewal theory. This method can be used, because the embedded Markov chain is of special type, more exactly, it is an integer-valued random walk on a straight line with delaying barrier at zero (see Section 1.5). The merits of this method will soon be clear (see Section 5.5).

As already mentioned, we assume, for the sake of simplicity, that the system is empty at the initial instant 0.

For a Markov chain $\{\nu_n, n \geq 0\}$, the instants τ_n after which the system becomes empty are referred to as the *renewal epochs*. Obviously, the renewal epochs (more exactly, the numbers of steps at which renewal occurred) are non-negative random integers. We also assume that the first renewal instant is the origin 0.

The number of steps between neighbouring renewals, i.e., neighbouring instants when the system becomes empty, is called the *busy period*. Obviously, busy periods are independent, identically distributed, positive, integer-valued random variables. Let g_n, $n \geq 1$, denote the probability that the busy period is equal to n. In order to find g_n, we introduce auxiliary variables $g_n(m)$, $n \geq 1, m \geq 1$,—the probabilities that the busy period is n, under the condition that there were m customers in the system at the initial instant 0 (in this case, the busy period is said to begin with m customers). Obviously, $g_n(1) = g_n$.

Assume that

$$G(z) = \sum_{n=1}^{\infty} g_n z^n, \qquad G(z \mid m) = \sum_{n=1}^{\infty} g_n(m) z^n.$$

We observe that in this case the embedded Markov chain has one more interesting property: its state cannot decrease by more than one in a step (such random walks are said to be continuous from below). Therefore, for a Markov chain to attain the state 0 from the state m, it must first attain the state $(m-1)$ from the state m, then attain the state $(m-2)$ from the state $(m-1)$, etc. Obviously, the times between neighbouring (first) attainments of the states $m-1, m-2, \dots, 0$ are independent random variables with the distribution $\{g_n,\ n \geq 1\}$. Hence, the total time to attain the state 0 is the sum of these attainment times, i.e., in of the generating function terms,

$$G(z \mid m) = (G(z))^m. \tag{5.1.4}$$

In order to calculate the distribution of the busy period, we must recall that the busy period terminates with the probability β_0 in the first step, the Markov chain returns with the probability β_1 to state 1 after the first step and goes to state 2 with the probability β_2, etc. Therefore,

$$g_1 = \beta_0,\, g_n = \beta_1 g_{n-1}(1) + \beta_2 g_{n-1}(2) + \cdots, \qquad n > 1.$$

Hence, using generating functions and applying (5.1.4), we obtain

$$G(z) = z\beta(\lambda - \lambda G(z)). \tag{5.1.5}$$

We will examine (5.1.5) in more detail in Section 5.5. Here we only say that it has a unique continuous solution for all z, $0 \le z \le 1$. Moreover, $0 \le G(z) \le 1$.

For further analysis, we need to introduce the renewal series $\{h_n, n \ge 0\}$, which is a set of probabilities that the system is empty after the instant τ_n. According to the renewal theory (see Section 1.3), the generating function

$$H(z) = \sum_{n=0}^{\infty} h_n z^n$$

is defined by the formula

$$H(z) = \frac{1}{1 - G(z)}. \tag{5.1.6}$$

Let a customer arrive at the instant 0 at an empty system. Let $f_i(n)$ denote the probability that the busy period does not terminate at the nth step and i, $i > 0$, customers are in the system after the nth step. Therefore,

$$f_i(1) = \beta_i,\, F_1(z) = \sum_{i=1}^{\infty} f_i(1) z^i = \beta(\lambda - \lambda z) - \beta_0,$$

and for $f_i(n)$, $n > 1$, we have the relation

$$f_i(n) = \sum_{j=1}^{i+1} f_j(n-1)\beta_{i-j+1}, \tag{5.1.7}$$

which is derived as follows: i customers are in the system after the nth step and the busy period does not terminate if the busy period does not terminate after the $(n-1)$th step, there are j customers in the system, and $(i - j + 1)$ more customers arrive in the course of service of the routine customer. Using the double generating function

$$F(z_1, z_2) = \sum_{n=1}^{\infty} \sum_{i=1}^{\infty} f_i(n) z_1^n z_2^i,$$

from (5.1.7) we obtain

$$\frac{1}{z_1}[F(z_1, z_2) - z_1 F_1(z_2)] = \frac{1}{z_2}[F(z_1, z_2)\beta(\lambda - \lambda z_2) - z_2 f(z_1)],$$

where

$$f(z) = \beta_0 \sum_{n=1}^{\infty} f_1(n) z^n.$$

Therefore,

$$F(z_1, z_2) = z_2 \frac{z_1 \beta(\lambda - \lambda z_2) - z_1 \beta_0 - z_1 f(z_1)}{z_2 - z_1 \beta(\lambda - \lambda z_2)}. \tag{5.1.8}$$

Before determining $f(z)$, we observe that $F(z_1, z_2)$ is a continuous function of z_1 and z_2 in the domains $0 \leq z_1 < 1$ and $0 \leq z_2 \leq 1$ respectively. Therefore, at the points (z_1, z_2) at which the denominator of (5.1.8) vanishes, the numerator must also vanish. But, according to (5.1.5), the denominator of (5.1.8) zeroises at the point $(z_1, G(z_1))$. Hence, $z_1\beta_0 + z_1 f(z_1) = G(z_1)$. We finally obtain

$$F(z_1, z_2) = z_2 \frac{z_1 \beta(\lambda - \lambda z_2) - G(z_1)}{z_2 - z_1 \beta(\lambda - \lambda z_2)}. \tag{5.1.9}$$

Now we are able to find $p_i(n)$. For this purpose, we observe that the system is empty after the instant τ_n if a renewal has occurred at the instant τ_n. Therefore,

$$p_0(n) = h_n.$$

At the instant $\tau_n + 0$, the system would contain i, $i > 0$, customers if a renewal occurred at some intermediate instant τ_m and in the remaining $(n - m)$ steps the system was never emptied and the Markov chain passed to the state i, i.e.,

$$p_i(n) = \sum_{m=0}^{n-1} h_m f_i(n - m).$$

Applying the double transformation $P(z_1, z_2)$, we obtain, by virtue of (5.1.6) and (5.1.9), the final formula

$$P(z_1, z_2) = \frac{z_2}{1 - G(z_1)} \frac{z_1 \beta(\lambda - \lambda z_2) - G(z_1)}{z_2 - z_1 \beta(\lambda - \lambda z_2)}.$$

Finally, let us re-examine the condition for the existence of stationary probabilities of states for an embedded Markov chain (see Section 1.2) and derive it from other considerations. According to the general theory of Markov chains (see Section 1.4), for an irreducible non-periodic Markov chain (and, as easily verified, such is the embedded Markov chain of the system $M/G/1/\infty$) to have a (proper) limiting distribution

$$p_i = \lim_{n \to \infty} p_i(n),$$

it is necessary and sufficient that the mean recurrence time to some (consequently, to any) state be finite; otherwise $p_i(n) \xrightarrow[n \to \infty]{} 0$ for any i. Applying this result to the state 0 and since the time of recurrence to the state 0 is equal to the busy period, for a stationary state to exist, it is necessary and sufficient that the mean busy period be finite; otherwise

$p_i(n) \xrightarrow[n\to\infty]{} 0$ for all i or, which is the same, the number of customers in the system tends to infinity in probability.

Differentiating (5.1.5) with respect to z and then passing to the limit as $z \to 1$, we find that the mean length of the busy period $g = 1/(1-\rho)$ is finite as soon as $\rho = \lambda b < 1$. Hence, the stationary state exists. But if $\rho = 1$, then, though the busy period is finite with the probability one, its mean length is infinite (see Section 5.5) and the queue length at the departure instants grows indefinitely. Finally, if $\rho > 1$ (see also Section 5.5), the busy period may be infinite (never ends) with a positive probability. Detailed analysis shows that the queue in this case is infinite even with the probability one.

Analogous queue properties also hold for the time probability $p_i(t)$.

5.1.5. Embedded Markov chain: description in terms of random walk

Now we consider a somewhat different interpretation of the embedded Markov chain $\{\nu_n,\ n \geq 0\}$ in terms of random walk with delaying barrier at zero. Moreover, using the results of this section, we will apply another method to finding, practically without computing, the stationary probability p_0^* and the mean number N of customers in the $M/G/1/\infty$ system.

Let ξ_n denote the service time of the nth customer and let η_n denote the number of customers that arrive at the system during this time. Clearly, the value of ν_n upon completion of service of the $(n+1)$th customer, on the one hand, decreases by one due to the departure of the served customer and, on the other hand, increases by η_{n+1}—the number of customers that arrived in time ξ_{n+1}. But this is not the case if a customer arrives at an empty system, because the customer is taken up for service immediately and the number of customers in the system after his departure is simply equal to η_{n+1}. Therefore, ν_n satisfies the recurrence relation

$$\nu_{n+1} = \begin{cases} \nu_n - 1 + \eta_{n+1}, & \nu_n \geq 1, \\ \eta_{n+1}, & \nu_n = 0. \end{cases}$$

Using the Heaviside function $u(x)$, we rewrite this relation as

$$\nu_{n+1} = \nu_n - u(\nu_n) + \eta_{n+1}. \tag{5.1.10}$$

Let us examine the relationship between the embedded Markov chain $\{\nu_n, n \geq 0\}$ and the integer-valued random walk $\{S_n, n \geq 0\}$ generated by a sequence of independent random variables $\{\tilde{\eta}_n = \eta_n - 1,\ n \geq 0\}$ and defined by the relation

$$S_0 = 0, \quad S_{n+1} = S_n + \tilde{\eta}_{n+1}, \quad n \geq 0.$$

For the time being, $\nu_n > 0$ and $S_n > 0$, and ν_n and S_n are defined by the formulas

$$\nu_{n+1} = \nu_n + \tilde{\eta}_{n+1}, \qquad S_{n+1} = S_n + \tilde{\eta}_{n+1}.$$

But the sequence S_n may also take negative values, unlike the sequence ν_n which cannot descend below zero due to the term $u(\nu_n)$. Thus, the embedded Markov chain $\{\nu_n, n \geq 0\}$ is a modification of the random walk $\{S_n, n \geq 0\}$, which can be called the random walk with delaying barrier at zero.

A specific property of the embedded Markov chain $\{\nu_n,\ n \geq 0\}$, like the random walk $\{S_n, n \geq 0\}$, is that their defining variables $\tilde{\eta}_n$ are such that $\mathbf{P}\{\tilde{\eta}_n < -1\} = 0$, i.e., the down-jump cannot be of amplitude exceeding one. Such a random walk, as already mentioned, is called a below-continuous integral random walk. Due to the below-continuity, it is possible to design, as demonstrated for the $M/G/1/\infty$ system, quite simple algorithms to compute steady-state and even transient distributions of random walks.

Now, using (5.1.10), we derive certain formulas that have already been derived for the embedded Markov chain $\{\nu_n, n \geq 0\}$.

Since $u(\nu_n)$ is a random variable, which is zero for $\nu_n = 0$ and one for $\nu_n > 0$, we have

$$\mathbf{E}u(\nu_n) = \mathbf{P}\{\nu_n > 0\} = 1 - p_0^*(n). \tag{5.1.11}$$

Taking the expectation of both sides of (5.1.10), we obtain

$$\mathbf{E}\nu_{n+1} = \mathbf{E}\nu_n - 1 + p_0^*(n) + \mathbf{E}\eta_{n+1}. \tag{5.1.12}$$

Recalling that

$$\mathbf{P}\{\eta_n = k\} = \beta_k$$

for any n and the generating function $\sum_{k=0}^{\infty} \beta_k z^k$ of the random variable η_n is $\beta(\lambda - \lambda z)$, we see that

$$\mathbf{E}(\eta_n)_i = \beta_z^{(i)}(\lambda - \lambda z)\Big|_{z=1} = (-\lambda)^i \beta^{(i)}(0) = \lambda^i b^{(i)}, \qquad i \geq 1,$$

where

$$b^{(i)} = \int_0^\infty t^i \, dB(t)$$

is the ith moment of the service time. Hence, in particular, we obtain

$$\begin{aligned} \mathbf{E}\eta_n &= \lambda b = \rho, \\ \mathbf{E}\eta_n^2 &= \mathbf{E}(\eta_n)_2 + \mathbf{E}\eta_n = \lambda^2 b^{(2)} + \rho. \end{aligned} \tag{5.1.13}$$

Using this fact, we rewrite relation (5.1.12) as

$$\mathbf{E}\nu_{n+1} = \mathbf{E}\nu_n - 1 + p_0^*(n) + \rho.$$

Now assuming that the system operates in steady state, we obtain $\mathbf{E}\nu_{n+1} = \mathbf{E}\nu_n = N$ and, therefore,

$$p_0^* = 1 - \rho. \tag{5.1.14}$$

Using (5.1.10), we arrive at the expression for the stationary probability p_0^* derived above.

Moments of any order of the stationary distribution of the number of customers in the system can be found similarly. In particular, to in order to find the stationary mean number of customers in the system, we square relation (5.1.10) and obtain

$$\nu_{n+1}^2 = \nu_n^2 - 2\nu_n u(\nu_n) + 2\nu_n\eta_{n+1} + u^2(\nu_n) - 2u(\nu_n)\eta_{n+1} + \eta_{n+1}^2. \tag{5.1.15}$$

Using the obvious equalities

$$\nu_n u(\nu_n) = \nu_n, \qquad \mathbf{E}(\nu_n \eta_{n+1}) = \mathbf{E}\nu_n \mathbf{E}\eta_{n+1},$$
$$u^2(\nu_n) = u(\nu_n), \qquad \mathbf{E}(u(\nu_n)\eta_{n+1}) = \mathbf{E}u(\nu_n)\mathbf{E}\eta_{n+1}$$

and taking the expectation of both sides of (5.1.15), we obtain

$$\mathbf{E}\nu_{n+1}^2 = \mathbf{E}\nu_n^2 - 2\mathbf{E}\nu_n + 2\mathbf{E}\nu_n \mathbf{E}\eta_{n+1} + \mathbf{E}u(\nu_n) - 2\mathbf{E}u(\nu_n)\mathbf{E}\eta_{n+1} + \mathbf{E}\eta_{n+1}^2.$$

Passing to the steady state and recalling that relations (5.1.11), (5.1.13), and (5.1.14) for this case yield the equality

$$\mathbf{E}u(\nu_n) = \rho,$$

we obtain

$$2N(1-\rho) = \rho - 2\rho^2 + \rho + \lambda^2 b^{(2)},$$

or finally,

$$N = \rho + \frac{\lambda^2 b^{(2)}}{2(1-\rho)}.$$

This is the Pollaczek–Khinchin formula for the mean number N of customers in the system in steady state derived above.

It must, however, be noted that we have derived the formula for p_0^* under the assumption that the stationary mean number N of customers in the system exists, whereas the formula for N, in its turn, was derived under the assumption that the stationary second moment of the number of customers in the system exists. Therefore, the method described here is not fully equivalent to the approach applied earlier in Section 5.1.2.

5.2. Virtual waiting time

We study the customer sojourn time with the use of a different Markov process. As will be soon clear, such an approach yields simple integro-differential equations, which can be solved easily in terms of Laplace and Laplace–Stieltjes transforms. For a reader acquainted with the general theory of continuous-time and continuous-state-space Markov processes, we note that the process examined below is the generalised Poisson process with negative drift and delaying barrier at zero.

This method has a drawback: it cannot be used for computing queue characteristics.

5.2.1. The Takács equation

Let $\xi(t)$ denote the virtual waiting time at instant t. Recall that the *virtual waiting time* $\xi(t)$ is the time during which a customer would have to wait for service, provided he had arrived at the instant t or, in other words, the 'work' that must be done by the server beginning from the instant t if no customer arrives after the instant t (Fig. 5.3). Obviously, $\{\xi(t),\ t \geq 0\}$ is a random process. The state set $\mathscr{X}$ of the process $\{\xi(t), t \geq 0\}$ is the half-line $[0, \infty)$. For the sake of simplicity, we assume that $\xi(0) = 0$, i.e., the system is empty at the initial instant 0. By the definition of the process $\{\xi(t),\ t \geq 0\}$, its state is re-calculated at the instant

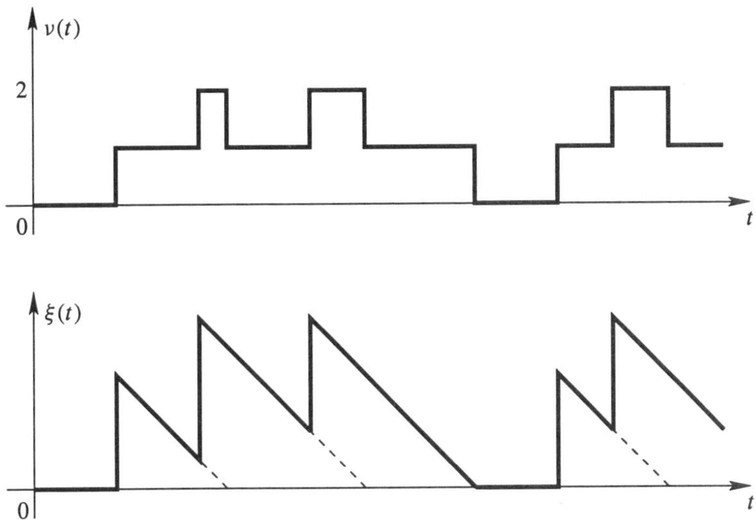

Figure 5.3.

of arrival of every customer, i.e., the service time of an arriving customer is known at this instant and this time is added to the 'work' already in the system.

Let us examine the process $\{\xi(t), t \geq 0\}$. If $\xi(t_0) = 0$, then $\xi(t)$ remains zero until the next instant of arrival of a customer. But if $\xi(t_0) > 0$, then $\xi(t)$ decreases at unit rate until it vanishes or a new customer arrives. From the instant when $\xi(t)$ vanishes, it remains zero until the instant of arrival of a new customer. At the instant of arrival of a new customer, the value of $\xi(t)$ increases by the service time of this new customer. Since the input flow is Poisson (i.e., the arrival process after the instant t_0 depends neither on the behaviour of the process $\xi(t)$ before the instant t_0 nor on the value of $\xi(t_0)$), and the service time of every customer depends neither on its arrival instant nor on operation of the system up to this instant, and has the same distribution $B(x)$, we find that $\{\xi(t), t \geq 0\}$ is a (time-) homogeneous Markov process.

Let

$$W(x, t) = \mathbf{P}\{\xi(t) < x\}$$

denote the distribution function of the process $\xi(t)$. To derive the main equation (5.1.2) in this section, it is more convenient to partition the state space $\mathscr{X}$ into two parts: state 0 corresponding to the empty system and the remaining half-line $(0, \infty)$. So, we set

$$p_0(t) = W(0+, t) = \mathbf{P}\{\xi(t) = 0\},$$

which is the probability that the system is empty at the instant t and

$$\widetilde{W}(x, t) = W(x, t) - p_0(t) = \mathbf{P}\{0 < \xi(t) < x\}, \qquad x > 0,$$

is the probability that customers are in the system at the instant t and the total service time of customers (including the customer at the server) is no greater than x. Obviously,

$$W(x,t) = \begin{cases} 0, & x \le 0; \\ p_0(t) + \widetilde{W}(x,t), & x > 0. \end{cases}$$

We now show that the functions $p_0(t)$ and $\widetilde{W}(x,t)$ have derivatives $p_0'(t)$ and $\partial \widetilde{W}(x,t)/\partial t$, respectively (which are also bounded).

Let us begin with $p_0(t)$. Consider the event A_t, i.e., the system is empty at the instant t. Obviously,

$$\mathbf{P}(A_t) = \mathbf{P}\{\xi(t) = 0\} = p_0(t).$$

We now represent the event $A_{t+\Delta}$ as

$$A_{t+\Delta} = (D_0 + D_1)A_{t+\Delta} = D_0 A_{t+\Delta} + D_1 A_{t+\Delta},$$

where the event D_0 means that the system is empty at the instant Δ and the event D_1 denotes that customers exist in the system at the instant Δ. Since the events D_0 and D_1 are mutually incompatible, from the total probability formula we obtain

$$p_0(t+\Delta) = \mathbf{P}(D_0)\mathbf{P}(A_{t+\Delta} \mid D_0) + \mathbf{P}(D_1)\mathbf{P}(A_{t+\Delta} \mid D_1).$$

Since the process $\{\xi(t), t \ge 0\}$ is Markov and homogeneous in time, we see that

$$\mathbf{P}(A_{t+\Delta} \mid D_0) = \mathbf{P}(A_t) = p_0(t).$$

The probability of the event D_1 is no greater than the probability of arrival of at least one customer at the system in the time Δ. The latter probability, in turn, is no greater than $\lambda\Delta$, i.e.,

$$\mathbf{P}(D_1) \le \lambda\Delta.$$

Moreover,

$$\mathbf{P}(D_0) = 1 - \mathbf{P}(D_1) \ge 1 - \lambda\Delta.$$

Therefore,

$$\begin{aligned} p_0(t+\Delta) &\ge (1-\lambda\Delta)p_0(t) \ge p_0(t) - \lambda\Delta, \\ p_0(t+\Delta) &\le p_0(t) + \lambda\Delta. \end{aligned}$$

Hence we obtain

$$-\lambda\Delta \le p_0(t+\Delta) - p_0(t) \le \lambda\Delta.$$

It is known that the last inequality implies the existence of the derivative $p_0'(t)$ bounded by λ. Applying a slightly more rigorous reasoning, we can also show that $p_0'(t)$ is a continuous function if the distribution function $B(x)$ of the service time is continuous.

Similarly we establish the existence of $\partial \widetilde{W}(x,t)/\partial t$, using A_t to denote the event that the process $\xi(t)$ at the instant t takes a value between zero and x. The proof of this assertion is left as an exercise to the reader.

The proof of the existence of the derivatives $p_0'(t)$ and $\partial\widetilde{W}(x,t)/\partial t$ given above is more of methodological interest than of any practical value: in the modern theory of generalised functions, there is no need to assume the existence of derivatives in the ordinary sense for deriving and solving linear differential equations. So, in the sequel we will not pay much attention to the proof of the existence of the respective derivatives.

The functions $p_0(t)$ and $\widetilde{W}(x,t)$ satisfy the relations

$$\begin{aligned} p_0(t+\Delta) &= [p_0(t)+\widetilde{W}(\Delta,t)](1-\lambda\Delta)+o(\Delta), \\ \widetilde{W}(x,t+\Delta) &= [\widetilde{W}(x+\Delta,t)-\widetilde{W}(\Delta,t)](1-\lambda\Delta) \\ &\quad +\lambda\Delta\int_0^x \widetilde{W}(x-y,t)\,dB(y)+\lambda\Delta B(x)p_0(t)+o(\Delta), \end{aligned} \tag{5.2.1}$$

which are derived as follows. First, no customer arrives with the probability $1-\lambda\Delta+o(\Delta)$ in the time Δ and the virtual waiting time decreases by Δ if it was greater than zero, or it remains equal to zero if it was zero. Second, the virtual waiting time may also vanish if it was less than Δ at the instant t and no customer had arrived in time Δ. This happens with the probability

$$\mathbf{P}\{0<\xi(t)<\Delta\}(1-\lambda\Delta)+o(\Delta)=\widetilde{W}(\Delta,t)+o(\Delta).$$

The virtual waiting time may be less than x at the instant $t+\Delta$ if at the instant t it was less than $(x-y)$ and a customer with service time y had arrived in time Δ. This case, along with the total probability formula, is taken into account by the second term in relation (5.2.1). Finally, the virtual waiting time at the instant $t+\Delta$ is less than x if a customer with service time less than x arrives at an empty system. The probability of other events is $o(\Delta)$.

Moving $p_0(t)$ or $\widetilde{W}(x+\Delta,t)$ to the left-hand side, dividing by Δ, and making Δ tend to zero, from (5.2.1) we obtain

$$\begin{aligned} p_0'(t) &= -\lambda p_0(t)+\frac{\partial}{\partial x}\widetilde{W}(x,t)\Big|_{x=0}, \\ \frac{\partial}{\partial t}\widetilde{W}(x,t)-\frac{\partial}{\partial x}\widetilde{W}(x,t) &= -\lambda\widetilde{W}(x,t)-\frac{\partial}{\partial x}\widetilde{W}(x,t)\Big|_{x=0} \\ &\quad +\lambda\int_0^x \widetilde{W}(x-y,t)\,dB(y)+\lambda B(x)p_0(t). \end{aligned} \tag{5.2.2}$$

The existence of the derivatives $\partial\widetilde{W}(x,t)/\partial x$ and $\partial\widetilde{W}(x,t)/\partial x\big|_{x=0}$ is implied by the existence of the derivatives $p_0'(t)$ and $\partial\widetilde{W}(x,t)/\partial t$, respectively. Equation (5.2.2) is called the *Takács equation.*

5.2.2. Stationary distribution of the customer sojourn time

Let us assume that the limit (stationary) distribution

$$W(x)=\lim_{t\to\infty}W(x,t)$$

exists. As in the previous section, for the sake of convenience we set

$$p_0=W(0+)=\lim_{t\to\infty}p_0(t), \qquad \widetilde{W}(x)=W(x)-p_0=\lim_{t\to\infty}\widetilde{W}(x,t).$$

Recall that the distribution of the virtual waiting time for the stationary case is the same as the distribution of the ordinary waiting time (see Section 2.1.2). Equations for p_0 and $\widetilde{W}(x)$ can be obtained by equating $p_0'(t)$ and $\partial \widetilde{W}(x,t)/\partial t$ to zero in (5.2.2):

$$
\begin{aligned}
0 &= -\lambda p_0 + \widetilde{W}'(0), \\
-\widetilde{W}'(x) &= -\lambda \widetilde{W}(x) - \widetilde{W}'(0) + \lambda \int_0^x \widetilde{W}(x-y)\, dB(y) + \lambda B(x) p_0.
\end{aligned}
$$

We solve these equations with the help of the Laplace–Stieltjes transform

$$
\omega(s) = \int_0^\infty e^{-sx} dW(x) = p_0 + \int_0^\infty e^{-sx} d\tilde{W}(x), \beta(s) = \int_0^\infty e^{-sx} dB(x).
$$

We thus obtain

$$
\begin{aligned}
\widetilde{W}'(0) &= \lambda p_0, \\
-s[\omega(s) - p_0] + \widetilde{W}'(0) &= -\lambda[\omega(s) - p_0] + \lambda \omega(s) \beta(s).
\end{aligned}
$$

Hence

$$
\omega(s) = p_0 \frac{s}{s - \lambda + \lambda \beta(s)}. \tag{5.2.3}
$$

The constant p_0 is determined from the normalisation condition $W(\infty) = \omega(0) = 1$. Applying the L'Hospital rule, we obtain $p_0 = 1 - \rho$, which was derived in the previous section. This means that the necessary condition for the existence of a stationary distribution $W(x)$, like the stationary distribution of the number of customers for an embedded Markov chain, is $\rho < 1$. The sufficiency of this condition for $W(x)$ will be shown in Section 5.5.

Since the customer sojourn time is the sum of the waiting and service times, the stationary distribution $V(x)$ of the sojourn time has the Laplace–Stieltjes transform

$$
\varphi(s) = \int_0^\infty e^{-sx} dV(x) = \omega(s)\beta(s). \tag{5.2.4}
$$

The stationary mean waiting time w and sojourn time v are given by the relations

$$
w = -\omega'(0) = \frac{\lambda b^{(2)}}{2(1-\rho)}, \tag{5.2.5}
$$

$$
v = -\varphi'(0) = b + w = b + \frac{\lambda b^{(2)}}{2(1-\rho)}, \tag{5.2.6}
$$

which can also be derived from the Little formulas.

5.2.3. Transient distribution of the virtual waiting time

Earlier we solved the Takács equation (5.2.2) for the stationary case. Now we derive its solution for the non-stationary case through double transformation.

Let

$$\omega(s_1, s_2) = \int_0^\infty e^{-s_1 t} \mathbf{E} e^{-s_2 \xi(t)} dt$$
$$= \int_0^\infty \int_0^\infty e^{-s_1 t - s_2 x} \widetilde{W}(dx, t)\, dt + \int_0^\infty e^{-s_1 t} p_0(t)\, dt$$

denote the double transformation of the virtual waiting time $\xi(t)$ at the instant t. Since the system is assumed to be empty at the initial instant 0, using the properties of the Laplace and Laplace–Stieltjes transforms, from the second equation of (5.2.2) we obtain

$$s_1 \left[\omega(s_1, s_2) - \int_0^\infty e^{-s_1 t} p_0(t) dt \right] - s_2 \left[\omega(s_1, s_2) - \int_0^\infty e^{-s_1 t} p_0(t) dt \right]$$
$$+ \int_0^\infty \widetilde{W}'(x, t)\Big|_{x=0} e^{-s_1 t} dt = -\lambda \left[\omega(s_1, s_2) - \int_0^\infty e^{-s_1 t} p_0(t) dt \right] + \lambda \beta(s_2) \omega(s_1, s_2).$$

Hence

$$\omega(s_1, s_2)[s_1 - s_2 + \lambda - \lambda\beta(s_2)] = \pi_0(s_1)[\varkappa(s_1) - s_2],$$
$$\omega(s_1, s_2) = \frac{\pi_0(s_1)[\varkappa(s_1) - s_2]}{s_1 + \lambda - \lambda\beta(s_2) - s_2}, \tag{5.2.7}$$

where

$$\pi_0(s_1) = \int_0^\infty e^{-s_1 t} p_0(t)\, dt,$$
$$\varkappa(s_1) = s_1 + \lambda - \frac{1}{\pi_0(s_1)} \int_0^\infty e^{-s_1 t} \widetilde{W}'(x, t)\Big|_{x=0} dt.$$

Before evaluating $\varkappa(s_1)$, we observe that the function $\omega(s_1, s_2)$ is defined and continuous for all $s_1 > 0$ and $s_2 \geq 0$ (furthermore, it is an analytic function of s_1 and s_2 in the domains $\Re s_1 > 0$ and $\Re s_2 > 0$, respectively). Therefore, the numerator of (5.2.7) must vanish at the points (s_1, s_2) where the denominator vanishes. Let us consider the equation

$$\beta(s_2) = \frac{1}{\lambda}(s_1 - s_2 + \lambda).$$

Assuming that $\gamma = (s_1 - s_2 + \lambda)/\lambda$, we can rewrite this equation as

$$\beta(s_1 + \lambda - \lambda\gamma) = \gamma. \tag{5.2.8}$$

In Section 5.5, we will analyse (5.2.8) in more detail. Here we observe only that this equation for every $s_1 > 0$ has a unique solution $\gamma = \gamma(s_1)$, $0 < \gamma < 1$. But then, as already mentioned, the numerator of (5.2.7) must vanish at the points $(s_1, s_2 = s_1 + \lambda - \lambda\gamma(s_1))$. Therefore,

$$\varkappa(s_1) = s_1 + \lambda - \lambda\gamma(s_1).$$

Furthermore,

$$\mathbf{E} e^{0\xi(t)} = 1.$$

Therefore,

$$\omega(s_1, 0) = 1/s_1, \qquad \pi_0(s_1) = \frac{1}{s_1 + \lambda - \lambda\gamma(s_1)}.$$

Substituting the values of $\pi_0(s_1)$ and $\varkappa(s_1)$ into formula (5.2.7), we obtain the final expression for the double transformation $\omega(s_1, s_2)$ of the virtual waiting time $\xi(t)$:

$$\omega(s_1, s_2) = \frac{s_1 + \lambda - \lambda\gamma(s_1) - s_2}{[s_1 + \lambda - \lambda\beta(s_2) - s_2][s_1 + \lambda - \lambda\gamma(s_1)]}.$$

This result is rather inconvenient in practical application, because the double transformation must be inverted.

5.3. Residual service time

In this and succeeding sections, we describe one more method for studying the $M/G/1/\infty$ system with a supplementary variable.

To understand the method and why it was developed, let us recall that the approaches described in the previous sections are one-sided: they are helpful only to find either the distribution of the number of customers or the distribution of the sojourn time. Obviously, a need arises for combining these two approaches and constructing a quite simple Markov process $\{\eta(t) = (\nu(t), \xi(t)), t \geq 0\}$ having a more complex state space consisting of two, (discrete and continuous) coordinates, but relatively simple in investigation.

Obviously, as the discrete coordinate $\nu(t)$ for such a process, it is convenient to choose, as before, the number of customers in the system. Let us examine which supplementary coordinate $\xi(t)$ must be adjoined to $\nu(t)$ for the resultant process $\{\eta(t) = (\nu(t), \xi(t)), t \geq 0\}$ be Markov. Clearly, this coordinate must not be related with the customers in the queue if we assume that the service time is known only at the instant when the customer is taken up for service. Hence $\xi(t)$ must depend only on the customer at the server.

Now there are two alternatives for the supplementary variable $\xi(t)$: either the time during which the customer at the server must be served to complete his service (residual service time) or the time during which the customer at the server has already been served (elapsed service time). In this section, we study the first option, choosing the residual service time as the supplementary variable $\xi(t)$.

5.3.1. Markov process describing the system functioning

Let $\nu(t)$ denote, as usual, the number of customers in the system and let $\xi(t)$ be the residual service time, i.e., the time during which the customer at the server at the instant t must stay in the system. None of these processes are Markov. But with the use of these processes we define a new process $\{\eta(t), t \geq 0\}$ that would be Markov.

If the system at the instant t is empty ($\nu(t) = 0$), then there is no need to introduce the supplementary variable $\xi(t)$ and in this case it is reasonable to set $\eta(t) = \nu(t)$. But if a customer is being served at the instant t ($\nu(t) > 0$), then, along with the number of customers in the system at this instant, we must also specify how much time is required to complete his service, i.e., we introduce a supplementary variable $\xi(t)$, and thereby define the process $\eta(t) = (\nu(t), \xi(t))$. The state set $\mathscr{X}$ of process admits the representation

$$\mathscr{X} = \{(0);\ (i, x),\ i = 1, 2, \dots, x > 0\},$$

where the state (0) corresponds to the empty system and the state (i, x) denotes that i customers exist in the system and the time x is required to complete the service of the customer at the server. In this case, the service time of a customer is known only at the instant when he is taken up for service, but not at the instant when he arrived at the system, as in case of virtual waiting time.

We now show that the process $\{\eta(t),\ t \geq 0\}$ thus defined is Markov. Indeed, if $\eta(t_0) = (0)$ at some instant t_0, i.e., the system is empty, then, as already noted, the behaviour of the process $\eta(t)$ after the instant t_0 does not depend on how the process evolved before the instant t_0. But if $\eta(t_0) = (i, x)$, then the future behaviour of the process $\eta(t)$ is determined not only by the number of customers i in the system at the instant t_0, but also by their service times. But the customers in the queue at the instant t_0 are yet to be served, so their service times do not depend on the prehistory of process $\eta(t)$. For the customer at the server, his (residual) service time x is known. Therefore, the process $\{\eta(t), t \geq 0\}$ is (time-homogenous) Markov and, as can be seen immediately, linearwise (see Sections1.6).

We set

$$p_0(t) = \mathbf{P}\{\nu(t) = 0\}, \qquad P_i(x, t) = \mathbf{P}\{\nu(t) = i, \xi(t) < x\}.$$

As usual, we assume that the system is empty at the initial instant 0, i.e., $\nu(0) = 0$, and $p_0(0) = 1$, $P_i(x, 0) = 0$.

Let us consider the state of the system at the instants t and $t + \Delta$. For the system to be empty at the instant $t + \Delta$, it is necessary that, accurate to probability $o(\Delta)$, either the system be empty at the instant t and no customer arrives in the time Δ or only one customer with residual service time less than Δ exist in the system at the instant t. Hence

$$p_0(t + \Delta) = p_0(t)(1 - \lambda\Delta) + P_1(\Delta, t) + o(\Delta).$$

The case $i \geq 2$ is investigated similarly. lines. At the instant $t + \Delta$ for the system to contain i customers and the residual service time of the customer at the server to be less than x, it is necessary that at the instant t

- either the system contains i customers, the residual service time of the customer at the server lies in the range from Δ to $x + \Delta$, and no customer arrives in time Δ; or
- the system contains $(i - 1)$ customers, the residual service time of the customer at the server lies in the range from Δ to $x + \Delta$, and one more customer arrives in time Δ; or
- the system contains $(i + 1)$ customers, the residual service time of the customer at the server is less than Δ, and after his departure a customer with service time less than x is taken for service.

Since all other events are of the probability $o(\Delta)$, we obtain

$$\begin{aligned} P_i(x, t + \Delta) &= [P_i(x + \Delta, t) - P_i(\Delta, t)](1 - \lambda\Delta) \\ &+ \lambda\Delta[P_{i-1}(x + \Delta, t) - P_{i-1}(\Delta, t)] + B(x)P_{i+1}(\Delta, t) + o(\Delta), \qquad i \geq 2. \end{aligned}$$

Finally, we give (without proof) the relation

$$P_1(x,t+\Delta)=[P_1(x+\Delta,t)-P_1(\Delta,t)](1-\lambda\Delta)+\lambda\Delta B(x)p_0(t)+B(x)P_2(\Delta,t)+o(\Delta).$$

Moving $p_0(t)$ or $P_i(x+\Delta,t)$ to the left-hand side, dividing by Δ, making Δ tend to zero, we obtain the simultaneous differential equations

$$\begin{aligned}
\frac{d}{dt}p_0(t) &= -\lambda p_0(t)+\frac{\partial}{\partial x}P_1(x,t)\Big|_{x=0},\\
\frac{\partial}{\partial t}P_1(x,t)-\frac{\partial}{\partial x}P_1(x,t) &= -\lambda P_1(x,t)-\frac{\partial}{\partial x}P_1(x,t)\Big|_{x=0}\\
&\quad+\lambda B(x)p_0(t)+B(x)\frac{\partial}{\partial x}P_2(x,t)\Big|_{x=0},\\
\frac{\partial}{\partial t}P_i(x,t)-\frac{\partial}{\partial x}P_i(x,t) &= -\lambda P_i(x,t)-\frac{\partial}{\partial x}P_i(x,t)\Big|_{x=0}\\
&\quad+\lambda P_{i-1}(x,t)+B(x)\frac{\partial}{\partial x}P_{i+1}(x,t)\Big|_{x=0},\qquad i\geq 2,
\end{aligned}\tag{5.3.1}$$

with the initial condition

$$p_0(0)=1,\quad P_i(x,0)=0,\quad i\geq 1.\tag{5.3.2}$$

Existence of the derivatives in (5.3.1) is shown in the preceding section.

5.3.2. Stationary state probabilities

Let us assume that the stationary probabilities

$$p_0=\lim_{t\to\infty}p_0(t),\qquad P_i(x)=\lim_{t\to\infty}P_i(x,t)$$

exist. They satisfy the simultaneous equations

$$\begin{aligned}
0&=-\lambda p_0+P_1'(0),\\
-P_1'(x)&=-\lambda P_1(x)-P_1'(0)+\lambda B(x)p_0+B(x)P_2'(0),\\
-P_i'(x)&=-\lambda P_i(x)-P_i'(0)+\lambda P_{i-1}(x)+B(x)P_{i+1}'(0),\qquad i\geq 2,
\end{aligned}\tag{5.3.3}$$

derived from (5.3.1) by equating the time derivatives to zero.

The term $P_{i+1}'(0)$ in the ith equation causes some inconvenience in solving (5.3.3). But this inconvenience can be easily surmounted. Indeed, let $p_i=P_i(\infty)$ denote the stationary probability that i customers exist in the system. Making x in (5.3.3) tend to infinity, we find that the limit

$$-P_i'(\infty)=\lim_{x\to\infty}(-P_i'(x))=-\lambda p_i-P_i'(0)+\lambda p_{i-1}+P_{i+1}'(0),\qquad i\geq 1,\tag{5.3.4}$$

exists. Since

$$p_i=P_i(\infty)=\int_0^\infty P_i'(x)\,dx\leq 1,$$

this limit can only be zero. Therefore, from (5.3.3) and (5.3.4) we obtain

$$0 = -\lambda p_0 + P_1'(0),$$
$$0 = -\lambda p_i + P_{i+1}'(0) + \lambda p_{i-1} - P_i'(0), \qquad i \geq 1.$$

Hence we finally obtain

$$P_{i+1}'(0) = \lambda p_i, \qquad i \geq 0.$$

Substituting the last equality into (5.3.3), after certain elementary transformations, we obtain the simultaneous equations

$$P_1'(x) + \lambda[p_1 - P_1(x)] = \lambda[1 - B(x)]p_0 + \lambda[1 - B(x)]p_1,$$
$$P_i'(x) + \lambda[p_i - P_i(x)] = \lambda[p_{i-1} - P_{i-1}(x)] + \lambda[1 - B(x)]p_i, \qquad i \geq 2. \tag{5.3.5}$$

Equations (5.3.5) can be solved recursively. Setting

$$R_i(x) = e^{-\lambda x}[p_i - P_i(x)], \tag{5.3.6}$$

from (5.3.5) we arrive at

$$-R_1'(x) = \lambda[1 - B(x)]e^{-\lambda x}p_0 + \lambda[1 - B(x)]e^{-\lambda x}p_1,$$
$$-R_i'(x) = \lambda R_{i-1}(x) + \lambda[1 - B(x)]e^{-\lambda x}p_i, \qquad i \geq 2,$$

with the obvious initial condition $R_i(\infty) = 0$. The solution is found recursively by the formula

$$R_1(x) = \lambda p_0 \int_x^\infty [1 - B(y)]e^{-\lambda y}dy + \lambda p_1 \int_x^\infty [1 - B(y)]e^{-\lambda y}dy,$$
$$R_i(x) = \lambda \int_x^\infty R_{i-1}(y)\,dy + \lambda p_i \int_x^\infty [1 - B(y)]e^{-\lambda y}dy, \qquad i \geq 2,$$

or, which is the same,

$$R_i(x) = Q_{i-1}(x)p_0 + \sum_{k=1}^{i} Q_{i-k}(x)p_k, \qquad i \geq 1, \tag{5.3.7}$$

where $Q_i(x)$, in its turn, is determined recursively by the formula

$$Q_0(x) = \lambda \int_x^\infty [1 - B(y)]e^{-\lambda y}dy,$$
$$Q_i(x) = \lambda \int_x^\infty Q_{i-1}(y)\,dy, \qquad i \geq 1.$$

Integrating the last relation by parts, we obtain

$$Q_i(x) = \frac{\lambda^{i+1}}{i!} \int_x^\infty (y - x)^i[1 - B(y)]e^{-\lambda y}dy, \qquad i \geq 0. \tag{5.3.8}$$

Although the probabilities p_i were computed earlier (see Section 5.1), let us compute them once again. Setting $x = 0$ in (5.3.6), we obtain $R_i(0) = p_i$. By virtue of this equality, from (5.3.7) and (5.3.8) we obtain

$$p_i = Q_{i-1} p_0 + \sum_{k=1}^{i} Q_{i-k} p_k, \qquad i \geq 1, \tag{5.3.9}$$

where

$$Q_i = \frac{\lambda^{i+1}}{i!} \int_0^\infty x^i [1 - B(x)] e^{-\lambda x} dx, \qquad i \geq 0. \tag{5.3.10}$$

Simultaneous equations (5.3.9) differ from (5.1.1) used earlier for finding $p_i^* = p_i$. Using the method of generating functions, let us prove the equivalence of (5.3.9) and (5.1.1). Setting

$$P(z) = \sum_{i=0}^{\infty} p_i z^i, \qquad Q(z) = \sum_{i=0}^{\infty} Q_i z^i,$$

from (5.3.9) we obtain

$$P(z) - p_0 = z p_0 Q(z) + \sum_{i=1}^{\infty} z^i \sum_{k=1}^{i} Q_{i-k} p_k = [z p_0 + P(z) - p_0] Q(z).$$

From (5.3.10) we obtain

$$\begin{aligned} Q(z) &= \lambda \int_0^\infty e^{\lambda x z} [1 - B(x)] e^{-\lambda x} dx \\ &= \lambda \int_0^\infty [1 - B(x)] e^{-\lambda x (1-z)} dx = \frac{1 - \beta(\lambda - \lambda z)}{1 - z}. \end{aligned}$$

Therefore,

$$P(z) - p_0 = [P(z) - p_0(1 - z)] \frac{1 - \beta(\lambda - \lambda z)}{1 - z},$$

and, consequently,

$$P(z) = \frac{(1 - z)\beta(\lambda - \lambda z)}{\beta(\lambda - \lambda z) - z} p_0.$$

Hence we immediately see that $p_0 = 1 - \rho$. Thus, we have derived the Pollaczek–Khinchin formula derived in Section 5.1.

Equations (5.3.5) can also be solved with the help of a generating function. Assuming that

$$P(z, x) = \sum_{i=1}^{\infty} P_i(x) z^i,$$

from (5.3.5) we obtain the equation

$$\frac{\partial}{\partial x} P(z, x) = \lambda(1 - z) P(z, x) + \lambda[z - B(x)] P(z) + \lambda B(x)(1 - z) p_0.$$

Its solution is

$$P(z,x) = \lambda e^{\lambda(1-z)x} \int_0^x [(z - B(y))P(z) + (1-z)B(y)p_0]e^{-\lambda(1-z)y}dy.$$

In turn, it is more convenient to express the last formula in terms of the double transformation

$$\begin{aligned}\pi(z,s) &= \int_0^\infty e^{-sx}P(z,dx) = \sum_{i=1}^\infty z^i \int_0^\infty e^{-sx}dP_i(x)\\ &= \lambda^2(1-z)\int_0^\infty e^{-(s-\lambda(1-z))x}dx \int_0^x [(z-B(y))P(z)\\ &\quad + (1-z)B(y)p_0]e^{-\lambda(1-z)y}dy\\ &\quad + \lambda \int_0^\infty e^{-sx}[(z-B(x))P(z) + (1-z)B(x)p_0]\,dx.\end{aligned}$$

Changing the order of integration in the first integral, we obtain

$$\begin{aligned}\pi(z,s) &= \lambda^2(1-z)\int_0^\infty dy \int_y^\infty e^{-(s-\lambda(1-z))x}e^{-\lambda(1-z)y}[(z-B(y))P(z)\\ &\quad + (1-z)B(y)p_0]\,dx\\ &\quad + \lambda\int_0^\infty e^{-sx}[(z-B(x))P(z) + (1-z)B(x)p_0]\,dx\\ &= \left[\frac{\lambda^2(1-z)}{s-\lambda(1-z)} + \lambda\right]\int_0^\infty e^{-sx}[(z-B(x))P(z) + (1-z)B(x)p_0]\,dx\\ &= \left[\frac{\lambda^2(1-z)}{s-\lambda(1-z)} + \lambda\right]\left[\frac{z-\beta(s)}{s}P(z) + \frac{(1-z)\beta(s)}{s}p_0\right]\end{aligned}$$

or finally,

$$\pi(z,s) = \frac{\lambda(1-z)}{s-\lambda(1-z)}\frac{z[\beta(\lambda-\lambda z) - \beta(s)]}{\beta(\lambda-\lambda z) - z}p_0. \tag{5.3.11}$$

According to the results of this section, for the stationary probabilities p_0 and $P_i(x)$ to exist, it is necessary that $\rho = \lambda b < 1$. The sufficiency of this condition will be shown in Section 5.5. Moreover, the necessity and sufficiency can be derived from Theorem 1.6.5 and results of Section 5.1.

5.3.3. Joint stationary distribution of the number of customers in the system and the waiting time

A customer arrived at a system in steady state finds with the probability p_0 that the system is empty. Obviously, the waiting time for service in this case is zero.

A customer arrived at the system fins with probability $dP_i(x)$ that the system contains i customers and the residual service time of the customer at the server lies in the range from

x to $x + dx$. But the distribution of the service times of $(i - 1)$ customers in the queue is the convolution $B^{(*(i-1))}(x)$ of $(i - 1)$ distribution functions $B(x)$. Therefore, the total stationary probability $W_i(x)$ that a new customer finds that the system contains i customers and has to wait for service for a time less than x is given by the expression

$$W_i(x) = \int_0^x B^{(*(i-1))}(x - y)dP_i(y), \qquad x > 0.$$

Hence, applying the double transformation

$$\omega(z, s) = p_0 + \sum_{i=1}^{\infty} z^i \int_0^{\infty} e^{-sx} dW_i(x),$$

by virtue of (5.3.11), we obtain

$$\begin{aligned} \omega(z, s) &= p_0 + \frac{\pi(z\beta(s), s)}{\beta(s)} \\ &= (1 - \rho)\left\{1 + \lambda z \frac{[z\beta(s) - 1][\beta(s) - \beta(\lambda - \lambda z\beta(s))]}{[z\beta(s) - \beta(\lambda - \lambda z\beta(s))][\lambda - \lambda z\beta(s) - s]}\right\}. \end{aligned} \tag{5.3.12}$$

In particular, setting $s = 0$ or $z = 1$, from the last expression we can easily derive the Pollaczek–Khinchin formula (5.1.2) for the stationary distribution of the number of customers in the system or formula (5.2.3) for the stationary distribution of the waiting time for service.

5.3.4. Transient characteristics

Solving simultaneous equations (5.3.1) for the transient case is more of a demonstration of the potentialities of the modern mathematical methods than of any practical value. We have already encountered this situation in Chapter 3, where we had to take recourse to the Laplace time-transformation to find determining the non-stationary probabilities of the states of the $M/M/1/\infty$ system. Nevertheless, we briefly sketch the possible routes to finding $P_i(x, t)$ for an inquisitive reader.

The first route, as usual, consists of applying the generating functions and Laplace and Laplace–Stieltjes transformations. Setting

$$\pi_0(s) = \int_0^{\infty} e^{-st} p_0(t)\, dt,$$

$$\pi(s_1, z, s_2) = \pi_0(s_1) + \sum_{i=1}^{\infty} z^i \int_0^{\infty} e^{-s_1 t} \int_0^{\infty} e^{-s_2 x} P_i(dx, t)\, dt,$$

$$\tilde{\pi}(s, z) = \sum_{i=1}^{\infty} z^i \int_0^{\infty} e^{-st} \frac{\partial}{\partial x} P_i(x, t)\bigg|_{x=0} dt$$

and applying these transformations to (5.3.1), after certain simple computations under initial conditions (5.3.2) we obtain

$$\pi(s_1, z, s_2) = \frac{\beta(s_2) + \{(s_1 + \lambda - \lambda z)[1 - \beta(s_2)] - s_2\}\pi_0(s_1) - \left[1 - \frac{\beta(s_2)}{z}\right]\tilde{\pi}(s_1, z)}{s_1 - s_2 + \lambda(1 - z)}. \tag{5.3.13}$$

While determining $\pi_0(s)$ and $\tilde{\pi}(s, z)$ we must keep in mind that the numerator of (5.3.13) vanishes for those $s_2 = s_1+\lambda(1-z)$ for which the denominator of (5.3.13) vanishes. Therefore,

$$\beta(s + \lambda - \lambda z)[1 - (s + \lambda - \lambda z)\pi_0(s)] = \left[1 - \frac{\beta(s + \lambda - \lambda z)}{z}\right]\tilde{\pi}(s, z). \tag{5.3.14}$$

Now consider the equation

$$\gamma = \beta(s + \lambda - \lambda\gamma).$$

We have encountered this equation earlier and we know that it has a unique continuous solution $\gamma = \gamma(s)$ for $s \geq 0$. Let $z = \gamma(s)$. Then

$$1 - \frac{\beta(s + \lambda - \lambda\gamma(s))}{\gamma(s)} = 0.$$

Therefore, the right-hand side of (5.3.14) is zero and hence,

$$\pi_0(s) = \frac{1}{s + \lambda - \lambda\gamma(s)}. \tag{5.3.15}$$

Substituting the above $\pi_0(s)$ into (5.3.14) and (5.3.13), we obtain

$$\pi(s_1, z, s_2) = \frac{1}{s_1 + \lambda - \lambda\gamma(s_1)}\left(1 + \lambda z \frac{[z - \gamma(s_1)][\beta(s_2) - \beta(s_1 + \lambda - \lambda z)]}{[s_1 - s_2 + \lambda - \lambda z][z - \beta(s_1 + \lambda - \lambda z)]}\right). \tag{5.3.16}$$

In particular, the joint non-stationary distribution $W_i(x, t)$ of the number of customers in the system and virtual waiting time at instant t can be easily found from formulas (5.3.16) in terms of the transformation of $\omega(s_1, z, s_2)$. Indeed, doing as in Section 5.3.4, we obtain

$$\begin{aligned}\omega(s_1, z, s_2) &= \pi_0(s_1) + \sum_{i=1}^{\infty} z^i \int_0^\infty \int_0^\infty e^{-s_1 t} e^{-s_2 t} W_i(dx, t)\, dt \\ &= \frac{1}{\lambda + s_1 - \lambda\gamma(s_1)} \\ &\quad \times \left\{1 + \lambda z \frac{[z\beta(s_2) - \gamma(s_1)][\beta(s_2) - \beta(s_1 + \lambda - \lambda z\beta(s_2))]}{[z\beta(s_2) - \beta(s_1 + \lambda - \lambda z\beta(s_2))][s_1 + \lambda - \lambda z\beta(s_2) - s_2]}\right\}.\end{aligned} \tag{5.3.17}$$

The second method consists of the following. Looking at equations (5.3.1), we see that the ith equation for $P_i(x, t)$, $i \geq 1$, is a non-homogeneous linear first-order partial differential equation, which, as known from the theory of partial differential equations, can be solved in explicit form.

But the unknowns in (5.3.1) are $\partial P_i(x,t)/\partial x|_{x=0}$, $i \geq 1$. To find them, as in the steady case, we must make x tend to infinity. Since $\partial P_i(x,t)/\partial x \xrightarrow[x\to\infty]{} 0$, from (5.3.1) we obtain

$$p_0'(t) = -\lambda p_0(t) + \frac{\partial}{\partial x} P_1(x,t)\bigg|_{x=0},$$

$$p_i'(t) = -\lambda p_i(t) - \frac{\partial}{\partial x} P_i(x,t)\bigg|_{x=0} + \lambda p_{i-1}(t) + \frac{\partial}{\partial x} P_{i+1}(x,t)\bigg|_{x=0}, \qquad i \geq 1,$$

where

$$p_i(t) = P_i(\infty, t) = \mathbf{P}\{\nu(t) = i\}.$$

The last can be applied to finding $\partial P_i(x,t)/\partial x|_{x=0}$ recursively for known $p_i(t)$. Thus, calculation of the transient probabilities $p_0(t)$ and $P_i(x,t)$, $i \geq 1$, is reduced to designing an efficient algorithm for $p_i(t)$, $i \geq 0$, which we do not consider here.

The functions $\partial P_i(x,t)/\partial x|_{x=0}$ can also be found from equalities (5.3.14) and (5.3.15) by means of double transformation.

5.4. Elapsed waiting time

In this section, we turn back to the method of supplementary variables. In contrast with the previous section, here we use the elapsed service time as the supplementary variable $\xi(t)$. The reader will see that the equations derived in these two sections closely resemble each another and yield the same final result.

Rigorous mathematical justification for using the equilibrium equation in applying the elapsed service time, as will be clear soon, is far more complicated than in the previous case. To free the reader of cumbersome computations, first we restrict ourselves to the steady case and, second, formulate certain assertions without proof. Detailed derivations are given in (Belyaev, 1962) (see also Section 1.6).

5.4.1. Markov process describing the system functioning

First we define the process $\{\eta(t),\ t \geq 0\}$ that describes the system functioning. Let, as before, $\nu(t)$ be the number of customers in the system at the instant t. If $\nu(t) = 0$, then no supplementary variable is required and hence, $\eta(t) = \nu(t)$. But if $\nu(t) > 0$, then the process $\eta(t)$ is defined by $\eta(t) = (\nu(t), \xi(t))$, where, by our notational convention, the supplementary coordinate $\xi(t)$ is the elapsed service time, i.e., the time during which the customer at the server at the instant t has already been served. Thus, the set of states of the process $\{\eta(t),\ t \geq 0\}$ can be expressed as

$$\mathscr{X} = \{(0);\ (i,x),\ i = 1,2,\ldots,\ x \geq 0\}.$$

Here, if the process $\eta(t)$ is in the state (0) at any instant t, it means that the system is empty; if the process finds itself in the state (i,x), then that the system contains i customers and the service of the customer at the server has been continuing for time x. Note that, in this approach, the service time of a customer is known only at the instant when he quits the system.

We now show that the process $\{\eta(t),\ t \geq 0\}$ is Markov. Indeed, if the system is empty at the instant t_0, i.e., $\eta(t_0) = (0)$, then the behaviour of the process $\eta(t)$ at any $t > t_0$, as we have already shown, does not depend on the trajectory of the process $\eta(t)$ prior to the instant t_0. But if $\eta(t_0) = (i, x)$, then the future behaviour of the process $\eta(t)$ is determined not only by the input flow and service times of the customers in the queue at instant t_0 and the new arrivals after the instant t_0, which, as before, do not depend on the prehistory of the operation of the system prior to the instant t_0, but also on the residual service time of the customer at the server at the instant t_0. Since the elapsed service time x of this customer is known, the distribution of his residual service time, denoted by ζ_x, is completely determined. Indeed, if γ is the total time of service of the customer at the server at the instant t_0 and since he has been served for a time x by the instant t_0, then, by the conditional probability formula, we obtain

$$\mathbf{P}\{\zeta_x < y\} = \mathbf{P}\{\gamma - x < y \mid \gamma > x\} = \frac{B(x+y) - B(x)}{1 - B(x)}.$$

Thus, the behaviour of the process $\eta(t)$ prior to the instant t_0 has no influence on the probability of the events that may occur after the instant t_0 and the process $\{\eta(t),\ t \geq 0\}$ is Markov and, as can be verified easily, linearwise (see Section 1.6).

In order to describe the Markov process $\{\eta(t),\ t \geq 0\}$, we must also know its initial state $\eta(0)$, i.e., the joint distribution of the number of customers in the system at the instant 0 and the elapsed service time of the customer at the server at the instant 0. But, in this section, we study only the limiting (steady-state) characteristics of a system and, as will be shown in what follows, limit distributions do not depend on the initial state of a system. Therefore, we will not specify the initial state $\eta(0)$. But a rigorous reader may assume that the system is empty at the initial instant 0, i.e., $\eta(0) = (0)$.

5.4.2. Equations for the stationary probabilities of the Markov process

We use the notation

$$\begin{aligned} p_0(t) &= \mathbf{P}\{\nu(t) = 0\}, \\ P_i(x,t) &= \mathbf{P}\{\nu(t) = i, \xi(t) < x\}, \\ p_i(x,t) &= \frac{\partial}{\partial x} P_i(x,t), \qquad i \geq 1. \end{aligned}$$

Existence of the densities $p_i(x,t)$ is shown as in the previous sections (for details, see, e.g., (Kleinrock, 1984)). For the sake of simplicity, we assume that the inequality $B(x) < 1$ holds for all x and the distribution function $B(x)$ is continuous, and hence, $p_i(x,t)$ is also continuous.

Let us derive the equations. For this, as usual, we consider the instants t and $t + \Delta$. For the process $\eta(t)$ to exist in the state $(i, x + \Delta)$ at the instant $t + \Delta$, the following conditions must hold:

- at the instant t the process $\eta(t)$ exists in state (i, x), and no customer arrives in the time interval Δ (with the probability $1 - \lambda\Delta$) and the service of the customer at the server is not completed (the probability of this event, as is clear from what has been said above, is $[1 - B(x + \Delta)]/[1 - B(x)]$);

- at the instant t the process $\eta(t)$ is in the state $(i-1, x)$, one customer (with the probability $\lambda\Delta$) arrives in the time interval Δ, and the service of the customer at the service is not completed. Note that for $i = 1$, the last event cannot occur, because the elapsed service time of a customer arriving at an empty system is zero.

Since all other events are of probability $o(\Delta)$, using the total probability formula, we write

$$p_i(x+\Delta, t+\Delta) = p_i(x,t)(1-\lambda\Delta)\frac{1-B(x+\Delta)}{1-B(x)} + u(i-1)p_{i-1}(x,t)\lambda\Delta\frac{1-B(x+\Delta)}{1-B(x)} + o(\Delta), \quad (5.4.1)$$

where $u(x)$ is the Heaviside function.

The next step consists of passing to the limit as $\Delta \to 0$. But before doing so, we assume that the limit (stationary) probabilities

$$p_0 = \lim_{t\to\infty} p_0(t), \qquad P_i(x) = \lim_{t\to\infty} P_i(x,t)$$

exist. Soon we will state the condition under which $p_0 > 0$ and $P_i(x) > 0$ exist (though, the reader may easily guess that this condition is $\rho = \lambda b < 1$). For the time being, we only state that existence of the densities $p_i(x) = P_i'(x)$ can indeed be proved. For the steady case, equality (5.4.1) is rewritten as

$$p_i(x+\Delta) = p_i(x)(1-\lambda\Delta)\frac{1-B(x+\Delta)}{1-B(x)} + u(i-1)p_{i-1}(x)\lambda\Delta\frac{1-B(x+\Delta)}{1-B(x)} + o(\Delta). \quad (5.4.2)$$

Moreover, in what follows it would be more convenient to deal with the functions

$$q_i(x) = \frac{p_i(x)}{1-B(x)}$$

rather than the densities $p_i(x)$. After some simple transformations, from (5.4.2) we obtain

$$q_i(x+\Delta) - q_i(x) = -\lambda\Delta q_i(x) + u(i-1)\lambda\Delta q_{i-1}(x) + o(\Delta).$$

Now dividing by Δ and making Δ tend to zero, we arrive at the simultaneous equations

$$q_i'(x) = -\lambda q_i(x) + u(i-1)\lambda q_{i-1}(x), \qquad i \geq 1. \quad (5.4.3)$$

5.4.3. Boundary conditions

Set of equations (5.4.3) is simpler than (5.3.3) because it does not contain any derivatives at the point $x = 0$. But in order to solve it, we require certain additional (boundary) conditions for p_0 and $q_i(0)$. We now derive precisely these conditions.

We begin with the probability p_0. Let us turn to the transient case and consider, as before, the state of the system at the instants t and $t+\Delta$. For the process $\eta(t)$ to exist in the state (0) at the instant $t + \Delta$, it is necessary that either the system was empty at the instant t and no customer had arrived in the time interval Δ (with the probability $1 - \lambda\Delta$), or the process $\eta(t)$ was in the state $(1, x)$ at the instant t and the service of the customer at the server was completed in time Δ (this occurs with the probability $[B(x + \Delta) - B(x)]/[1 - B(x)]$). The probability of all other events is $o(\Delta)$. With the use of the total probability formula we obtain

$$p_0(t + \Delta) = (1 - \lambda\Delta) p_0(t) + \int_0^\infty p_1(x, t) \frac{B(x + \Delta) - B(x)}{1 - B(x)} \, dx + o(\Delta).$$

Passing now to the steady state and recalling the definition of the function $q_1(x)$, we arrive at the equality

$$\lambda\Delta p_0 = \int_0^\infty q_1(x)[B(x + \Delta) - B(x)] \, dx + o(\Delta). \tag{5.4.4}$$

Let us show that

$$p_i(x) \leq \lambda[1 - B(x)].$$

Hence it follows, in particular, that $q_i(x)$ are bounded. Since $q_i(x)$ are continuous and bounded (see (Sevastyanov, 1957)),

$$\lim_{\Delta \to 0} \int_0^\infty q_1(x) \frac{B(x + \Delta) - B(x)}{\Delta} \, dx = \int_0^\infty q_1(x) \, dB(x).$$

Therefore, dividing both sides of (5.4.4) by Δ and making Δ tend to zero, we obtain the first boundary condition

$$\lambda p_0 = \int_0^\infty q_1(x) \, dB(x). \tag{5.4.5}$$

Similar reasoning yields the boundary conditions for $q_i(0)$. If $i \geq 2$, then for the event $\{\nu(t + \Delta) = i, \xi(t + \Delta) < \Delta\}$ to take place, it is necessary that the process $\eta(t)$ exist in the state $(i + 1, x)$ at the instant t and the service of the customer at the server be completed in time Δ. All other possible transitions are of the probability $o(\Delta)$. Therefore,

$$P_i(\Delta, t + \Delta) = \int_0^\infty p_{i+1}(x, t) \frac{B(x + \Delta) - B(x)}{1 - B(x)} \, dx + o(\Delta).$$

For the steady case, applying the well-known mean value theorem, we obtain

$$\begin{aligned} P_i(\Delta) &= \int_0^\Delta p_i(x) \, dx = p_i(\theta\Delta)\Delta \\ &= \int_0^\infty q_{i+1}(x)[B(x + \Delta) - B(x)] \, dx + o(\Delta), \qquad 0 < \theta < 1. \end{aligned}$$

Dividing by Δ and passing to the limit as $\Delta \to 0$, we obtain

$$q_i(0) = \int_0^\infty q_{i+1}(x) \, dB(x), \qquad i \geq 2. \tag{5.4.6}$$

Finally, for $i = 1$, the event $\{\nu(t+\Delta) = 1, \xi(t+\Delta) < \Delta\}$ might also occur if the system had been empty at the instant t but a customer arrived in time Δ. Then

$$q_1(0) = \int_0^\infty q_2(x)\,dB(x) + \lambda p_0. \tag{5.4.7}$$

Boundary conditions (5.4.6) and (5.4.7) can be combined into one condition, using the Heaviside function:

$$q_i(0) = \int_0^\infty q_{i+1}(x)\,dB(x) + u(2-i)\lambda p_0, \qquad i \geq 1. \tag{5.4.8}$$

5.4.4. Stationary state probabilities

Thus, we have shown that the stationary probabilities p_0 and $P_i(x)$ are determined by solving recursive differential equations (5.4.3) under boundary conditions (5.4.5) and (5.4.8), and then applying the substitution

$$p_i(x) = P_i'(x) = [1 - B(x)]q_i(x). \tag{5.4.9}$$

Solving (5.4.3) for $i = 1, 2$ and then using induction, we easily obtain

$$q_i(x) = e^{-\lambda x} \sum_{k=0}^{i-1} \frac{(\lambda x)^k}{k!} q_{i-k}(0), \qquad i \geq 1. \tag{5.4.10}$$

It is easy to validate (5.4.10) by substituting $q_i(x)$ into (5.4.3) and thereby converting them into simultaneous identities.

From (5.4.9) and (5.4.10), in particular, we obtain the expression

$$p_i(x) = [1 - B(x)]e^{-\lambda x} \sum_{k=0}^{i-1} \frac{(\lambda x)^k}{k!} q_{i-k}(0), \qquad i \geq 1, \tag{5.4.11}$$

of the density $p_i(x)$ of the distribution function $P_i(x)$—the probability that i customers are in the system and the service time of the customer at the server is less than x. It is worthwhile to note that the stationary density $p_i(x)$ of the elapsed service time is determined in a somewhat simpler manner than the respective density of the residual service time (cf. formulas (5.3.6)–(5.3.8)). Derivation of formulas interrelating these two densities is left as an exercise to the reader.

In order to find the stationary probabilities $p_i = P_i(\infty)$, $i \geq 1$, that i customers exist in the system, we need β_i defined in Section 5.1. Moreover, we introduce

$$B_i = 1 - \sum_{k=0}^{i} \beta_k,$$

which is the probability that more than i, $i \geq 0$, new customers arrive at the system during the time of service of a current customer. For known β_i, we find B_i with the use of the recurrence relations

$$B_0 = 1 - \beta_0, \quad B_i = B_{i-1} - \beta_i, \quad i \geq 1.$$

Recalling that

$$p_i = \int_0^\infty p_i(x)\,dx,$$

from (5.4.11) we obtain

$$p_i = \sum_{k=0}^{i-1} \frac{\lambda^k}{k!} q_{i-k}(0) \int_0^\infty x^k e^{-\lambda x}[1 - B(x)]\,dx, \qquad i \geq 1.$$

Integrating the last integral by parts and applying the formula

$$\int x^k e^{\lambda x} dx = e^{\lambda x} \sum_{j=0}^{k} (-1)^j \frac{(k)_j}{\lambda^{j+1}} x^{k-j} + C,$$

after simple transformations we arrive at

$$p_i = \frac{1}{\lambda} \sum_{k=0}^{i-1} B_k q_{i-k}(0), \qquad i \geq 1. \tag{5.4.12}$$

What remains is to find $q_i(0)$; to do this, we substitute $q_i(x)$ from (5.4.10) into the equations for boundary conditions (5.4.5) and (5.4.8). We thus obtain

$$\begin{aligned} \lambda p_0 &= \beta_0 q_1(0), \\ q_i(0) &= \sum_{k=0}^{i} \beta_k q_{i-k+1}(0) + \lambda u(2-i) p_0, \qquad i \geq 1. \end{aligned} \tag{5.4.13}$$

Assuming that

$$\tilde{q}_i = \frac{q_i(0)}{p_0}, \qquad i \geq 1, \tag{5.4.14}$$

after simple transformations of equalities (5.4.13), we obtain recurrence relations for $\tilde{q}_i$:

$$\begin{aligned} \tilde{q}_1 &= \frac{\lambda}{\beta_0}, \\ \tilde{q}_i &= \frac{1}{\beta_0} \left[\tilde{q}_{i-1} - \sum_{k=1}^{i-1} \beta_k \tilde{q}_{i-k} - \lambda u(3-i) \right], \qquad i \geq 2. \end{aligned}$$

Without calculating p_0 right now, we give other formulas for $\tilde{q}_i$ and p_i, which are often more convenient in numerical computations.

Summing the leading i equations in (5.4.13), we obtain the recurrence relations

$$\tilde{q}_i = \frac{1}{\beta_0} \sum_{k=1}^{i-1} B_{i-k} \tilde{q}_k, \qquad i \geq 2. \tag{5.4.15}$$

This formula is preferable in computations since the expressions in the right-hand side are sums of positive addends.

For the stationary probabilities p_i as well, we are able to derive simpler relations than (5.4.12). Indeed, multiplying both sides of differential equations (5.4.3) by $(1 - B(x))$ and then integrating with respect to x from 0 to ∞, we obtain

$$-q_i(0) + \int_0^\infty q_i(x)\, dB(x) = -\lambda p_i + u(i-1)\lambda p_{i-1}, \qquad i \geq 1.$$

Adding this equality with $i = 1$ to (5.4.5) and for $i = 2, 3, \ldots$, to (5.4.8), we obtain

$$-q_1(0) + \lambda p_0 = -\lambda p_1,$$
$$-q_i(0) + q_{i-1}(0) - u(3-i)\lambda p_0 = -\lambda p_i + \lambda p_{i-1}, \qquad i \geq 2.$$

Hence, successively adding these relations, we obtain

$$p_1 = \frac{q_1(0)}{\lambda} - p_0,$$
$$p_i = \frac{q_i(0)}{\lambda}, i \geq 2. \tag{5.4.16}$$

Obviously, formula (5.4.16), due to its simplicity, is more attractive for computations than formula (5.4.12), although (5.4.12) can also be used, for example, to check calculations.

We now calculate the probability p_0 as follows. Summing equations (5.4.3) for $i = 1, 2, \ldots$, we obtain the equation

$$q_{\cdot}'(x) = 0,$$

where the dot, as before, denotes summation over all possible values of the discrete argument. Solving this equation, we obtain

$$q_{\cdot}(x) = q_{\cdot}(0).$$

Multiplying both sides of the last relation by $(1 - B(x))$ and then integrating it from 0 to ∞, by virtue of (5.4.9) and the normalisation condition, we obtain

$$1 - p_0 = bq_{\cdot}(0). \tag{5.4.17}$$

Summing equalities (5.4.16) and using the normalisation condition, we obtain

$$q_{\cdot}(0) = \lambda.$$

Hence, from (5.4.17) we obtain the relation

$$p_0 = 1 - \rho,$$

where $\rho = \lambda b$, which is what should be expected.

Thus, we have once again arrived at the conclusion that if the system is in steady state, then the traffic intensity ρ is less than one, i.e., the condition $\rho < 1$ is necessary for the system to be in stationary state. The sufficiency of this condition will be demonstrated in Section 5.5 (see also Theorem 1.6.6 and Section 5.1).

Finally, as usual, we derive the generating function of the stationary distribution of the number of customers in the system. Setting

$$P(z) = \sum_{i=0}^{\infty} p_i z^i, \qquad Q(z) = \sum_{i=1}^{\infty} q_i(0) z^i,$$

multiplying (5.4.16) by z^i, and summing over i, we obtain

$$P(z) = \frac{1}{\lambda} Q(z) + (1 - z) p_0. \tag{5.4.18}$$

In order to find the generating function $Q(z)$, we make use of formulas (5.4.13)–(5.4.15) after reducing them to the form

$$q_i(0) = \sum_{k=1}^{i} B_{i-k} q_k(0) + u(2 - i)\lambda p_0, \qquad i \geq 1.$$

Multiplying both sides of this equality by z^i and then summing over i, by virtue of the convolution property of generating functions, we obtain

$$Q(z) = p_0 \frac{\lambda z}{1 - B(z)},$$

where

$$B(z) = \sum_{i=0}^{\infty} B_i z^i.$$

For this generating function,

$$B(z) = \sum_{i=0}^{\infty} \left(1 - \sum_{k=0}^{i} \beta_k \right) z^i = \frac{1}{1 - z} - \frac{\beta(\lambda - \lambda z)}{1 - z}.$$

Thus,

$$Q(z) = p_0 \frac{\lambda z (1 - z)}{\beta(\lambda - \lambda z) - z}$$

and, by virtue of (5.4.18), we write the final expression for $P(z)$ as

$$P(z) = (1 - \rho) \frac{(1 - z)\beta(\lambda - \lambda z)}{\beta(\lambda - \lambda z) - z},$$

which is the well-known Pollaczek–Khinchin formula.

5.4.5. Other characteristics of the system

Using the results derived thus far, we can study other characteristics of the $M/G/1/\infty$ system, including transient ones. However, we omit them for the mere reason that they are calculated in the same way as in the preceding section. The only exceptions are the distributions of waiting and sojourn times, because the distribution of elapsed service time must be expressed through the distribution of the residual waiting time.

5.5. Use of renewal processes

Using the $M/G/1/\infty$ system, we now analyse the general approach to investigating the characteristics of a wide class of queueing systems. The approach is based on the use of regeneration (renewal) instants at which the process that describes the operation of the system completely forgets its 'past' (see Section 1.3). As a result, we can apply the tools of the renewal theory and reduce the analysis of the characteristics of the system on a given time interval $[0, t]$ to a study of these characteristics on an individual regeneration period, i.e., the period between renewals. This approach is helpful in expressing (mostly in terms of transforms) the joint distribution of almost any characteristic of the system both for steady and transient states.

Obviously, most convenient for the analysis of renewal instants of the $M/G/1/\infty$ system are the instants of arrivals of customers at an empty system (or, which is virtually the same, customer departure instants after which the system becomes empty). The interval between neighbouring renewal instants is called a cycle. Clearly, cycles are mutually independent and each cycle consists of two independent parts: a busy period, i.e., the time interval from the instant of arrival of a customer at an empty system to the succeeding instant when the system becomes empty, and an idle period, i.e., the time interval from the instant when the system becomes empty and the succeeding instant when a customer arrives at the empty system. Since the input flow is Poisson, the idle period is distributed by the exponential law with parameter λ.

We proceed as follows. First, we study the busy period to find the distribution of the main characteristics of the system in a busy period. We also find transient characteristics of the system and, finally, show how to derive steady-state characteristics from transient ones.

Although, as already mentioned, the approach applied here is adequate to find the joint transient distribution of virtually any system characteristic, we restrict ourselves, for the sake of simplicity, only to the number of served customers at any instant t, number of customers in the system, and the virtual waiting time at any instant t.

5.5.1. Busy period

Let $G_n(t)$, $n \geq 1$, denote the joint distribution of the busy period and the number of customers served in this period, i.e., the probability that the busy period lasts for a time less than t and exactly n customers are served in this period.

In order to find the distribution $G_n(t)$, let us change the service discipline and assume that customers are served according to the LCFS discipline, i.e., in the reverse order of arrival, and a newly arriving customer is immediately taken up for service by interrupting the service of the customer at the server. The service of the interrupted customer is resumed later. Obviously, such a change in the service discipline has no effect on the busy period and number of customers served in this period. But the busy period now coincides with the sojourn time of the tagged customer arrived at the empty system and the number of customers served in the busy period is the same as the number of customers served in period of sojourn in the system of that tagged customer arrived at the empty system, including the newly arrived customer. In what follows, it is more convenient to refer to the service time of a customer (i.e., the time spent at the server) as the 'length' of this customer.

Assuming that the length of a newly arriving tagged customer is x, let us calculate how much total time he would stay in the system and how many customers would be served in this period. Since the input flow is Poisson, the service of the tagged customer is interrupted exactly n times in the period x with the probability $(\lambda x)^n e^{-\lambda x}/n!$. In turn, since the service discipline is LCFS, the duration of every interruption is equal to the sojourn period of the interrupting customer in the system, i.e., to the busy period, and the number of customers served during the interruption period coincides with the number of customers served in a busy period. Let $G_n(t \mid x)$ denote the conditional distribution of the busy period and the number of customers served in this period under the condition that the length of the (tagged) customer starting this period is x, and let

$$\gamma(s, z) = \sum_{n=1}^{\infty} z^n \int_0^{\infty} e^{-st} dG_n(t),$$

$$\gamma(s, z \mid x) = \sum_{n=1}^{\infty} z^n \int_0^{\infty} e^{-st} G_n(dt \mid x)$$

denote the double transformations of the distributions $G_n(t)$ and $G_n(t \mid x)$. Using the property of Laplace–Stieltjes transforms and generating functions, namely, the total length of n busy periods and the total number of customers served in these busy periods admit the double transformation $[\gamma(s, z)]^n$, and applying the total probability formula, we obtain

$$\gamma(s, z \mid x) = ze^{-sx} \sum_{n=0}^{\infty} \frac{(\lambda x)^n}{n!} e^{-\lambda x} (\gamma(s, z))^n = ze^{-[s+\lambda-\lambda\gamma(s,z)]x}. \tag{5.5.1}$$

The factor z in formula (5.5.1) indicates that the customer of length x opening a busy period is also served in this period, whereas the factor e^{-sx} indicates that the busy period, along with interruption durations, also includes the length x of the customer opening this period. Now recalling that the length of any customer, even the customer that opens a busy period, is distributed by the law $B(x)$ and using the total probability formula, we obtain

$$\begin{aligned}\gamma(s, z) &= \int_0^{\infty} \gamma(s, z \mid x)\, dB(x) \\ &= z \int_0^{\infty} e^{-[s+\lambda-\lambda\gamma(s,z)]x} dB(x) = z\beta(s + \lambda - \lambda\gamma(s, z)).\end{aligned}$$

Finally, we find that the double transform $\gamma(s, z)$ of the busy period and number of customers served in this period satisfies the equation

$$\gamma(s, z) = z\beta(s + \lambda - \lambda\gamma(s, z)). \tag{5.5.2}$$

Like the double transform, the function $\gamma(s, z)$ is defined in the complex domain $\Re s \geq 0$, $|z| \leq 1$, is continuous, satisfies the inequality $|\gamma(s, z)| \leq 1$ inside this domain, and is also analytic in the domain $\Re s > 0$, $|z| < 1$. Furthermore, $\gamma(s, 1) = \gamma(s)$ is the Laplace–Stieltjes transform of the busy period, whereas $\gamma(0, z)$ is the generating function of the number of customers served in a busy period.

We now study the solution of (5.5.2). We restrict ourselves only to the Laplace–Stieltjes transform $\gamma(s) = \gamma(s, 1)$, leaving the general transform of $\gamma(s, z)$ and generating function of $\gamma(0, z)$ to the reader as a topic for self-study.

We need the following properties of $\beta(s + \lambda - \lambda\gamma)$ as a function of γ, $0 \leq \gamma \leq 1$, and s, $s \geq 0$:

(A) For a fixed s, $s \in [0, \infty)$, the function $\beta(s + \lambda - \lambda\gamma)$ is strictly increasing and strictly downward convex function of γ, $\gamma \subset [0, 1]$. It takes the values $\beta(s + \lambda) > 0$ and $\beta(s) \leq 1$ at the points $\gamma = 0$ and $\gamma = 1$ respectively.

(B) For a fixed γ, $\gamma \in [0, 1]$, the function $\beta(s + \lambda - \lambda\gamma)$ is a strictly decreasing function of s, $s \in [0, \infty)$.

(C) The equality $(\beta(\lambda - \lambda\gamma))'_{\gamma}\Big|_{\gamma=1} = \lambda b = \rho$ holds.

Properties A, B, and C are obvious corollaries to the definition of the Laplace–Stieltjes transform

$$\beta(s) = \int_0^{\infty} e^{-sx} dB(x)$$

and its derivative

$$\beta'(s) = -\int_0^{\infty} x e^{-sx} dB(x).$$

We now study three possible cases separately.

TRAFFIC INTENSITY $\rho < 1$. In this case (see Fig. 5.4a), the equation

$$\gamma = \beta(s + \lambda - \lambda\gamma) \tag{5.5.3}$$

for any $s \geq 0$ has a unique solution $\gamma = \gamma(s)$, $0 < \gamma(s) \leq 1$, where $\gamma(0) = 1$ and $\gamma(s) \to 1$ as $s \to 0$. Therefore, the busy period is finite with the probability one. Differentiating (5.5.3) with respect to s, we obtain the equation for the mean busy period g:

$$g = -\gamma'(0) = -\beta'(0)[1 - \lambda\gamma'(0)] = b(1 + \lambda g).$$

Hence $g = b/(1 - \rho)$. Thus, for $\rho < 1$, the busy period is not only finite with the probability one, but also has a finite mean value. As we know from the previous sections, the condition $\rho < 1$ is necessary for the existence of stationary characteristics of the queue and time of sojourn of a customer in the system. In Section 5.5.4, we will show that for $\rho < 1$, the system will necessarily pass to the steady state with time.

TRAFFIC INTENSITY $\rho = 1$. In this case (Fig. 5.4b), (5.5.3) has a unique solution $\gamma(s)$ and $\gamma(s) \to 1$ as $s \to 0$. But, as can be easily verified, the mean busy period is infinite.

TRAFFIC INTENSITY $\rho > 1$ (INCLUDING $\rho = \infty$). This case (Fig. 5.4c) significantly differs from the previous cases. Equation (5.5.3) for $s = 0$ has two solutions: $\gamma_1(0) = 1$ and $\gamma_2(0) = \gamma(0) < 1$. But for $s > 0$, its solution $\gamma(s)$ is unique and $\gamma(s) \to \gamma(0)$ as $s \to 0$. Hence the busy period is $+\infty$ with probability $1 - \gamma(0)$, i.e., never terminates.

In conclusion, let us write the expression for the distribution $G_n^*(t)$ of cycles and the number of customers served in a cycle. Since each cycle consists of a busy period and an idle period, the idle period is exponentially distributed with parameter λ, and no customer

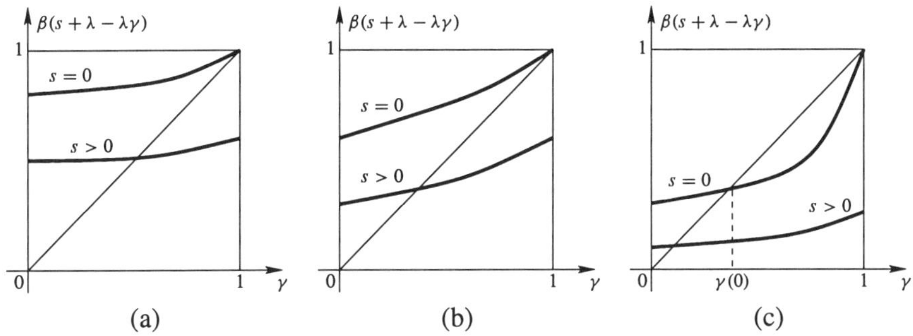

Figure 5.4.

is served in an idle period, the double transform $\gamma^*(s, z)$ of the distribution $G_n^*(t)$ is given by the expression

$$\gamma^*(s, z) = \frac{\lambda \gamma(s, z)}{\lambda + s}.$$

The mean cycle length g^* is defined by the formula

$$g^* = g + \frac{1}{\lambda} = \frac{b}{1-\rho} + \frac{1}{\lambda} = \frac{1}{\lambda(1-\rho)}.$$

5.5.2. Characteristics of the system in a busy period

Let a customer arrive at the system at the instant 0. In this section, we study the characteristics of the system at instant t, assuming that the first busy period starting at the instant 0 does not terminate at the instant t.

First, we find $F_m(n, t)$, $n \geq 1$, $m \geq 1$, $t > 0$, the probability that the following events occur simultaneously: the nth customer is served up to the instant t, the system never becomes free until the service of the nth customer is completed (the busy period that started at the instant 0 is not completed), and m, $m \geq 1$, customers remain in the system immediately upon completion of the service of the nth customer. Here it is reasonable to take $F_1(0, t) = 1$ and $F_m(0, t) = 0$, $m \geq 2$, for all $t > 0$. We use the notation $f_m(n, t) = F_m'(n, t)$, $n \geq 0$, $m \geq 1$ ($f_1(0, t)$ is the generalised derivative of the Heaviside function). For the sake of simplicity, we assume that the density of the distribution function $B(x)$ of the customer length is $b(x) = B'(x)$. Since for the service of the nth customer to be completed at the instant t and the system to contain m customers immediately after this instant, it is necessary that the service of the $(n-1)$th customer be completed at some intermediate instant y, the busy period be not terminated prior to the instant y, i, $i \geq 1$, customers remain in the system immediately after the instant y, the service time of the nth customer be $(t-y)$, and $(m-i+1)$ customers arrive in the period $(t-y)$ of service of the nth customer, we can, using the total probability formula, express $f_m(n, t)$ as follows:

$$f_m(n, t) = \sum_{i=1}^{m+1} \int_0^t \frac{[\lambda(t-y)]^{m-i+1}}{(m-i+1)!} e^{-\lambda(t-y)} f_i(n-1, y) b(t-y)\, dy, \quad n \geq 1, \quad m \geq 1. \tag{5.5.4}$$

Applying the triple transformation

$$\phi(z_1, s, z_2) = \sum_{n=0}^{\infty}\sum_{m=1}^{\infty} z_1^n z_2^m \int_0^\infty e^{-st} f_m(n,t)\,dt,$$

from (5.5.4) we obtain

$$\phi(z_1, s, z_2) = z_2 + z_1 \frac{\beta(s+\lambda-\lambda z_2)}{z_2}\phi(z_1, s, z_2) - z_1\beta(s+\lambda)\left.\frac{\partial}{\partial z_2}\phi(z_1, s, z_2)\right|_{z_2=0},$$

or

$$\phi(z_1, s, z_2) = z_2 \frac{z_2 - z_1\beta(s+\lambda)\frac{\partial}{\partial z_2}\phi(z_1, s, z_2)|_{z_2=0}}{z_2 - z_1\beta(s+\lambda-\lambda z_2)}. \tag{5.5.5}$$

In order to find the numerator of the fraction in (5.5.5), we recall that the equation

$$z_2 - z_1\beta(s+\lambda-\lambda z_2) = 0$$

has a unique solution $z_2 = \gamma(s, z_1)$ (see the preceding section) for all z_1, $0 \le z_1 \le 1$, and $s \ge 0$ (for $\rho > 1$, we take the minimum root of the equation for $z_1 = 1$ and $s = 0$). Since the numerator of the fraction in (5.5.5) must vanish at the point $(z_1, s, \gamma(s, z_1))$, i.e.,

$$\gamma(s, z_1) = z_1\beta(s+\lambda)\left.\frac{\partial}{\partial z_2}\phi(z_1, s, z_2)\right|_{z_2=0},$$

we obtain

$$\phi(z_1, s, z_2) = z_2 \frac{z_2 - \gamma(s, z_1)}{z_2 - z_1\beta(s+\lambda-\lambda z_2)}. \tag{5.5.6}$$

Let $R_m^*(x, n, t)$, $n \ge 0$, $m \ge 1$, denote the probability of the following event: the busy period does not terminate up to the instant t (as before, we assume that a customer arrives at the instant 0 at an empty system), n customers were served up to the instant t, m customers are at the instant t in the system, and the $(n+1)$th customer is served up to the instant $(t+x)$. Using the total probability formula, for

$$r_m^*(x, n, t) = \frac{\partial}{\partial x} R_m^*(x, n, t)$$

we obtain the expression

$$r_m^*(x, n, t) = \sum_{i=1}^{m}\int_0^t f_i(n, y)\frac{[\lambda(t-y)]^{m-i}}{(m-i)!}e^{-\lambda(t-y)}b(t-y+x)\,dy, \quad n \ge 0, \quad m \ge 1. \tag{5.5.7}$$

Applying the transformation

$$\rho^*(z_1, s_1, z_2, s_2) = \sum_{n=0}^{\infty}\sum_{m=1}^{\infty} z_1^n z_2^m \int_0^\infty \int_0^\infty e^{-s_1 t} e^{-s_2 x} r_m^*(x, n, t)\,dx\,dt$$

and using the equality

$$\int_0^\infty \int_0^\infty e^{-s_1 t} e^{-s_2 x} b(t+x)\,dx\,dt = \int_0^\infty dt \int_t^\infty e^{-s_1 t} e^{-s_2 (y-t)} b(y)\,dy$$
$$= \int_0^\infty e^{-s_2 y} b(y)\,dy \int_0^y e^{-(s_1-s_2)t}\,dt$$
$$= \int_0^\infty \frac{1-e^{-(s_1-s_2)y}}{s_1-s_2} e^{-s_2 y} b(y)\,dy$$
$$= \frac{1}{s_1-s_2}\int_0^\infty (e^{-s_2 y} - e^{-s_1 y}) b(y)\,dy$$
$$= \frac{\beta(s_2)-\beta(s_1)}{s_1-s_2},$$

from (5.5.7) we obtain

$$\rho^*(z_1, s_1, z_2, s_2) = \phi(z_1, s_1, z_2)\frac{\beta(s_2) - \beta(s_1 + \lambda - \lambda z_2)}{s_1 + \lambda(1-z_2) - s_2}. \tag{5.5.8}$$

One more function $R_m(x, n, t)$ which we need is the probability that the busy period does not terminate up to instant t, n customers were served up to the instant t, there are m customers in the system at the instant t, and the virtual waiting time at the instant t is less than x. Therefore, $R_m(x, n, t)$ differs from $R_m^*(x, n, t)$ in that the residual service time of the customer at the server at the instant t is replaced by the virtual waiting time at this instant. Since the virtual waiting time at the instant t is the sum of two independent terms, namely, the residual service time of the customer at the server and the service times of $(m-1)$ customers in the queue, then, using the total probability formula, we find that $r_m(x, n, t) = (R_m(x, n, t))'$ is given by the expression

$$r_m(x, n, t) = \int_0^x r_m^*(y, n, t) b^{(*(m-1))}(x-y)\,dy,$$

where $b^{(*m)}(x)$ is the distribution density of the total service time of m customers (m-fold convolution of $b(x)$). Using the transformation

$$\rho(z_1, s_1, z_2, s_2) = \sum_{n=0}^\infty \sum_{m=1}^\infty z_1^n z_2^m \int_0^\infty \int_0^\infty e^{-s_1 t} e^{-s_2 x} r_m(x, n, t)\,dx\,dt,$$

we obtain

$$\rho(z_1, s_1, z_2, s_2) = \frac{\rho^*(z_1, s_1, z_2\beta(s_2), s_2)}{\beta(s_2)},$$

or, by virtue of formulas (5.5.8) and (5.5.6),

$$\rho(z_1, s_1, z_2, s_2) = z_2 \frac{z_2\beta(s_2) - \gamma(s_1, z_1)}{z_2\beta(s_2) - z_1\beta(s_1 + \lambda - z_2\lambda\beta(s_2))} \times \frac{\beta(s_2) - \beta(s_1 + \lambda - z_2\lambda\beta(s_2))}{s_1 + \lambda - z_2\lambda\beta(s_2) - s_2}. \tag{5.5.9}$$

The last formula expresses the joint distribution of the key service characteristics in a busy period in terms of transforms.

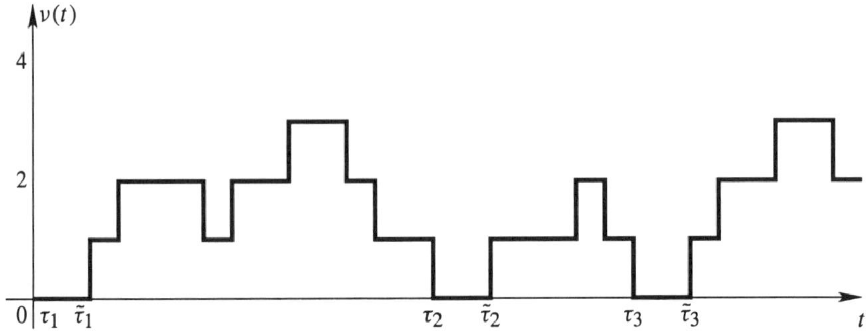

Figure 5.5.

5.5.3. Transient characteristics

Let us find the joint transient distribution of base characteristics of system functioning at instant t in terms of transforms, namely, the number of customers served by instant t, the number of customers in the system at instant t, and the virtual waiting time at instant t. For the sake of simplicity, we assume that the system is empty at the initial instant 0, though all results of this section can, obviously, be extended to the case of an arbitrary initial instant.

Let us consider two renewal processes generated by cycles.

The first renewal process consists of instants $\tau_1, \tau_2, \dots$ at which the system becomes empty. We assume that the initial instant 0 is a also a renewal instant with number 1 (Fig. 5.5). Let $H^{(1)}(t,n)$, $n \geq 0$, denote the mean number of renewals that occurred up to the instant t, and the probability that n customers were served in the renewal periods (cycles) ending by the instant t. Then

$$h^{(1)}(t,n) = \frac{\partial}{\partial t} H^{(1)}(t,n)$$

has a simple physical meaning: $h^{(1)}(t,n)\,dt$ is the probability that (some) busy period (cycle) had ended in time $(t, t+dt)$ and n customers were served by the instant $t+dt$. As is known from renewal theory, the transform

$$\varkappa^{(1)}(s,z) = \sum_{n=0}^{\infty} z^n \int_0^{\infty} e^{-st} H^{(1)}(dt,n)$$

is defined by

$$\varkappa^{(1)}(s,z) = \frac{1}{1-\gamma^*(s,z)} = \frac{\lambda+s}{\lambda+s-\lambda\gamma(s,z)}. \tag{5.5.10}$$

The second renewal process, which consists of the instants $\tilde{\tau}_1, \tilde{\tau}_2, \dots$ of arrival of customers at an empty system (see Fig. 5.5), differs from the first renewal process from the viewpoint of renewal theory in that the first renewal instant is shifted by an exponentially distributed random variable (the idle period) with parameter λ. Introducing $H^{(2)}(t,n)$,

$h^{(2)}(t,n)$, and $\varkappa^{(2)}(s,z)$ in analogy with $H^{(1)}(t,n)$, $h^{(1)}(t,n)$, and $\varkappa^{(1)}(s,z)$, respectively, we obtain

$$\varkappa^{(2)}(s,z) = \frac{\frac{\lambda}{\lambda+s}}{1-\gamma^*(s,z)} = \frac{\lambda}{\lambda+s-\lambda\gamma(s,z)}. \tag{5.5.11}$$

Now we can find $W_m(n,x,t)$—the joint distribution of the main characteristics of the system at any instant t, i.e., the probability that n customers have been served, there are m customers in the system, and the virtual waiting time is less than x.

For the system to be empty at instant t, it is necessary that the system be empty (the first renewal process is renewed) at some intermediate instant y $(0 < y < t)$ and no customer arrive in the residual time $(t-y)$ at the empty system (recall that the virtual waiting time in this case is zero). Therefore,

$$W_0(n,x,t) = \int_0^t e^{-\lambda(t-y)} h^{(1)}(y,n)\,dy, \quad n \geq 0, \quad x > 0. \tag{5.5.12}$$

Similarly, for n customers to be served by the instant t, the system to contain m customers at the instant t, and the virtual waiting time at the instant t to be less than x, it is necessary that a customer arrive at the empty system at an intermediate instant y (the first renewal process be renewed), i, $i = 0,\ldots,n$, customers be served by the instant y, the busy period that started at the instant y be not completed in the residual time $(t-y)$, $(n-i)$ customers be served, m customers remain in the system at the instant t, and the virtual waiting time be less than x. Hence,

$$W_m(n,x,t) = \sum_{i=0}^{n} \int_0^t R_m(x,n-i,t-y) h^{(2)}(y,i)\,dt, \quad m \geq 1, \quad n \geq 0. \tag{5.5.13}$$

Using the transformation

$$\omega(z_1,s_1,z_2,s_2) = \sum_{n=0}^{\infty}\sum_{m=0}^{\infty} z_1^n z_2^m \int_0^\infty \int_0^\infty e^{-s_1 t} e^{-s_2 x} W_m(n,dx,t)\,dt,$$

from (5.5.12) and (5.5.13) we obtain

$$\omega(z_1,s_1,z_2,s_2) = \frac{1}{s_1+\lambda}\varkappa^{(1)}(s_1,z_1) + \rho(z_1,s_1,z_2,s_2)\varkappa^{(2)}(s_1,z_1),$$

or, by virtue of (5.5.9)–(5.5.11),

$$\begin{aligned}\omega(z_1,s_1,z_2,s_2) &= \frac{1}{\lambda+s_1-\lambda\gamma(s_1,z_1)} \\ &\times \left\{1+\lambda z_2 \frac{[z_2\beta(s_2)-\gamma(s_1,z_1)][\beta(s_2)-\beta(s_1+\lambda-z_2\lambda\beta(s_2))]}{[z_2\beta(s_2)-z_1\beta(s_1+\lambda-z_2\lambda\beta(s_2))][s_1+\lambda-z_2\lambda\beta(s_2)-s_2]}\right\}.\end{aligned} \tag{5.5.14}$$

In particular, setting $z_2 = z$ and $z_1 = 1$ in (5.5.14), we arrive at formula (5.3.17) via a different route.

5.5.4. Stationary joint distribution of the number of customers in the system and the waiting time

We now assume that the traffic intensity is $\rho < 1$. Let

$$W_m(x,t) = \sum_{n=0}^{\infty} W_m(n,x,t)$$

stand for the joint distribution of the number of customers in the system and virtual waiting time at the instant t, and let

$$W_m(x) = \lim_{t\to\infty} W_m(x,t)$$

denote the limit (stationary) joint distribution of the number of customers in the system and waiting time for service. Obviously, here we are not considering the number of served customers, because this characteristic tends to infinity.

Prior to finding $W_m(x)$, we prove (what was skipped earlier) that the condition $\rho < 1$ is sufficient for existence of $W_m(x)$ (and hence, for existence of marginal limit distributions of the number of customers in the system and virtual waiting time). We will soon see that this fact is a trivial corollary to the key renewal theorem 1.3.4. For the sake of simplicity, let the distribution function $B(t)$ be continuous (existence of the distribution density $b(t)$ is not assumed). If this assumption is broken, then first we must to prove the key renewal theorem under conditions that are somewhat more general than those usually used.

First, let us rewrite (5.5.12) and (5.5.13) in terms of renewal functions, omitting the number of customers that had been served by the instant t. Setting

$$H^{(1)}(t) = \sum_{n=0}^{\infty} H^{(1)}(t,n), \qquad H^{(2)}(t) = \sum_{n=0}^{\infty} H^{(2)}(t,n),$$

$$R_m(x,t) = \sum_{n=0}^{\infty} R_m(x,n,t),$$

from (5.5.12) and (5.5.13) we obtain

$$W_0(x,t) = \int_0^t e^{-\lambda(t-y)} dH^{(1)}(y), \qquad x > 0, \tag{5.5.15}$$

$$W_m(x,t) = \int_0^t R_m(x,t-y)\, dH^{(2)}(y), \quad x > 0, \quad m \ge 1. \tag{5.5.16}$$

Now we assert a few properties of the function $R_m(x,t)$. By the definition of the function $R_m(x,t)$, we see that

$$\sum_{m=0}^{\infty} R_m(x,t) = 1 - G(t).$$

Since $\rho < 1$, we obtain the inequality

$$\int_0^{\infty} R_m(x,t)\, dt \le \int_0^{\infty} [1 - G(t)]\, dt = g = \frac{b}{1-\rho} < \infty.$$

Therefore, the function $R_m(x, t)$ is integrable on $(0, \infty)$. By virtue of continuity of $B(t)$, we see that the function $R_m(x, t)$ is also continuous in t.

Finally, since $\rho < 1$, the mean-renewal time (mean cycle length) $g^* = 1/(\lambda(1-\rho))$ is finite for the renewal processes generated by the instants $\tau_1, \tau_2, \ldots$ and $\tilde{\tau}_1, \tilde{\tau}_2, \ldots$. Moreover, the distribution function of the time between renewals is continuous (and even has a bounded density), because it is the convolution of the distribution function $G(t)$ of the busy period and exponential distribution function of the idle period with parameter λ.

Thus, we can apply the key renewal theorem to relations (5.5.15) and (5.5.16) as $t \to \infty$:

$$p_0 = W_0(x) = \lim_{t\to\infty} W_0(x, t) = \frac{1}{g^*}\int_0^\infty e^{-\lambda t}\,dt = \frac{1}{\lambda g^*} = 1 - \rho, \qquad x > 0,$$

$$W_m(x) = \lim_{t\to\infty} W_m(x, t) = \frac{1}{g^*}\int_0^\infty R_m(x, t)\,dt.$$

These equalities prove the existence of $W_m(x)$.

The fact that $W_m(x)$ does not depend on the initial state of the system follows from the fact that the initial state affects only the first renewal period of the renewal processes generated by the instants $\tau_1, \tau_2, \ldots$ and $\tilde{\tau}_1, \tilde{\tau}_2, \ldots$, and, as is known, the key renewal theorem holds for general renewal processes.

This result is a particular case of the ergodic theorem 1.3.5 for regenerative processes.

To find the limit distribution $W_m(x)$ by means of the double transformation

$$\omega(z, s) = \sum_{m=0}^{\infty} z^m \int_0^\infty e^{-sx}\,dW_m(x),$$

it suffices to recall the definition of the Laplace (time) transformation

$$\omega(1, s_1, z, s_2) = \int_0^\infty e^{-s_1 t}\left[\sum_{n=0}^{\infty} 1^n \sum_{m=0}^{\infty} z^m \int_0^\infty e^{-s_2 x} W_m(n, dx, t)\right] dt$$

$$= \int_0^\infty e^{-s_1 t}\left[\sum_{m=0}^{\infty} z^m \int_0^\infty e^{-s_2 x} W_m(dx, t)\right] dt.$$

Since $W_m(x, t) \xrightarrow[t\to\infty]{} W_m(x)$, we find that $\omega(1, s_1, z, s_2) \sim \omega(z, s_2)/s_1$ as $s_1 \to \infty$. Therefore,

$$\omega(z, s_2) = \lim_{s_1\to 0} s_1\omega(1, s_1, z, s_2).$$

Using (5.5.14) and the L'Hospital rule

$$\lim_{s\to 0}\frac{s}{s + \lambda[1 - \gamma(s)]} = \frac{1}{1 + \lambda g} = 1 - \rho,$$

we obtain

$$\omega(z, s) = (1 - \rho)\left\{1 + \lambda z\frac{[z\beta(s) - 1][\beta(s) - \beta(\lambda - z\lambda\beta(s))]}{[z\beta(s) - \beta(\lambda - z\lambda\beta(s))][\lambda - z\lambda\beta(s) - s]}\right\}.$$

Thus, we have again arrived at formula (5.3.12).

In conclusion, we give without proof the following assertion.

If $\rho = 1$, then the number of customers in the system and the virtual waiting time at instant t as $t \to \infty$ tend to infinity in probability. If $\rho > 1$, then these characteristics as $t \to \infty$ tend to infinity even with the probability one.

6

Other simple non-Markov models

In this chapter, we continue the study of simple non-Markov models and show that the methods applied to study the $M/G/1/\infty$ system can also be applied, after certain slight modification, to study several other queueing systems. However, not all methods developed for the $M/G/1/\infty$ system in the preceding chapter are suitable for other systems. For example, formulas for the $M/D/n/\infty$ system can be derived only for state probabilities, and moreover, only through the method of embedded Markov chains. For the $M/G/n/0$ system, the situation is opposite: stationary state probabilities can be computed only by the method of supplementary variables.

Truly speaking, the objective of this chapter is, in essence, the converse of the problem formulated and solved in the previous chapter. Here we will not go into details, but restrict ourselves to a demonstration which method is suited best to a particular system. A reader, if he has mastered the tools of the previous chapter, can himself study the system in depth that interests him most.

As in the whole book, the topics in this chapter are arranged so that their research methods increase in complexity. First we study the infinite-server $M/G/\infty$ and $G/G/\infty$ systems; their investigation requires, indeed, nothing but only the concept of the input flow. For the $M/D/n/\infty$ and $G/M/1/\infty$ systems, we apply the embedded Markov chain technique, and, finally, we analyse the $M/G/1/r$ and $M/G/n/0$ systems by the method of supplementary variables.

For the sake of definiteness, we assume that the customers are served in the order of their arrival, i.e., according to the FCFS discipline.

However, it must be emphasised that the queueing systems we study constitute only a small part of the whole range of models of real-world technical systems. Nevertheless, we hope that our book will stimulate interest for independent study.

6.1. $M/G/\infty$ system

In the infinite-server $M/G/\infty$ queueing system, each customer is taken up for service immediately upon arrival. Therefore, it is reasonable to restrict ourselves only to the characteristics related to the number of customers in the system. The arrival intensity, as usual, is denoted by λ. The service time of a customer is distributed by some law $B(x)$.

There are different methods to study the $M/G/\infty$ system. Here we deal with the simplest method based on the description of state probabilities in terms of a non-stationary Poisson flow. This method is helpful in determining, via simple reasoning, both stationary and transient distributions of the number of customers in the system.

6.1.1. Transient distribution of the number of customers

Let $\nu(t)$ denote the number of customers in the system at instant t and let

$$p_i(t) = \mathbf{P}\{\nu(t) = i\}$$

denote the probability that i customers exist in the system at instant t. Obviously, the state set of the process $\{\nu(t), t \geq 0\}$ is a set of non-negative integers $\mathcal{X} = \{0, 1, \dots\}$. For the sake of simplicity, we assume that the system is empty at the initial instant 0 ($\nu(0) = 0$), though the result can be trivially extended to an arbitrary initial instant.

In this case, the process $\{\nu(t),\ t \geq 0\}$ is obviously not Markov. But, in reality using only a non-stationary Poisson flow, we can easily find the transient distribution $\{p_i(t),\ i \geq 0\}$ of the process $\{\nu(t),\ t \geq 0\}$. We recommend the reader to apply the method described here to find the joint multivariate distribution

$$\mathbf{P}\{\nu(t_1) = i_1, \dots, \nu(t_n) = i_n\}$$

of the process $\{\nu(t), t \geq 0\}$ and the joint multivariate distribution of $\nu(t)$ at the instants $t_1, \dots, t_n$ and residual (or elapsed) service times at these instants.

On the interval $(0, t)$, let us consider a flow of customers whose service is not completed by the instant t (the current instant is denoted by s). Clearly, the total number of unserved customers at the instant t is equal to the value of the process $\nu(t)$ at the instant t.

This flow has no memory, because, since the input is Poisson and service times are independent of each other and of the input, the numbers of unserved customers at the instant t that had arrived in non-overlapping time intervals $(0, t)$ are independent random variables.

Furthermore, this flow is ordinary, because the probability that more than one customer arrives in an interval of length Δ that will not be served by the instant t is no greater than the probability that more than one customer arrives in this time interval, i.e., is $o(\Delta)$.

But the flow is not stationary, because the distribution of the number of customers arriving in time intervals of identical length, obviously, depends on the location of intervals: the farther the interval from the instant t, the smaller the number of customers that would remain in the system by the instant t out of those that had arrived in this interval. Hence the flow of unserved customers at the instant t is a non-stationary Poisson flow with intensity $\lambda(s)$, which depends on time s. It is easy to find the intensity $\lambda(s)$: it suffices to multiply the initial input intensity λ by the probability $(1 - B(t - s))$ that a customer is not served in time $(t - s)$, i.e.,

$$\lambda(s) = \lambda[1 - B(t - s)].$$

Thus, we conclude that the number of customers $\nu(t)$ in the $M/G/\infty$ system at the instant t is equal to the number of customers arrived in the interval $(0, t)$ from a non-stationary

Poisson flow of intensity $\lambda(s) = \lambda[1 - B(t - s)]$. Hence (see Section 2.1) the number $\nu(t)$ is distributed by the Poisson law with parameter

$$\Lambda(t) = \int_0^t \lambda[1 - B(t - s)]\, ds = \lambda \int_0^t [1 - B(s)]\, ds, \tag{6.1.1}$$

and the probability $p_i(t)$ that there are i customers at the instant t is

$$p_i(t) = \frac{[\Lambda(t)]^i}{i!} e^{-\Lambda(t)}. \tag{6.1.2}$$

6.1.2. Stationary distribution of the number of customers

In order to find the stationary probabilities

$$p_i = \lim_{t\to\infty} p_i(t),$$

it suffices to pass to the limit in formulas (6.1.1) and (6.1.2) as $t \to \infty$.

If the mean service time

$$b = \int_0^\infty x\, dB(x) = \int_0^\infty [1 - B(x)]\, dx$$

is finite, then

$$\Lambda(t) \to \lambda b = \rho, \qquad t \to \infty.$$

Therefore, the stationary distribution of the number of the customers is Poisson with parameter ρ:

$$p_i = \frac{\rho^i}{i!} e^{-\rho}, \qquad i = 0, 1, \dots.$$

But if $b = \infty$, then $\Lambda(t) \to \infty$. Therefore, $p_i(t) \xrightarrow[t\to\infty]{} 0$, $i \geq 0$, i.e., the number $\nu(t)$ of customers tends to infinity in probability as time grows. Rigorous analysis shows that $\nu(t)$ tends to infinity even with probability one.

6.1.3. Output flow

First we study the output flow for the case where the service time of each customer is constant, i.e., the service time takes only one value b with probability one. Obviously, the output flow in this case is shifted in time from the input flow by b. But this means that the output is an ordinary non-stationary Poisson flow of intensity $\lambda^*(t) = 0$ for $t < b$ and $\lambda^*(t) = \lambda$ for $t \geq b$.

Now we assume that the service time of each incoming customer takes one of two values b_1 and b_2 (for definiteness, assume that $b_1 < b_2$) with some probabilities β_1 and β_2 respectively. Then the input flow (see Theorems 2.1.1 and 2.1.2) can be represented as the superposition (union) of two independent Poisson flows: a customer flow of intensity $\beta_1\lambda$ with service time b_1 and a customer flow of intensity $\beta_2\lambda$ with service time b_2. The output flow is also a superposition of two independent but non-stationary Poisson flows. The first output flow is obtained from the first input flow by shift by the service time b_1, i.e., it has a time-dependent intensity $\lambda_1^*(t)$ equal to 0 for $t < b_1$ and $\beta_1\lambda$ for $t \geq b_1$. The second output flow is obtained from the second input flow by shifting by the service time b_2; its intensity $\lambda_2^*(t)$ is 0 for $t < b_2$ and $\beta_2\lambda$ for $t \geq b_2$. Thus, the total output flow is a non-stationary Poisson flow of intensity $\lambda^*(t) = \lambda_1^*(t) + \lambda_2^*(t)$, i.e., $\lambda^*(t) = 0$ for $0 < t < b_1$, $\lambda^*(t) = \beta_1\lambda$ for $b_1 \leq t < b_2$, and $\lambda^*(t) = \lambda$ for $t \geq b_2$.

Similarly, we can show that the output flow is also Poisson even if the service time takes any finite or denumerable values $b_1, b_2, \ldots$

For an arbitrary service time distribution $B(x)$, the input flow admits the representation as a superposition of infinitely many independent Poisson flows of intensity $\lambda dB(x)$, and each individual flow is retarded in time by x. So, the output flow is also Poisson (but non-stationary) and its intensity is

$$\lambda^*(t) = \int_0^t \lambda\, dB(x) = \lambda B(t).$$

Obviously, $\lambda^*(t) \to \lambda$ as $t \to \infty$. Therefore, the output in the limit (in the steady state) is a stationary Poisson flow of intensity λ. No condition is imposed on the distribution function $B(x)$ of the service time.

6.2. $G/G/\infty$ system

The $G/G/\infty$ system is the most general of infinite-server systems in the Kendall classification, because the input is an arbitrary recurrent flow and the service time of each customer is distributed by an arbitrary law. As before, $A(x)$ denotes the distribution function of the inter-arrival time and $B(x)$, the distribution function of service times. As in the case of $M/G/\infty$ systems, we restrict ourselves only to the number $\nu(t)$ of customers in the system at instant t or, in other words, the number of busy servers at instant t.

As will be clear soon, although this system is apparently complicated, satisfactory calculation formulas for the distribution of the number of customers in the system at an arbitrary instant can be found, because the number of servers is infinite. These formulas are expressed as recursive procedures for the Laplace time transforms of binomial moments of the number of customers in the system at an instant t.

6.2.1. Simultaneous equations for the state probabilities

For the sake of simplicity, we assume that the first customer arrives at the instant 0 at an empty system. Let $p_i(t)$ denote the probability that there are i customers in the system at the instant t. Then $p_i(t)$ satisfies the recursive equations

$$\begin{aligned}
p_0(t) &= B(t)\left[1 - A(t) + \int_0^t p_0(t-x)\, dA(x)\right], \\
p_1(t) &= B(t)\int_0^t p_1(t-x)\, dA(x) + [1 - B(t)]\left[1 - A(t) + \int_0^t p_0(t-x)\, dA(x)\right], \\
p_i(t) &= B(t)\int_0^t p_i(t-x)\, dA(x) + [1 - B(t)]\int_0^t p_{i-1}(t-x)\, dA(x), \qquad i \geq 2.
\end{aligned} \tag{6.2.1}$$

System (6.2.1) is derived as follows. Let us consider the arrival instant x next to 0. Then the system is empty at the instant t only if the service of the customer that had arrived at the instant 0 is completed before the instant t (this occurs with probability $B(t)$) and either x is greater than t (with the probability $1 - A(t)$) or x is less than t, but the customers that had arrived by the instant x and those that had arrived in the interval (x, t) are served before the instant t. Obviously, the probability that the customers that have arrived by the instant x and those that had arrived in the interval (x, t) are served before the instant t equals $p_0(t - x)$. Hence, using the total probability formula, we arrive at the first equation in (6.2.1).

Other equations in (6.2.1) are derived similarly. Thus, there are i, $i \geq 2$, customers at the instant t if the service of the first customer is completed before the instant t and at the instant t there remain i of the customers that had arrived in the interval $(0, t)$, or if the service of the first customer is completed before the instant t and at the instant t there remain $(i - 1)$ of customers that had arrived in the interval $(0, t)$.

Using the generating function

$$P(z, t) = \sum_{i=0}^{\infty} p_i(t) z^i,$$

we rewrite (6.2.1) as

$$P(z, t) = [z + B(t) - zB(t)] \left[1 - A(t) + \int_0^t P(z, t - x)\, dA(x) \right]. \tag{6.2.2}$$

6.2.2. Transient state probabilities

Although the integral part of (6.2.2) is the convolution of the functions $P(z, t)$ and $A(t)$, (6.2.2) cannot be solved with the help of the Laplace transform because the integral has a t-dependent factor, $B(t)$. Since this factor at the point $z = 1$ equals one, the Laplace time transform at this point can be used not only for the function $P(z, t)$, but also for all derivatives $\partial^i P(z, t)/\partial z^i$. We can show that these derivatives at the point $z = 1$ exist and do not grow too fast with i, thereby guaranteeing the convergence of all series encountered in what follows in some neighbourhood of the point $z = 1$.

Thus, we set

$$r_i(t) = \frac{1}{i!} \frac{\partial^i}{\partial z^i} P(z, t) \bigg|_{z=1} = \sum_{k=i}^{\infty} \binom{k}{i} p_k(t), \qquad i \geq 1,$$

$r_i(t)$ is the ith binomial moment of the distribution $\{p_k(t), k \geq 0\}$. For $i = 0$, $r_0(t) = 1$ by the normalisation condition $\sum_{i=0}^{\infty} p_i(t) = 1$. Differentiating (6.2.2) with respect to z at the point $z = 1$ and using the Leibniz formula, we obtain

$$\begin{aligned} r_1(t) &= \int_0^t r_1(t - x)\, dA(x) + 1 - B(t), \\ r_i(t) &= \int_0^t r_i(t - x)\, dA(x) + [1 - B(t)] \int_0^t r_{i-1}(t - x)\, dA(x), \qquad i \geq 2. \end{aligned} \tag{6.2.3}$$

The functions $r_i(t)$, like $p_i(t)$, are defined by recursive equations, which in our case are (6.2.3). But now, in distinction to (6.2.1), we may apply the Laplace transform. Assuming that

$$\rho_i(s) = \int_0^\infty e^{-st} r_i(t)\,dt, \qquad i \geq 1,$$

$$\psi_1(s) = \int_0^\infty e^{-st}[1 - B(t)]\,dt = \frac{1 - \beta(s)}{s}$$

and

$$\psi_i(s) = \int_0^\infty e^{-st}[1 - B(t)] \int_0^t r_{i-1}(t - x) dA(x)\,dt, \qquad i \geq 2, \tag{6.2.4}$$

from (6.2.3) we obtain

$$\rho_i(s) = \rho_i(s)\alpha(s) + \psi_i(s), \qquad i \geq 1,$$

or

$$\rho_i(s) = \frac{\psi_i(s)}{1 - \alpha(s)}, \qquad i \geq 1. \tag{6.2.5}$$

Formulas (6.2.4) and (6.2.5), in principle, are helpful in recursive calculation of $\rho_i(s)$, beginning with $\rho_1(s)$. But $\rho_{i-1}(s)$ in $\psi_i(s)$ must be re-calculated each time. For this purpose, we must return from the Laplace transform $\rho_{i-1}(s)$ to the original function $r_{i-1}(t)$ and then apply formula (6.2.4) to finding $\psi_i(s)$.

In order to find $p_k(t)$, $k \geq 0$, it suffices to recall that $r_k(t)$ are the coefficients in the expansion of the function $P(z, t)$ into powers of $(z - 1)$, i.e.,

$$P(z, t) = \sum_{k=0}^\infty r_k(t)(z - 1)^k.$$

In particular, if its convergence radius exceeds one for some t, then

$$p_k(t) = \sum_{l=k}^\infty \binom{l}{k} (-1)^{l-k} r_l(t), \qquad k \geq 0.$$

The functions $r_i(t)$ themselves are helpful in calculating all moments (all moments are indeed finite) of the number of customers in the system at instant t. In particular, $r_1(t)$ is the mean number of customers in the system at instant t.

6.2.3. Stationary state probabilities

We assume that the mean inter-arrival time

$$a = \int_0^\infty x\,dA(x)$$

and the mean service time

$$b = \int_0^\infty x\, dB(x)$$

are finite. We also assume that one more technical condition is fulfilled, namely, the inter-arrival time cannot take the values kx_0, $k = 0, 1, 2, \ldots, x_0 > 0$. We can show that the limit probabilities

$$p_i = \lim_{t\to\infty} p_i(t), \qquad i \geq 0,$$

exist under these assumptions. Then the generating function

$$P(z) = \sum_{i=0}^{\infty} p_i z^i$$

can be expressed as

$$P(z) = \sum_{i=0}^{\infty} r_i (z-1)^i, \qquad r_i = \lim_{t\to\infty} r_i(t).$$

In turn, using the properties of the Laplace transform, from (6.2.5) we obtain

$$r_i = \lim_{s\to 0} s\rho_i(s) = \lim_{s\to 0} \frac{s\psi_i(s)}{1-\alpha(s)} = \frac{\psi_i(0)}{a}, \qquad i \geq 1.$$

The stationary probabilities p_k, $k \geq 0$, and the moments of the stationary distribution of the number of customers in the system are found from this formula exactly as in the transient case. In particular, since $\psi_1(0) = b$, the stationary mean number of customers in the system is

$$N = \frac{b}{a}.$$

This equality can also be derived from the Little formula, because the arrival intensity in the stationary state is $1/a$ and the sojourn time is the same as his service time and has the mean value b.

6.3. $M/D/n/\infty$ system

The $M/D/n/\infty$ system is a special case of the n-server $M/G/n/\infty$ queueing system. Each customer is served for a constant (not random) time, which, for the sake of definiteness, is assumed to be 1. As before, the intensity of the Poisson input flow is denoted by λ.

Thus far, there is no method to study multiserver M/G/n/∞ systems (let alone even more complicated G/G/n/∞ systems) that might be of help even in determining stationary performance indices in a convenient form for analysis. But, as will be clear soon, the $M/D/n/\infty$ system is one of the few exceptions, though only certain characteristics related to the number of customers in the system can be determined for it.

6.3.1. Embedded Markov chain

As usual, let $\nu(t)$ denote the number of customers in the system at an instant t. Certainly, the process $\{\nu(t), t \geq 0\}$ is not Markov. Therefore, we study the values $\nu_0 = \nu(0)$, $\nu_1 = \nu(1)$, $\ldots, \nu_m = \nu(m), \ldots$ of the process $\nu(t)$ only at integer instants $0, 1, \ldots, m, \ldots$

If a customer at the instant m was at the server, then he would certainly quit the system by the instant $m+1$. Out of other customers that were in the system at the instant m and that had arrived in the time interval $(m, m+1)$, exactly n (or less than n customers if the total number of customers in the queue at the instant m and of those arrived in the time interval $(m, m+1)$ was less than n) would be under service at the instant $m+1$, and others, if any, would wait for service in the queue. Therefore, since the input flow is Poisson, the sequence $\{\nu_m, m \geq 0\}$ is, obviously, a homogeneous Markov chain.

Let us write the transition probabilities p_{ij} of the embedded Markov chain $\{\nu_m, m \geq 0\}$.

If $i \leq n$, then all customers at the instant m quit the system by the instant $m+1$ and the system contains those and only those customers that arrived in the time interval $(m, m+1)$. Since the number of such customers is j with the probability $\beta_j = \lambda^j e^{-\lambda}/j!$, we obtain

$$p_{ij} = \beta_j, \quad i = 0, \ldots, n, \quad j \geq 0.$$

But if $i > n$, then at the instant $m+1$ the system contains $(i-n)$ out of i customers that were in the system at the instant m and $(j-(i-n))$ more customers must arrive. Therefore,

$$p_{ij} = \beta_{j-i+n}, \quad i > n, \quad j \geq i-n.$$

Thus, the matrix of transition probabilities of the embedded Markov chain $\{\nu_m,\ m \geq 0\}$ is

$$P = (p_{ij}) = \begin{pmatrix} \beta_0 & \beta_1 & \beta_2 & \beta_3 & \cdots \\ \beta_0 & \beta_1 & \beta_2 & \beta_3 & \cdots \\ \cdots & \cdots & \cdots & \cdots & \cdots \\ \beta_0 & \beta_1 & \beta_2 & \beta_3 & \cdots \\ 0 & \beta_0 & \beta_1 & \beta_2 & \cdots \\ 0 & 0 & \beta_0 & \beta_1 & \cdots \\ \cdots & \cdots & \cdots & \cdots & \cdots \end{pmatrix}.$$

6.3.2. Stationary state probabilities

Let us assume that the condition $\rho = \lambda/n < 1$ is satisfied (of course, ρ can be referred to as the traffic intensity of the system, because it defines the ratio of the arrival intensity λ to the mean number of customers n that the system can serve in unit time). Therefore, the limit (stationary) state probabilities

$$p_i = \lim_{m\to\infty} \mathbf{P}\{\nu_m = i\}$$

exist, and they are also the limit (time-stationary) state probabilities

$$p_i = \lim_{t\to\infty} \mathbf{P}\{\nu(t) = i\}$$

(necessity and sufficiency of the condition $\rho < 1$ can be proved as before). The stationary probabilities p_i are found from the equilibrium equations

$$p_i = \beta_i \sum_{k=0}^{n-1} p_k + \sum_{k=0}^{i} \beta_k p_{n+i-k}, \qquad i \geq 0. \tag{6.3.1}$$

If p_i, $i = 0, \ldots, n-1$, are known, then we are able to find all other probabilities p_i, $i \geq n$, from (6.3.1). Indeed,

$$p_{n+i} = \frac{1}{\beta_0}\left(p_i - \beta_i \sum_{k=0}^{n-1} p_k - \sum_{k=1}^{i} \beta_k p_{n+i-k}\right), \qquad i \geq 0.$$

In order to find p_i, $i = 0, \ldots, n-1$, let us solve (6.3.1) in terms of the generating function

$$P(z) = \sum_{i=0}^{\infty} p_i z^i.$$

Multiplying the ith equation in (6.3.1) by z^i, summing over i, and recalling the definition of β_i, we obtain

$$P(z) = e^{-\lambda(1-z)} \sum_{k=0}^{n-1} p_k + \frac{1}{z^n}\left[P(z) - \sum_{k=0}^{n-1} p_k z^k\right] e^{-\lambda(1-z)}.$$

Hence

$$P(z) = \frac{z^n \sum_{k=0}^{n-1} p_k - \sum_{k=0}^{n-1} p_k z^k}{z^n e^{\lambda(1-z)} - 1}. \tag{6.3.2}$$

We now make use of the Rouché theorem, which asserts that if two functions $f(z)$ and $g(z)$ are analytic inside a closed domain D, continuous up to the boundary of D, and satisfy the condition $|f(z)| > |g(z)|$ on the boundary, then the functions $f(z)$ and $f(z) - g(z)$ have identical number of zeros (with regard for their multiplicities) inside the domain D. Setting

$$f(z) = z^n e^{\lambda(1-z)}, \qquad g(z) = 1,$$

by virtue of this theorem we conclude that $|f(z)| > |g(z)|$ on the circle $|z| = 1$, except for the point $z = 1$, at which $f(1) = g(1) = 1$. Moreover, by the condition $\rho < 1$,

$$[z^n e^{\lambda(1-z)} - 1]'\Big|_{z=1} = n - \lambda > 0.$$

Therefore, there exists $\varepsilon_1 > 0$ such that for any ε, $0 < \varepsilon < \varepsilon_1$, the inequality $z^n e^{\lambda(1-z)} > 1$ holds true for all z lying on the circle $|z| = 1 + \varepsilon$. But, by virtue of the Rouché theorem, the function $z^n e^{\lambda(1-z)} - 1$ has as many zeros as the function $z^n e^{\lambda(1-z)}$ (with regard for their multiplicities) inside the circle $|z| < 1 + \varepsilon$ (and, since ε is arbitrary, inside the circle $|z| \leq 1$). But the function $z^n e^{\lambda(1-z)}$ has one zero of multiplicity n at the point $z = 0$.

Therefore, $z^n e^{\lambda(1-z)} - 1$ also has n zeros $z_0, z_1, \ldots, z_{n-1}$ (it is not hard to see that all zeros are simple). The only zero on the circle $|z| = 1$ (for definiteness, we denote it by z_0) is, obviously, $z_0 = 1$. To find other zeros, we must find all $(n-1)$ solutions of the functional equation

$$z^n e^{\lambda(1-z)} - 1 = 0$$

lying inside the circle $|z| < 1$. Recall that $P(z)$ is an analytic function of the complex variable z inside the circle $|z| < 1$. Therefore, the numerator of (6.3.2) must vanish at the points $z_1, \ldots, z_{n-1}$. Thus, in order to find p_i, $i = 0, \ldots, n-1$, we have $(n-1)$ linear equations

$$z_i^n \sum_{k=0}^{n-1} p_k - \sum_{k=0}^{n-1} p_k z_i^k = 0, \qquad i = 1, \ldots, n-1. \tag{6.3.3}$$

The last root $z_0 = 1$ is not of any use in calculating $p_i, i = 0, \ldots, n-1$, because an obvious identity is obtained on its substitution into the numerator of (6.3.2). The missing equation is obtained from the normalisation condition $\sum_{i=0}^{\infty} p_i = P(1) = 1$ and the L'Hospital rule. Thus, we obtain

$$\frac{1}{n-\lambda}\left(n \sum_{k=0}^{n-1} p_k - \sum_{k=0}^{n-1} k p_k\right) = 1.$$

This equality will be self-evident in view of the fact that $\sum_{k=0}^{n-1}(n-k)p_k/n$ is the stationary probability $1 - \rho = (n-\lambda)/n$ that any server is idle.

Instead of solving (6.3.3), we can also use the representation

$$z^n - c \sum_{k=0}^{n-1} p_k z^k = (z-1)(z-z_1)\cdots(z-z_{n-1}),$$

where

$$c = \left(\sum_{k=0}^{n-1} p_k\right)^{-1},$$

which can be used to calculate the probability p_i, $i = 0, \ldots, n-1$, up to a normalisation factor without solving equations (6.3.3).

6.4. $G/M/1/\infty$ system

For the single-server $G/M/1/\infty$ system, the input is a recurrent flow and the inter-arrival times are distributed by some law $A(x)$. The service times are exponentially distributed with parameter μ.

As we will see in this section, the $G/M/1/\infty$ system is described by random processes that closely resemble the processes arisen in studying the $M/G/1/\infty$ system, but a close look shows that in fact they are the same but turned upside-down. Nevertheless, the calculation formulas appear to be quite different.

Though the $G/M/1/\infty$ system can be investigated by any of the methods described in the preceding chapter, we restrict our consideration to the simplest method of embedded Markov chains, which (with slight modification) can be applied to studying in more detail all characteristics (including transient characteristics, which we will not consider here). The reason for such a universality of this method consists of the following. First, the Markov chain is much simpler for the $G/M/1/\infty$ system than for the $M/G/1/\infty$ system, because there is no need for the lapse of any additional time after the completion of service for the next customer to arrive at an empty system. Second, the arrival instants, i.e., the instants at which the waiting times and sojourn times are determined, are precisely the Markov instants of the embedded Markov chain. Finally, the stationary state probabilities for the $G/M/1/\infty$ system can be represented in a simple geometric form.

6.4.1. Embedded Markov chain

We begin the study with finding the stationary state probabilities. Let $\nu(t)$ be the number of customers in the system at instant t. As in the case of the $M/G/1/\infty$ system, the process $\{\nu(t), t \geq 0\}$ is not Markov and there are instants $\tau_0, \tau_1, \dots, \tau_n, \dots$ at which the values of the process $\{\nu(t),\ t \geq 0\}$ form a Markov chain. But, as contrasted to the $M/G/1/\infty$ system, these instants are not service termination instants but customer arrival ones. For definiteness, we assume that the value of the embedded Markov chain ν_n is the number of customers in the system immediately upon the arrival of the next (the nth) customer, i.e., $\nu_n = \nu(\tau_n - 0)$. We also assume that at the initial instant 0 a customer arrives at an empty system, i.e., $\tau_0 = 0$ and $\nu_0 = \nu(\tau_0 - 0) = 0$.

We now find the transition probabilities p_{ij} of the embedded Markov chain $\{\nu_n, n \geq 0\}$. Let

$$\alpha_k = \int_0^\infty \frac{(\mu x)^k}{k!} e^{-\mu x} dA(x), \qquad k \geq 0,$$

denote the probability that exactly k customers are served in the time between arrivals of two successive customers (provided that there were more than k customers in the system after arrival of the first customer). Furthermore, we set

$$A_k = \sum_{i=k+1}^{\infty} \alpha_i = 1 - \sum_{i=0}^{k} \alpha_i, \qquad k \geq 0.$$

Let $\nu_n = i$ (i.e., i customers were in the system at the instant of arrival of the nth customer). Then, for ν_{n+1} to be equal to j, $j = 1, \dots, i+1$ (for the system to contain j customers at the instant of arrival of the $(n+1)$th customer), exactly $i - j + 1$ customers must be served in the time interval $\tau_{n+1} - \tau_n$, i.e., $p_{ij} = \alpha_{i-j+1}$. But if $j = 0$, then the $i + 1$ customers remaining in the system after the instant τ_n must be served. Therefore,

$$p_{i0} = 1 - \sum_{k=0}^{i} \alpha_k = A_i.$$

Hence we obtain the matrix $P = (p_{ij})$ of transition probabilities of the embedded Markov chain:

$$P = (p_{ij}) = \begin{pmatrix} A_0 & \alpha_0 & 0 & 0 & \cdots \\ A_1 & \alpha_1 & \alpha_0 & 0 & \cdots \\ A_2 & \alpha_2 & \alpha_1 & \alpha_0 & \cdots \\ \vdots & \vdots & \vdots & \vdots & \ddots \end{pmatrix}.$$

6.4.2. Stationary state probabilities for the embedded Markov chain

We set $\rho = 1/(\mu a)$, where

$$a = \int_0^\infty x\, dA(x)$$

is the mean inter-arrival time. Obviously, ρ has the same meaning as before, i.e., is the traffic intensity of the system. In order to find the steady-state probabilities, we reasonably assume that $\rho < 1$. Soon we will see that the condition $\rho < 1$ is necessary and sufficient for the existence of a steady state.

For the time being, we assume that the limit probabilities

$$p_i^* = \lim_{n\to\infty} \mathbf{P}\{\nu_n = i\}, \qquad i \geq 0,$$

i.e., the stationary probabilities for the embedded Markov chain $\{\nu_n,\ n \geq 0\}$, exist. But, as contrasted to the $M/G/1/\infty$ system, the probabilities p_i^* for the system under consideration, by definition, are also the stationary probabilities that a customer upon arrival finds i other customers in the system, because customers arrive only at the instants τ_n.

The stationary probabilities p_i^* satisfy the equilibrium equations

$$p_0^* = \sum_{k=0}^{\infty} A_k p_k^*,$$
$$p_i^* = \sum_{k=0}^{\infty} \alpha_k p_{i+k-1}^*, \qquad i \geq 1.$$

To solve this system, we consider the generating function

$$P^*(z) = \sum_{i=0}^{\infty} p_i^* z^i.$$

Multiplying the ith equation, $i = 1, 2, \ldots$, by z^i and then summing, we obtain, for any complex z, $|z| \le 1$,

$$\begin{aligned}P^*(z) - p_0^* &= \sum_{i=1}^{\infty} z^i \sum_{k=0}^{\infty} \alpha_k p_{i+k-1}^* = z \sum_{k=0}^{\infty} \frac{\alpha_k}{z^k} \sum_{i=k}^{\infty} p_i^* z^i \\ &= z \left(P^*(z) \sum_{k=0}^{\infty} \frac{\alpha_k}{z^k} - \sum_{k=1}^{\infty} \frac{\alpha_k}{z^k} \sum_{i=0}^{k-1} p_i^* z^i \right) \\ &= z P^*(z) \sum_{k=0}^{\infty} \frac{\alpha_k}{z^k} - \sum_{i=0}^{\infty} p_i^* \sum_{k=i}^{\infty} \frac{\alpha_{k+1}}{z^{k-i}} \\ &= z P^*(z) A\left(\frac{1}{z}\right) + f(z) - p_0^*,\end{aligned}$$

where

$$f(z) = p_0^* - \sum_{i=0}^{\infty} p_i^* \sum_{k=i}^{\infty} \frac{\alpha_{k+1}}{z^{k-i}},$$

$$A(z) = \sum_{k=0}^{\infty} \alpha_k z^k = \int_0^{\infty} e^{-\mu(1-z)x} dA(x) = \alpha(\mu - \mu z).$$

Then

$$P^*(z) \left[1 - z\alpha\left(\mu - \frac{\mu}{z}\right) \right] = f(z), \tag{6.4.1}$$

where the function $f(z)$ is analytic outside the circle $|z| \le 1$ (at infinity as well).

Let us consider the function $1 - z\alpha(\mu - \mu/z)$. Obviously, it is analytic outside the circle $|z| \le 1$, except at the point $z = \infty$ with a simple pole at this point, and continuous. Furthermore, $|z\alpha(\mu - \mu/z)| < 1$ on the circle $|z| = 1$, except at the point $z = 1$. Differentiating, we obtain

$$\left(z\alpha\left(\mu - \frac{\mu}{z}\right) \right)' \bigg|_{z=1} = 1 - \mu a = 1 - \frac{1}{\rho}.$$

By virtue of the condition $\rho < 1$, this derivative is negative. Therefore, there exists $\varepsilon_1 > 0$ such that for any ε, $0 < \varepsilon < \varepsilon_1$, we have $\Re(1 - z\alpha(\mu - \mu/z)) > 0$ for $|z| = 1 + \varepsilon$. Therefore, by the theorem of logarithmic residues, the number of zeros minus the number of poles of the function $1 - z\alpha(\mu - \mu/z)$ in the domain $|z| \ge 1 + \varepsilon$ and hence, outside the circle $|z| \le 1$ as well, is zero. But the function $1 - z\alpha(\mu - \mu/z)$ has a unique simple pole at $z = \infty$. Thus, the equation

$$\frac{1}{z} - \alpha\left(\mu - \frac{\mu}{z}\right) = 0 \tag{6.4.2}$$

has a unique solution z_0, whose absolute value exceeds one. As shown in studying the $M/G/1/\infty$ system, this solution is positive.

We now represent the function $1 - z\alpha(\mu - \mu/z)$ as

$$1 - z\alpha\left(\mu - \frac{\mu}{z}\right) = (z - z_0) g(z).$$

The function $g(z)$ is, obviously, analytic outside the circle $|z| \leq 1$ (at $z = \infty$ as well) and does not vanish in this domain. Hence (6.4.1) can be rewritten as

$$P^*(z)(z - z_0) = \frac{f(z)}{g(z)}. \tag{6.4.3}$$

The left-hand side of (6.4.3), which is the product of the generating function $P^*(z)$ and the function $(z - z_0)$, is analytic inside the circle $|z| < 1$ and continuous up to the circle $|z| = 1$. The right-hand side of (6.4.3) is also analytic and continuous, but outside the circle $|z| \leq 1$. Therefore, the left-hand and right-hand sides of (6.4.3) are the analytic continuations of the same function, which, being analytic in the whole extended plane, must equal a constant C. Thus,

$$P^*(z)(z - z_0) = C, \qquad P^*(z) = \frac{C}{z - z_0}.$$

The constant C is, as usual, found from the normalisation condition $P^*(1) = 1$. We thus obtain

$$P^*(z) = \frac{1 - z_0}{z - z_0}.$$

Expanding the generating function $P^*(z)$ in powers of z, we easily obtain

$$p_i^* = \left(1 - \frac{1}{z_0}\right)(z_0)^{-i}. \tag{6.4.4}$$

Thus, the stationary distribution of the number of customers in the system for the embedded Markov chain is a geometric distribution with parameter $1/z_0$.

In conclusion, let us consider the conditions for existence of limit (stationary) probabilities p_i^*, $i \geq 0$, i.e., ergodicity of the embedded Markov chain $\{\nu_n, n \geq 0\}$. Sufficiency of the condition $\rho < 1$ is implied by the Foster theorem 1.4.3, because the probabilities p_i^*, $i \geq 0$, define precisely the solution of the equilibrium equations stated in the Foster theorem. It is somewhat more difficult to establish necessity of the condition $\rho < 1$. For this purpose, considering, say, the instant τ_0 of recurrence of the Markov chain $\{\nu_n,\ n \geq 0\}$ to the state 0 (see Section 5.1), we see that $\mathbf{E}\tau_0 < \infty$ for $\rho < 1$ and $\mathbf{E}\tau_0 = \infty$ for $\rho = 1$ and $\tau_0 = \infty$ with non-zero probability if $\rho > 1$. This, according to the Feller theorem 1.4.2, proves necessity (and sufficiency) of the condition $\rho < 1$ for the ergodicity of the embedded Markov chain.

6.4.3. Time-stationary distribution of the number of customers in the system

We assume that the inter-arrival time cannot take the values kx_0.

We now find the time-stationary probabilities of the number of customers in the system

$$p_i = \lim_{t \to \infty} \mathbf{P}\{\nu(t) = i\},$$

using the renewal theory apparatus. We fix an instant for determining the probability p_i. According to the key renewal theorem 1.3.4, the probability that the last customer in the stationary state arrives in the interval $[t, t + dt)$ before the instant under consideration is

$[1 - A(t)]\,dt/a$. Since there were j customers with the probability $p_j^* = (1 - 1/z_0)(z_0)^{-j}$ in the system prior to the last arrival (see formula (6.4.4)) and k customers were served in time t with the probability $(\mu t)^k e^{-\mu t}/k!$, by the total probability formula, the probability that i, $i \geq 1$, customers exist in the system at the instant under consideration is given by the expression

$$\begin{aligned}
p_i &= \sum_{j=i-1}^{\infty} p_j^* \int_0^\infty \frac{(\mu x)^{j-i+1}}{(j-i+1)!} e^{-\mu x} \frac{1 - A(x)}{a} dx \\
&= \left(1 - \frac{1}{z_0}\right)\left(\frac{1}{z_0}\right)^{i-1} \int_0^\infty \sum_{j=0}^{\infty} \left(\frac{1}{z_0}\right)^j \frac{(\mu x)^j}{j!} e^{-\mu x} \frac{1 - A(x)}{a} dx \\
&= \left(1 - \frac{1}{z_0}\right)\left(\frac{1}{z_0}\right)^{i-1} \int_0^\infty e^{-\mu x(1-1/z_0)} \frac{1 - A(x)}{a} dx \\
&= \left(\frac{1}{z_0}\right)^{i-1} \frac{1 - \alpha(\mu - \mu/z_0)}{a\mu}.
\end{aligned}$$

By virtue of (6.4.2),

$$\alpha\left(\mu - \frac{\mu}{z_0}\right) = \frac{1}{z_0}, \qquad a\mu = \frac{1}{\rho};$$

therefore,

$$p_i = \rho(z_0 - 1)(z_0)^{-i}, \qquad i \geq 1. \tag{6.4.5}$$

Hence, beginning with $i = 1$, the time-stationary distribution of the number of customers is a geometric distribution with the same parameter $1/z_0$.

The most convenient way to find p_0 is to apply the normalisation condition $\sum_{i=0}^{\infty} p_i = 1$. Then we obtain

$$p_0 = 1 - \sum_{i=1}^{\infty} p_i = 1 - (z_0 - 1)\rho \sum_{i=1}^{\infty} (z_0)^{-i} = 1 - \rho, \tag{6.4.6}$$

which is the familiar formula $p_0 = 1 - \rho$ for an infinite-capacity system.

The condition $\rho < 1$ is also necessary and sufficient for existence of a time-stationary distribution for the number of customers (this is demonstrated in the same manner as in Section 5.1 for the $M/G/1/\infty$ system). Here an interesting point deserves attention: for this system, the case of $a = \infty$, i.e., of infinite mean inter-arrival time, is not excluded. Then, as can be easily verified with formulas (6.4.5) and (6.4.6), $p_0 = 1$ and $p_i = 0$, $i \geq 1$. Hence the system is empty as $t \to \infty$.

6.4.4. Stationary distribution of sojourn times

In the steady state, as we have seen, a customer upon arrival finds i other customers in the system with the probability $p_i^* = (1 - 1/z_0)(z_0)^{-i}$. Since customers are served in order of arrival and service time is exponential, the waiting time for service of a customer is Erlang-distributed with parameters i and μ. Passing to the Laplace–Stieltjes transform, from the total probability formula we obtain the expression for the Laplace–Stieltjes transform $\omega(s)$ of the stationary distribution $W(x)$ of the waiting time:

$$\omega(s) = \sum_{i=0}^{\infty} p_i^* \left(\frac{\mu}{\mu+s}\right)^i = \frac{(\mu+s)(z_0-1)}{(\mu+s)z_0-\mu}$$
$$= 1 - \frac{1}{z_0} + \frac{\mu(z_0-1)}{z_0[(\mu+s)z_0-\mu]}.$$

Turing back to the original distribution, we obtain

$$W(x) = 1 - \frac{1}{z_0} + \frac{1}{z_0}\left(1 - e^{-\mu(1-1/z_0)x}\right)$$
$$= 1 - \frac{1}{z_0} e^{-\mu(1-1/z_0)x}, \qquad x > 0.$$

In order to find the stationary distribution $V(x)$ of the customer sojourn time, the waiting time must be added to the exponentially distributed service time or, in terms of the Laplace–Stieltjes transform,

$$\varphi(s) = \frac{\mu\omega(s)}{\mu+s}.$$

The reader can easily invert the last Laplace–Stieltjes transform.

6.5. $M/G/1/r$ system

In this section, we study a single-server system with Poisson input of intensity λ and arbitrary service time distribution $B(x)$. But, in contrast to Chapter 5, we assume that the waiting buffer r is finite.

As in Section 5.4, we restrict ourselves only to determination of stationary state probabilities for the $M/G/1/r$ system with the use of the method of supplementary variable, using elapsed service time as the supplementary variable. We will show that the stationary state probabilities for the $M/G/1/r$ system (except the stationary probability p_{r+1} that the buffer is full) are the same, up to a constant, as those for the $M/G/1/\infty$ system. For the sake of brevity, as before, the total capacity of the system is denoted by $R = r+1$. Furthermore, to leave needlessly complicated particular cases out of consideration, we take $r \geq 2$.

6.5.1. Markov process describing the system

Let $\nu(t)$ be the number of customers in the system. If $\nu(t) = 0$ at an instant t, then we take $\eta(t) = \nu(t)$. But if $\nu(t) \geq 1$, then, along with $\nu(t)$, we use a supplementary component $\xi(t)$ equal to the elapsed service time of the customer at the instant t at the server, and introduce $\eta(t) = (\nu(t), \xi(t))$. The process $\{\eta(t), t \geq 0\}$ is a (linearwise) Markov one, but, as distinct from Section 5.4, its state set $\mathscr{X}$ is $\mathscr{X} = \{(0); (i,x), i = 1, \ldots, R, x \geq 0\}$.

Let

$$p_0(t) = \mathbf{P}\{\eta(t) = 0\}, \quad P_i(x,t) = \mathbf{P}\{\nu(t) = i, \xi(t) < x\}, \quad i = 1, \ldots, R.$$

We assume that the expectation

$$b = \int_0^\infty x\,dB(x)$$

of service time is finite. Therefore (see Theorem 1.6.6), the limit (stationary) probabilities

$$p_0 = \lim_{t\to\infty} p_0(t), \qquad P_i(x) = \lim_{t\to\infty} P_i(x,t)$$

exist, as well as the densities $p_i(x) = P_i'(x)$.

The stationary probability that there exist i customers without regard for the elapsed service time, as in Section 5.4, is denoted by p_i:

$$p_i = \int_0^\infty p_i(x)\,dx.$$

Let us introduce the functions

$$q_i(x) = \frac{p_i(x)}{1 - B(x)}, \qquad i = 1, \dots, R.$$

They satisfy the equations

$$q_i'(x) = -\lambda q_i(x) + u(i-1)\lambda q_{i-1}(x), \qquad i = 1, \dots, r, \tag{6.5.1}$$
$$q_R'(x) = \lambda q_r(x) \tag{6.5.2}$$

under the boundary conditions

$$\begin{aligned} \lambda p_0 &= \int_0^\infty q_1(x)\,dB(x), \\ q_i(0) &= \int_0^\infty q_{i+1}(x)\,dB(x) + u(2-i)\lambda p_0, \qquad i = 1, \dots, r, \\ q_R(0) &= 0, \end{aligned} \tag{6.5.3}$$

where $u(x)$ is the Heaviside function.

Equations (6.5.1) and (6.5.2) and boundary conditions (6.5.3) are derived exactly as those for the $M/G/1/\infty$ system. Only the relations for $q_R(x)$ and $q_R(0)$ need to be explained. They are easily understood if we recall that it is not possible to pass from the state (R, x) to the state $(R+1, x)$ due to absence of such a state (this explains (6.5.2)). For a similar reason, it is impossible to pass from the state $(R+1, x)$ to the state $(R, 0)$.

6.5.2. Stationary state probabilities

The solution of (6.5.1), (6.5.2) under the boundary conditions (6.5.3) is

$$q_i(x) = e^{-\lambda x} \sum_{k=0}^{i-1} \frac{(\lambda x)^k}{k!} q_{i-k}(0), \qquad i = 1, \dots, r, \tag{6.5.4}$$

$$q_R(x) = \sum_{k=1}^{r} [q_k(0) - q_k(x)]. \tag{6.5.5}$$

It is easy to validate relations (6.5.4) and (6.5.5) by substituting them into (6.5.1) and (6.5.2); thereby (6.5.1) is transformed into an identity and (6.5.2) reduces to

$$\sum_{k=1}^{r} q_k'(x) = -\lambda q_r(x).$$

In turn, the last relation can be obtained by summing (6.5.1) over $i = 1, 2, \ldots, r$. Thus, the functions $q_i(x)$, $i = 1, \ldots, R$, defined by formulas (6.5.4) (6.5.5) identically satisfy (6.5.1) and (6.5.2).

Turning back to the densities $p_i(x)$, $i = 1, \ldots, R$, we find that

$$p_i(x) = [1 - B(x)]e^{-\lambda x} \sum_{k=0}^{i-1} \frac{(\lambda x)^k}{k!} q_{i-k}(0), \qquad i = 1, \ldots, r,$$

$$p_R(x) = [1 - B(x)] \sum_{k=1}^{r} [q_k(0) - q_k(x)].$$

Integrating both relations with respect to x from zero to ∞, we obtain

$$p_i = \frac{1}{\lambda} \sum_{k=0}^{i-1} B_k q_{i-k}(0), \qquad i = 1, \ldots, r, \tag{6.5.6}$$

$$p_R = -\sum_{k=1}^{r} p_k + b \sum_{k=1}^{r} q_k(0), \tag{6.5.7}$$

where B_k is defined in Section 5.4, and b is the mean service time. Relation (6.5.6) is proved exactly (5.4.12), and (6.5.7) is obvious.

In order to determine the boundary conditions $q_i(0)$, we substitute formula (6.5.4) into equations (6.5.3). We thus obtain a set of R equations for p_0 and $q_i(0)$, $i = 1, \ldots, r$:

$$\lambda p_0 = \beta_0 q_1(0),$$

$$q_i(0) = \sum_{k=0}^{i} \beta_k q_{i-k+1}(0) + \lambda u(2 - i) p_0, \qquad i = 1, \ldots, r - 1, \tag{6.5.8}$$

where β_k are defined in Section 5.4.

As in Section 5.4, introducing $\tilde{q}_i = q_i(0)/p_0$, from (6.5.8) we obtain an expression for $\tilde{q}_i$:

$$\tilde{q}_1 = \frac{\lambda}{\beta_0},$$

$$\tilde{q}_i = \frac{1}{\beta_0}\left(\tilde{q}_{i-1} - \sum_{k=1}^{i-1} \beta_k \tilde{q}_{i-k} - \lambda u(3 - i)\right), \qquad i = 2, \ldots, r. \tag{6.5.9}$$

In the same manner as in Section 5.4, we obtain the relations

$$p_1 = \frac{1}{\lambda} q_1(0) - p_0,$$

$$p_i = \frac{1}{\lambda} q_i(0), \qquad i = 2, \ldots, r, \tag{6.5.10}$$

which relate the probabilities p_i and the boundary values $q_i(0)$.

In order to find p_0, let us apply normalisation condition $\sum_{k=0}^{R} p_i = 1$ unused yet and relation (6.5.7). From (6.5.7) it follows that

$$\sum_{k=1}^{R} p_i = bq.(0),$$

where the dot, as before, means summation over all values of the discrete argument. Hence, from the normalisation condition we obtain

$$1 - p_0 = bq.(0). \tag{6.5.11}$$

Expressing $q_i(0)$ in terms of $\tilde{q}_i$, we obtain the formula

$$p_0 = [1 + b\tilde{q}.]^{-1}. \tag{6.5.12}$$

Thus, we have found the stationary densities $p_i(x)$, $i = 1, \dots, R$, and stationary probabilities p_i, $i = 0, \dots, R$.

We now derive one more relation between the probabilities p_0 and p_R. Summing (6.5.10) over $i = 1, 2, \dots, r$ and recalling the normalisation condition, we obtain

$$1 - p_R = \frac{1}{\lambda} q.(0).$$

Hence from (6.5.11) we obtain

$$\lambda(1 - p_R) = \mu(1 - p_0), \tag{6.5.13}$$

where $\mu = b^{-1}$ is the service intensity.

Using the traffic intensity $\rho = \lambda/\mu$, we rewrite (6.5.13) as

$$\rho(1 - p_R) = 1 - p_0. \tag{6.5.14}$$

Since p_R is the customer loss probability, relation (6.5.13) shows that the intensities of accepted and served flows in steady state are equal. Indeed, the left-hand and right-hand sides of (6.5.13) yield two equivalent expressions for the output flow intensity λ_D:

$$\lambda_D = \mu(1 - p_0),$$
$$\lambda_D = \lambda(1 - p_R).$$

Recall that equalities of type (6.5.13) and (6.5.14) were derived in Chapters 3 and 4 for certain finite-capacity queueing systems. So, they hold true for a wider class of finite-capacity systems. This can be demonstrated using the ergodicity of the processes describing the behaviour of the respective systems.

Finally, we explain the assertion that 'stationary probabilities of states of the $M/G/1/r$ system are virtually the same as the respective probabilities for the $M/G/1/\infty$ system.' Setting $\tilde{p}_i = p_i/p_0$, we consider $\tilde{p}_i^{(\infty)}$, $\tilde{q}_i^{(\infty)}$, and $\tilde{p}_i^{(r)}$, $\tilde{q}_i^{(r)}$ for the $M/G/1/\infty$ and $M/G/1/r$ systems respectively. Comparing the results obtained in Section 5.4 and in this section, we find that

$$\tilde{q}_i^{(\infty)} = \tilde{q}_i^{(r)}, \quad \tilde{p}_i^{(\infty)} = \tilde{p}_i^{(r)}, \quad i = 1, \dots, r.$$

Therefore, knowing the state probabilities for one system, we can easily find the corresponding probabilities for the other system.

6.6. $M/G/n/0$ system

In analogy with the $M/M/n/0$ system, the multiserver $M/G/n/0$ system with losses is sometimes called the Erlang system (the $M/M/n/0$ system, a particular case of the $M/G/n/0$ system, is investigated in Section 3.2). It is a non-Markov system; its operation cannot be described by a continuous-time Markov process with a discrete state set.

The main practical interest in $M/G/n/0$ system is the probability of loss of a customer. Thus far, there is no method for finding this characteristic in the transient state. Nevertheless, as will be clear soon, the steady-state probabilities of states of this system are defined by surprisingly simple formulas.

As usual, we assume that the Poisson arrival flow is of intensity λ and the service time of each customer at any server is distributed by an arbitrary law $B(x)$.

6.6.1. Equations for the Markov process

In order to find the stationary distribution of the number of customers in the system, we make use of the method of supplementary variables, which yields a piecewise-linear Markov process (see Section 1.6). As supplementary variables, along with the number $\nu(t)$ of customers in the system at instant t, we can use, for example, the residual service times $\xi_1(t), \dots, \xi_{\nu(t)}(t)$ of customers at busy servers at the instant t in order of their enumeration. Supplementing these coordinates to $\nu(t)$, we obtain a piecewise-linear Markov process $\{\eta(t) = (\nu(t), \xi_1(t), \dots, \xi_{\nu(t)}(t)), t \geq 0\}$, whose state set $\mathcal{X}$ consists of the vectors $(i, x_1, \dots, x_i)$, $i = 0, \dots, n$, $x_1, \dots, x_i \geq 0$, of varying dimensionality $i+1$.

We set

$$p_0(t) = \mathbf{P}\{\nu(t) = 0\},$$
$$P_i(x_1, \dots, x_i, t) = \mathbf{P}\{\nu(t) = i, \xi_1(t) < x_1, \dots, \xi_i(t) < x_i\}, \quad i = 1, \dots, n.$$

In (Borovkov, 1976; Sevastyanov, 1957), it is shown that if the mean service time $b = \int_0^\infty x\, dB(x)$ is finite, then there exist stationary probabilities

$$p_0 = \lim_{t\to\infty} p_0(t), \quad P_i(x_1, \dots, x_i) = \lim_{t\to\infty} P_i(x_1, \dots, x_i, t), \quad i = 1, \dots, n,$$

having the densities

$$p_i(x_1, \dots, x_i) = \frac{\partial^i}{\partial x_1 \cdots \partial x_i} P_i(x_1, \dots, x_i).$$

For the sake of simplicity, we assume that the distribution function $B(x)$ of the service time also has the density $b(x) = B'(x)$.

Let us derive equations which the stationary probability p_0 and the densities $p_i(x_1, \dots, x_i)$, $i = 1, \dots, n$, satisfy.

Let us consider all possible transitions from the state $(i, x_1, \dots, x_i)$ in some 'small' time Δ. In this time interval, the residual service time of each customer decreases by Δ, a customer with probability $\lambda\Delta + o(\Delta)$ may arrive, and the service of the kth customer having a residual service time less than Δ may be completed with the probability $p_i(x_1, \dots, x_{k-1}, 0, x_{k+1}, \dots, x_i)\Delta + o(\Delta)$. In turn, a newly arriving customer has a service time y with the distribution density $b(y)$. Since we are considering the steady state, the distributions of the piecewise-linear process $\{\eta(t), t \geq 0\}$ at the instants t and $t+\Delta$ are identical and have the distribution density $p_i(x_1, \dots, x_i)$. We assume that the new arrival may join any idle server with equal probability. Therefore, $p_i(x_1, \dots, x_i)$, $i = 1, \dots, n-1$, satisfy the relations

$$\begin{aligned} p_i(x_1, \dots, x_i) = p_i(x_1 + \Delta, \dots, x_i + \Delta)(1 - \lambda\Delta) \\ + \lambda\Delta\frac{1}{i}\sum_{k=1}^{i} p_{i-1}(x_1, \dots, x_{k-1}, x_{k+1}, \dots, x_i)b(x_k) \\ + \Delta\sum_{k=1}^{i+1} p_{i+1}(x_1, \dots, x_{k-1}, 0, x_k, \dots, x_i) + o(\Delta). \end{aligned}$$

The stationary probability p_0 that the system is empty is

$$p_0 = p_0(1 - \lambda\Delta) + P_1(\Delta) + o(\Delta)$$

or, if $P_1(\Delta)$ is replaced by $p_1(0)\Delta + o(\Delta)$,

$$p_0 = p_0(1 - \lambda\Delta) + p_1(0)\Delta + o(\Delta).$$

Finally, if the system contains n customers, the new arrival is lost. This is helpful in writing the last equality:

$$\begin{aligned} p_n(x_1, \dots, x_n) = p_n(x_1 + \Delta, \dots, x_n + \Delta) \\ + \lambda\Delta\frac{1}{n}\sum_{k=1}^{n} p_{n-1}(x_1, \dots, x_{k-1}, x_{k+1}, \dots, x_n)b(x_k) + o(\Delta). \end{aligned}$$

Subtracting p_0, $p_i(x_1 + \Delta, \dots, x_i + \Delta)$, and $p_n(x_1 + \Delta, \dots, x_n + \Delta)$ from both sides of the respective relations, dividing by Δ, and making Δ tend to zero, we arrive at the equations

$$\begin{aligned} &\lambda p_0 = p_1(0), \\ &-\left(\frac{\partial}{\partial x_1} + \dots + \frac{\partial}{\partial x_i}\right) p_i(x_1, \dots, x_i) \\ &\quad = -\lambda p_i(x_1, \dots, x_i) + \sum_{k=1}^{i+1} p_{i+1}(x_1, \dots, x_{k-1}, 0, x_k, \dots, x_i) \\ &\qquad + \lambda\frac{1}{i}\sum_{k=1}^{i} p_{i-1}(x_1, \dots, x_{k-1}, x_{k+1}, \dots, x_i)b(x_k), \qquad i = 1, \dots, n-1, \\ &-\left(\frac{\partial}{\partial x_1} + \dots + \frac{\partial}{\partial x_n}\right) p_n(x_1, \dots, x_n) \\ &\quad = \lambda\frac{1}{n}\sum_{k=1}^{n} p_{n-1}(x_1, \dots, x_{k-1}, x_{k+1}, \dots, x_n)b(x_k). \end{aligned} \tag{6.6.1}$$

6.6.2. Stationary distribution of the number of customers in the system

By means of direct checking, we find that the functions

$$p_i(x_1, \ldots, x_i) = \frac{\lambda^i}{i!} p_0 \prod_{k=1}^{i} [1 - B(x_k)], \qquad i = 1, \ldots, n, \tag{6.6.2}$$

solve (6.6.1), i.e., are the densities of the stationary distribution of the process $\{\eta(t), t \geq 0\}$.

The stationary probability p_i that the system contains exactly i customers is

$$\begin{aligned} p_i &= P_i(\infty, \ldots, \infty) \\ &= \int_0^\infty \cdots \int_0^\infty p_i(x_1, \ldots, x_i)\, dx_1 \cdots dx_i = \frac{\rho^i}{i!} p_0, \qquad i = 1, \ldots, n, \end{aligned}$$

where $\rho = \lambda b$ and p_0 is found from the normalisation condition $\sum_{i=0}^n p_i = 1$, i.e.,

$$p_0 = \left(\sum_{i=0}^{n} \frac{\rho^i}{i!} \right)^{-1}.$$

A vital role in practical computations is played by the stationary probability π that a customer is lost, which coincides with the stationary probability p_n that all servers are busy.

Obviously, stationary probabilities p_i for the $M/G/n/0$ system are defined by the same expressions as for the stationary probabilities for the $M/M/n/0$ system with exponentially distributed service time with parameter $\mu = 1/b$. Thus, we have found an interesting fact, known as the invariance property:

The stationary state probabilities p_i of the $M/G/n/0$ system depend only on the mean service time, but not on the type of the distribution $B(x)$.

6.6.3. Output flow

We define the output flow of the $M/G/n/0$ system to be the total output consisting of all departures and lost customers. To study the structure of the output flow, we make use of the time inversion method (Sections 1.5 and 3.1).

Let us consider the reversed Markov process $\{\tilde{\eta}(t) = \eta(-t), t \leq 0\}$ obtained from the Markov process $\{\eta(t),\ t \geq 0\}$ by time reversion. We assume that the initial system operates in steady state, i.e., the processes $\{\eta(t), -\infty < t < \infty\}$ and $\{\tilde{\eta}(t),\ -\infty < t < \infty\}$ are stationary. Let us now describe the reversed system that corresponds to the reversed process $\{\tilde{\eta}(t), -\infty < t < \infty\}$. The coordinates $\nu(t)$ and $\tilde{\nu}(t)$ are the numbers of customers at instant t in the initial and reversed systems respectively. But the customer arrival instant τ for the reversed system corresponds to the customer departure instant $-\tau$ for the initial system and vice versa, i.e., the departure instant τ for reversed system corresponds to the arrival instant $-\tau$ for the initial system. Obviously, if $\xi_j(t)$ for the initial system denotes the residual service time of the customer at instant t at the jth server, then $\tilde{\xi}_j(-t)$ is the elapsed service time of the customer in the reversed system at instant $-t$ at the jth server.

The intensity of the arrival flow for the reversed system, in general, must depend on the state of the process $\{\tilde{\eta}(t), -\infty < t < \infty\}$: $\tilde{\lambda}_i = \tilde{\lambda}_i(x_1, \ldots, x_i)$, where i is the value of the coordinate of $\tilde{\nu}(t)$ and x_j is the value of the coordinate of $\tilde{\xi}_j(t)$. Recalling that an arrival of a customer at the reversed system in the interval $(t, t+\Delta)$ corresponds to the completion of service in the interval $(-t-\Delta, -t)$ of a customer at one of the $j+1$ servers in the initial system and applying the definitions of flow intensity and conditional probability, we obtain

$$\begin{aligned}
&\tilde{\lambda}_i(x_1, \ldots, x_i) \\
&= \lim_{\Delta \to 0} \frac{\mathbf{P}\left(\{\tilde{\nu}(t) = i, \tilde{\nu}(t+\Delta) = i+1\} \bigcap_{j=1}^{i} \{x_j \le \tilde{\xi}_j(t) < x_j + dx_j\}\right)}{\Delta \mathbf{P}\left(\{\tilde{\nu}(t) = i\} \bigcap_{j=1}^{i} \{x_j \le \tilde{\xi}_j(t) < x_j + dx_j\}\right)} \\
&= \lim_{\Delta \to 0} \sum_{j=1}^{i+1} \frac{1}{\Delta \mathbf{P}\left(\{\nu(-t) = i\} \bigcap_{j=1}^{i} \{x_j \le \xi_j(-t) < x_j + dx_j\}\right)} \\
&\quad \times \mathbf{P}\left(\{\nu(-t-\Delta) = i+1\} \bigcap_{k=1}^{j-1} \{x_k + \Delta \le \xi_k(-t-\Delta) < x_k + \Delta + dx_k\}\right. \\
&\quad \left. \cap \{0 \le \xi_j(-t-\Delta) < \Delta\} \bigcap_{k=j}^{i} \{x_k + \Delta \le \xi_{k+1}(-t-\Delta) < x_k + \Delta + dx_k\}\right).
\end{aligned}$$

Since, by the assumption, the system is in steady state and formula (6.6.2) holds for the steady-state probability densities, we finally obtain

$$\tilde{\lambda}_i(x_1, \ldots, x_i) = \sum_{j=1}^{i+1} \frac{\frac{\lambda^{i+1}}{(i+1)!} \prod_{k=1}^{j-1} [1 - B(x_k)][1 - B(0)] \prod_{k=j}^{i} [1 - B(x_k)]}{\frac{\lambda^i}{i!} \prod_{k=1}^{i} [1 - B(x_k)]} = \lambda.$$

If $i = n$ (the initial and reversed systems are busy), then for any value of $\tilde{\xi}_1(t), \ldots, \tilde{\xi}_n(t)$, by our assumption that an arrival at the reversed system corresponds to a departure from the initial system, i.e., arrival of a lost customer at the initial system, since the input flow of the initial system is Poisson, we again obtain $\tilde{\lambda}_n(x_1, \ldots, x_n) = \lambda$.

Thus, the arrival intensity for the reversed system does not depend on the state of the process $\{\tilde{\eta}(t), -\infty < t < \infty\}$. Hence, the output flow (including lost customers) from the $M/G/n/0$ system in steady state is Poisson with parameter λ.

We leave as an exercise to prove the following to the reader: if $\tilde{\xi}_j(t) = x_j$ at instant t (i.e., there is a customer with residual service time x_j at instant $-t$ at the jth server in the initial system), then, irrespective of the state of other components $\tilde{\xi}_i(t)$ of the process $\{\tilde{\eta}(t), -\infty < t < \infty\}$, the probability that the customer at the jth server in the reversed system quits in the time interval $(t, t+dt)$ (or this customer arrives at the initial system in the time interval $(-t-dt, -t)$ immediately before the instant $-t$) is equal to the conditional probability that the service is completed in the interval $(x_j, x_j + dt)$, provided that the service has been continuing for time x_j. This probability is given by the formula $b(x_j)dt/[1 - B(x_j)]$. Along with the constant parameter λ, this means that the reversed system in steady state is also an $M/G/n/0$ system, the only difference is that its piecewise-linear process $\{\tilde{\eta}(t), -\infty < t < \infty\}$ is constructed using the elapsed service time but not the residual service time as the supplementary variable $\tilde{\xi}_j(t)$.

7

$MAP/G/1/r$ system

In this chapter, we study the $MAP/G/1/r$ system. The abbreviation MAP, according to the Kendall classification, denotes that the input is a Markov arrival process.

The Markov flow generalises phase-type flows but is not recurrent. Therefore, a queueing model with Markov flow, on the one hand, is attractive due to its generality and, on the other hand, deterrent due to its complexity. Nevertheless, as shown in Section 4.6, a queueing model in many cases is easily generalised to Markov flows. A similar situation is also observed for the system studied in this section. The methods that were applied in Chapters 5 and 6 to $M/G/1/r$ and $M/G/1/\infty$ systems can also be applied, with appropriate modification, to $MAP/G/1/r$ and $MAP/G/1/\infty$ systems respectively. Methodologically, however, it is reasonable to begin with a finite-capacity system. Moreover, along with the traditional FCFS discipline, we will also study the LCFS discipline with service interruption and resumption, because it is helpful in obtaining certain results, in particular, that are related to the sojourn time, by a rather simple, illustrative method.

Thus, we begin with a single-server queueing system of finite buffer capacity r, $r < \infty$. Its input is a Markov arrival flow, which, as before, is characterised by matrices Λ and N of order l. We assume that the matrix $\Lambda^* = \Lambda + N$ is indecomposable and the matrix N is non-zero. So, the matrix Λ is non-degenerate.

In what follows, $\boldsymbol{\pi} = (\pi_1, \dots, \pi_l)^T$ for systems with Markov input flow denotes the vector of stationary probabilities of states (phases) of the customer generation process $\{\xi(t), t \geq 0\}$ for this flow (recall that $\{\xi(t), t \geq 0\}$ is a homogeneous Markov process with state set $\{1, \dots, l\}$ and infinitesimal matrix Λ^*), i.e., the vector determined from the equilibrium equations

$$\boldsymbol{\pi}^T \Lambda^* = \mathbf{0}^T.$$

By virtue of these assumptions, the solution of these simultaneous equilibrium equations under the normalisation condition

$$\boldsymbol{\pi}^T \mathbf{1} = \pi_{\cdot} = 1$$

is unique (here, as before, the dot in the subscript denotes that summation is taken over all possible values of this subscript). Let $\boldsymbol{\lambda} = N\mathbf{1}$ denote the vector of intensities of the Markov flow, depending on the phase of the customer generation process, i.e., a vector whose coordinate $(\boldsymbol{\lambda})_i$ is equal to the intensity of the Markov flow, provided the generation process stays in phase i. Let $\lambda = \boldsymbol{\pi}^T \boldsymbol{\lambda}$ denote the intensity of the Markov flow in the steady state. By virtue of the relation $\Lambda^* \mathbf{1} = (\Lambda + N)\mathbf{1} = \mathbf{0}$, the vector $\boldsymbol{\lambda}$ can also be expressed as $\boldsymbol{\lambda} = -\Lambda \mathbf{1}$.

Service times are independent and have a common distribution function $B(x)$ with finite mean $b = \int_0^\infty [1 - B(x)]\, dx < \infty$.

A customer is lost if he arrives at the instant when the buffer is full.

As before, let $R = r + 1$ be the total capacity of the system.

7.1. Embedded Markov chain: FCFS discipline

We begin with the method of embedded Markov chains. Methodologically, it is the simplest technique that employs only the Markov chain apparatus. Applying it to queueing systems, we can find the stationary state probabilities at chosen instants. For our case, these instants are service completion instants. The stationary state probabilities of the queueing system under study at these instants are used as the starting point in future analysis. In particular, applying the methods of renewal theory, we will determine stationary state probabilities at arbitrary instants and the stationary distribution of the linearwise Markov process describing the behaviour of the system.

7.1.1. Embedded Markov chain: Construction

Let us consider the successive service completion instants τ_n, $n \geq 0$, for the $MAP/G/1/r$/FCFS system. If there are customers in queue at the instant of completion of service of a customer, then that instant is also the instant of commencement of service of the succeeding customer. Only exceptions are the busy period termination instants after which the system is empty.

Let $\xi(t)$ be the phase of the customer generation process at instant t and let $\nu(t)$ be the number of customers in the system at this instant.

Let the random variables $\xi_n = \xi(\tau_n + 0)$ and $\nu_n = \nu(\tau_n + 0)$ define the phase of the customer generation process and the number of customers in the system, respectively, immediately after the instant of completion of service of the nth customer. Setting $\zeta_n = (\xi_n, \nu_n)$, we find that the sequence $\{\zeta_n,\ n \geq 0\}$ is a homogeneous Markov chain, called the embedded chain. Obviously, its state set is $\mathcal{X} = \{(i,k), i = 1, \ldots, l, k = 0, \ldots, r\}$, where i and k denote the phase of the generation process and the number of customers in the system after the service completion instant respectively. The sequence of random variables $\{\zeta_n, n \geq 0\}$ is a Markov chain, which can be demonstrated as in Section 5.1 where the embedded Markov chain was constructed for the $M/G/1/\infty$ system. This is left to the reader as an exercise.

In order to find the transition probability matrix for the embedded Markov chain $\{\zeta_n,\ n \geq 0\}$, first we determine the following matrices:

$F_k(x)$ is the matrix whose element $(F_k(x))_{ij}$ is the probability that exactly k customers arrive in time x and the generation process passes from phase i to phase j;

$F_k^*(x)$ is the matrix whose element $(F_k^*(x))_{ij}$ is the probability that at least k customers arrive in time x and the generation process passes from phase i to phase j;

B_k is the matrix, whose element $(B_k)_{ij}$ is the probability that exactly k customers arrive in the course of service of a customer and the generation process passes from phase i to phase j; and

B_k^* is the matrix whose element $(B_k^*)_{ij}$ is the probability that at least k customers arrive in the course of service of a customer and the generation process passes from phase i to phase j.

The functions $F_k(x)$ and $F_k^*(x)$ satisfy the recurrence relations

$$F_0(x) = e^{\Lambda x}, \tag{7.1.1}$$

$$F_k(x) = \int_0^x F_{k-1}(y) N F_0(x-y)\,dy, \qquad k \geq 1, \tag{7.1.2}$$

$$F_0^*(x) = e^{\Lambda^* x}, \tag{7.1.3}$$

$$F_k^*(x) = \int_0^x F_{k-1}(y) N F_0^*(x-y)\,dy, \qquad k \geq 1. \tag{7.1.4}$$

These relations can be easily derived with the use of the probabilistic meaning of the elements of the matrices Λ, N, and Λ^* (recall that the elements Λ_{ij} and N_{ij} of the matrices Λ and N are the intensities of transition of the process $\xi(t)$ from phase i to phase j without and with a new arrival, respectively, and the element Λ_{ij}^* of the matrix Λ^* is the total intensity of transition of the process $\xi(t)$ from phase i to phase j). Indeed, by definition, the matrix $F_0^*(x)$ is merely the transition probability matrix of the generation process $\xi(t)$ for time x. According to formula (1.5.14), it is of form (7.1.3). Formula (7.1.1) is derived along similar lines with a terminating Markov process, regarding an arrival as a termination of the generation process (see Section 1.5.4). In (7.1.2), the expressions for $F_{k-1}(y)$ and $F_0(x-y)$ are the matrix representation of the probability that $k-1$ customers arrive in time y and no customer arrives in time $x-y$ (with account for the corresponding changes in the phases of the generation process), respectively, whereas $N\,dx$ is the matrix expression for the probability of a new arrival in a small time interval dx (again with account for the corresponding changes in the generation phases). Expression (7.1.4) differs from (7.1.2) in that any number of customers may arrive in time $x-y$.

The matrix $F_0^*(x)$ for any x is stochastic.

Using the total probability formula, we obtain

$$B_k = \int_0^\infty F_k(x)\,dB(x), \qquad k \geq 0, \tag{7.1.5}$$

$$B_k^* = \int_0^\infty F_k^*(x)\,dB(x), \qquad k \geq 0. \tag{7.1.6}$$

We observe that

$$B_0^* = B_0 + B_1^* = B_0 + B_1 + B_2^* = \cdots$$

These equalities follow from the obvious fact that the probability of arrival of no less than 0 customers is the sum of the probabilities of arrivals of 0 customers and, at least, one customer. This sum, in turn, is the sum of probabilities of arrivals of 0 customers, one customer, and, at least, two customers, etc. Moreover, since the matrix Λ^* satisfies the condition $\Lambda^*\mathbf{1} = \mathbf{0}$, the matrix B_0^* is stochastic. These observations are required in what follows. Furthermore, they are helpful while checking calculations.

Since the matrix functions B_k and B_k^* are the generalisations of exponential moments, we call them the matrix exponential moments. Due to their importance for implementation of the algorithms derived below, we will describe numerical methods of finding these matrices in Section 7.4.

In what follows, we also need the matrix B whose element B_{ij} is the probability that the generation process stays in phase j at the completion of an idle period, provided the process was in phase i at the commencement of the idle period. It is easily seen that this matrix is defined by the formula

$$B = \int_0^\infty e^{\Lambda x} N \, dx = -\Lambda^{-1} N. \tag{7.1.7}$$

The matrix B is stochastic. Indeed, from (7.1.7) we obtain

$$B\mathbf{1} = -\Lambda^{-1} N\mathbf{1} = \Lambda^{-1}\Lambda\mathbf{1} = \mathbf{1}.$$

Let $p_{(i,k),(j,m)}$ denote the probability that the embedded Markov chain passes from state (i, k) to state (j, m) and let us introduce matrices P_{km}, $k, m = 0, \ldots, r$, of order l consisting of these probabilities.

The matrices P_{km} are defined by the relations

$$P_{0m} = BB_m, \qquad m = 0, \ldots, r-1, \tag{7.1.8}$$

$$P_{km} = B_{m-k+1}, \qquad k = 1, \ldots, r, \qquad m = k-1, \ldots, r-1, \tag{7.1.9}$$

$$P_{0r} = BB_r^*, \tag{7.1.10}$$

$$P_{kr} = B_{R-k}^*, \qquad k = 1, \ldots, r, \tag{7.1.11}$$

$$P_{km} = 0, \qquad k = 2, \ldots, r, \qquad m = 0, \ldots, k-2. \tag{7.1.12}$$

Relations (7.1.8)–(7.1.12) are derived similarly to the relations for the elements of the transition probability matrix of the embedded Markov chain for the $M/G/1/\infty$ system (see Section 5.1). Here we only make a few remarks concerning equality (7.1.10). Indeed, let the system be empty after the completion of service of a customer and let the customer generation process be in phase i. Then, for r customers to remain in the system after the completion of service of the next customer and the generation process to pass to phase j, it is necessary that the generation process pass during the idle period from phase i to some intermediate phase k (with the probability B_{ik}) and in the course of service of the next customer at least r customers arrive (if more than r customers arrive, then the 'excessive' customers are lost) and the generation process pass from the phase k to phase j (with the probability $(B_r^*)_{kj}$).

Thus, the transition probability matrix P of the embedded Markov chain in the block form $P = (P_{km})_{k,m=0,\dots,r}$ can be expressed as

$$P = \begin{array}{c} \mathcal{X}_0 \\ \mathcal{X}_1 \\ \mathcal{X}_2 \\ \mathcal{X}_3 \\ \vdots \\ \mathcal{X}_r \end{array} \overset{\begin{array}{cccccc} \mathcal{X}_0 & \mathcal{X}_1 & \mathcal{X}_2 & \dots & \mathcal{X}_{r-1} & \mathcal{X}_r \end{array}}{\begin{pmatrix} BB_0 & BB_1 & BB_2 & \dots & BB_{r-1} & BB_r^* \\ B_0 & B_1 & B_2 & \dots & B_{r-1} & B_r^* \\ 0 & B_0 & B_1 & \dots & B_{r-2} & B_{r-1}^* \\ 0 & 0 & B_0 & \dots & B_{r-3} & B_{r-2}^* \\ \vdots & \vdots & \vdots & \ddots & \vdots & \vdots \\ 0 & 0 & 0 & \dots & B_0 & B_1^* \end{pmatrix}},$$

where, for the sake of clarity, we use the subsets

$$\mathcal{X}_k = \{(1,k), (2,k), \dots, (l,k)\}, \qquad k = 0, \dots, r,$$

of states of the embedded Markov chain. The subset $\mathcal{X}_k$ contains states with k customers in the system. As will be shown in the next section, since the matrix P is of rather peculiar form, stationary state probabilities can be found with the help of algorithms whose complexity is proportional to the squared buffer capacity r, but not to the cubed capacity, as in the case of general matrices.

7.1.2. Stationary probabilities of the embedded Markov chain

The embedded Markov chain $\{\zeta_n,\ n \ge 0\}$ is finite; furthermore, it is irreducible. Indeed, it can pass from the state (j, r) to any state (j, k), $k = 0, \dots, r-1$, with non-zero probability, because during the progress of service of $r-k$ customers ($r-k$ steps of the Markov chain) the phase j of the customer generation process does not change with non-zero probability. Furthermore, by the assumption stipulated at the beginning of this chapter, namely the matrix Λ^* is indecomposable, during the time of service of a customer (in one step of the Markov chain) the generation process may pass with non-zero probability from phase i to phase j for any i and j, and no less than r customers arrive at the system. Hence the probability of reaching the state (j, r) from the state (i, k), $k = 1, \dots, r$, is non-zero. Similarly, we can show that the probability of reaching the state (j, r) from the state $(i, 0)$ is non-zero. All these facts imply the irreducibility of the embedded Markov chain.

Since, as shown just above, the probability of transition in one step from the state (j, r) to the same state (j, r) is non-zero, the embedded Markov chain is non-periodic. Therefore, this chain is ergodic, and the limit (stationary) probabilities

$$p_{ik}^* = \lim_{n\to\infty} \mathbf{P}\{\zeta_n = (i,k)\} > 0, \qquad (i,k) \in \mathcal{X},$$

exist. We set $\boldsymbol{p}_k^* = (p_{1k}^*, \dots, p_{lk}^*)^T$ and denote the vector of stationary probabilities of the embedded Markov chain by $\boldsymbol{p}^* = (\boldsymbol{p}_0^{*T}, \boldsymbol{p}_1^{*T}, \dots, \boldsymbol{p}_r^{*T})^T$.

The vector $\boldsymbol{p}^*$ is the only solution of the simultaneous equilibrium equations

$$\boldsymbol{p}^{*T} = \boldsymbol{p}^{*T} P,$$

or, in the scalar form,

$$p_k^{*T} = p_0^{*T} B B_k + \sum_{m=1}^{k+1} p_m^{*T} B_{k+1-m}, \qquad k = 0, \dots, r-1,$$

$$p_r^{*T} = p_0^{*T} B B_r^* + \sum_{m=1}^{r} p_m^{*T} B_{R-m}^*, \tag{7.1.13}$$

under the normalisation condition

$$p_{\cdot,\cdot}^* = 1. \tag{7.1.14}$$

In Section 1.4.5, we described two algorithms to solve simultaneous equilibrium equations. Since the matrix P is of special type, we can apply any matrix modification of either of these algorithms. In this section, we solve equations (7.1.13) with the use of the matrix version of the first algorithm with account for the specific form of the matrix P. Furthermore, as in Section 1.4.5, we construct the algorithm by a probabilistic technique, using supplementary Markov chains.

Let us represent the matrix P as

$$P = \begin{pmatrix} BB_0 & BB_1 & BB_2 & \dots & BB_{r-2} & BB_{r-1} & P_{0r} \\ B_0 & B_1 & B_2 & \dots & B_{r-2} & B_{r-1} & P_{1r} \\ 0 & B_0 & B_1 & \dots & B_{r-3} & B_{r-2} & P_{2r} \\ 0 & 0 & B_0 & \dots & B_{r-4} & B_{r-3} & P_{3r} \\ \vdots & \vdots & \vdots & \ddots & \vdots & \vdots & \vdots \\ 0 & 0 & 0 & \dots & B_0 & B_1 & P_{r-1,r} \\ 0 & 0 & 0 & \dots & 0 & B_0 & P_{rr} \end{pmatrix}.$$

Since the embedded Markov chain $\{\zeta_n, n \geq 0\}$ is irreducible, the matrix $P_{rr} = B_1^*$ is substochastic. Therefore (Kemeny and Snell, 1960), the matrix

$$P_{rr}^{(1)} = (I - P_{rr})^{-1} = (I - B_1^*)^{-1}$$

exists.

In view of the structure of the matrix P, the Markov chain $\{\zeta_n,\ n \geq 0\}$ can pass from the states of the set $\mathcal{X}_r$ to only one of the states of the set $\mathcal{X}_{r-1}$ in one step. Moreover, the matrix $P_{r,r-1}^{(1)}$ consisting of the probabilities $p_{(i,r),(j,r-1)}^{(1)}$ that the Markov chain $\{\zeta_n, n \geq 0\}$ passes from the state $(i, r) \in \mathcal{X}_r$ to the state $(j, r-1) \in \mathcal{X}_{r-1}$ is

$$P_{r,r-1}^{(1)} = B_0 + P_{rr} B_0 + (P_{rr})^2 B_0 + \cdots = (I - P_{rr})^{-1} B_0 = P_{rr}^{(1)} B_0.$$

Now consider one more Markov chain $\{\zeta_n^{(1)},\ n \geq 0\}$ with the state set

$$\mathcal{X}^{(1)} = \{(i, k), i = 1, \dots, l, k = 0, \dots, r-1\}$$

generated by the instants when the initial Markov chain $\{\zeta_n, n \geq 0\}$ passes to the state set $\mathcal{X}^{(1)}$.

It is easily seen that, for all states (i,k), $i=1,\dots,l$, $k=0,\dots,r-1$, and for all (j,m), $j=1,\dots,l$, $m=0,\dots,r-2$, the Markov chain $\{\zeta_n^{(1)}, n\geq 0\}$ passes from the state (i,k) to the state (j,m) if and only if so does the Markov chain $\{\zeta_n,\ n\geq 0\}$. experiences Therefore, the matrices $P_{km}^{(1)}$, $k=0,\dots,r-1$, $m=0,\dots,r-2$, for the Markov chain $\{\zeta_n^{(1)},\ n\geq 0\}$ are the same as the corresponding matrices for the Markov chain $\{\zeta_n, n\geq 0\}$.

When the Markov chain $\{\zeta_n^{(1)},\ n\geq 0\}$ passes from the state (i,k) to the state $(j,r-1)$, the Markov chain $\{\zeta_n,\ n\geq 0\}$

- with probability $p_{(i,k),(j,r-1)}$ may pass from the state (i,k) directly to the state $(j,r-1)$; or
- with probability $p_{(i,k),(m,r)}$ may first pass from the state (i,k) to the state (m,r) and then with probability $p_{(m,r),(j,r-1)}^{(1)}$ pass from the state (m,r) to the state $(j,r-1)$.

Therefore,

$$P_{k,r-1}^{(1)} = P_{k,r-1} + P_{kr}P_{r,r-1}^{(1)}.$$

The Markov chain $\{\zeta_n^{(1)}, n\geq 0\}$ is irreducible and non-periodic.

We introduce the matrix

$$\hat{P}^{(1)} = \begin{pmatrix} BB_0 & BB_1 & BB_2 & \dots & BB_{r-2} & P_{0,r-1}^{(1)} & P_{0r} \\ B_0 & B_1 & B_2 & \dots & B_{r-2} & P_{1,r-1}^{(1)} & P_{1r} \\ 0 & B_0 & B_1 & \dots & B_{r-3} & P_{2,r-1}^{(1)} & P_{2r} \\ 0 & 0 & B_0 & \dots & B_{r-4} & P_{3,r-1}^{(1)} & P_{3r} \\ \vdots & \vdots & \vdots & \ddots & \vdots & \vdots & \vdots \\ 0 & 0 & 0 & \dots & B_0 & P_{r-1,r-1}^{(1)} & P_{r-1,r} \\ 0 & 0 & 0 & \dots & 0 & P_{r,r-1}^{(1)} & P_{r,r}^{(1)} \end{pmatrix}.$$

It is easy to see that its principal minor

$$P^{(1)} = \begin{pmatrix} BB_0 & BB_1 & BB_2 & \dots & BB_{r-3} & BB_{r-2} & P_{0,r-1}^{(1)} \\ B_0 & B_1 & B_2 & \dots & B_{r-3} & B_{r-2} & P_{1,r-1}^{(1)} \\ 0 & B_0 & B_1 & \dots & B_{r-4} & B_{r-3} & P_{2,r-1}^{(1)} \\ 0 & 0 & B_0 & \dots & B_{r-5} & B_{r-4} & P_{3,r-1}^{(1)} \\ \vdots & \vdots & \vdots & \ddots & \vdots & \vdots & \vdots \\ 0 & 0 & 0 & \dots & B_0 & B_1 & P_{r-2,r-1}^{(1)} \\ 0 & 0 & 0 & \dots & 0 & B_0 & P_{r-1,r-1}^{(1)} \end{pmatrix},$$

which is the transition probability matrix of the Markov chain $\{\zeta_n^{(1)},\ n\geq 0\}$, has the same structure as the matrix P. Therefore, the same procedure can be applied to it, and thus we obtain the matrix $P^{(2)}$, etc., until we arrive at the Markov chain $\{\zeta_n^{(r)},\ n\geq 0\}$ with the state set $\mathscr{X}^{(r)} = \mathscr{X}_0 = \{(i,0), i=1,\dots,l\}$.

It remains to calculate the stationary probabilities $\boldsymbol{p}_m^*$, $m = 0, \ldots, r$, of the embedded Markov chain, applying Corollary 1.4.2, which asserts that these probabilities are identical to the stationary probabilities $\boldsymbol{p}_m^{(k)*}$ of any Markov chain $\{\zeta_n^{(k)}, n \geq 0\}$, $k = 0, \ldots, r-1$, up to a constant factor. Therefore, we first find $\boldsymbol{p}_0^*$ from the equilibrium equations for the Markov chain $\{\zeta_n^{(r)}, n \geq 0\}$

$$\boldsymbol{p}_0^{*T} = \boldsymbol{p}_0^{*T} P_{00}^{(r)}.$$

These equations have a unique solution up to a constant. (This constant is determined from the normalisation condition (7.1.14), which is to be applied after finding all $\boldsymbol{p}_k^*$). Then $\boldsymbol{p}_k^*$, $k = 1, \ldots, r$, are found sequentially from the last equilibrium matrix equation for the Markov chain $\{\zeta_n^{(r-k)}, n \geq 0\}$ (here the initial Markov chain is assumed to be the chain $\{\zeta_n^{(0)},\ n \geq 0\}$):

$$\boldsymbol{p}_k^{*T} = \left(\sum_{m=0}^{k-1} \boldsymbol{p}_m^{*T} P_{mk}^{(r-k)} \right) P_{kk}^{(r-k+1)}. \tag{7.1.15}$$

Finally, the probabilities $\boldsymbol{p}_k^*$, $k = 0, \ldots, r$, thus obtained are normalised by means of (7.1.14).

7.1.3. Time-stationary queue distribution

Let us calculate the time-stationary distribution of the number of customers in the $MAP/G/1/r$ system.

First, we introduce the following matrices:

$T_k(x)$, $k = 0, \ldots, r-1$, is the matrix whose element $(T_k(x))_{ij}$ is the mean sojourn time of the system in the time interval $[0, x)$ in a state in which there are exactly k customers in the system and the customer generation process is in phase j, provided at the initial instant it was in phase i and the service of the customer that had arrived at an empty system was immediately started and not completed by the instant x;

$T_k^*(x)$, $k = 0, \ldots, r$, is the matrix whose element $(T_k^*(x))_{ij}$ is the mean sojourn time in the time interval $[0, x)$ in a state in which there are at least k customers in the system and the customer generation process stays in phase j, provided at the initial instant it was in phase i and the service of the customer that had arrived at an empty system was immediately started and not completed by the instant x;

T_k, $k = 0, \ldots, r-1$, is the matrix whose element $(T_k)_{ij}$ is the mean sojourn time of the system during the service time of a customer in a state in which there are exactly k customers in the system and the customer generation process is in phase j, provided at the initial instant it was in phase i and the service of the customer that had arrived at an empty system was commenced; and

T_k^*, $k = 0, \ldots, r$, is a matrix, whose element $(T_k^*)_{ij}$ is the mean sojourn time of the system during the service time of a customer in a state in which there are at least k customers in the system and the customer generation process is in phase j, provided at the initial instant it was in phase i and the service of the customer that had arrived at an empty system was commenced.

In order to find the defining relation for $T_k(x)$, we partition the interval $[0, x)$ into subintervals $[y, y + dy)$ of 'small' length dy. In each subinterval, the system stays with probability $(F_k(y))_{ij}$ in state (j, k) or with some additional probability does not stay in this state. Therefore, in the subinterval $[y, y + dy)$, the mean sojourn time in state (j, k) increases by $(F_k(y))_{ij}dy$. Since the total sojourn time of the system in the state (j, k) in the interval $[0, x)$ is the sum of sojourn times in this state on individual subintervals $[y, y + dy)$, by the property of the expectation of the sum of random variables, we obtain

$$T_k(x) = \int_0^x F_k(y)\,dy, \qquad k = 0, \dots, r-1.$$

Similarly, we derive the relation

$$T_k^*(x) = \int_0^x F_k^*(y)\,dy, \qquad k = 0, \dots, r.$$

By the total expectation formula,

$$T_k = \int_0^\infty T_k(x)\,dB(x) = \int_0^\infty (1 - B(y))F_k(y)\,dy, \qquad k = 0, \dots, r-1,$$
$$T_k^* = \int_0^\infty T_k^*(x)dB(x) = \int_0^\infty (1 - B(y))F_k^*(y)dy, \qquad k = 0, \dots, r.$$

In what follows, we need the following obvious relations:

$$T_0^* = T_0 + T_1^* = T_0 + T_1 + T_2^* = \cdots, \qquad T_0^*\mathbf{1} = b\mathbf{1}.$$

They are also helpful while checking calculations.

Moreover, let T^* denote the matrix whose element T_{ij}^* is the mean sojourn time of the system in an idle period in a state in which the customer generation process is in phase j, provided at the initial instant of the idle period it was in phase i, i.e.,

$$T^* = \int_0^\infty e^{\Lambda x}dx = -\Lambda^{-1}.$$

At the instant $\tau_n + 0$, let the coordinate ν_n of the embedded Markov chain $\{\zeta_n,\ n \geq 0\}$ be $k \geq 1$. Then the mean time $\mathbf{E}(\tau_{n+1} - \tau_n)$ to the next instant τ_{n+1}, irrespective of the value of the coordinate ξ_n, is simply equal to the mean service time b.

But if $\nu_n = 0$, then the time $\mathbf{E}(\tau_{n+1} - \tau_n)$ is the sum of the mean idle period $T^*\mathbf{1}$ (with account for the value of the coordinate ξ_n) and the mean service time b.

Therefore, in the steady state, the mean time T between successive instants generating the embedded Markov chain is

$$T = \left(\boldsymbol{p}_0^{*T}(T^* + bI) + \sum_{k=1}^{r} b\boldsymbol{p}_k^{*T}\right)\mathbf{1} = \boldsymbol{p}_0^{*T}T^*\mathbf{1} + b. \tag{7.1.16}$$

Now, we set $\boldsymbol{p}_k = (p_{1k}, \dots, p_{lk})^T$, where p_{ik} is the probability that at any instant there are k customers in the system in steady state and the customer generation process is in phase i.

Using the theory of semi-Markov processes and reasoning as in Section 5.1.3, since the busy period following the completion of the idle period that started when the customer generation process was in phase i begins with probability B_{ij} in phase j of the customer generation process, we find, by Theorem 1.6.2, that the stationary state probabilities of the system at arbitrary instants are given by the expressions

$$\begin{aligned} p_0^T &= \frac{1}{T} p_0^{*T} T^*, \\ p_k^T &= \frac{1}{T}\left(p_0^{*T} B T_{k-1} + \sum_{m=1}^{k} p_m^{*T} T_{k-m} \right), \qquad k = 1, \dots, r, \\ p_R^T &= \frac{1}{T}\left(p_0^{*T} B T_r^* + \sum_{m=1}^{r} p_m^{*T} T_{R-m}^* \right). \end{aligned} \tag{7.1.17}$$

7.1.4. Stationary distribution of the Markov process describing the system functioning

Using Markov renewal (semi-Markov) processes and basing on the stationary distribution of the embedded Marked chain, we can refine the result of the previous section and find the time-stationary distribution of the Markov process describing the operation of the system under study.

Let $\nu(t)$ denote the number of customers in the system at instant t and let $\xi(t)$ be the phase of the customer generation process. Let $\eta(t)$ be the elapsed service time of the customer at the server at instant t (if there are customers in the system). We set $\zeta(t) = (\xi(t), \nu(t))$ if there are no customers in the system ($\nu(t) = 0$) and $\zeta(t) = (\xi(t), \nu(t), \eta(t))$ otherwise ($\nu(t) > 0$). Then $\{\zeta(t), t \geq 0\}$ is a linearwise Markov process, as can be easily verified reasoning as in Section 5.4 for the $M/G/1/\infty$ system and taking account for the fact that the sojourn times of the generation process in all phases are exponentially distributed.

The state set of the process $\{\zeta(t), t \geq 0\}$ is

$$\mathscr{X} = \{(i, 0), i = 1, \dots, l; (i, k, x), i = 1, \dots, l, k = 1, \dots, R, x \geq 0\}.$$

The states of the process $\{\zeta(t),\ t \geq 0\}$ are interpreted as follows. If $\zeta(t) = (i, 0)$ at some instant t, then the customer generation process is in phase i and the system is empty. If $\zeta(t) = (i, k, x)$, then the customer generation process is in phase i, there are k customers in the system, and the elapsed service time of the customer at the server is x.

Setting

$$p_{i0}(t) = \mathbf{P}\{(\xi(t), \nu(t)) = (i, 0)\}, \qquad P_{ik}(x, t) = \mathbf{P}\{(\xi(t) = i, \nu(t) = k, \eta(t) < x)\},$$
$$i = 1, \dots, l, \qquad k = 1, \dots, R,$$

as usual, we introduce $\boldsymbol{p}_0(t)$ and $\boldsymbol{P}_k(x, t)$, $k = 1, \dots, R$.

Since the embedded Markov chain $\{\zeta_n,\ n \geq 0\}$ is irreducible and the input flow is a Markov arrival process, Theorem 1.6.6 is true. Therefore, the process $\{\zeta(t), t \geq 0\}$ is ergodic. Its limit (stationary) distribution hence exists:

$$\boldsymbol{p}_0 = \lim_{t\to\infty} \boldsymbol{p}_0(t),$$
$$\boldsymbol{P}_k(x) = \lim_{t\to\infty} \boldsymbol{P}_k(x,t), \quad k = 1,\dots,R, \quad x \geq 0.$$

Moreover, the density $\boldsymbol{p}_k(x)$ of the stationary probability $\boldsymbol{P}_k(x)$ exists and $\boldsymbol{p}_k(x) = P_k'(x)$. Existence of the density $\boldsymbol{p}_k(x)$ is proved as in Section 5.4 for the $M/G/1/\infty$ queueing system.

Now note that for the coordinate $\nu(t)$ of the process $\{\zeta(t), t \geq 0\}$ to be zero at any instant (let this instant be 0) of operation of the system in steady state, it is necessary that in some time interval $(-x - dx, -x]$ the coordinate ν_n of the embedded Markov chain $\{\zeta_n,\ n \geq 0\}$ become 0 (by Theorem 1.6.1, this occurs with the probability $\boldsymbol{p}_0^* \, dx/T$) and no new customers arrive in time x (with account for the phase changes of the customer generation process, the probability of this event is $e^{\Lambda x}$). Therefore, using the total probability formula, we obtain

$$\boldsymbol{p}_0^T = \frac{1}{T}\boldsymbol{p}_0^{*T}\int_0^\infty e^{\Lambda x} dx = \frac{1}{T}\boldsymbol{p}_0^{*T} T^*.$$

Obviously, this result for $\boldsymbol{p}_0$ is the same as formula (7.1.17).

For the system in steady state at instant 0, the process $\{\zeta(t), t \geq 0\}$ stays in state subset $(i, k, [x, x + dx))$, $k = 1, \dots, r$, in two cases, namely

(1) If in the interval $(-x - dx, -x]$, the embedded Markov chain passes to the state (j, m), $m = 1, \dots, k$ (with probability $p_{jm}^* \, dx/T$), and in time x the service of the customer at the server is not completed (with the probability $1 - B(x)$), the customer generation process passes to phase i, and $k - m$ customers arrive (with the probability $(F_{k-m}(x))_{ji}$).

(2) If in the interval $(-x - dx, -x]$, a customer arrives at an empty system and the customer generation process passes to phase j (with the probability $(\boldsymbol{p}_0^T N)_j \, dx = (\boldsymbol{p}_0^{*T} T^* N/T)_j \, dx = (\boldsymbol{p}_0^{*T} B/T)_j \, dx$, and in time x the service of the customer at the server is not completed (with the probability $1 - B(x)$), the customer generation process passes to phase i, and $(k-1)$ customers arrive (with probability $(F_{k-1}(x))_{ji}$).

Hence

$$\boldsymbol{p}_k^T(x) = \frac{1 - B(x)}{T}\left(\boldsymbol{p}_0^{*T} B F_{k-1}(x) + \sum_{m=1}^{k} \boldsymbol{p}_m^{*T} F_{k-m}(x)\right), \qquad k = 1, \dots, r. \quad (7.1.18)$$

Similarly, replacing F_k by F_k^*, we find the vector $\boldsymbol{p}_R(x)$:

$$\boldsymbol{p}_R^T(x) = \frac{1 - B(x)}{T}\left(\boldsymbol{p}_0^{*T} B F_r^*(x) + \sum_{m=1}^{r} \boldsymbol{p}_m^{*T} F_{R-m}^*(x)\right). \quad (7.1.19)$$

From these formulas we can easily derive formulas (7.1.17) for $k = 1, \dots, R$ using the relation

$$\boldsymbol{p}_k = \int_0^\infty \boldsymbol{p}_k(x)\, dx.$$

7.1.5. Stationary state distribution of the system at customer arrival instants

Let us find the stationary distribution of the states at the instants when customers arrive at the system in steady state.

Let $p_{ik}^{-}(x)$ be the stationary probability density of the state (i, k, x), $k = 1, \dots, R$, and let p_{i0}^{-}, p_{ik}^{-} be the stationary probabilities of the states $(i, 0)$ and $(i, k) = \bigcup_{x \geq 0}(i, k, x)$, respectively, just before customer arrival instants. Let us introduce two vectors $\boldsymbol{p}_k^{-}(x)$ and $\boldsymbol{p}_k^{-}$, which are analogous in structure and dimensionality to the vectors $\boldsymbol{p}_k(x)$ and $\boldsymbol{p}_k$ respectively.

Let us show that the following equalities hold:

$$\boldsymbol{p}_k^{-T}(x) = \frac{1}{\lambda}\boldsymbol{p}_k^T(x)N, \qquad k = 1, \dots, R, \tag{7.1.20}$$

$$\boldsymbol{p}_k^{-T} = \frac{1}{\lambda}\boldsymbol{p}_k^T N, \qquad k = 0, \dots, R. \tag{7.1.21}$$

To prove (7.1.20), it suffices to note that $N_{ij}\,\Delta$ is the probability that a customer arrives in a small interval $(t, t + \Delta)$ (to within $o(\Delta)$) and the process $\{\zeta(t), t \geq 0\}$ at the instant $t + \Delta$ remains in states in which the customer generation process is in phase j, provided that at the instant t the process was in the same state but in phase i. In the steady state of the system, it is seen that

$$\boldsymbol{p}_k^{-T}(x) = \lim_{\Delta \to 0} \frac{\boldsymbol{p}_k^T(x)N\Delta}{\left(\boldsymbol{p}_0^T + \int_0^\infty \boldsymbol{p}_{\cdot}^T(x)\,dx\right)N\mathbf{1}\Delta} = \frac{\boldsymbol{p}_k^T(x)N}{\boldsymbol{p}_{\cdot}^T\boldsymbol{\lambda}}. \tag{7.1.22}$$

We observe that $\boldsymbol{p}_{\cdot}$ is the same as the vector $\boldsymbol{\pi}$ defining the stationary distribution of the customer generation process $\{\xi(t), t \geq 0\}$. Therefore,

$$\boldsymbol{p}_{\cdot}^T\boldsymbol{\lambda} = \boldsymbol{\pi}^T\boldsymbol{\lambda} = \lambda, \tag{7.1.23}$$

where λ is the intensity of the Markov input flow. From (7.1.22), by virtue of (7.1.23), we obtain (7.1.20).

Equality (7.1.21) for $k = 0$ is proved similarly, whereas equality (7.1.21) for $k = 1, \dots, R$ follows directly from relations (7.1.20) after integration with respect to x from 0 to ∞.

Let $p_k^{-}(x)$ be the stationary probability density of the state $(k, x) = \bigcup_{i=1}^{l}(i, k, x)$ of the system immediately before the arrival instant of a customer, and let p_k^{-} be the stationary probability that a customer upon arrival finds k other customers in the system. Then, from (7.1.20) and (7.1.21) we obtain

$$p_k^{-}(x) = \frac{1}{\lambda}\boldsymbol{p}_k^T(x)\boldsymbol{\lambda}, \qquad k = 1, \dots, R, \tag{7.1.24}$$

$$p_k^{-} = \frac{1}{\lambda}\boldsymbol{p}_k^T\boldsymbol{\lambda}, \qquad k = 0, \dots, R. \tag{7.1.25}$$

The probability p_R^{-} defined by (7.1.25) for $k = R$ is merely the stationary loss probability. Therefore, denoting the loss probability by π, we obtain

$$\pi = \frac{1}{\lambda}\boldsymbol{p}_R^T\boldsymbol{\lambda}. \tag{7.1.26}$$

7.1.6. Stationary distribution of the waiting time

As usual, considering the steady state of the queueing system under study, we find the waiting time distribution of a customer accepted by the system. Let us recall that we are considering the service of customers in the order of their arrival, i.e., the FCFS discipline.

Let $W(t \mid (i, k, x))$ denote the conditional distribution function of the waiting time for service of a customer arriving at the system in state (i, k, x) and let $\omega(s \mid (i, k, x))$ be the Laplace–Stieltjes transform of this distribution function. The waiting time of the customer arriving at the system in state (i, k, x) is the sum of service time of $(k - 1)$ preceding customers plus the time required to complete the service of the customer at the server at the instant of arrival of that customer. Hence, since service times are independent, we have

$$\omega(s \mid (i, k, x)) = \beta^{k-1}(s) \int_0^\infty e^{-su} \frac{d_u B(x+u)}{1 - B(x)}, \quad x \geq 0, \quad i = 1, \ldots, l, \quad k = 1, \ldots, r. \tag{7.1.27}$$

Moreover, a customer may arrive at the system in state $(i, 0)$. Let $W(t \mid (i, 0))$ be the conditional distribution function of this event and $\omega(s \mid (i, 0))$ its Laplace–Stieltjes transform. Since, in this case, the customer arrives at an empty system, it is immediately taken up for service. Therefore,

$$\omega(s \mid (i, 0)) = 1, \qquad i = 1, \ldots, l. \tag{7.1.28}$$

Let a customer arrive at the system in steady state. Let $W_i(t)$ denote the probability that his waiting time is less than t and the customer generation process is in phase i at the instant of arrival of this customer, and let $\omega_i(s)$ be the Laplace–Stieltjes transform of $W_i(t)$. Furthermore, we introduce a vector $\boldsymbol{\omega}(s) = (\omega_1(s), \ldots, \omega_l(s))^T$.

Now, using the total probability formula, by virtue of (7.1.27) and (7.1.28), we obtain the expression for $\boldsymbol{\omega}(s)$:

$$\boldsymbol{\omega}^T(s) = \frac{1}{1 - \pi} \left[\boldsymbol{p}_0^{-T} + \sum_{k=1}^{r} \beta^{k-1}(s) \int_0^\infty \boldsymbol{p}_k^{-T}(x)\, dx \int_0^\infty e^{-su} \frac{d_u B(x+u)}{1 - B(x)} \right].$$

Using formulas (7.1.20) and (7.1.21), we reduce this expression to

$$\boldsymbol{\omega}^T(s) = \frac{1}{\lambda(1 - \pi)} \left[\boldsymbol{p}_0^T + \sum_{k=1}^{r} \beta^{k-1}(s) \int_0^\infty \boldsymbol{p}_k^T(x)\, dx \int_0^\infty e^{-su} \frac{d_u B(x+u)}{1 - B(x)} \right] N. \tag{7.1.29}$$

If $\omega(s)$ is the Laplace–Stieltjes transform of the conditional distribution function of the waiting time for service of a customer, provided it was accepted by the system, then

$$\omega(s) = \boldsymbol{\omega}^T(s)\mathbf{1}. \tag{7.1.30}$$

By virtue of (7.1.29), (7.1.30), (7.1.17), and (7.1.18), the expression for $\omega(s)$ can be rewritten as

$$\omega(s) = \frac{1}{\lambda(1-\pi)}\left(\frac{1}{T}\boldsymbol{p}_0^{*T}T^* + \frac{1}{T}\sum_{k=1}^{r}\beta^{k-1}(s)\left(\boldsymbol{p}_0^{*T}B\int_0^\infty F_{k-1}(x)\,dx\int_0^\infty e^{-su}d_uB(x+u) + \sum_{m=1}^{k}\boldsymbol{p}_m^{*T}\int_0^\infty F_{k-m}(x)\,dx\int_0^\infty e^{-su}d_uB(x+u)\right)\boldsymbol{\lambda}\right). \quad (7.1.31)$$

Formula (7.1.31) is hardly useful in calculating the stationary distribution of the waiting time (even with supplementary transformations, see Section 7.2.5). But it is of use in determining the moments of this distribution. Let us calculate the mean waiting time w.

Differentiating (7.1.31) at the point $s = 0$, we obtain

$$w = -\omega'(0) = \frac{1}{\lambda(1-\pi)}\left(\sum_{k=1}^{r}(k-1)b\left(\boldsymbol{p}_0^{*T}BT_{k-1} + \sum_{m=1}^{k}\boldsymbol{p}_m^{*T}T_{k-m}\right)\boldsymbol{\lambda} + \sum_{k=1}^{r}\left(\boldsymbol{p}_0^{*T}B\int_0^\infty F_{k-1}(x)\,dx\int_0^\infty u\,d_uB(x+u) + \sum_{m=1}^{k}\boldsymbol{p}_m^{*T}\int_0^\infty F_{k-m}(x)\,dx\int_0^\infty u\,d_uB(x+u)\right)\boldsymbol{\lambda}\right).$$

Since the matrix $F_0^*(x)$ is stochastic for any x, i.e., $F_0^*(x)\mathbf{1} = \mathbf{1}$, we obtain

$$T_{k-1}\boldsymbol{\lambda} = \int_0^\infty dB(x)T_{k-1}(x)\boldsymbol{\lambda} = \int_0^\infty dB(x)\int_0^x F_{k-1}(y)\boldsymbol{\lambda}\,dy = \int_0^\infty dB(x)\int_0^x F_{k-1}(y)NF_0^*(x-y)\mathbf{1}\,dy = B_k^*\mathbf{1}.$$

Furthermore, integrating by parts, we obtain

$$\int_0^\infty F_k(x)\,dx\int_0^\infty u\,d_uB(x+u) = \int_0^\infty F_k(x)\,dx\int_x^\infty (1-B(u))\,du = \int_0^\infty (1-B(u))\,du\int_0^u F_k(x)\,dx = T_k.$$

With this reasoning, we express w as

$$w = \frac{1}{\lambda(1-\pi)}\left(\sum_{k=1}^{r}(k-1)b\left(\boldsymbol{p}_0^{*T}BB_k^* + \sum_{m=1}^{k}\boldsymbol{p}_m^{*T}B_{k-m+1}^*\right)\mathbf{1} + \sum_{k=1}^{r}\left(\boldsymbol{p}_0^{*T}BT_k^*\mathbf{1} + \sum_{m=1}^{k}\boldsymbol{p}_m^{*T}T_{k-m+1}^*\mathbf{1}\right)\right). \quad (7.1.32)$$

This formula is quite adequate for numerical computations.

7.1.7. Performance indices of the system

Using expression (7.1.17) for the stationary probabilities $\boldsymbol{p}_k$, $k = 0, \ldots, R$, we, in particular, find easily the stationary probability $p_k = p_{\cdot,k}$, $k = 0, \ldots, R$, that there are k customers in the system at any instant (without account for the state of the customer generation process), the stationary mean number of customers in the system

$$\overline{N} = \sum_{k=1}^{R} k p_k,$$

and the stationary mean queue length

$$Q = \sum_{k=2}^{R} (k-1) p_k.$$

Let λ_A denote the stationary intensity of the input flow accepted for service. The ith component of the vector $\boldsymbol{p}_{\cdot} - \boldsymbol{p}_R$ is the stationary probability that the customer generation process is in phase i and there is at least one free entry in the waiting buffer (in this case, the system is accessible to the arriving customers). The intensity of the accepted input flow is

$$\lambda_A = \sum_{k=0}^{r} \boldsymbol{p}_k^T \boldsymbol{\lambda}. \tag{7.1.33}$$

By virtue of (7.1.17), formula (7.1.33) can expressed as

$$\lambda_A = \frac{1}{T}\left(\boldsymbol{p}_0^{*T} T^* \boldsymbol{\lambda} + \sum_{k=1}^{r}\left(\boldsymbol{p}_0^{*T} B T_{k-1} \boldsymbol{\lambda} + \sum_{m=1}^{k} \boldsymbol{p}_m^{*T} T_{k-m} \boldsymbol{\lambda}\right)\right).$$

Obviously,

$$T^* \boldsymbol{\lambda} = T^* N \mathbf{1} = B \mathbf{1} = \mathbf{1}.$$

Moreover, using the equality $F_0^*(x)\mathbf{1} = \mathbf{1}$, we obtain

$$T_k(x)\boldsymbol{\lambda} = T_k(x) N \mathbf{1} = \int_0^x F_k(y) N F_0^*(x-y)\mathbf{1}\, dx = F_{k+1}^*(x)\mathbf{1}$$

and

$$T_k \boldsymbol{\lambda} = B_{k+1}^* \mathbf{1}. \tag{7.1.34}$$

Therefore,

$$\lambda_A = \frac{1}{T}\left(\boldsymbol{p}_0^{*T} B \mathbf{1} + \sum_{k=1}^{r}\left(\boldsymbol{p}_0^{*T} B B_k^* \mathbf{1} + \sum_{m=1}^{k} \boldsymbol{p}_m^{*T} B_{k+1-m}^* \mathbf{1}\right)\right).$$

Consider

$$G = \boldsymbol{p}_0^{*T} B B_k^* + \sum_{m=1}^{k} \boldsymbol{p}_m^{*T} B_{k+1-m}^*.$$

Using the relations

$$B_k^* = \sum_{m=k}^{r-1} B_m + B_r^*, \qquad B_{k+1-m}^* = \sum_{l=k+1-m}^{r-m} B_l + B_{r+1-m}^*$$

the above expression reduces to

$$G = \sum_{m=k}^{r-1}\left(\boldsymbol{p}_0^{*T} B B_m + \sum_{l=1}^{m+1} \boldsymbol{p}_l^{*T} B_{m+1-l}\right) + \boldsymbol{p}_0^{*T} B B_r^* + \sum_{l=1}^{r} \boldsymbol{p}_l^{*T} B_{r+1-l}^* - \sum_{m=k+1}^{r} \boldsymbol{p}_m^{*T} B_0^*.$$

Then from equilibrium equations (7.1.14), equalities $B_0^*\mathbf{1} = \mathbf{1}$ and $B\mathbf{1} = \mathbf{1}$, and the normalisation condition (7.1.15) we obtain

$$G\mathbf{1} = \sum_{m=k}^{r} \boldsymbol{p}_m^{*T}\mathbf{1} - \sum_{m=k+1}^{r} \boldsymbol{p}_m^{*T}\mathbf{1} = \boldsymbol{p}_k^{*T}\mathbf{1}$$

and

$$\lambda_A = \frac{1}{T}. \tag{7.1.35}$$

Expression (7.1.16) of T, by virtue of (7.1.15), reduces to

$$T = \boldsymbol{p}_0^{*T} T^*\mathbf{1} + b = p_0 T + b.$$

Hence

$$T = \frac{b}{1 - p_0}.$$

Therefore, expression (7.1.35) of λ_A can be rewritten as

$$\lambda_A = \frac{1 - p_0}{b} = \mu(1 - p_0), \tag{7.1.36}$$

where $\mu = b^{-1}$ is the service intensity.

Let λ_L denote the stationary intensity of the lost customer flow. The ith component of the vector $\boldsymbol{p}_R$ is the stationary probability that the customer generation process is in phase i and the buffer is full. Then the intensity of the lost customer flow is

$$\lambda_L = \boldsymbol{p}_R^T \boldsymbol{\lambda}. \tag{7.1.37}$$

As shown above, the total intensity λ of the Markov input flow can be expressed as

$$\lambda = \sum_{k=0}^{R} \boldsymbol{p}_k^T \boldsymbol{\lambda} = \lambda_A + \lambda_L.$$

Hence, by virtue of (7.1.37), we obtain

$$\frac{\lambda_L}{\lambda} = 1 - \frac{\lambda_A}{\lambda} = \frac{1}{\lambda} \boldsymbol{p}_R^T \boldsymbol{\lambda} = \pi,$$

where π is the loss probability defined by (7.1.26).

Thus, we have arrived at a different but equivalent definition of the loss probability, namely,

$$\pi = \frac{\lambda_L}{\lambda} \tag{7.1.38}$$

or

$$\pi = 1 - \frac{\lambda_A}{\lambda}. \tag{7.1.39}$$

From the last equality, we obtain one more expression for λ_A:

$$\lambda_A = \lambda(1 - \pi). \tag{7.1.40}$$

In turn, by virtue of (7.1.36), we obtain another expression for the loss probability:

$$\pi = 1 - \frac{1 - p_0}{\rho}, \tag{7.1.41}$$

where $\rho = \lambda b$ is the traffic intensity of the system.

Finally, let λ_D denote the stationary intensity of the flow of served customers—the output intensity. Since the linearwise Markov process $\{\zeta(t), t \geq 0\}$ is ergodic, we express λ_D as

$$\lambda_D = \mu(1 - p_0). \tag{7.1.42}$$

Since a system in steady state satisfies the equality $\lambda_A = \lambda_D$, from (7.1.36) and (7.1.41) we obtain

$$\lambda(1 - \pi) = \mu(1 - p_0). \tag{7.1.43}$$

Let us turn our attention to expression (7.1.32) of the mean waiting time.

Using (7.1.40) and the equality $b\mathbf{1} = T_0^*\mathbf{1}$, we obtain

$$\begin{aligned} w &= \frac{1}{\lambda_A T}\left(\sum_{k=1}^{r}(k-1)\boldsymbol{p}_k^{*T}T_0^*\mathbf{1} + \sum_{k=1}^{r}\left(\boldsymbol{p}_0^{*T}BT_k^*\mathbf{1} + \sum_{m=1}^{k}\boldsymbol{p}_m^{*T}T_{k-m+1}^*\mathbf{1}\right)\right) \\ &= \frac{1}{\lambda_A T}\left(\sum_{k=1}^{r}(k-1)\boldsymbol{p}_k^{*T}\left(\sum_{l=0}^{r-k}T_l + T_{R-k}^*\right) + \sum_{k=1}^{r}\left(\boldsymbol{p}_0^{*T}B\left(\sum_{l=k}^{r-1}T_l + T_R^*\right)\right.\right. \\ &\qquad \left.\left. + \sum_{m=1}^{k}\boldsymbol{p}_m^{*T}\left(\sum_{l=k-m+1}^{r-m}T_l + T_{R-m}^*\right)\right)\right)\mathbf{1} \\ &= \frac{1}{\lambda_A T}\left(\sum_{k=1}^{r}(k-1)\left(\boldsymbol{p}_0^{*T}BT_{k-1} + \sum_{m=1}^{k}\boldsymbol{p}_m^{*T}T_{k-m}\right)\right. \\ &\qquad \left. + r\left(\boldsymbol{p}_0^{*T}BT_r^* + \sum_{m=1}^{r}\boldsymbol{p}_m^{*T}T_{R-m}^*\right)\right)\mathbf{1} \\ &= \frac{1}{\lambda_A}\sum_{k=1}^{R}(k-1)\boldsymbol{p}_k^T\mathbf{1} = \frac{1}{\lambda_A}\sum_{k=1}^{R}(k-1)p_k = \frac{Q}{\lambda_A}. \end{aligned}$$

Thus, we have once again verified the validity of the Little formula

$$Q = \lambda_A w. \tag{7.1.44}$$

Obviously, the Little formula for this case contains the intensity of flow of customers accepted for service.

Equalities (7.1.43) and (7.1.44), as well as some other relations derived in this section, can be used to check calculations implemented via the algorithms described in this section.

7.1.8. Other definitions of characteristics via arrival instants

In what follows, by the limit (stationary) probability of an event for customer arrival instants is meant the probability defined as the conditional probability of occurrence of this event in a small time interval Δ under the condition that a customer arrives in this interval at the system in steady state.

But other definitions are also possible. For example, for a fixed initial state of the system, we can study successive instants of arrival of customers. At the arrival of the nth customer, an event may take place with probability P_n. Then, if $\lim_{n\to\infty} P_n$ exists, it is also called the limit (stationary) probability of the event at the customer arrival instants.

For stochastic simulation, it is more convenient to adopt the following procedure. Consider a large time interval T and introduce a random variable $\nu_A(T)$, the number of occurrences of an event at the customer arrival instants, and a random variable $\nu(T)$—the number of customers that had arrived in this interval. Then, if $\lim_{T\to\infty} \nu_A(T)/\nu(T)$ exists, (it is usually assumed to exist with probability one), it is precisely the limit (stationary) probability of the event at the customer arrival instants.

For the systems studied in previous chapters, these definitions all lead to the same result. But for more complicated systems, these definitions, in general, are not equivalent.

Let us give an example.

Consider the $MAP/M/1/0$ system with matrices

$$\Lambda = \begin{pmatrix} -\lambda_1 & 0 \\ 0 & -\lambda_2 \end{pmatrix}, \qquad N = \begin{pmatrix} 0 & \lambda_1 \\ \lambda_2 & 0 \end{pmatrix},$$

and service intensity μ. For this system, the sequence $\{\zeta_n^* = (\xi_n, \nu_n), n \geq 0\}$ is a Markov chain, where ξ_n is the phase of the customer generation process and ν_n is the number of customers in the system just before the instant of arrival of a customer. Its state set is $\mathcal{X} = \{(i,k), i = 1,2; k = 0,1\}$, where i is the generation phase and k is the number of customers in the system. Obviously, the Markov chain passes from the state $(1,0)$ only to the state $(2,1)$, from the state $(2,0)$ only to the state $(1,1)$, from the state $(1,1)$ only to the states $(2,1)$ and $(2,0)$, and from the state $(2,1)$ only to the states $(1,1)$ and $(1,0)$.

Therefore, the Markov chain $\{\zeta_n^*, n \geq 0\}$ is periodic of period two. Though the stationary distribution for this chain exists, its limit probabilities do not. This, in turn, leads to a situation where the loss probabilities for customers with even and odd numbers may tend to different limits.

This example demonstrates, as said in Section 1.4, that the concepts of stationary and limit probabilities may not coincide.

Nevertheless, for the systems studied in this and succeeding sections with Markov flow, the limit (properly stationary) probabilities for the arrival instants always exist and are identical in the sense of the first and third definitions. Furthermore, if the limit probability exists in the sense of the second definition, it coincides with the stationary probability in the sense of other two definitions. These facts are corollaries to the presence of renewal instants, which can be taken to be, for example, the instants at which the system becomes empty when the generation process stays in some fixed phase i.

7.2. Supplementary variables: FCFS discipline

In this section, we apply a different method to analysing the $MAP/G/1/r$ system—the supplementary variable technique, using elapsed service time as the supplementary variable. Applying this method, we studied the $M/G/1/\infty$ system in Section 5.4, and the $M/G/1/r$ system in Section 6.5. Here we show that it can be easily applied to systems with Markov input flows.

Along with the assumptions made at the beginning of this chapter, we, in addition, assume for the sake of simplicity, as in Section 5.4, that the distribution function $B(x)$ is continuous and satisfies the condition $B(x) < 1$ for all x.

Furthermore, we apply one more technical condition that all eigenvalues $\{\sigma_i\}$ of the matrix Λ satisfy the condition $\beta(-\sigma_i) \neq 0$, where $\beta(s)$ is the Laplace–Stieltjes transform of the distribution function $B(x)$.

7.2.1. Simultaneous equations

Let us study the linearwise Markov process $\{\zeta(t),\ t \geq 0\}$ described in Section 7.1.4, using the notation introduced therein $p_{i0}(t)$, $P_{ik}(x,t)$, p_{i0}, $P_{ik}(x)$, and $p_{ik}(x)$ for probabilities and $\boldsymbol{p}_0(t)$, $\boldsymbol{P}_k(x,t)$, $\boldsymbol{p}_0$, $\boldsymbol{P}_k(x)$, and $\boldsymbol{p}_k(x)$ for their vector analogues, and introduce the derivative $p_{ik}(x,t) = \partial P_{ik}(x,t)/\partial x$, $k = 1, \dots, R$ (existence of the derivative $p_{ik}(x,t)$ is demonstrated exactly as in Section 5.4).

Let us introduce functions $q_{ik}(x)$, defining them by the relations

$$p_{ik}(x) = [1 - B(x)]q_{ik}(x), \tag{7.2.1}$$

and vectors $\boldsymbol{q}_k(x) = (q_{1k}(x), \dots, q_{lk}(x))^T$.

We show that $\boldsymbol{q}_k(x)$, $k = 1, \dots, R$, satisfy the simultaneous differential equations

$$\frac{d}{dx}\boldsymbol{q}_k^T(x) = \boldsymbol{q}_k^T(x)\Lambda + u(k-1)\boldsymbol{q}_{k-1}^T(x)N, \qquad k = 1, \dots, r, \tag{7.2.2}$$

$$\frac{d}{dx}\boldsymbol{q}_R^T(x) = \boldsymbol{q}_R^T(x)\Lambda^* + \boldsymbol{q}_r^T(x)N \tag{7.2.3}$$

under the boundary conditions

$$\boldsymbol{0}^T = \boldsymbol{p}_0^T\Lambda + \int_0^\infty \boldsymbol{q}_1^T(x)\,dB(x), \tag{7.2.4}$$

$$\boldsymbol{q}_k^T(0) = \int_0^\infty \boldsymbol{q}_{k+1}^T(x)\,dB(x) + u(2-k)\boldsymbol{p}_0^T N, \qquad k = 1, \dots, r, \tag{7.2.5}$$

$$\boldsymbol{q}_R(0) = \boldsymbol{0}, \tag{7.2.6}$$

where $u(x)$ is the Heaviside function.

As usual, we consider the instants t and $t+\Delta$, where Δ is a small increment of time. For the process $\zeta(t)$ to stay in state $(i,k,x+\Delta)$, $i=1,\dots,l$, $k=1,\dots,r$, at instant $t+\Delta$, it is necessary that

(1) at instant t, the process $\zeta(t)$ be in state (i,k,x), and in time Δ, the customer generation process do not change its phase i (with the probability $1+\Lambda_{ii}\Delta+o(\Delta)$) and the service of the customer at the server be not completed (with the probability $[1-B(x+\Delta)]/[1-B(x)]$);

(2) at instant t, the process $\zeta(t)$ be in state (j,k,x), $j=1,\dots,l$, $j\neq i$; in time Δ, the customer generation process $\xi(t)$ pass from phase j to phase i without any new arrival (with the probability $\Lambda_{ji}\Delta+o(\Delta)$), and the service of the customer at the server be not completed;

(3) at the instant t, the process $\zeta(t)$ be in state $(j,k-1,x)$, $j=1,\dots,l$; in time Δ, the customer generation process $\xi(t)$ pass from phase j to phase i with a new arrival (with the probability $N_{ji}\Delta+o(\Delta)$), and the service of the customer at the server be not completed. Note that if $k=1$, then this event may not occur, because the elapsed service time of a customer arriving at an empty system is zero.

Since the probability of all other transitions is $o(\Delta)$, with the use of the total probability formula we obtain

$$\begin{aligned} p_{ik}(x+\Delta,t+\Delta) = p_{ik}(x,t)(1+\Lambda_{ii}\Delta)\frac{1-B(x+\Delta)}{1-B(x)} \\ + \sum_{\substack{j=1\\ j\neq i}}^{l} p_{jk}(x,t)\Lambda_{ji}\Delta\frac{1-B(x+\Delta)}{1-B(x)} \\ + u(k-1)\sum_{j=1}^{l} p_{j,k-1}(x,t)N_{ji}\Delta\frac{1-B(x+\Delta)}{1-B(x)} + o(\Delta). \end{aligned}$$

Passing to the stationary state in this equality and using relation (7.2.1), we obtain

$$\begin{aligned} q_{ik}(x+\Delta)-q_{ik}(x) = \sum_{j=1}^{l} q_{jk}(x)\Lambda_{ji}\Delta \\ + u(k-1)\sum_{j=1}^{l} q_{j,k-1}(x)N_{ji}\Delta + o(\Delta), \quad i=1,\dots,l, \quad k=1,\dots,r. \end{aligned}$$

Hence, dividing by Δ and making Δ tend to zero, we finally obtain the simultaneous differential equations

$$q'_{ik}(x) = \sum_{j=1}^{l} q_{jk}(x)\Lambda_{ji} + u(k-1)\sum_{j=1}^{l} q_{j,k-1}(x)N_{ji}, \quad i=1,\dots,l, \quad k=1,\dots,r.$$

They in matrix form are equivalent to (7.2.2).

In order to derive (7.2.3), we use the following condition: for the process $\zeta(t)$ to be in state (i, R, x) at the instant $t + \Delta$, it is necessary, along with conditions for $k \neq R$ that are analogous to the conditions defined for $k = R$, that

(4) at instant t, the process $\zeta(t)$ be in state (j, R, x), $j = 1, \ldots, l$, and in time Δ, a customer arrive and be lost, the customer generation process $\xi(t)$ pass from state j to phase i (the probability of this event is $N_{ji}\Delta + o(\Delta)$), and the service of the customer at the server be not completed.

All other steps in deriving (7.2.3) are similar to the steps involved in deriving (7.2.2).

We now examine boundary conditions. We begin with the condition for $\boldsymbol{p}_0$. Turning back to the transient case, we consider the state of the process $\zeta(t)$ at instants t and $t + \Delta$. Now for the process $\zeta(t)$ to be in state $(i, 0)$, $i = 1, \ldots, l$, at the instant $t + \Delta$, it is necessary that

(1) at instant t, the process $\zeta(t)$ be in state $(i, 0)$, and in time Δ, the customer generation process $\xi(t)$ do not change its phase i (with probability $1 + \Lambda_{ii}\Delta + o(\Delta)$);

(2) at instant t, the process $\zeta(t)$ be in state $(j, 0)$, $j = 1, \ldots, l$, $j \neq i$, and in time Δ, the customer generation process $\xi(t)$ pass from phase j to phase i and no customer arrive (this occurs with the probability $\Lambda_{ji}\Delta + o(\Delta)$); and

(3) at instant t, the process $\zeta(t)$ be in state $(i, 1, x)$, and in time Δ the service of the customer at the server be completed (with the probability $[B(x + \Delta) - B(x)]/[1 - B(x)]$).

The probability of all other possible transitions is $o(\Delta)$.

Now, using the total probability formula, we obtain

$$p_{i0}(t + \Delta) = p_{i0}(t)(1 + \Lambda_{ii}\Delta) + \sum_{\substack{j=1 \\ j \neq i}}^{l} p_{j0}(t)\Lambda_{ji}\Delta + \int_0^\infty p_{i1}(x, t)\frac{B(x + \Delta) + B(x)}{1 - B(x)}dx + o(\Delta), \qquad i = 1, \ldots, l.$$

Hence, passing to the steady state and using relation (7.2.1), we obtain

$$0 = \sum_{j=1}^{l} p_{j0}\Lambda_{ji}\Delta + \int_0^\infty q_{i1}(x)[B(x + \Delta) - B(x)]\,dx + o(\Delta), \qquad i = 1, \ldots, l.$$

In fact, $p_{ik}(x)$ and hence, $q_{ik}(x)$ are bounded functions (for the proof, see (Belyaev, 1962)). Therefore, by the continuity and boundedness of $q_{ik}(x)$, we obtain (see (Sevastyanov, 1957))

$$\lim_{\Delta \to 0} \int_0^\infty q_{ik}(x)\frac{B(x + \Delta) - B(x)}{\Delta}dx = \int_0^\infty q_{ik}(x)\,dB(x).$$

Now dividing both sides of the equations for p_{i0}, by Δ and passing to the limit as $\Delta \to 0$, we obtain thc boundary conditions

$$0 = \sum_{j=1}^{l} p_{j0}\Lambda_{ji} + \int_0^\infty q_{i1}(x)\,dB(x), \qquad i = 1, \dots, l.$$

This equation in matrix form corresponds to (7.2.4).

We now find boundary conditions for the vector $\boldsymbol{q}_k(0)$. Consider the event $\{\xi(t+\Delta) = i, \nu(t+\Delta) = k, \eta(t+\Delta) < \Delta\}$, $i = 1, \dots, l$, $k = 2, \dots, r$. For this event to occur, it is necessary that at instant t the process $\zeta(t)$ stay in state $(i, k+1, x)$ and the service of the customer at the server be completed in time Δ. The probability of all other possible events is $o(\Delta)$. Therefore,

$$P_{ik}(\Delta, t+\Delta) = \int_0^\infty p_{i,k+1}(x,t)\frac{B(x+\Delta) - B(x)}{1 - B(x)}dx + o(\Delta),$$

where $i = 1, \dots, l$, $k = 2, \dots, r$. Hence, passing to the stationary state, applying the mean value theorem, and using (7.2.1), we obtain

$$\begin{aligned} P_{ik}(\Delta) &= \int_0^\Delta p_{ik}(x)\,dx = p_{ik}(\theta\Delta)\Delta \\ &= \int_0^\infty q_{i,k+1}(x)[B(x+\Delta) - B(x)]\,dx + o(\Delta), \end{aligned}$$

where $0 < \theta < 1$, $i = 1, \dots, l$, $k = 2, \dots, r$. Dividing the last relation by Δ and passing to the limit as $\Delta \to 0$, in view of the fact that $p_{ik}(0) = q_{ik}(0)$, we obtain

$$q_{ik}(0) = \int_0^\infty q_{i,k+1}(x)\,dB(x), \quad i = 1, \dots, l, \quad k = 2, \dots, r,$$

which in matrix form yields (7.2.5) for $k = 2, \dots, r$.

While deriving (7.2.5) for $k = 1$, we must additionally take into account that the event $\{\xi(t+\Delta) = i, \nu(t+\Delta) = 1, \eta(t+\Delta) < \Delta\}$ may occur not only if the service of the customer at the server at instant t is completed, but also if the process $\zeta(t)$ is in state $(j, 0)$ at instant t and the customer generation process in time Δ passes from phase j to phase i and a customer arrives (with the probability $N_{ji}\Delta + o(\Delta)$). Other steps in deriving (7.2.5) for $k = 1$ are similar to the steps used in deriving the equation for $k \neq 1$.

Equality (7.2.6) is self-evident, because it is not possible to reach the state $(i, R, 0)$ by completing the service for the reason that the system cannot contain more than R customers.

Let us now introduce the stationary probability p_{ik} of the state (i, k, x) with no regard for the elapsed service time:

$$p_{ik} = \int_0^\infty p_{ik}(x)\,dx,$$

and let $\boldsymbol{p}_k = (p_{1k}, \dots, p_{lk})^T$. Then the normalisation condition takes the form

$$\boldsymbol{p}_{\cdot}^T \mathbf{1} = 1. \tag{7.2.7}$$

Note that $p_{ik}(0) = q_{ik}(0)$ has a clear physical meaning. It is the stationary intensity of occurrence of instants at which service begins and after each such instant the system contains k customers and the generation process is in phase i. As in the previous section, we can show that a Markov chain is obtained for the number of customers in the system and phases of the generation process at service commencement instants. But this chain, which is naturally called the embedded Markov chain for the service commencement instants, is not helpful (because it may have non-essential states).

7.2.2. Solution of equations: a recurrent matrix algorithm

In order to solve differential equations (7.2.2), (7.2.3), let us consider the matrix functions $F_k(x)$, $k \geq 0$, defined by formulas (7.1.1) and (7.1.2).

Let us derive a relation for the matrix functions $F_k(x)$ which we need in what follows. Differentiating (7.1.1) and (7.1.2), we obtain differential equations for $F_k(x)$:

$$F'_k(x) = F_k(x)\Lambda + u(k)F_{k-1}(x)N, \qquad k \geq 0. \tag{7.2.8}$$

Let us solve differential equations (7.2.2). Solving first for $k = 1$ and 2 with the use of (7.1.1) and (7.1.2) and then applying induction, we obtain

$$\boldsymbol{q}_k^T(x) = \sum_{j=0}^{k-1} \boldsymbol{q}_{k-j}^T F_j(x), \qquad k = 1, \ldots, r, \tag{7.2.9}$$

where, for brevity, we use the notation $\boldsymbol{q}_k = \boldsymbol{q}_k(0)$.

Validity of formula (7.2.9) can be checked by substituting it into (7.2.2) with account for (7.2.8): it becomes an identity.

In order to find $\boldsymbol{q}_R(x)$, summing (7.2.2) and (7.2.3) over $k = 1, 2, \ldots, R$, we obtain

$$\frac{d}{dx}\boldsymbol{q}_{\cdot}^T(x) = \boldsymbol{q}_{\cdot}^T(x)\Lambda^*.$$

Hence

$$\boldsymbol{q}_{\cdot}^T(x) = \boldsymbol{q}_{\cdot}^T e^{\Lambda^* x}. \tag{7.2.10}$$

From this equality we obtain

$$\boldsymbol{q}_R^T(x) = \boldsymbol{q}_{\cdot}^T e^{\Lambda^* x} - \sum_{k=1}^{r} \boldsymbol{q}_k^T(x). \tag{7.2.11}$$

The vectors $\boldsymbol{q}_k$, $k = 1, \ldots, r$, in (7.2.9) and (7.2.11) are unknown yet. To find them, substituting (7.2.9) into (7.2.4) and (7.2.5) for $r = 1, \ldots, r-1$, we obtain the simultaneous algebraic equations in unknowns $\boldsymbol{p}_0$ and $\boldsymbol{q}_k$, $k = 1, \ldots, r$:

$$\begin{aligned} \boldsymbol{0}^T &= \boldsymbol{p}_0^T \Lambda + \boldsymbol{q}_1^T B_0, \\ \boldsymbol{q}_k^T &= \sum_{j=0}^{k} \boldsymbol{q}_{k-j+1}^T B_j + u(2-k)\boldsymbol{p}_0^T N, \qquad k = 1, \ldots, r-1, \end{aligned} \tag{7.2.12}$$

where

$$B_k = \int_0^\infty F_k(x)\, dB(x).$$

Recall that the matrix B_k is the matrix analogue of the kth-order exponential moment of the distribution function $B(x)$, and its element $(B_k)_{ij}$ is the probability that k customers arrive during the service of a customer and the generation process upon completion of service passes to phase j, provided it was in phase i at the commencement of service.

Note that (7.2.12) literally defines the first r equations in a set of equations which differs from the set of equilibrium equations for the embedded Markov chain stated in the previous section in that it contains the stationary probabilities p_{i0} of states $(i, 0)$ at arbitrary instants, but not the probabilities p_{i0}^* for the embedded Markov chain, as in the previous section. But here to solve (7.2.12) we use the matrix analogue of the second algorithm described in Section 1.4.5.

Transforming equations (7.2.12) so that they represent a equations in unknowns $\boldsymbol{q}_k$, $k = 1, \dots, r$, under the assumption that the vector $\boldsymbol{p}_0$ is known, we obtain

$$(\boldsymbol{q}_1^T, \dots, \boldsymbol{q}_r^T) \begin{pmatrix} B_0 & B_1 - I & B_2 & \dots & B_{r-1} \\ 0 & B_0 & B_1 - I & \dots & B_{r-2} \\ 0 & 0 & B_0 & \dots & B_{r-3} \\ \vdots & \vdots & \vdots & \ddots & \vdots \\ 0 & 0 & 0 & \dots & B_0 \end{pmatrix} = (-\boldsymbol{p}_0^T \Lambda, -\boldsymbol{p}_0^T N, \boldsymbol{0}^T, \dots, \boldsymbol{0}^T). \tag{7.2.13}$$

Since the matrix of coefficients of (7.2.13) is a block-triangular matrix, all unknowns $\boldsymbol{q}_k$, $k = 1, \dots, r$, can be expressed in terms of the vector $\boldsymbol{p}_0$ if the matrix B_0 is invertible.

We observe that

$$B_0 = \int_0^\infty e^{\Lambda x} dB(x) = \beta(-\Lambda),$$

where $\beta(s)$ is the Laplace–Stieltjes transform of the distribution function $B(x)$. As we know (Kato, 1966), if σ_i is an eigenvalue of the matrix Λ, then $\beta(-\sigma_i)$ is an eigenvalue of the matrix $\beta(-\Lambda) = B_0$. By the assumption in the beginning of this section, namely, $\beta(-\sigma_i) \neq 0$ for all $i = 1, \dots, l$, we see that the matrix B_0 is non-degenerate and hence, its inverse B_0^{-1} exists.

According to the abovesaid, we will seek for the solution of (7.2.13) in the form

$$\boldsymbol{q}_k^T = \boldsymbol{p}_0^T Q_k, \qquad k = 1, \dots, r, \tag{7.2.14}$$

where the matrices Q_k are defined by the recurrence relations

$$Q_1 = -\Lambda B_0^{-1},$$
$$Q_k = \left[Q_{k-1} - \sum_{j=1}^{k-1} Q_{k-j} B_j - u(3-k) N \right] B_0^{-1}, \qquad k = 2, \dots, r. \tag{7.2.15}$$

In order to verify formula (7.2.14), we substitute it into (7.2.13). In view of (7.2.15), it becomes an identity.

The vector $\boldsymbol{p}_0$, which is unknown yet, can be found (up to a constant) by substituting expressions (7.2.9) and (7.2.11) of $\boldsymbol{q}_k(x)$, $k = 1, \dots, r$, and expression (7.2.14) of $\boldsymbol{q}_k$, $k = 1, \dots, r$, into the last relation (with $k = r$) in (7.2.5). Indeed, homogeneous equations (7.2.12), along with the relation for $\boldsymbol{p}_0$, contain lR unknowns $\boldsymbol{p}_0$ and $\boldsymbol{q}_k$, $k = 1, \dots, r$, and as many equations. Since the stationary distribution of the process $\{\zeta(t),\ t \geq 0\}$ is unique, the rank of such a set of equations is $lR - 1$. Therefore, their solution can be found up to a constant, which, in its turn, is found from the (additional) normalisation condition.

We will, however, find $\boldsymbol{p}_0$ not from the equations thus obtained, but from equivalent ones. For this, adding together all equations in (7.2.4) and (7.2.5), we obtain

$$\boldsymbol{q}_{\cdot}^T = \int_0^\infty \boldsymbol{q}_{\cdot}^T(x)\, dB(x) + \boldsymbol{p}_0^T \Lambda^*.$$

Substituting expression (7.2.10) of $\boldsymbol{q}_{\cdot}(x)$ into this relation, we obtain

$$\boldsymbol{q}_{\cdot}^T = \boldsymbol{q}_{\cdot}^T B_0^* + \boldsymbol{p}_0^T \Lambda^*. \tag{7.2.16}$$

where

$$B_0^* = \int_0^\infty e^{\Lambda^* x} dB(x).$$

Finally, substituting expression (7.2.14) of $\boldsymbol{q}_k$, $k = 1, \dots, r$, into (7.2.16), we obtain the equations for the vector $\boldsymbol{p}_0$

$$\boldsymbol{p}_0^T Q = \boldsymbol{0}^T, \tag{7.2.17}$$

where $Q = Q.(I - B_0^*) - \Lambda^*$.

As we seen, the vector $\boldsymbol{p}_0$ is determined by (7.2.17) up to a constant. To find the constant, we derive an equation equivalent to the normalisation condition (7.2.7). For this purpose, multiplying both sides of (7.2.10) from the right by the vector $\boldsymbol{1}$, we obtain

$$q_{\cdot}(x) = q_{\cdot}, \tag{7.2.18}$$

where $q_{\cdot}(x) = \boldsymbol{q}_{\cdot}^T(x)\boldsymbol{1}$ and $q_{\cdot} = \boldsymbol{q}_{\cdot}^T\boldsymbol{1}$. Multiplying both sides of (7.2.18) by $(1 - B(x))$, integrating them with respect to x from zero to ∞, and using the normalisation condition (7.2.7), we obtain

$$1 - p_0 = bq_{\cdot}. \tag{7.2.19}$$

Thus, equations (7.2.17), along with relation (7.2.19) and expressions (7.2.14) of $\boldsymbol{q}_k$, $k = 1, \dots, r$, uniquely define the vector $\boldsymbol{p}_0$.

Let us find the vectors $\boldsymbol{p}_k$, $k = 1, \dots, R$. For this purpose, multiplying both sides of differential equations (7.2.2) and (7.2.3) by $1 - B(x)$ and then integrating with respect to x from zero to ∞, after certain rather elementary calculations involving equalities (7.2.4) and (7.2.5) we arrive at the relations

$$-\boldsymbol{q}_1^T = (\boldsymbol{p}_0^T + \boldsymbol{p}_1^T)\Lambda, \tag{7.2.20}$$

$$-\boldsymbol{q}_k^T + \boldsymbol{q}_{k-1}^T - u(3-k)\boldsymbol{p}_0^T N = \boldsymbol{p}_k^T \Lambda + \boldsymbol{p}_{k-1}^T N, \qquad k = 2, \dots, r, \tag{7.2.21}$$

$$\boldsymbol{q}_r^T = \boldsymbol{p}_R^T \Lambda^* + \boldsymbol{p}_r^T N. \tag{7.2.22}$$

From (7.2.20) and (7.2.21) we obtain

$$p_1^T = -(q_1^T \Lambda^{-1} + p_0^T),$$
$$p_k^T = -\left(q_k^T + \sum_{j=0}^{k-1} p_j^T \Lambda^* \right) \Lambda^{-1}, \qquad k = 2, \dots, r. \tag{7.2.23}$$

I remains to find the vector p_R. Multiplying both sides of (7.2.23) from the right by the matrix Λ, and then adding (7.2.23) for $k = r$ to (7.2.22), we obtain the equations for p_R

$$p_.^T \Lambda^* = 0^T. \tag{7.2.24}$$

By the definition of the Markov arrival flow, the matrix Λ^* is the transition intensity matrix of the Markov process $\{\xi(t), t \geq 0\}$. Therefore, equations (7.2.24), along with normalisation condition (7.2.7), are used to determine uniquely the vector $p_. = \pi$, which, in its turn, is used to find the vector p_R from the relation

$$p_R = p_. - \sum_{k=0}^{r} p_k.$$

7.2.3. Main performance indices of the system

The matrix algorithm derived above can be applied to finding the stationary probability densities $\{p_k(x),\ k = 1, \dots, R\}$ and stationary probability distribution $\{p_k, k = 0, \dots, R\}$ of the states of the $MAP/G/1/r$ queueing system at arbitrary instants. In a sense, these characteristics are basic parameters, because they can be used to find many other stationary performance indices of the system. In particular, reasoning as in Section 7.1.5, we find the stationary distribution of the process $\{\zeta(t), t \geq 0\}$ at customer arrival instants, and hence, formula (7.1.26) for the loss probability. We also find the stationary distribution of the process $\{\zeta(t),\ t \geq 0\}$ at service completion instants. In the concluding section, we will describe this distribution.

In this section, we derive expressions for certain macrocharacteristics of the quality of operation of the system under study, as well as some useful relations.

Some of the performance indices and their relations derived below have already been found in the preceding section by the method of embedded Markov chains. For methodological reasons, using the results of this section, we will once again find the main performance indices derived in the previous section.

First we find the intensity λ_A of the flow of accepted customers, the intensity λ_D of the flow of served customers, and the intensity λ_L of the flow of lost customers.

The input flow intensity, by virtue of (7.2.24), admits the expression

$$\lambda = p_.^T \boldsymbol{\lambda}, \tag{7.2.25}$$

and the intensity of accepted flow is expressed as

$$\lambda_A = (p_.^T - p_R^T)\boldsymbol{\lambda}. \tag{7.2.26}$$

In order to calculate the intensity λ_D of flow of accepted customers, we observe that the output flow intensity $\mu(x)$ is $dB(x)/[(1 - B(x))dx]$ if the process $\{\zeta(t), t \geq 0\}$ is in state (i, k, x), $k = 1, \ldots, R$, and zero otherwise. Therefore, by the total probability formula,

$$\lambda_D = \sum_{k=1}^{R} \int_0^\infty \mu(x) p_{\cdot,k}(x)\, dx = \sum_{k=1}^{R} \int_0^\infty \frac{p_{\cdot,k}(x)\, dB(x)}{1 - B(x)},$$

or, by virtue of (7.2.1), (7.2.18), and (7.2.19),

$$\lambda_D = \frac{1 - p_0}{b} = \mu(1 - p_0). \tag{7.2.27}$$

Since $\lambda_A = \lambda_D$ in the steady state, we see that

$$1 - p_0 = b(\boldsymbol{p}_\cdot^T - \boldsymbol{p}_R^T)\boldsymbol{\lambda}. \tag{7.2.28}$$

Finally, since $\lambda = \lambda_A + \lambda_L$, from (7.2.25) and (7.2.26) we immediately obtain the equality

$$\lambda_L = \boldsymbol{p}_R^T \boldsymbol{\lambda}, \tag{7.2.29}$$

and another expression for the loss probability

$$\pi = \frac{\lambda_L}{\lambda},$$

which was derived earlier in Section 7.1.

Multiplying both sides of equalities (7.2.23) from the right by $\boldsymbol{\lambda}$ and summing them over $k = 1, 2, \ldots, r$, by virtue (7.2.25) we obtain

$$q_\cdot = \lambda - \boldsymbol{p}_R^T \boldsymbol{\lambda}.$$

Hence, with the use of (7.2.19) and (7.1.26) we obtain the relation

$$\lambda(1 - \pi) = \frac{1 - p_0}{b} = \mu(1 - p_0), \tag{7.2.30}$$

which was also derived in Section 7.1.

Equalities (7.2.28) and (7.2.30) are helpful while checking the calculations carried out by the matrix algorithm described above.

7.2.4. Stationary state distribution at departure instants

Let τ_n be the instant of completion of service of the nth customer. The sequence $\{\zeta_n, n \geq 0\}$ consisting of the first two coordinates $\xi(t)$ and $\nu(t)$ of the process $\{\zeta(t), t \geq 0\}$ at the instants $\tau_n + 0$ (obviously, the third coordinate $\eta(t)$ vanishes at these instants) is the embedded Markov chain introduced in the preceding section.

The vectors $\boldsymbol{p}_k^*$ of stationary state probabilities of the embedded Markov chain can be calculated by the formula

$$\boldsymbol{p}_k^{*T} = \begin{cases} -\frac{1}{\lambda_D}\boldsymbol{p}_0^T\Lambda, & k=0; \\ \frac{1}{\lambda_D}(\boldsymbol{q}_1^T - \boldsymbol{p}_0^T N), & k=1; \\ \frac{1}{\lambda_D}\boldsymbol{q}_k^T, & k=2,\dots,r. \end{cases} \qquad (7.2.31)$$

In order to prove formula (7.2.31), observe first that the probability of completion of service of a customer in a small time interval dx is $dB(x)/[1-B(x)]$ if the service was not completed in the time x. Therefore, for the steady state of the queueing system, we see that

$$\boldsymbol{p}_k^{*T} = \frac{\int_0^\infty \boldsymbol{q}_{k+1}^T(x)\,dB(x)}{\int_0^\infty q.(x)\,dB(x)}, \qquad k=0,\dots,r.$$

Hence, by virtue of (7.2.18), (7.2.19), and (7.2.27), we obtain

$$\boldsymbol{p}_k^{*T} = \frac{1}{\lambda_D}\int_0^\infty \boldsymbol{q}_{k+1}^T(x)\,dB(x), \qquad k=0,\dots,r,$$

which, combined with (7.2.4) and (7.2.5), yields (7.2.31).

Setting $p_k^* = \boldsymbol{p}_k^{*T}\mathbf{1}$, we see that p_k^*—the stationary probability that k customers are in the system at the instant of completion of service—is given by the formula

$$p_k^* = \frac{1}{\lambda_D}\boldsymbol{p}_k^T\boldsymbol{\lambda}, \qquad k=0,\dots,r, \qquad (7.2.32)$$

which follows from (7.2.31) by virtue of (7.2.23).

Formulas (7.2.31) and (7.2.32) are of a simple physical meaning. Indeed, $\lambda_D p_{ik}^*$ is the intensity of occurrence of service completion instants after which k customers remain in the system and the customer generation process is in phase i. For $k=2,\dots,r$, they are the instants of commencement of service of successive customers. This proves the last equality in (7.2.31). Furthermore, by virtue of (7.2.12), we obtain

$$-\boldsymbol{p}_0^T\Lambda = \boldsymbol{q}_1^T B_0.$$

Therefore, the first equality in (7.2.31) can be rewritten as

$$\lambda_D \boldsymbol{p}_0^{*T} = \boldsymbol{q}_1^T B_0.$$

The left-hand side of this equality is the intensity of occurrence of service completion instants after which the system is empty. On the other hand, this intensity is the product of the intensity of occurrence of commencement of service at instants at which there is no queue and the probability that no customer arrives in the course of service of a customer (this probability in matrix form is B_0).

Finally, the second equality in (7.2.31) implies that every service commencement instant in the absence of queue is either a service completion instant after which only one customer remains in the system, or an instant when a customer arrives at an empty system.

7.2.5. Stationary distribution of waiting time

In Section 7.1.6, we studied the waiting time for service of a customer accepted by the system. We also derived expression (7.1.29) of $\boldsymbol{\omega}(s) = (\omega_1(s), \dots, \omega_l(s))^T$, where $\omega_i(s)$ is the Laplace–Stieltjes transform of the probability $W_i(t)$ that the waiting time is less than t and the customer generation process at the instant of arrival of a customer is in phase i.

Equality (7.1.29), by virtue of formula (7.2.1), reduces to

$$\boldsymbol{\omega}^T(s) = \frac{1}{\lambda(1-\pi)} \left[\boldsymbol{p}_0^T + \sum_{k=1}^{r} \beta^{k-1}(s) \int_0^\infty dx \int_0^\infty \boldsymbol{q}_k^T(x) e^{-su} d_u B(x+u) \right] N.$$

Hence, in view of expressions (7.2.9) of $\boldsymbol{q}_k(x)$ and relations (7.2.8), we obtain the expression of $\boldsymbol{\omega}(s)$

$$\boldsymbol{\omega}^T(s) = \frac{1}{\lambda_D} \Bigg\{ \boldsymbol{p}_0^T + \sum_{k=1}^{r} \beta^{k-1}(s) \sum_{j=0}^{k-1} \boldsymbol{q}_{k-j}^T [(-1)^{j+1} \beta(s)(sI+\Lambda)^{-1}(N(sI+\Lambda)^{-1})^j + \sum_{m=0}^{j} (-1)^m B_{j-m} (sI+\Lambda)^{-1}(N(sI+\Lambda)^{-1})^m] \Bigg\} N. \quad (7.2.33)$$

Note that the matrix $sI + \Lambda$ is invertible due to the assumptions we made concerning the matrices Λ and N.

From (7.2.33), using formula (7.1.30), we easily find the Laplace–Stieltjes transform $\omega(s)$ of the stationary distribution function of waiting time.

7.2.6. Stationary distribution of inter-departure times of served customers

Let us find the stationary distribution of intervals between the departures of served customers.

Let $D(t \mid (i,k))$ denote the conditional distribution function of the interval between the departures of successive served customers, provided that the system was in state (i,k) immediately after the completion of service of the first of two successive customers, and let $\delta(s \mid (i,k))$ be the Laplace–Stieltjes transform of this distribution function. The Laplace–Stieltjes transform $\delta(s)$ of the distribution function of the inter-departure times is given by the relation

$$\delta(s) = \sum_{k=0}^{r} \boldsymbol{p}_k^{*T} \boldsymbol{\delta}_k(s), \quad (7.2.34)$$

where $\boldsymbol{\delta}_k(s) = (\delta(s \mid (1,k)), \dots, \delta(s \mid (l,k)))^T$.

If the system is at the instant of departure of a served customer in the state (i,k), $i = 1, \dots, l$, $k = 1, \dots, r$, then the next customer quits the system in a time equal to his service time. Therefore,

$$\boldsymbol{\delta}_k(s) = \beta(s)\mathbf{1}, \qquad k = 1, \dots, r. \quad (7.2.35)$$

But if the system at the instant of departure stays in the state $(i, 0)$, $i = 1, \dots, l$, i.e., no customer is in the queue, then the time period up to the departure of a customer is the sum of two independent random variables: the time up to the arrival of a customer at an empty system and his service time. Let us find the Laplace–Stieltjes transform of this total time. To do this, we use (7.1.1). Setting $\boldsymbol{f}_0(x) = \mathbf{1} - F_0(x)\mathbf{1}$ and denoting by $\boldsymbol{\varphi}(s)$ the Laplace–Stieltjes transform of $\boldsymbol{f}_0(x)$, from (7.1.1) we obtain

$$\boldsymbol{\varphi}_0(s) = (sI - \Lambda)^{-1}\boldsymbol{\lambda}.$$

Hence

$$\boldsymbol{\delta}_0(s) = (sI - \Lambda)^{-1}\boldsymbol{\lambda}\beta(s). \tag{7.2.36}$$

Now, combining formulas (7.2.34)–(7.2.36), we obtain the final expression for $\delta(s)$

$$\delta(s) = [1 - p_0^* + \boldsymbol{p}_0^{*T}(sI - \Lambda)^{-1}\boldsymbol{\lambda}]\beta(s). \tag{7.2.37}$$

Formula (7.2.37) is identical to formula (4.5.53) for the Laplace–Stieltjes transform of the inter-departure intervals for the $PH/PH/1/r$ system (see Section 4.5). Therefore, formula (4.5.54) can be used to calculate the moments δ_j of order j for the distribution function $D(t)$.

7.3. LCFS discipline

Now we examine the $MAP/G/1/r$ system under the LCFS discipline with service interruption and resumption. Under this discipline, any customer, if there is at least one waiting place, is immediately taken up for service, superseding the customer under service, who is placed at the head of the queue. The service of the interrupted customer is resumed only after the departure of all customers that had arrived later.

Note that under the LCFS discipline, the customer sojourn time is the duration from the instant of his arrival at the system to the instant of completion of his service. This sojourn time, since the service may be interrupted, is not equal to the sojourn time of the customer at the server, i.e., the time during which he might have been served if no other customers had arrived at the system. In this case, it is convenient to refer to the sojourn time of a customer at the server as the customer length. As usual, the length of a customer does not depend on the lengths of other customers, arrival and service processes, and has some distribution function $B(x)$ with finite mean b. Furthermore, for the sake of simplicity, we assume that the density $b(x) = B'(x)$ exists.

The LCFS discipline is somewhat artificial. Nonetheless, it reveals potentialities envisaged in Lemma 1.4.1. The advantages of this discipline will be clear in the next chapter concerned with the analysis of the $MAP/G/1/\infty$ system.

7.3.1. Busy periods

In this section, we study $MAP/G/1/k$ systems with different buffer capacities k, $k = 0, \dots, r$.

Let $\Gamma(s \mid k)$ denote the matrix whose element $\Gamma_{ij}(s \mid k)$ is the Laplace–Stieltjes transform of the busy period of a system of buffer capacity k and the probability that the customer generation process is in phase j at the instant of termination of the busy period, provided that it was in phase i at the instant of commencement of the busy period. Let $\Gamma(s \mid x, k)$ denote a similar matrix but under the additional condition that the busy period begins with a customer of length x. For the sake of brevity, we refer to $\Gamma(s \mid k)$ and $\Gamma(s \mid x, k)$ simply as the Laplace–Stieltjes transform and will not mention the respective probability where this will not lead to ambiguity. For the sake of uniformity, we write

$$\Gamma(s \mid -1) = I. \tag{7.3.1}$$

In order to compute $\Gamma(s \mid x, k)$, we consider, along with a customer of length x, a customer of length $x + \Delta$. According to the LCFS discipline, the busy period that begins with a customer of length $x + \Delta$ consists of the busy period that begins with a customer of length x and the busy period that begins with a customer of length Δ, provided the customer generation process at the beginning of the latter busy period is in the phase in which it was after the completion of the busy period opened by the customer of length x.

During the time of service of the customer of length Δ, the busy period necessarily increases by Δ—the length of this customer (the Laplace–Stieltjes transform of such an increase is $e^{-s\Delta}$). Furthermore, during this time period the customer generation process with the probability $1 + \Lambda_{ii}\Delta + o(\Delta)$ does not change its phase i, with the probability $\Lambda_{ij}\Delta + o(\Delta)$, $j \neq i$, changes its phase from i to phase j without any new arrival, and with the probability $N_{ij}\Delta + o(\Delta)$ changes its phase from i to phase j with a new arrival. In the last case, the new arrival interrupts the service for the time equal to the busy period of the $MAP/G/1/k-1$ system (with Laplace–Stieltjes transform $\Gamma(s \mid k-1)$), and this interruption time is added to the length of the customer Δ.

Since the probability of all other events is of order $o(\Delta)$ and Laplace–Stieltjes transforms are multiplied when independent random variables are summed, by the total probability formula we obtain, the in matrix form,

$$\Gamma(s \mid x + \Delta, k) = \Gamma(s \mid x, k)e^{-s\Delta}(I + \Lambda\Delta + N\Gamma(s \mid k-1)\Delta) + o(\Delta).$$

Hence, expanding $e^{-s\Delta}$ into a series up to $o(\Delta)$, we obtain the differential equation

$$\Gamma'_x(s \mid x, k) = \Gamma(s \mid x, k)(\Lambda + N\Gamma(s \mid k-1) - sI), \qquad k = 0, \dots, r, \tag{7.3.2}$$

under the obvious initial condition

$$\Gamma(s \mid 0, k) = I. \tag{7.3.3}$$

Solving (7.3.2) under initial condition (7.3.3) yields

$$\Gamma(s \mid x, k) = e^{(\Lambda + N\Gamma(s \mid k-1) - sI)x}, \qquad k = 0, \dots, r. \tag{7.3.4}$$

Obviously, the Laplace–Stieltjes transform of $\Gamma(s \mid k)$ is defined by the formula

$$\Gamma(s \mid k) = \int_0^\infty \Gamma(s \mid x, k)\, dB(x), \qquad k = 0, \dots, r. \tag{7.3.5}$$

Formulas (7.3.1), (7.3.4), and (7.3.5) allow us to compute the matrix $\Gamma(s \mid k)$ recursively.

Let

$$F(x,k)=\Gamma(0\mid x,k), \qquad F(k)-\Gamma(0\mid k).$$

The element $F_{ij}(x,k)$ of the matrix $F(x,k)$ is the probability that the customer generation process is in phase j at the termination of the busy period of a system of buffer capacity k, provided it was in phase i at the commencement of the busy period and the busy period began with a customer of length x. The element $F_{ij}(k)$ of the matrix $F(k)$ has a similar meaning, but without the condition that the busy period begins with a customer of length x. So, from (7.3.1), (7.3.4), and (7.3.5) we obtain

$$F(x,k)=e^{(\Lambda+NF(k-1))x}, \qquad k=0,\dots,r,$$

$$F(-1)=I, \quad F(k)=\int_0^\infty F(x,k)\,dB(x), \quad k=0,\dots,r.$$

Expressions (7.3.1), (7.3.4), and (7.3.5) are also useful in calculating the moments of the busy period in the matrix form as

$$S^{(m)}(x,k)=(-1)^m\frac{\partial^m}{\partial s^m}\Gamma(s\mid x,k)\bigg|_{s=0}, \qquad S^{(m)}(k)=(-1)^m\frac{\partial^m}{\partial s^m}\Gamma(s\mid k)\bigg|_{s=0}.$$

Here the element $(S^{(1)}(x,k))_{ij}$ of the matrix $S^{(1)}(x,k)$ is the mean busy period of a system of buffer capacity k and probability that the customer generation process stays in phase j at the end of this busy period, provided it began with a customer of length x and the generation process was in phase i.

Indeed, for $k=0$ from (7.3.1), (7.3.4), and (7.3.5) we obtain

$$S^{(m)}(x,0)=x^m e^{\Lambda^* x}, \qquad S^{(m)}(0)=\int_0^\infty x^m e^{\Lambda^* x}\,dB(x).$$

Since the matrix $e^{\Lambda^* x}$ is stochastic, we have $e^{\Lambda^* x}\mathbf{1}=\mathbf{1}$ and

$$S^{(m)}(0)\mathbf{1}=\int_0^\infty x^m\,dB(x)\mathbf{1}=b^{(m)}\mathbf{1},$$

where $b^{(m)}$ is the mth moment of the customer length. Hence, in particular, the mth moment of the busy period of a system with no waiting room is finite if and only if the mth moment of the customer length is finite.

In order to find the moment $(S^{(m)}(x,k))$, $k=1,\dots,r$ (for this moment to exist, it is necessary and sufficient that the mth moment of the customer length be finite), we must differentiate (7.3.2) (with respect to s) and solve the differential equation (in x) thus obtained at the point $s=0$. Using $(S^{(1)}(x,k))$ as example, let us demonstrate how this is done. Carrying out the operations mentioned above, we obtain

$$(S^{(1)}(x,k))'=S^{(1)}(x,k)(\Lambda+NF(k-1))+F(x,k)(I+NS^{(1)}(k-1)).$$

Its solution is

$$S^{(1)}(x,k)=Ce^{(\Lambda+NF(k-1))x}+\int_0^x F(y,k)(I+NS^{(1)}(k-1))e^{(\Lambda+NF(k-1))(x-y)}dy$$

or, in view of the obvious initial condition $S^{(1)}(0,k)=0$,

$$S^{(1)}(x,k)=\int_0^x F(y,k)(I+NS^{(1)}(k-1))e^{(\Lambda+NF(k-1))(x-y)}dy.$$

Obviously, the matrix $S^{(m)}(k)$ is related to the matrix $S^{(m)}(x,k)$ by the formula

$$S^{(m)}(k)=\int_0^\infty S^{(m)}(x,k)\,dB(x).$$

Now let us introduce two matrices $T(k)$ and $T^*(k)$. The element $T_{ij}(k)$ of the matrix $T(k)$ is the mean busy time of a system of buffer capacity k staying in states when the customer generation process is in phase j, provided it was in phase i at the commencement of the busy period, and the element $T^*_{ij}(k)$ of matrix $T^*(k)$ is the mean busy period of a system of buffer capacity k in states when there are no customers in the queue and the customer generation process is in phase j, provided it was in phase i at the commencement of the busy period. A similar meaning is attributed to the matrices $T(x,k)$ and $T^*(x,k)$, but under an additional condition that the busy period begins with a customer of length x.

To find the matrix $T^*(x,k)$, we observe that under the LCFS discipline there is no queue during a busy period if and only if the customer that started the busy period is under service. By the instant when his served length reaches a value y, the customer generation process with the probability $F_{ij}(y,k)$ passes from state i to state j. Therefore, by the total expectation formula, we obtain

$$T^*(x,k)=\int_0^x F(y,k)\,dy,\qquad k=0,\dots,r. \tag{7.3.6}$$

The mean busy time of a system of buffer capacity k in states when the customer generation process is in phase j, provided it was in phase i at the commencement of the busy period, is the sum of the mean busy time of this system in states when no customers were in the queue and the customer generation process was in phase j and the mean busy time of the system in states when at least one customer was in the queue and the customer generation process was in phase j. At the instant when the served length of the customer that began the busy period reaches a value y, let the customer generation process be in phase m. Then in the time period in which the served length of this customer increases from y to $y+dy$ with the probability $N_{mj}dy$, a new customer may arrive interrupting the service of the first customer for a time equal to the busy period of a system of buffer capacity $k-1$. Since the mean time spent by the customer generation process in phase j in the busy periods of a system of waiting capacity $k-1$ is given by the matrix $T(k-1)$, using the total expectation formula and equality (7.3.6) we obtain

$$\begin{aligned}T(x,k)&=T^*(x,k)+\int_0^x F(y,k)NT(k-1)\,dy\\&=T^*(x,k)(I+NT(k-1)),\qquad k=0,\dots,r,\end{aligned} \tag{7.3.7}$$

where $T(-1)=0$. From (7.3.6) and (7.3.7) we easily derive the formulas

$$\begin{aligned}T^*(k)&=\int_0^\infty (1-B(x))F(x,k)\,dx,\qquad k=0,\dots,r,\\T(k)&=T^*(k)(I+NT(k-1)),\qquad k=0,\dots,r.\end{aligned} \tag{7.3.8}$$

7.3.2. The Markov process

As in the preceding section, we continue the study of our queueing system by the method of supplementary variables. But here we use the residual lengths of customers as supplementary variables. Furthermore, there can be several supplementary variables, as many as the number of customers in the system.

Let $\xi(t)$ be the phase of the customer generation process at instant t, and let $\nu(t)$ be the number of customers at this instant. Let the supplementary variables be $\eta_1(t), \eta_2(t), \ldots, \eta_{\nu(t)}(t)$, where $\eta_1(t)$ is the residual length of the customer at the server at instant t, $\eta_2(t)$ is the residual length of the customer at the first place in the queue, etc., and $\eta_{\nu(t)}(t)$ is the residual length of the customer at the last, $(\nu(t)-1)$th, place in the queue.

We set $\zeta(t) = (\xi(t), 0)$ if there are no customers in the system at instant t $(\nu(t) = 0)$ and $\zeta(t) = (\xi(t), \nu(t), \eta_1(t), \ldots, \eta_{\nu(t)}(t))$ if there are $\nu(t)$ customers, $\nu(t) \geq 1$.

Reasoning as in Section 5.3 for the $M/G/1/\infty$ system and recalling that the customer generation process is Markov, we easily see that the process $\{\zeta(t), t \geq 0)$ is Markov.

By the definition of the process $\{\zeta(t), t \geq 0)$, its state set is

$$\begin{aligned}\mathscr{X} = \{&(i, 0),\ i = 1, \ldots, l;\\ &(i, k, x_1, \ldots, x_k),\ x_j \geq 0,\ i = 1, \ldots, l,\ j = 1, \ldots, k,\ k = 1, \ldots, R\}.\end{aligned}$$

The states of the process $\{\zeta(t), t \geq 0)$ are interpreted as follows. In the state $(i, 0)$, the customer generation process is in phase i and no customers are in the system. In the state $(i, k, x_1, x_2, \ldots, x_k)$, the customer generation is in phase i, k customers exist in the system, the residual length of the customer at the server is x_1, and the residual lengths of the customers at the first, second, ... , and last, $(k-1)$th, place in the queue are $x_2, x_3, \ldots,$ x_k respectively.

We set

$$\begin{aligned}p_{i0}(t) &= \mathbf{P}\{\zeta(t) = (i, 0)\},\\ P_{ik}(x_1, \ldots, x_k, t) &= \mathbf{P}\{\xi(t) = i, \nu(t) = k, \eta_1(t) < x_1, \ldots, \eta_k(t) < x_k\}\end{aligned}$$

and introduce the vectors $\boldsymbol{p}_0(t)$, $\boldsymbol{P}_k(x_1, \ldots, x_k, t)$, $k = 1, \ldots, R$, whose coordinates are the probabilities $p_{i0}(t)$ and $P_{ik}(x_1, \ldots, x_k, t), i = 1, \ldots, l$.

It is known (Sevastyanov, 1957) that if the system is empty at the initial instant, then probability densities exist and can be expressed in vector form as

$$\boldsymbol{p}_k(x_1, \ldots, x_k, t) = \frac{\partial^k \boldsymbol{P}_k(x_1, \ldots, x_k, t)}{\partial x_1 \cdots \partial x_k}.$$

Since the mean customer length is finite, then, by the results of Section 7.1, the mean busy period is also finite. Hence, by Theorem 1.3.5, the process $\{\zeta(t), t \geq 0)$ is ergodic, i.e., the limit (stationary) probabilities exist and they are expressed in the vector form as

$$\begin{aligned}\boldsymbol{p}_0 &= \lim_{t\to\infty} \boldsymbol{p}_0(t),\\ \boldsymbol{P}_k(x_1, \ldots, x_k) &= \lim_{t\to\infty} \boldsymbol{P}_k(x_1, \ldots, x_k, t).\end{aligned}$$

Moreover, the stationary probabilities $P_{ik}(x_1, \dots, x_k)$ have probability densities which are expressed in vector form as

$$\boldsymbol{p}_k(x_1, \dots, x_k) = \frac{\partial^k \boldsymbol{P}_k(x_1, \dots, x_k)}{\partial x_1 \cdots \partial x_k}.$$

7.3.3. Auxiliary systems. Differential equations

We now introduce $R+1$ auxiliary systems. The k-system, $k = 1, \dots, R$, is defined as follows. In this system, the buffer capacity is $k-1$. So long as the number of customers in the system is less than k, the k-system coincides with the initial system. But if the number of customers is k (the server is busy and there is no free waiting places in the buffer), the infinitesimal matrix of the customer generation process is $\Lambda + NF(r-k)$. In view of (7.3.1), $F(-1) = \Gamma(0 \mid -1) = I$; therefore the R-system coincides with the initial system.

Obviously, the k-system is also described by the Markov process $\{\zeta^{(k)}(t),\ t \geq 0)$ with the state set

$$\mathscr{X}^{(k)} = \{(i,0),\ i = 1, \dots, l,\\ (i, m, x_1, \dots, x_m),\ x_j \geq 0,\ i = 1, \dots, l,\ j = 1, \dots, m,\ m = 1, \dots, k\}.$$

The processes $\{\zeta^{(k)}(t), t \geq 0)$ for some fixed $k = 1, \dots, r$ and $\{\zeta(t), t \geq 0)$ are related as follows. Let $t^{(k)}$ be the first instant when the process $\{\zeta(t),\ t \geq 0)$ attains the state in which the initial system contains k customers ($\nu(t^{(k)}) = k$), let $t_a^{(k)}$ be the first instant after $t^{(k)}$ when a new customer arrives at the initial system, and let $t_s^{(k)}$ be the first instant after $t_a^{(k)}$ when the process $\{\zeta(t),\ t \geq 0)$ attains the state in which k are customers in the system. According to the LCFS discipline, the service of the customer interrupted at the instant $t_a^{(k)}$ is resumed at the instant $t_s^{(k)}$. The time interval $(t_a^{(k)}, t_s^{(k)})$ is the busy period for an analogous system of buffer capacity $(r-k)$. Discarding this time interval, i.e., 'gluing' the instants $t_a^{(k)}$ and $t_s^{(k)}$, we consider only the possible change in the state of the customer generation process by the instant $t_s^{(k)}$, provided that the state of the process at the instant $t_a^{(k)}$ is known. The probabilities of such changes in the customer generation phases are defined by the matrix $F(r-k)$ described in the preceding section. Since $\{\zeta(t), t \geq 0\}$ and $\{\xi(t), t \geq 0\}$ are (homogeneous) Markov processes, the gluing yields a Markov process $\{\zeta'(t), t \geq 0\}$ which is stochastically equivalent to the process $\{\zeta^{(k)}(t),\ t \geq 0\}$ describing the behaviour of the k-system defined above.

We refer to a fictitious system described by a Markov process with state set $\{1, 2, \dots, l\}$ and infinitesimal matrix $\Lambda + NF(r)$ as the 0-system.

Let $\boldsymbol{p}_0^{(k)}(t)$, $\boldsymbol{p}_0^{(k)}$, $\boldsymbol{p}_m^{(k)}(x_1, \dots, x_m, t)$, $\boldsymbol{p}_m^{(k)}(x_1, \dots, x_m)$, $m = 1, \dots, k$, denote vectors that are similar to the vectors $\boldsymbol{p}_0(t)$, $\boldsymbol{p}_0$, $\boldsymbol{p}_k(x_1, \dots, x_k, t)$, $\boldsymbol{p}_k(x_1, \dots, x_k)$ of the initial system.

Obviously, the stationary probability vector $\boldsymbol{p}_0^{(0)}$ of the 0-system is found from the equilibrium equations

$$\boldsymbol{p}_0^{(0)T}(\Lambda + NF(r)) = \boldsymbol{0}^T. \tag{7.3.9}$$

Let us derive the differential equation for $\boldsymbol{p}_k^{(k)}(x_1, \dots, x_k, t)$, $k = 1, \dots, R$. It will be clear soon why we restrict ourselves to the equation for $\boldsymbol{p}_k^{(k)}(x_1, \dots, x_k, t)$.

Let us study the state of the k-system at the instant $t + \Delta$. The k-system at the instant $t + \Delta$ stays in the state $(i, k, x_1, \dots, x_k)$ in two cases:

(1) if it was at instant t in the state $(j, k, x_1 + \Delta, x_2, \dots, x_k)$ (with the probability density $p_{jk}^{(k)}(x_1 + \Delta, \dots, x_k, t - \Delta)$), and in time Δ the phase of the customer generation process did not change for $i = j$ (with probability $1 + (\Lambda + NF(r-k))_{ii}\Delta + o(\Delta)$) or the generation process passed from phase j to phase i for $i \neq j$ (with the probability $(\Lambda + NF(r-k))_{ji}\Delta + o(\Delta)$);

(2) if it was at instant t in the state $(j, k-1, x_2, \dots, x_k)$ (with the probability density $p_{j,k-1}^{(k)}(x_2, \dots, x_k, t - \Delta)$), and in time Δ the customer generation process passed from phase j to phase i and a customer of length x_1 arrived at the system (with the probability $b(x_1)N_{ji}\Delta + o(\Delta)$).

Since the probability of all other transitions is of order of magnitude $o(\Delta)$, we obtain

$$\boldsymbol{p}_k^{(k)T}(x_1, x_2, \dots, x_k, t) = \boldsymbol{p}_k^{(k)T}(x_1 + \Delta, \dots, x_k, t - \Delta)(I + [\Lambda + NF(r-k)]\Delta) + \boldsymbol{p}_{k-1}^{(k)T}(x_2, \dots, x_k, t - \Delta)b(x_1)N\Delta + o(\Delta).$$

Hence we arrive at the partial differential equation

$$\frac{\partial}{\partial t}\boldsymbol{p}_k^{(k)T}(x_1, \dots, x_k, t) - \frac{\partial}{\partial x_1}\boldsymbol{p}_k^{(k)T}(x_1, \dots, x_k, t) = \boldsymbol{p}_k^{(k)T}(x_1, \dots, x_k, t)[\Lambda + NF(r-k)] + b(x_1)\boldsymbol{p}_{k-1}^{(k)T}(x_2, \dots, x_k, t)N.$$

For the steady case, this equation reduces to an ordinary differential equation

$$-\frac{d}{dx_1}\boldsymbol{p}_k^{(k)T}(x_1, \dots, x_k) = \boldsymbol{p}_k^{(k)T}(x_1, \dots, x_k)[\Lambda + NF(r-k)] + b(x_1)\boldsymbol{p}_{k-1}^{(k)T}(x_2, \dots, x_k)N. \tag{7.3.10}$$

Recall that the process $\{\zeta^{(k)}(t),\ t \geq 0\}$ is stochastically equivalent to the process obtained from the process $\{\zeta(t) = \zeta^{(R)}(t),\ t \geq 0\}$ by discarding all instants at which the system contains more than k customers ($\nu(t) > k$) and gluing the residual parts of the process $\zeta(t)$. Therefore, by virtue of Corollary 1.4.1, the vectors of stationary state probabilities $\boldsymbol{p}_m^{(k)}(x_1, \dots, x_m)$ and $\boldsymbol{p}_m(x_1, \dots, x_m)$ for the processes $\{\zeta^{(k)}(t), t \geq 0\}$ and $\{\zeta(t), t \geq 0\}$ coincide for all $m = 0, \dots, k$ within a constant.

In particular, (7.3.9) and (7.3.10) hold true for $\boldsymbol{p}_0$ and $\boldsymbol{p}_k(x_1, \dots, x_k)$:

$$\boldsymbol{p}_0^T(\Lambda + NF(r)) = \boldsymbol{0}^T, \tag{7.3.11}$$

$$-\frac{d}{dx_1}\boldsymbol{p}_k^T(x_1, \dots, x_k) = \boldsymbol{p}_k^T(x_1, \dots, x_k)[\Lambda + NF(r-k)] + b(x_1)\boldsymbol{p}_{k-1}^T(x_2, \dots, x_k)N. \tag{7.3.12}$$

7.3.4. Stationary state probabilities

Equations (7.3.11) and (7.3.12) help to calculate recursively the stationary state probabilities for the Markov process $\{\zeta(t), t \geq 0)\}$.

Indeed, first the vector $\boldsymbol{p}_0$ is found from (7.3.11) up to a normalisation constant. Then (7.3.12) is solved in succession in k from $k = 1$ to $k = R$ up to a constant:

$$\begin{aligned} \boldsymbol{p}_k^T(x_1, \ldots, x_k) &= \boldsymbol{c}_k^T(x_2, \ldots, x_k) e^{-(\Lambda + NF(r-k))x_1} \\ &\quad + \boldsymbol{p}_{k-1}^T(x_2, \ldots, x_k) N \int_{x_1}^{\infty} e^{(\Lambda + NF(r-k))(x - x_1)} dB(x). \end{aligned} \quad (7.3.13)$$

The second term in the right-hand side of (7.3.13) tends to zero vector as $x_1 \to \infty$. The eigenvalues of the matrix $-(\Lambda + NF(r - k))$ have non-negative real parts; hence, for the vector $\boldsymbol{p}_k(x_1, \ldots, x_k)$ defined by (7.3.13) to be a probability density vector it is necessary and sufficient that $\boldsymbol{c}_k(x_2, \ldots, x_k) = \boldsymbol{0}$. Therefore,

$$\begin{aligned} \boldsymbol{p}_k^T(x_1, \ldots, x_k) &= \boldsymbol{p}_{k-1}^T(x_2, \ldots, x_k) N \int_{x_1}^{\infty} e^{(\Lambda + NF(r-k))(x - x_1)} dB(x) \\ &= \boldsymbol{p}_{k-1}^T(x_2, \ldots, x_k) N \int_0^{\infty} F(y, R-k) B(x_1 + dy). \end{aligned} \quad (7.3.14)$$

Thus, we have found $\boldsymbol{p}_0$ and $\boldsymbol{p}_k(x_1, \ldots, x_k)$ up to a constant which is found from the normalisation condition formulated below.

In actual practice, it usually suffices to know $\boldsymbol{p}_k(x) = \partial \boldsymbol{P}_k(x, \infty, \ldots, \infty)/\partial x$ or $\boldsymbol{p}_k = \boldsymbol{P}_k(\infty, \ldots, \infty)$. Integrating (7.3.14), in view of (7.3.8) we obtain

$$\begin{aligned} \boldsymbol{p}_k^T(x) &= \boldsymbol{p}_{k-1}^T N \int_0^{\infty} F(y, R-k) B(x_1 + dy), \\ \boldsymbol{p}_k^T &= \int_0^{\infty} \boldsymbol{p}_k^T(x)\, dx = \boldsymbol{p}_{k-1}^T N \int_0^{\infty} (1 - B(y)) F(y, R-k)\, dy = \boldsymbol{p}_{k-1}^T N T^*(R-k), \end{aligned}$$

which yields

$$\boldsymbol{p}_k^T = \boldsymbol{p}_0^T (NT^*(r))(NT^*(r-1)) \cdots (NT^*(R-k)), \qquad k = 1, \ldots, R. \quad (7.3.15)$$

The unknown constant is found from the normalisation condition

$$\boldsymbol{p}_{\cdot}^T \boldsymbol{1} = 1.$$

From the formulas thus obtained we can find different performance indices of the system. In particular, from (7.3.15) we easily find that the mean number $\overline{N}$ of customers in the system is

$$\begin{aligned} \overline{N} = \sum_{k=1}^{R} k \boldsymbol{p}_k^T \boldsymbol{1} = \boldsymbol{p}_0^T N T^*(r)(I + NT^*(r-1)(2I + NT^*(r-2)(3I + \cdots \\ + NT^*(1)(rI + RNT^*(0)) \ldots)))\boldsymbol{1}. \end{aligned}$$

Reasoning as in section 7.1, we prove formulas (7.1.21) and (7.1.25) for the stationary state probabilities $\boldsymbol{p}_k^-$, $k = 0, \ldots, R$, of the system at arrival instants. In particular, the stationary loss probability, as in Section 7.1, is defined by the formula

$$\pi = \frac{1}{\lambda} \boldsymbol{p}_R^T \boldsymbol{\lambda}.$$

In conclusion, we observe that formula (7.3.15) is the matrix analogue of the well-known formula for the stationary distribution of the number of customers in the $M/G/1/r$ system under the LCFS discipline with interruption and resumption of service

$$p_k = p_0 \rho^k, \qquad k = 1, \ldots, R,$$

which, in its turn, is the same as the analogous formula for the $M/M/1/r$ system with the FCFS discipline.

7.3.5. Stationary sojourn time distribution. Total work

Let $\boldsymbol{\varphi}_k(s)$ denote the vector whose coordinate $\varphi_{ik}(s)$ is the Laplace–Stieltjes transform of the sojourn time of a customer arriving at an instant when k customers are in the system in steady state and the customer generation process stays in phase i. Let $\boldsymbol{\varphi}_k(s \mid x)$ denote a similar vector but under the additional condition that this customer is of length x.

We easily see that under the LCFS discipline the sojourn time of a customer arriving at a system of buffer capacity r at an instant when k other customers exist in the system is the same as the busy period of a similar system of buffer capacity $(r-k)$. Therefore,

$$\boldsymbol{\varphi}_k(s \mid x) = \Gamma(s \mid x, r-k)\mathbf{1},$$
$$\boldsymbol{\varphi}_k(s) = \Gamma(s \mid r-k)\mathbf{1}.$$

As already noted, the stationary state distribution at the arrival instant of a customer is defined by formula (7.1.25). Therefore, the Laplace–Stieltjes transform $\varphi(s)$ of the stationary distribution of the sojourn time and the Laplace–Stieltjes transform $\varphi(s \mid x)$ of the analogous characteristic under the additional condition that the customer is of length x are

$$\varphi(s \mid x) = \frac{1}{1-\pi} \sum_{k=0}^{r} \boldsymbol{p}_k^- \Gamma(s \mid x, r-k)\mathbf{1} = \frac{1}{\lambda(1-\pi)} \sum_{k=0}^{r} \boldsymbol{p}_k^T N \Gamma(s \mid x, r-k)\mathbf{1},$$

$$\varphi(s) = \frac{1}{\lambda(1-\pi)} \sum_{k=0}^{r} \boldsymbol{p}_k^T N \Gamma(s \mid r-k)\mathbf{1}.$$

In particular, the mean sojourn time of a customer of length x is

$$v(x) = \frac{1}{\lambda(1-\pi)} \sum_{k=0}^{r} \boldsymbol{p}_k^T N T(x, r-k)\mathbf{1},$$

and that of an arbitrary customer is

$$v = \frac{1}{\lambda(1-\pi)} \sum_{k=0}^{r} \boldsymbol{p}_k^T N T(r-k)\mathbf{1}.$$

In this formula, replacing $\boldsymbol{p}_k$ and $T(k)$ by their expressions (7.3.15) and (7.3.8), we obtain

$$v = \frac{1}{\lambda(1-\pi)} \sum_{k=1}^{R} k \boldsymbol{p}_k^T \mathbf{1} = \frac{\overline{N}}{\lambda(1-\pi)};$$

hence we obtain the Little formula

$$\overline{N} = \lambda(1-\pi)v.$$

Using the results of Section 7.3.1, we find moments of any order of the customer sojourn time.

Let $\boldsymbol{W}_k(x)$, $k = 0, \ldots, R$, denote the vector whose coordinate $W_{ik}(x)$ is the stationary probability that the system contains k customers, whose total work (i.e., the total length of all customers in the system) is less than x and the customer generation process is in phase i. Let $\boldsymbol{\omega}_k(s)$ denote the Laplace–Stieltjes transform of the vector $\boldsymbol{W}_k(x)$. Obviously,

$$\begin{aligned}\boldsymbol{\omega}_0(s) &= \boldsymbol{p}_0^T, \\ \boldsymbol{\omega}_k(s) &= \underset{\mathbb{R}^{k+}}{\int \cdots \int} e^{-(x_1+\cdots+x_k)s} \boldsymbol{p}_k(x_1, \ldots, x_k)\, dx_1 \cdots dx_k.\end{aligned}$$

Using formula (7.3.14) and then integrating in succession, we obtain

$$\begin{aligned}\boldsymbol{\omega}_k^T(s) &= \underset{\mathbb{R}^{(k-1)+}}{\int \cdots \int} e^{-(x_2+\cdots+x_k)s} \boldsymbol{p}_{k-1}^T(x_2, \ldots, x_k)\, dx_2 \cdots dx_k R_k(s) \\ &= \cdots = \boldsymbol{p}_0^T R_1(s) \cdots R_k(s), \qquad k = 1, \ldots, R, \end{aligned} \tag{7.3.16}$$

where

$$R_k(s) = N \int_0^\infty e^{-sx} dx \int_x^\infty e^{(\Lambda + NF(r-k))(y-x)} dB(y). \tag{7.3.17}$$

For the sake of definiteness, we set $R_0(s) = I$.

Relation (7.3.17) admits another representation. Indeed, changing the order of integration, we obtain

$$\begin{aligned}R_k(s) &= N \int_0^\infty e^{(\Lambda+NF(r-k))y} dB(y) \int_0^y e^{-(sI+\Lambda+NF(r-k))x} dx \\ &= N \int_0^\infty e^{(\Lambda+NF(r-k))y} (I - e^{-(sI+\Lambda+NF(r-k))y})\, dB(y)(sI + \Lambda + NF(r-k))^{-1} \\ &= N \int_0^\infty (e^{(\Lambda+NF(r-k))y} - e^{-sy} I)\, dB(y)(sI + \Lambda + NF(r-k))^{-1} \\ &= N(\beta(-(\Lambda + NF(r-k)) - \beta(s)I)(sI + \Lambda + NF(r-k))^{-1}.\end{aligned}$$

Furthermore, from (7.3.17) we immediately derive the equalities

$$
\begin{aligned}
R_k(0) &= N\int_0^\infty dB(y)\int_0^y e^{(\Lambda+NF(r-k))(y-x)}dx \\
&= N\int_0^\infty dB(y)\int_0^y e^{(\Lambda+NF(r-k))t}dt = NT^*(R-k), \qquad k=1,\dots,R, \\
-R_k'(0) &= N\int_0^\infty dB(y)\int_0^y e^{(\Lambda+NF(r-k))(y-x)}dx, \qquad k=1,\dots,R.
\end{aligned}
\tag{7.3.18}
$$

Denoting the stationary probability that the total work is less than x (without account for the phase of the customer generation process and the number of customers in the system) by $W(x) = W_{\cdot,\cdot}(x)$, we derive its Laplace–Stieltjes transform $\omega(s)$. From (7.3.16) we obtain

$$
\begin{aligned}
\omega(s) &= \sum_{k=0}^{R}\boldsymbol{\omega}_k^T(s)\mathbf{1} = \sum_{k=0}^{R}\boldsymbol{p}_0^T R_1(s)\cdots R_k(s)\mathbf{1} \\
&= \boldsymbol{p}_0^T(I + R_1(s)(I + R_2(s)(I + \cdots + R_r(s)(I + R_R(s))\cdots)))\mathbf{1}.
\end{aligned}
$$

Using the last formula and (7.3.17), we find moments of any order of the stationary work in the system. Say, the mean stationary work is

$$w = -\omega'(0),$$

which, by virtue of (7.3.18), can be rewritten in a more convenient form as

$$w = \boldsymbol{p}_0^T Q_1\mathbf{1},$$

where Q_1 is defined by the recurrence relations

$$
\begin{aligned}
&D_R = I, && Q_R = -R_R'(0), \\
&D_k = I + NT^*(r-k)D_{k+1}, && Q_k = -R_k'(0)D_k + NT^*(R-k)Q_{k+1}, \quad k=1,\dots,r.
\end{aligned}
$$

7.4. Matrix exponential moments

An essential part of implementation of the algorithms described in the previous sections consists of finding the matrices B_k, B_k^*, T_k, T_k^*, $F(k)$, $T(k)$, $T^*(k)$, and other. In this section, we describe how to calculate them. Since these methods are almost identical for all matrices of such a kind, we restrict ourselves to the matrix exponential moments of B_k and B_0^* only. However, while computing the matrices entering formulas for moments of the mean sojourn time, it is necessary that service time have higher order moments.

Here we consider a number of methods and describe their merits and demerits.

7.4.1. Direct integration

The simplest method that does not require any explanation is direct integration of (differential or integral) equations which the matrices satisfy. Using the matrix B_0 as an example, we illustrate this method.

First, we consider the matrix $F_0(x)$. It solves the differential equation (7.2.8):

$$F_0'(x) = F_0(x)\Lambda.$$

There are a great body of numerical methods to solve first-order differential equations. The simplest is the computation of the values of the matrix $F(x)$ sequentially at the points $x_k = k\Delta$ by expanding it into a Taylor series up to the first term. In our case,

$$F_0(x_{k+1}) = F_0(x_k) + F_0(x_k)\Lambda\Delta,$$

where Δ is the integration step.

Let the matrix B_0 be computed by some numerical integration formula; the simplest among them is the rectangular formula

$$B_0 = \sum_{k=0}^{\infty}(B(x_{k+1}) - B(x_k))F_0(x_k).$$

Direct integration is satisfactory only if the service time is bounded by a number c not much greater than the mean service time (say, twice or thrice as large for uniform distribution). In the general case, the error is rather large and computations are cumbersome.

7.4.2. Power series expansion

The method consists of expanding the exponential function into a power series and changing the integration and summation order. Using the matrix B_0, let us illustrate this method.

Expanding the matrix $e^{\Lambda x}$ in the formula

$$B_0 = \int_0^{\infty} e^{\Lambda x}\, dB(x)$$

as a power series and changing the order of integration and summation, we obtain

$$B_0 = \int_0^{\infty}\sum_{k=0}^{\infty}\frac{(\Lambda x)^k}{k!}dB(x) = \sum_{k=0}^{\infty}\int_0^{\infty}\frac{(\Lambda x)^k}{k!}dB(x) = \sum_{k=0}^{\infty}\Lambda^k\frac{b^{(k)}}{k!},$$

where $b^{(k)}$ is the kth service moment.

But doing so is forbidden, because the series $\sum_{k=0}^{\infty}\Lambda^k b^{(k)}/k!$ may not converge absolutely for the simple reason that all moments, beginning from the second, of certain distributions may, generally speaking, be infinite (in the beginning of this chapter, we assumed that the first moment is finite). Therefore, we modify our reasoning and represent the matrix B_0 as

$$B_0 = \int_0^{\infty} e^{-ax}e^{aPx}dB(x),$$

where $P = I + a^{-1}\Lambda$ and a is positive. Now, if the number a is no less than the maximum of the absolute values of the diagonal elements of the matrix Λ, then the matrix P is substochastic. Hence the series

$$\sum_{k=0}^{\infty}\frac{a^k\|A\|^k}{k!}\int_0^{\infty} e^{-ax}x^k dB(x)$$

converges absolutely, where $\|A\|$ is some norm of the matrix A.

We observe that the larger the number a, the slower the convergence rate. Therefore, the most appropriate choice of a is the maximum of the absolute values of the diagonal elements of the matrix Λ.

We now describe a method to compute the matrices B_k and B_0^* in detail.

First, let us compute the matrices $F_k(x)$, $k \geq 0$. Now consider the matrix $F_0(x) = e^{\Lambda x}$. Let $a = \max_i |\Lambda_{ii}|$, and set

$$P = I + a^{-1}\Lambda. \tag{7.4.1}$$

Matrix (7.4.1) is the transition probability matrix of a terminating Markov chain.

From (7.1.1) and (7.4.1) we see that the matrix $F_0(x)$ admits the representation

$$F_0(x) = e^{-ax}\sum_{n=0}^{\infty}\frac{(ax)^n}{n!}F_{0n}, \tag{7.4.2}$$

where $F_{0n} = P^n$.

Substituting (7.4.2) into (7.1.2) for $k = 1$, we obtain

$$F_1(x) = e^{-ax}\sum_{n=1}^{\infty}\frac{(ax)^n}{n!}F_{1n},$$

where

$$F_{1n} = \sum_{l=0}^{n-1} P^l S P^{n-l-1}, \qquad S = a^{-1}N.$$

Repeating this procedure in the general case, with the use of induction we obtain

$$F_k(x) = e^{-ax}\sum_{n=k}^{\infty}\frac{(ax)^n}{n!}F_{kn}, \tag{7.4.3}$$

where

$$F_{kn} = \sum_{l=0}^{n-k} P^l S F_{k-1,n-l-1}, \qquad n \geq k \geq 1. \tag{7.4.4}$$

For the matrix F_{kn}, we derive from (7.4.2) and (7.4.4) the recurrence relations

$$\begin{aligned} F_{0n} &= P^n, && n \geq 0,\\ F_{kk} &= F_{k-1,k-1}S, && k \geq 1,\\ F_{kn} &= F_{k,n-1}P + F_{k-1,n-1}S, && n \geq k+1, \quad k \geq 1. \end{aligned} \tag{7.4.5}$$

Validity of relations (7.4.3) and (7.4.5) can be proved by substituting expressions (7.4.3) of $F_k(x)$, with account for (7.4.1), into (7.2.1).

Similarly, assuming that

$$a^* = \max_i |\Lambda^*_{ii}|, \qquad P^* = I + (a^*)^{-1}\Lambda^*,$$

we obtain

$$F^*_0(x) = e^{\Lambda^* x} = e^{-a^* x} \sum_{n=0}^{\infty} \frac{(a^* x)^n}{n!} F^*_{0n},$$

where

$$F^*_{0n} = (P^*)^n, \qquad n \geq 0.$$

Now we consider the matrix B_k. Substituting expression (7.4.3) of $F_k(x)$ into the equality

$$B_k = \int_0^\infty F_k(x)\, dB(x),$$

we conclude that the matrix B_k admits the representation

$$B_k = \sum_{n=k}^{\infty} F_{kn}\beta_n, \qquad k = 0, \ldots, r, \tag{7.4.6}$$

where β_n is the nth-order exponential moment of the distribution function $B(x)$:

$$\beta_n = \int_0^\infty e^{-ax} \frac{(ax)^n}{n!} dB(x), \qquad n \geq 0.$$

Similarly, for the matrix B^*_0 we obtain

$$B^*_0 = \sum_{n=0}^{\infty} F^*_{0n}\beta^*_n,$$

where

$$\beta^*_n = \int_0^\infty e^{-a^* x} \frac{(a^* x)^n}{n!} dB(x), \qquad n \geq 0.$$

Using series (7.4.6), we compute the matrix B_k as follows: for a given $\varepsilon > 0$, we first find n_0 such that

$$\sum_{n=0}^{n_0} \beta_n \geq 1 - \varepsilon.$$

Summation in the right-hand side of series (7.4.6) halts when n equals n_0 and the n_0th term is included in the partial sum.

In this case, the remainder term of the matrix series (7.4.6), by virtue of the probabilistic meaning of β_n and matrices P and S that define the matrices F_{kn} in terms of relations (7.4.5), is estimated as

$$\left\| \sum_{n=n_0+1}^{\infty} F_{kn}\beta_n \right\| < \varepsilon.$$

A similar procedure is used to find the matrix B_0^*.

The main disadvantage of the method is that the expansion into an infinite power series involves a large number of computations and (scalar) exponential moments of the distribution function $B(x)$ are to be computed.

7.4.3. Rational Laplace–Stieltjes transform

Imposing certain additional constraints on the distribution function $B(x)$, we can find explicit expressions of the matrices B_k.

Let the distribution function $B(x)$ have a rational Laplace–Stieltjes transform. Then (see Section 1.6.2) $B(x)$ admits the representation

$$B(x) = 1 - \boldsymbol{\beta}^T e^{-Gx}\mathbf{1}, \qquad x > 0, \tag{7.4.7}$$

where G is a matrix of order m with eigenvalues having positive real parts and $\boldsymbol{\beta}$ is a vector of dimensionality m. Let $M = -G$ and $\boldsymbol{\mu} = -M\mathbf{1}$. Then expression (7.4.7) of the distribution function $B(x)$ reduces to

$$B(x) = 1 - \boldsymbol{\beta}^T e^{Mx}\mathbf{1}, \qquad x > 0.$$

In this case,

$$\begin{aligned} B_k &= \int_0^\infty F_k(x)\,dB(x) = \int_0^\infty F_k(x)\boldsymbol{\beta}^T e^{Mx}\boldsymbol{\mu}\,dx \\ &= (I \otimes \boldsymbol{\beta}^T)\int_0^\infty (F_k(x) \otimes e^{Mx})\,dx(I \otimes \boldsymbol{\mu}), \end{aligned} \tag{7.4.8}$$

where $A \otimes B$ is the Kronecker product of the matrices A and B. We set

$$\varphi_k = \int_0^\infty (F_k(x) \otimes e^{Mx})\,dx, \qquad k \geq 0. \tag{7.4.9}$$

Therefore, by virtue of (7.4.8), we obtain

$$B_k = (I \otimes \boldsymbol{\beta}^T)\varphi_k(I \otimes \boldsymbol{\mu}). \tag{7.4.10}$$

Thus, in order to calculate the matrices B_k, we must find the matrix φ_k. For this purpose, consider (7.4.9) for $k = 0$. Then from (7.1.1) we obtain

$$\varphi_0 = \int_0^\infty (e^{\Lambda x} \otimes e^{Mx})\,dx = \int_0^\infty e^{(\Lambda \oplus M)x}dx = -(\Lambda \oplus M)^{-1}, \tag{7.4.11}$$

where

$$\Lambda \oplus M = \Lambda \otimes I + I \otimes M$$

is the Kronecker sum of the matrices Λ and M. Invertibility of the matrix $\Lambda \oplus M$ is implied by the properties of the Kronecker product of matrices and stability of the matrices Λ and M. Furthermore, for $k = 1$, from (7.4.9) and (7.1.2), by virtue of (7.4.11), we obtain

$$\begin{aligned} \varphi_1 &= \int_0^\infty dx \left(\int_0^x F_0(y) N e^{\Lambda(x-y)} dy \otimes e^{Mx} \right) \\ &= \int_0^\infty (F_0(y) \otimes e^{My})\,dy(N \otimes I) \int_0^\infty (e^{\Lambda x} \otimes e^{Mx})\,dx = \varphi_0(N \otimes I)\varphi_0. \end{aligned}$$

Similarly, by with the use of induction, we obtain

$$\varphi_k = \varphi_{k-1}(N \otimes I)\varphi_0, \qquad k \geq 1. \tag{7.4.12}$$

From (7.4.10)–(7.4.12) we easily derive explicit expressions for the matrices B_k:

$$B_k = (-1)^{k+1}(I \otimes \boldsymbol{\beta}^T)[(\Lambda \oplus M)^{-1}(N \otimes I)]^k(\Lambda \oplus M)^{-1}(I \otimes \boldsymbol{\mu}).$$

Similarly we obtain an explicit expression for B_0^*:

$$B_0^* = -(I \otimes \boldsymbol{\beta}^T)(\Lambda^* \oplus M)^{-1}(I \otimes \boldsymbol{\mu}).$$

Invertibility of the matrix $\Lambda^* \oplus M$ is again implied by the properties of the Kronecker product of the matrices, semi-stability of the matrix Λ^*, and stability of the matrix M.

The main disadvantage is that this method is applicable only to queueing systems in which the service time has a rational Laplace–Stieltjes transform. Moreover, the complexity of the method rapidly increases as m grows.

7.4.4. Reduction to diagonal form

In order to compute matrices of the type

$$B_0 = \int_0^\infty e^{\Lambda x} dB(x), \tag{7.4.13}$$

to which we actually restricted ourselves in the study of $MAP/G/1/r$ systems with LCFS discipline, we use the method based on the reduction of matrices to the diagonal form.

From linear algebra we know that any matrix Λ can be reduced with some non-degenerate linear transformation A to the diagonal form (in the general case, to the Jordan normal form, but we will not study such a case):

$$\Lambda = ADA^{-1},$$

where

$$D = \begin{pmatrix} \sigma_1 & 0 & \dots & 0 \\ 0 & \sigma_2 & \dots & 0 \\ \dots & \dots & \dots & \dots \\ 0 & 0 & \dots & \sigma_l \end{pmatrix},$$

and σ_k, $k = 1, \dots, l$, are the eigenvalues of the matrix Λ with negative real parts. Then

$$e^{\Lambda x} = A \begin{pmatrix} e^{\sigma_1 x} & 0 & \dots & 0 \\ 0 & e^{\sigma_2 x} & \dots & 0 \\ \dots & \dots & \dots & \dots \\ 0 & 0 & \dots & e^{\sigma_l x} \end{pmatrix} A^{-1}$$

and

$$B_0 = A \begin{pmatrix} \beta(-\sigma_1) & 0 & \dots & 0 \\ 0 & \beta(-\sigma_2) & \dots & 0 \\ \dots & \dots & \dots & \dots \\ 0 & 0 & \dots & \beta(-\sigma_l) \end{pmatrix} A^{-1}.$$

Thus, computation of the matrix B_0 reduces to finding the eigenvalues σ_k and the eigenvectors of thc matrix Λ and computing the Laplace–Stieltjes transform $\beta(s)$ at certain points σ_k, i.e., to problems that can be solved by well-known numerical methods.

The disadvantage is that this method is applicable only to matrices of type (7.4.13) and yields good results only if the matrices Λ are of moderate order.

8

$MAP/G/1/\infty$ system

In this chapter, we study the $MAP/G/1/\infty$ system, i.e., a queueing system with Markov input flow and infinite buffer. In some sense, it is easier to study this system than the $MAP/G/1/r$ system with a finite buffer. This is not surprising, because we had encountered such a situation in studying queueing systems with Poisson input flow.

We use the same notation and assumptions concerning the parameters of the system as in the preceding chapter. For stationary probabilities, we assume that the traffic intensity $\rho = b\lambda = b\pi^T\boldsymbol{\lambda} = b\pi^T N\mathbf{1}$ of the system is less than one. This constraint is necessary and sufficient for existence of a steady operation mode of the system. Although the proof of this intuitive fact will be given in Section 8.1 for an embedded Markov chain with the use of apparatus of Markov renewal processes theory (see Section 1.6.1), it can be easily extended to other processes considered in this chapter.

As in the preceding chapter, we consider two service disciplines, FCFS and LCFS with service interruption and resumption. We analyse the $MAP/G/1/\infty$ system with the LCFS discipline for the following reasons. As observed in the preceding chapter, for a similar system of finite buffer capacity, it is simple and easy to find characteristics associated with the customer sojourn time. But if the buffer capacity is infinite, the total work introduced in the preceding chapter for a system with the LCFS discipline is the same as the virtual waiting time in the system with the FCFS discipline. Therefore, using the LCFS discipline, we can illustrate the potentialities of an alternative approach to studying the time characteristics of operation of a system under the FCFS discipline.

Since most of the reasoning repeats that of the preceding chapter word for word, we will not present a full-depth analysis.

8.1. Embedded Markov chain

In this section, we study the method of embedded Markov chains. For the investigation purposes, a system of infinite buffer capacity differs from a system of finite buffer capacity in that the number of states of the former is denumerable and that of the latter is finite.

8.1.1. Construction of an embedded Markov chain

As in Section 7.1, an embedded Markov chain is formed by the sequence $\{\zeta_n, n \geq 0\}$, where $\zeta_n = (\xi_n, \nu_n)$, and ξ_n and ν_n are, respectively, the phase of the customer generation process and the number of customers in the system immediately after the instant of completion of service of the nth customer. Obviously, the state set $\mathscr{X}$ of the Markov chain for a system of infinite capacity is of the form $\mathscr{X} = \{(i, k), i = 1, \dots, l, k \geq 0\}$; the state (i, k) is interpreted as in Section 7.1.

Reasoning as in Section 7.1, and using the same notation, we easily obtain the expression

$$P = \begin{pmatrix} BB_0 & BB_1 & BB_2 & BB_3 & \dots \\ B_0 & B_1 & B_2 & B_3 & \dots \\ 0 & B_0 & B_1 & B_2 & \dots \\ 0 & 0 & B_0 & B_1 & \dots \\ \vdots & \vdots & \vdots & \vdots & \ddots \end{pmatrix}$$

for the transition probability matrix of the embedded Markov chain.

Here and in what follows, the matrices B_k^* will not be encountered, as a consequence of the infinite capacity of the buffer.

8.1.2. Stationary probabilities of the embedded Markov chain

We assume that the embedded Markov chain $\{\zeta_n, n \geq 0\}$ is ergodic. As in Chapter 7, let p_{ik}^* denote the stationary probability that the customer generation process at the instant of completion of service is in phase i and the system contains k customers, and let $\boldsymbol{p}_k^*$ denote the corresponding vector.

In order to find the stationary distribution of the probabilities $\boldsymbol{p}_k^*$, $k \geq 0$, we introduce the matrix G whose element G_{ij} is the probability that the customer generation process at the end of a busy period in the $MAP/G/1/\infty$ system is in phase j, provided that it was in phase i at the beginning of this busy period.

In order to calculate the matrix G, we first consider a busy period starting with k, $k \geq 2$, customers (the k-busy period). Let $G^{(k)}$ denote the matrix whose element $G_{ij}^{(k)}$ is the probability that the customer generation process at the end of the k-busy period is in phase j, provided that it was in phase i at the beginning of this busy period. Let τ_m, $m = 1, \dots, k$, denote successive instants at which $k - m$ customers remain in the system for the first time. Obviously, for a system of infinite buffer, the times $\tau_m - \tau_{m-1}$, where τ_0 is the instant of beginning of the k-busy period, are ordinary busy periods. Therefore, the k-busy period is the sum of k conditionally independent busy periods (provided that the phase of the generation process at the beginning of a busy period is given) and, in particular,

$$G^{(k)} = G^k. \tag{8.1.1}$$

In the course of service of a customer that opens a (usual) busy period, k new customers arrive, and the customer generation process passes from phase i to phase j with the probability $(B_k)_{ij}$. Therefore, the residual time to the end of this busy period is a k-busy period. Hence from (8.1.1) by the total probability formula we obtain

$$G = \tilde{B}(G), \tag{8.1.2}$$

where

$$\tilde{B}(A) = \sum_{k=0}^{\infty} B_k A^k \tag{8.1.3}$$

is the matrix generating function of the matrix sequence $\{B_k,\ k \geq 0\}$.

At the end of this section, we will describe a numerical method to solve (8.1.2) and prove that the matrix G is stochastic if $\rho < 1$.

In Section 8.4, another functional equation for the matrix G will be derived in terms of matrix exponential moments.

In order to calculate the stationary probabilities of the embedded Markov chain, as in Section 7.1, we introduce more Markov chains, now for all non-negative integers k. The Markov chain $\{\zeta_n^{(k)},\ n \geq 0\}$ with the state set $\mathscr{X}^{(k)} = \{(i, m), i = 1, \ldots, l, m = 0, \ldots, k\}$ is obtained from the embedded Markov chain $\{\zeta_n,\ n \geq 0\}$ by discarding the instants at which $\nu_n > k$ and then gluing the remaining segments of the trajectory of the chain. Note In distinction to Section 7.1, here we use a different (reverse) numeration for the Markov chains obtained after gluing.

Reasoning as in Section 7.1, we see that the transition probability matrix $P^{(k)}$ of the Markov chain $\{\zeta_n^{(k)},\ n \geq 0\}$ consists of the elements

$$\begin{aligned} P_{mn}^{(k)} &= P_{mn}, && k \geq 1, \quad m = 0, \ldots, k, \quad n = 0, \ldots, k-1, \\ P_{0k}^{(k)} &= B\tilde{B}_k, && k \geq 0, \\ P_{mk}^{(k)} &= \tilde{B}_{k-m+1}, && k \geq 1, \quad m = 1, \ldots, k, \end{aligned} \tag{8.1.4}$$

where

$$\tilde{B}_k = \sum_{i=0}^{\infty} B_{k+i} G^i, \qquad k \geq 0. \tag{8.1.5}$$

Hence, by virtue of Corollary 1.4.1, the stationary probabilities $\boldsymbol{p}_k^*$, $k \geq 0$, of the embedded Markov chain $\{\zeta_n,\ n \geq 0\}$ can be found up to a normalisation factor from the equations

$$\begin{aligned} \boldsymbol{p}_0^{*T} &= \boldsymbol{p}_0^{*T} P_{00}^{(0)}, \\ \boldsymbol{p}_k^{*T} &= \boldsymbol{p}_k^{*T} P_{kk}^{(k)} + \sum_{m=0}^{k-1} \boldsymbol{p}_m^{*T} P_{mk}^{(k)}, \qquad k \geq 1, \end{aligned} \tag{8.1.6}$$

which for fixed k are the equilibrium equations for the vector $\boldsymbol{p}_k^*$ of the Markov chain $\{\zeta_n^{(k)},\ n \geq 0\}$, $k \geq 0$. Hence, by virtue of (8.1.5), we obtain the following recurrence relations for $\boldsymbol{p}_k^*$, $k \geq 0$:

$$\begin{aligned} \boldsymbol{p}_0^{*T} &= \boldsymbol{p}_0^{*T} BG, \\ \boldsymbol{p}_k^{*T} &= \sum_{m=1}^{k-1} \boldsymbol{p}_m^{*T} \tilde{B}_{k-m+1} (I - \tilde{B}_1)^{-1}, \qquad k \geq 1. \end{aligned} \tag{8.1.7}$$

Invertibility of the matrix $I - \tilde{B}_1$ follows from the fact that $\tilde{B}_1$ is the principal minor of the indecomposable transition probability matrix $P^{(k)}$ (Kemeny and Snell, 1960).

The normalisation constant can be found from the normalisation condition. But we will see soon that, which is common for systems of infinite buffer capacity, the normalisation constant can be easily expressed in terms of the traffic intensity ρ.

8.1.3. Time-stationary distribution of the queue

As in Section 7.1.3, we find the matrices T_k and T_k^*, $k \geq 0$, whose elements $(T_k)_{ij}$ and $(T_k)_{ij}^*$ are the mean sojourn times of the system in the course of service of a customer in the states in which the buffer contains exactly k customers and at least k customers, respectively, and the customer generation process is in phase j, provided that at the initial instant it was in phase i and the service of the customer that had arrived at an empty system has begun. Let us also introduce the matrix T^* whose element T_{ij}^* is the mean sojourn time of the system in an idle period in a state in which the customer generation process exists in phase j, provided that the process at the beginning of the idle period was in phase i.

In the steady state of the system, the mean interval T between successive instants generating the embedded Markov chain is

$$T = \left(\boldsymbol{p}_0^{*T} T^* + \sum_{k=0}^{\infty} \boldsymbol{p}_k^{*T} T_0^* \right) \mathbf{1}.$$

This equality, by virtue of the properties of matrix T_0^* (see Section 7.1.3), reduces to

$$T = \boldsymbol{p}_0^{*T} T^* \mathbf{1} + b. \tag{8.1.8}$$

Let p_{ik} be the probability that at any instant in the system in steady state there are k customers and the customer generation process is in phase i. We set $\boldsymbol{p}_k = (p_{1k}, \ldots, p_{lk})^T$. We recall that in our notation the probabilities p_{ik} are referred to as the time-stationary state probabilities.

Using the apparatus of the theory of Markov renewal processes, we find that the time-stationary probabilities $\boldsymbol{p}_k$ are

$$\begin{aligned} \boldsymbol{p}_0^T &= \frac{1}{T} \boldsymbol{p}_0^{*T} T^*, \\ \boldsymbol{p}_k^T &= \frac{1}{T} \left(\boldsymbol{p}_0^{*T} B T_{k-1} + \sum_{m=1}^{k} \boldsymbol{p}_m^{*T} T_{k-m} \right), \qquad k \geq 1. \end{aligned} \tag{8.1.9}$$

For a system with infinite buffer in steady state, the output flow intensity λ coincides with the intensity of service completion instants, which, in turn, is $1/T$. Therefore, multiplying both sides of equality (8.1.8) by λ and recalling the expression of the vector $\boldsymbol{p}_0^{*T} T^*$ in terms of the vector $\boldsymbol{p}_0$ from formula (8.1.9), we obtain

$$p_0 = \boldsymbol{p}_0^T \mathbf{1} = 1 - \rho. \tag{8.1.10}$$

This relation is well known. In particular, it implies that the condition $\rho < 1$ is necessary for existence of the steady state of our system.

8.1.4. Stationary distribution of the Markov process describing the system functioning

Denoting the number of customers in the system at instant t by $\nu(t)$, the phase in which the customer generation process is at this instant by $\xi(t)$, and the time that has elapsed since the beginning of service of the customer at instant t by $\eta(t)$ (if there are customers in the system), we introduce, as in the preceding chapter, the linearwise Markov process $\{\zeta(t), t \geq 0\}$, where $\zeta(t) = (\xi(t), \nu(t))$ if there are no customers in the system, and the process $\zeta(t) = (\xi(t), \nu(t), \eta(t))$ otherwise.

The state set of the process $\{\zeta(t),\ t \geq 0\}$ is of the form $\mathscr{X} = \{(i, 0), i = 1, \ldots, l;\ (i, k, x), i = 1, \ldots, l, k \geq 0, x \geq 0\}$. Naturally, the states of the process $\{\zeta(t),\ t \geq 0\}$ are interpreted in the same way as in the preceding chapter. We also retain the notation of Chapter 7, i.e., p_{i0} for the stationary probabilities of the states $(i, 0)$, $p_{ik}(x)$ for the stationary probability densities of the states (i, k, x), and $\boldsymbol{p}_0$ and $\boldsymbol{p}_k(x)$ for the corresponding vectors. Then the stationary probability vector $\boldsymbol{p}_0$ for the process $\{\zeta(t), t \geq 0\}$ is defined by the first equality in (8.1.9) and the vectors of stationary probability densities $\boldsymbol{p}_k(x)$, $k \geq 1$, are defined by the formula

$$\boldsymbol{p}_k^T(x) = \frac{1 - B(x)}{T}\left(\boldsymbol{p}_0^{*T} B F_{k-1}(x) + \sum_{m=1}^{k} \boldsymbol{p}_m^{*T} F_{k-m}(x)\right), \qquad k \geq 1. \tag{8.1.11}$$

These relations are proved exactly as Section 7.1.4.

8.1.5. Generating functions and Laplace–Stieltjes transforms: applications

Let us demonstrate how to calculate the stationary probability distribution of the embedded Markov chain in terms of generating functions.

We introduce the generating functions

$$F(z, x) = \sum_{k=0}^{\infty} z^k F_k(x), \quad \tilde{B}(z) = \sum_{k=0}^{\infty} z^k B_k, \quad \boldsymbol{P}^*(z) = \sum_{k=0}^{\infty} z^k \boldsymbol{p}_k^*, \quad |z| \leq 1.$$

It is easily seen that the elements $F_{ij}(z, x)$ and $\tilde{B}_{ij}(z)$ of the matrices $F(z, x)$ and $\tilde{B}(z)$, respectively, are the generating functions of the number of customers arriving at the system in a time interval x and in the course of service of a customer with account for the fact that the phase of the customer generation process changes from i to j and the coordinate $P_i^*(z)$ of the vector $\boldsymbol{P}^*(z)$ is the generating function of the stationary distribution of the number of customers in the system at the instant of completion of service with account for the phase i of the customer generation process. We recall that that $F(1, x)$ and $\tilde{B}(1)$ are the transition probability matrices of the customer generation process in the time interval x and in the course of service of a customer. Therefore, they are stochastic and indecomposable, and $\boldsymbol{\pi}$ is their unique (left) eigenvector corresponding to the eigenvalue one.

Multiplying the kth equation in (7.2.8) by z^k and summing over $k = 0, 1 \ldots$, we obtain the differential equation for $F(z, x)$

$$\frac{\partial}{\partial x} F(z, x) = F(z, x)(\Lambda + zN). \tag{8.1.12}$$

Its solution, in view of the obvious initial condition

$$F(z,0) = I, \tag{8.1.13}$$

is

$$F(z,x) = e^{(\Lambda+zN)x}. \tag{8.1.14}$$

Hence we obtain the formula for $\tilde{B}(z)$

$$\tilde{B}(z) = \int_0^\infty F(z,x)\,dB(x) = \int_0^\infty e^{(\Lambda+zN)x}dB(x) = \beta(-\Lambda - zN). \tag{8.1.15}$$

Let us present the simultaneous equilibrium equations for the stationary probabilities $\boldsymbol{p}_k^*$, $k \geq 0$, of the embedded Markov chain:

$$\boldsymbol{p}_k^{*T} = \boldsymbol{p}_0^{*T} B B_k + \sum_{m=1}^{k+1} \boldsymbol{p}_m^{*T} B_{k+1-m}, \qquad k \geq 0. \tag{8.1.16}$$

Multiplying the kth equation in this set by z^k and then summing over $k = 0, 1 \ldots,$ we obtain

$$\boldsymbol{P}^{*T}(z) = \boldsymbol{p}_0^{*T} B \tilde{B}(z) + \frac{1}{z}(\boldsymbol{P}^{*T}(z) - \boldsymbol{p}_0^{*T})\tilde{B}(z),$$

or

$$\boldsymbol{P}^{*T}(z)(zI - \tilde{B}(z)) = \boldsymbol{p}_0^{*T}(zB - I)\tilde{B}(z), \tag{8.1.17}$$

where $B = -\Lambda^{-1}N$. From (8.1.17) we find that

$$\boldsymbol{P}^{*T}(z) = \boldsymbol{p}_0^{*T}(I - zB)\tilde{B}(z)(\tilde{B}(z) - zI)^{-1}. \tag{8.1.18}$$

Formula (8.1.18) is the matrix analogue of the Pollaczek–Khinchin formula (5.2.1).

We observe that the determinant of the matrix $\tilde{B}(z) - zI$ vanishes at certain points on the unit circle $|z| \leq 1$. This is of help in applying the Pollaczek–Khinchin formula to deriving one more method of calculation of the vector $\boldsymbol{p}_0^*$ based on the analytic properties of the generating function $\boldsymbol{P}^*(z)$. Since serious technical difficulties arise while applying this method, we restrict ourselves only to a brief description of the method under certain simplifying assumptions without rigorous mathematical proof.

The equation $\det(zI - \tilde{B}(z)) = 0$ has, within the circle $|z| \leq 1$, exactly l roots $z_1, \ldots, z_l$; furthermore, one of them (say, z_l) is equal to 1, whereas others are less than 1 in absolute value. Let us assume that z_i are distinct and a unique (right) vector $\boldsymbol{x}_i$ corresponds to each of them so that $(z_i I - \tilde{B}(z_i))\boldsymbol{x}_i = \boldsymbol{0}$.

Multiplying relation (8.1.17) from the right by the vector $\boldsymbol{x}_i$, we obtain

$$\boldsymbol{P}^{*T}(z_i)(z_i I - \tilde{B}(z_i))\boldsymbol{x}_i = \boldsymbol{p}_0^{*T}(z_i B - I)\tilde{B}(z_i)\boldsymbol{x}_i.$$

The left-hand side of this equality is zero, and hence, the right-hand side is zero as well. Therefore, the vector $\boldsymbol{p}_0^*$ solves the equation

$$\boldsymbol{p}_0^{*T}(z_i B - I)\tilde{B}(z_i)\boldsymbol{x}_i = 0.$$

There are $(l-1)$ such equations, because the vector $\boldsymbol{x}_l = \mathbf{1}$ corresponds to the solution $z_l = 1$ and it is an eigenvector (corresponding to the eigenvalue 1) for any stochastic matrix, even the matrices I, B, and $\tilde{B}(1)$. The additional condition that allows us to determine the vector $\boldsymbol{p}_0^*$ uniquely, as usual, is the normalisation condition.

Proceeding along similar lines, we derive formulas for the generating function of the time-stationary distribution $\{\boldsymbol{p}_k,\ k \geq 0\}$ and the generating function for the sequence $\{\boldsymbol{p}_k(x), k \geq 1\}$ of stationary densities of the Markov process describing the functioning of the system.

Indeed, introducing the generating functions

$$\tilde{T}(z) = \sum_{k=0}^{\infty} z^k T_k, \qquad \boldsymbol{P}(z) = \sum_{k=0}^{\infty} z^k \boldsymbol{p}_k$$

and making use of results of Section 7.1.3, we obtain

$$T_k = \int_0^\infty (1 - B(y)) F_k(y)\, dy, \qquad k \geq 0.$$

Multiplying both sides of this equality by z^k and summing over $k = 0, 1 \ldots$, in view of (8.1.14) we obtain

$$\tilde{T}(z) = \int_0^\infty e^{(\Lambda + zN)x}(1 - B(x))\, dx. \tag{8.1.19}$$

Integrating (8.1.19) by parts, we obtain the expression of $\tilde{T}(z)$

$$\tilde{T}(z) = (\tilde{B}(z) - I)(\Lambda + zN)^{-1}. \tag{8.1.20}$$

Multiplying the kth equation in (8.1.9) by z^k and summing over $k = 1, 2 \ldots$, by virtue of the first equation in (8.1.9) we obtain

$$\boldsymbol{P}^T(z) = \frac{1}{T}\left(\boldsymbol{p}_0^{*T}[(zB - I)\tilde{T}(z) + T^*] + \boldsymbol{P}^{*T}(z)\tilde{T}(z)\right).$$

This equality, by virtue of (8.1.17), (8.1.20), and the first equality in (8.1.9), reduces to

$$\boldsymbol{P}^T(z) = (1 - z)\boldsymbol{p}_0^T(\Lambda + zN)(\tilde{B}(z) - zI)^{-1}\tilde{B}(z)(\Lambda + zN)^{-1}$$

or, since the matrices in the last formula are commutative,

$$\boldsymbol{P}^T(z) = (1 - z)\boldsymbol{p}_0^T\tilde{B}(z)(\tilde{B}(z) - zI)^{-1}. \tag{8.1.21}$$

By virtue of formulas (8.1.18) and (8.1.21), for a queueing system with Markov input flow, the stationary distribution of the number of customers for the embedded Markov chain, in general, does not coincide with the time-stationary distribution of this number, i.e., the Khinchin stationary queue law is broken for such a system.

The generating function

$$\boldsymbol{P}(z, x) = \sum_{k=1}^{\infty} z^k \boldsymbol{p}_k(x)$$

is calculated similarly. Multiplying both sides of equality (8.1.11) by z^k and summing over $k = 1, 2 \ldots$, we obtain

$$\boldsymbol{P}^T(z,x) = \frac{1}{T}\Big((1 - B(x)\boldsymbol{p}_0^{*T}[(zB - I)F(z,x) + T^*] + \boldsymbol{P}^{*T}(z)F(z,x)\Big).$$

Substituting (8.1.14) and (8.1.18) into this equality, by virtue of the first equality in (8.1.9) we obtain

$$\boldsymbol{P}^T(z,x) = (1 - B(x)z\boldsymbol{p}_0^T(\Lambda + zN)(zI - \tilde{B})^{-1}e^{(\Lambda+zN)x}. \tag{8.1.22}$$

Finally, let us find the Laplace–Stieltjes transform $\omega(s)$ of the stationary distribution of the waiting time for service. By analogy with (7.1.29), we obtain

$$\begin{aligned}\omega(s) = \frac{1}{\lambda T}\Bigg[\boldsymbol{p}_0^{*T}T^* + \sum_{k=1}^{\infty}\beta^{k-1}(s)\Bigg(\boldsymbol{p}_0^{*T}B\int_0^\infty F_{k-1}(x)\,dx\int_0^\infty e^{-su}d_uB(x+u)\\ + \sum_{m=1}^{k}\boldsymbol{p}_m^{*T}\int_0^\infty F_{k-m}(x)\,dx\int_0^\infty e^{-su}d_uB(x+u)\Bigg)\Bigg]\boldsymbol{\lambda}\\ = \frac{1}{\lambda T}\Bigg[\boldsymbol{p}_0^{*T}T^* + \Bigg(\boldsymbol{p}_0^{*T}B + \frac{1}{\beta(s)}(\boldsymbol{P}^{*T}(\beta(s)) - \boldsymbol{p}_0^{*T})\Bigg)\\ \times\int_0^\infty F(\beta(s),x)\,dx\int_0^\infty e^{-su}d_uB(x+u)\Bigg]\boldsymbol{\lambda}.\end{aligned} \tag{8.1.23}$$

Let us consider the double integral in the right-hand side of (8.1.23). Changing the order of integration, we obtain

$$\int_0^\infty F(\beta(s),x)\,dx\int_0^\infty e^{-su}d_uB(x+u) = (\beta(-\Lambda-\beta(s)N)-\beta(s)I)(sI+\Lambda+\beta(s)N)^{-1}.$$

Substituting this expression into (8.1.23), we obtain

$$\begin{aligned}\omega(s) = \frac{1}{\lambda T}\Bigg[\boldsymbol{p}_0^{*T}T^* + \Bigg(\boldsymbol{p}_0^{*T}B + \frac{1}{\beta(s)}(\boldsymbol{P}^{*T}(\beta(s)) - \boldsymbol{p}_0^{*T})\Bigg)\\ \times[\beta(-\Lambda - \beta(s)N) - \beta(s)I][sI + \Lambda + \beta(s)N]^{-1}\Bigg]\boldsymbol{\lambda}.\end{aligned}$$

Hence, by virtue of (8.1.18) and the first equality in (8.1.9), after simple computations, which are left to the reader, we obtain the expression of $\omega(s)$

$$\omega(s) = \frac{s}{\lambda}\boldsymbol{p}_0^T[sI + \Lambda + \beta(s)N]^{-1}\boldsymbol{\lambda}. \tag{8.1.24}$$

Now we can find the Laplace–Stieltjes transform $\varphi(s)$ for the stationary distribution of the customer sojourn time. Since the customer sojourn time is the sum of the waiting time and the service time, and these times are independent, we obtain

$$\varphi(s) = \omega(s)\beta(s).$$

8.1.6. Mean characteristics of the queue

The results obtained above are helpful in finding the moments of stationary distribution of the number of customers in the system for the embedded Markov chain and of the time-stationary distribution of customers in the system. Using the mean values as example, let us illustrate how this is done.

Let us find the mean stationary number of customers for the embedded Markov chain. For this purpose, first we compute $\tilde{B}'(z)$. By the initial conditions (8.1.13),

$$\frac{\partial}{\partial z}F(z,x)\bigg|_{x=0} = 0. \tag{8.1.25}$$

Differentiating (8.1.12) with respect to z, we obtain the ordinary differential equation

$$\frac{\partial}{\partial x}\left(\frac{\partial}{\partial z}F(z,x)\right) = \frac{\partial}{\partial z}F(z,x)(\Lambda + zN) + F(z,x)N \tag{8.1.26}$$

in $\partial F(z,x)/\partial z$ as a function of x. Its solution under the initial condition (8.1.25), in view of (8.1.14), is

$$\frac{\partial}{\partial z}F(z,x) = \int_0^x e^{(\Lambda+zN)y}Ne^{(\Lambda+zN)(x-y)}dy.$$

From (8.1.15) we obtain

$$\tilde{B}'(z) = \int_0^\infty dB(x)\int_0^x e^{(\Lambda+zN)y}Ne^{(\Lambda+zN)(x-y)}dy.$$

In particular,

$$\begin{aligned}\tilde{B}'(1) &= \int_0^\infty dB(x)\int_0^x e^{\Lambda^* y}Ne^{\Lambda^*(x-y)}dy,\\ \tilde{B}'(1)\mathbf{1} &= \int_0^\infty dB(x)\int_0^x e^{\Lambda^* y}\boldsymbol{\lambda}\,dy.\end{aligned} \tag{8.1.27}$$

Now differentiating (8.1.17) with respect to z, we obtain

$$\boldsymbol{P}^{*\prime T}(z)(zI - \tilde{B}(z)) + \boldsymbol{P}^{*T}(z)(I - \tilde{B}'(z)) = \boldsymbol{p}_0^{*T}(B\tilde{B}(z) + (zB - I)\tilde{B}'(z)). \tag{8.1.28}$$

Taking $z = 1$ in (8.1.28), we obtain the simultaneous linear algebraic equations for $\boldsymbol{P}^{*\prime}(1)$

$$\boldsymbol{P}^{*\prime T}(1)(I - \tilde{B}(1)) = -\boldsymbol{P}^{*T}(1)(I - \tilde{B}'(1)) + \boldsymbol{p}_0^{*T}(B\tilde{B}(1) + (B - I)\tilde{B}'(1)). \tag{8.1.29}$$

Since the matrix $\tilde{B}(1)$ is the transition probability matrix of the irreducible Markov chain and also the transition probability matrix of the irreducible Markov process $\{\xi(t),\ t \geq 0\}$ during the service of a customer, the rank of the matrix $I - \tilde{B}(1)$ is $l - 1$, and $\boldsymbol{\pi}$, $\mathbf{1}$ are the left (unique up to a constant) and right eigenvectors corresponding to the zero eigenvalue. Hence, the vector $\boldsymbol{P}^{*\prime}(1)$ is defined by (8.1.29) up to a constant.

In order to derive one more equation defining this constant, we first find $\tilde{B}''(z)$. Differentiating (8.1.26) and (8.1.13) once again, we obtain the differential equation

$$\frac{\partial}{\partial x}\left(\frac{\partial^2}{\partial z^2}F(z,x)\right)=\frac{\partial^2}{\partial z^2}F(z,x)(\Lambda+zN)+2\frac{\partial}{\partial z}F(z,x)N$$

with the initial condition

$$\left.\frac{\partial^2}{\partial z^2}F(z,x)\right|_{x=0}=0.$$

Its solution is of the form

$$\frac{\partial^2}{\partial z^2}F(z,x)=2\int_0^x\frac{\partial}{\partial z}F(z,y)Ne^{(\Lambda+zN)(x-y)}dy.$$

Hence, by virtue of (8.1.28), we obtain

$$\begin{aligned}\tilde{B}''(1)&=2\int_0^\infty dB(x)\int_0^x\int_0^y e^{\Lambda^*t}Ne^{\Lambda^*(y-t)}Ne^{\Lambda^*(x-y)}\,dt\,dy,\\ \tilde{B}''(1)\mathbf{1}&=2\int_0^\infty dB(x)\int_0^x\int_0^y e^{\Lambda^*t}Ne^{\Lambda^*(y-t)}\boldsymbol{\lambda}\,dt\,dy.\end{aligned}\tag{8.1.30}$$

Differentiating (8.1.28) with respect to z and setting $z=1$, we obtain

$$\begin{aligned}\boldsymbol{P}^{*\prime\prime T}(1)(I-\tilde{B}(1))+2\boldsymbol{P}^{*\prime T}(1)(I-\tilde{B}'(1))-\boldsymbol{P}^{*T}(1)\tilde{B}''(1))&\\ =\boldsymbol{p}_0^{*T}(2B\tilde{B}'(1)+(B-I)\tilde{B}''(1)).&\end{aligned}\tag{8.1.31}$$

Multiplying (8.1.31) by $\mathbf{1}$, by virtue of the equality $(I-\tilde{B}(1))\mathbf{1}=\mathbf{0}$, we arrive at

$$2\boldsymbol{P}^{*\prime T}(1)(I-\tilde{B}'(1))\mathbf{1}=\boldsymbol{P}^{*T}(1)\tilde{B}''(1))\mathbf{1}+\boldsymbol{p}_0^{*T}(2B\tilde{B}'(1)+(B-I)\tilde{B}''(1))\mathbf{1}.\tag{8.1.32}$$

In view of (8.1.27),

$$\begin{aligned}\boldsymbol{\pi}^T(I-\tilde{B}'(1))\mathbf{1}&=1-\int_0^\infty dB(x)\int_0^x\boldsymbol{\pi}^Te^{\Lambda^*y}Ne^{\Lambda^*(x-y)}\mathbf{1}\,dy\\ &=1-\int_0^\infty dB(x)\int_0^x\boldsymbol{\pi}^TN\mathbf{1}\,dy\\ &=1-\int_0^\infty dB(x)\int_0^x\lambda\,dy=1-\rho>0.\end{aligned}\tag{8.1.33}$$

This means, first, that the vector $(I-\tilde{B}'(1))\mathbf{1}$ is non-zero and, second, (8.1.32) is linearly independent of equations (8.1.29).

Thus, equations (8.1.29), (8.1.32) is helpful in determining the vector $\boldsymbol{P}^{*\prime}(1)$ uniquely and applying this vector to finding $\bar{N}^*$—the stationary mean number of customers for the embedded Markov chain:

$$\bar{N}^*=\boldsymbol{P}^{*\prime T}(1)\mathbf{1}.$$

Let us now calculate the time-stationary mean number of customers in the system. For this purpose, we rewrite formula (8.1.21) as

$$\boldsymbol{P}^T(z)(\tilde{B}(z) - zI) = (1-z)\boldsymbol{p}_0^T\tilde{B}(z).$$

Differentiating this equality with respect to z, we obtain

$$\boldsymbol{P}'^T(z)(\tilde{B}(z) - zI) + \boldsymbol{P}^T(z)(\tilde{B}'(z) - I) = -\boldsymbol{p}_0^T\tilde{B}(z) + (1-z)\boldsymbol{p}_0^T\tilde{B}'(z). \qquad (8.1.34)$$

Hence, for $z = 1$, by virtue of the equality $\boldsymbol{P}(1) = \boldsymbol{\pi}$, we obtain simultaneous linear algebraic equations for $\boldsymbol{P}'(1)$:

$$\boldsymbol{P}'^T(1)(\tilde{B}(1) - I) = \boldsymbol{\pi}^T(I - \tilde{B}'(1)) - \boldsymbol{p}_0^T\tilde{B}(1). \qquad (8.1.35)$$

Reasoning as when we were finding the vector $\boldsymbol{P}^{*\prime}(1)$, we see that one more equation is required to calculate the vector $\boldsymbol{P}'(1)$. In order to find this missing equation, we differentiate (8.1.34) with respect to z and set $z = 1$. We thus obtain

$$\boldsymbol{P}''^T(1)(\tilde{B}(1) - I) + 2\boldsymbol{P}'^T(1)(\tilde{B}'(1) - I) + \boldsymbol{\pi}^T\tilde{B}''(1) = -2\boldsymbol{p}_0^T\tilde{B}'(1).$$

Multiplying this equality from the right by $\mathbf{1}$, we obtain

$$2\boldsymbol{P}'^T(1)(\tilde{B}'(1) - I)\mathbf{1} = -\boldsymbol{\pi}^T\tilde{B}''(1)\mathbf{1} - 2\boldsymbol{p}_0^T\tilde{B}'(1)\mathbf{1}. \qquad (8.1.36)$$

From (8.1.33) it follows that (8.1.36) can be used as a supplementary equation for equations (8.1.35).

Equations (8.1.35), (8.1.36) help to determine uniquely the vector $\boldsymbol{P}'(1)$ and then apply this vector to finding $\bar{N}$—the time-stationary mean number of customers in the system:

$$\bar{N} = \boldsymbol{P}'^T(1)\mathbf{1}.$$

While computing the right-hand side of (8.1.36), we can rewrite $\boldsymbol{\pi}^T\tilde{B}''(1)\mathbf{1}$, by virtue of (8.1.30), as

$$\boldsymbol{\pi}^T\tilde{B}''(1)\mathbf{1} = 2\int_0^\infty dB(x)\int_0^x\int_0^y \boldsymbol{\pi}^T N e^{\Lambda^*(y-t)}\boldsymbol{\lambda}\,dt\,dy.$$

Here we do not compute the stationary mean waiting time and sojourn time. They will be derived in Section 8.4.

8.1.7. Condition for ergodicity of the embedded Markov chain

As shown in Section 8.1.3, the condition $\rho < 1$ is necessary for ergodicity of the embedded Markov chain. We now show that this condition is also sufficient.

For this purpose, let us use the Moustafa criterion (Theorem 1.4.4), choosing x_{ik} so that they depend on the state (i, k) of the embedded Markov chain. We need the following lemma.

LEMMA 8.1.1. *If the matrix $\tilde{B}(1)$ is stochastic and indecomposable, and $\rho < 1$, then there exists a vector $\boldsymbol{\alpha} = (\alpha_1, \dots, \alpha_l)^T$ such that all coordinates of the vector $(\tilde{B}(1) - I)\boldsymbol{\alpha} + (\tilde{B}'(1) - I)\mathbf{1}$ are negative.*

The proof of the lemma is based on the algebraic properties of stochastic matrices and is given in (Dudin and Klimenok, 2000).

We now prove the sufficiency of the condition $\rho < 1$ for the ergodicity of the embedded Markov chain.

We have already proved that the matrix $\tilde{B}(1)$ is stochastic and indecomposable in Section 8.1.5. So, we can use Lemma 8.1.1.

We set

$$\boldsymbol{\varepsilon} = (\varepsilon_1, \dots, \varepsilon_l)^T = -(\tilde{B}(1) - I)\boldsymbol{\alpha} - (\tilde{B}'(1) - I)\mathbf{1},$$

where $\boldsymbol{\alpha}$ is the vector mentioned in Lemma 8.1.1. Let

$$\varepsilon = \min_{1 \le i \le l} \varepsilon_i, \qquad k_0 = \max_{1 \le i \le l} [|\alpha_i|] + 2 \tag{8.1.37}$$

($[x]$ is the integer part of x). We also assume that

$$x_{ik} = \begin{cases} 0, & k < k_0 - 1; \\ \alpha_i + k, & k \ge k_0 - 1. \end{cases}$$

Obviously, $\varepsilon > 0$ and $x_{ik} \ge 0$.

Let us show that conditions (1.4.3) and (1.4.4) of the Moustafa criterion are fulfilled.

First we consider the case $k \ge k_0$. Then

$$\begin{aligned} \sum_{m=0}^{\infty} \sum_{j=1}^{l} x_{jm}(P_{km})_{ij} - x_{ik} &= \sum_{m=k-1}^{\infty} \sum_{j=1}^{l} (\alpha_j + m)(B_{m-k+1})_{ij} - (\alpha_i + k) \\ &= \sum_{m=0}^{\infty} \sum_{j=1}^{l} (\alpha_j - \alpha_i)(B_m)_{ij} + \sum_{m=0}^{\infty} \sum_{j=1}^{l} (m-1)(B_m)_{ij}. \end{aligned} \tag{8.1.38}$$

We observe that the first and second terms in (8.1.38) are the ith coordinate of the vectors $(\tilde{B}(1) - I)\boldsymbol{\alpha}$ and $(\tilde{B}'(1) - I)\mathbf{1}$ respectively. Therefore, by Lemma 8.1.1 and (8.1.37), we obtain

$$\sum_{m=0}^{\infty} \sum_{j=1}^{l} (P_{km})_{ij} x_{jm} - x_{ik} \le -\varepsilon,$$

which proves (1.4.3).

If $1 \le k < k_0$, then, by virtue of (8.1.37), we see that

$$\begin{aligned} \sum_{m=0}^{\infty} \sum_{j=1}^{l} x_{jm}(P_{km})_{ij} &\le \sum_{m=k-1}^{\infty} \sum_{j=1}^{l} (k_0 + m)(B_{m-k+1})_{ij} \\ &\le \sum_{m=0}^{\infty} \sum_{j=1}^{l} (2k_0 + m)(B_m)_{ij} \\ &= 2k_0 + \sum_{m=0}^{\infty} \sum_{j=1}^{l} m(B_m)_{ij} = 2k_0 + (B'(1)\mathbf{1})_i, \end{aligned} \tag{8.1.39}$$

where $(B'(1)\mathbf{1})_i$ is the ith coordinate of the vector $B'(1)\mathbf{1}$. Bound (8.1.33) implies that the elements of the matrix $B'(1)$ are finite. Hence (8.1.39) implies validity of inequality (1.4.4).

Finally, for $k = 0$, inequality (1.4.4) is shown to be true exactly as in the previous case, except for the difference that here we use the matrices BB_m but not the matrices B_{m+1}, and this has no influence on computations.

Since there are only a finite number of states (i, k), $k < k_0$, all assumptions of the Moustafa criterion are satisfied and hence, the condition $\rho < 1$ is sufficient for ergodicity of the embedded Markov chain.

8.1.8. Computation of the matrix G

As becomes clear from the abovesaid, finding the vector $\boldsymbol{p}_0^*$ is the main difficulty in using the results in practice.

In turn, in order to find the vector $\boldsymbol{p}_0^*$ from equilibrium equations (8.1.6), we must calculate the matrix G. In this section, we consider an iterative procedure to solve this problem.

Before describing the method, we state certain properties of the matrix G and the matrix generating function $\tilde{B}(A)$.

We say that $A_1 \geq A_2$ if $(A_1)_{ij} \geq (A_2)_{ij}$ for all $i, j = 1, \ldots, l$.

LEMMA 8.1.2. *The matrix generating function $\tilde{B}(A)$ is a non-decreasing function of stochastic or substochastic matrices A. Moreover, the matrix G for $\rho < 1$ is stochastic, and the sequence of matrices $G^{(n)} = \tilde{B}(G^{(n-1)})$, $n = 1, 2, \ldots$, for any stochastic or substochastic matrix $G^{(0)}$ converges to the matrix G.*

PROOF. Since the elements of any stochastic and substochastic matrices A_1 and A_2 are non-negative, we see that $A_1^k \geq A_2^k$ for $A_1 \geq A_2$. Furthermore, the elements of the matrices B_k are also non-negative. Therefore, $\tilde{B}(A_1) \geq \tilde{B}(A_2)$ for $A_1 \geq A_2$, which proves the first assertion of the lemma.

For $\rho < 1$, the embedded Markov chain is ergodic. Therefore, starting from any state $(i, 1)$ the chain attains one of the states $(j, 0)$ in a finite number of steps with probability one. Hence, in particular, it follows that the matrix G is stochastic.

To prove the third assertion, along with the initial (embedded) Markov chain $\{\zeta_m, m \geq 0\}$, we us consider the sequence $\{\zeta_m^{(n)}, m \geq 0\}$, $n \geq 0$, of time-nonhomogeneous Markov chains with the same state set $\mathscr{X}$. These Markov chains are defined as follows.

We assume that for all Markov chains (including the initial chain), the states $(i, 0)$ are absorbing. Obviously, for the initial Markov chain the instant of reaching one of the absorbing states $(i, 0)$ is the instant of completion of the busy period of the system, and the coordinate i of the absorbing state is the phase of the customer generation process at this instant.

Immediately after the first step, the Markov chain $\{\zeta_m^{(0)},\ m \geq 0\}$ goes with the probability $(G^k(0))_{ij}$ from the state (i, k), $k \geq 1$, to the state $(j, 0)$. Furthermore, if the matrix $G(0)$ is substochastic, then the Markov chain terminates with the complementary probability $1 - \sum_{j=1}^{l}(G^k(0))_{ij}$. In particular, the elements $(G(0))_{ij}$ of the matrix $G(0)$ are the probabilities of reaching the absorbing state $(j, 0)$ from the state $(i, 1)$.

After the first step, the Markov chain $\{\zeta_m^{(1)},\ m \geq 0\}$ goes with the probability $(B_{r-k+1})_{is}$ from the state (i, k), $k \geq 1$, to the state (s, r), $r \geq k - 1$. If $k = 1$ and $r \geq 1$, then at the second step the chain goes with probability $(G^r(0))_{sj}$ from the state (s, r) to the state $(j, 0)$.

Clearly, in this case the probability of going from the state $(i, 1)$ to the absorbing state $(j, 0)$ is given by the matrix formula

$$G_1(1) = G(1) = \sum_{n=0}^{\infty} B_n G^n(0) = \tilde{B}(G(0)).$$

But if $k \geq 2$ and $r \geq k$, then at the second step the chain passes with the probability $(G^{r-k+1}(0))_{sj_1}$ from the state (s, r) to the state $(j_1, k-1)$. Thereafter, all steps are repeated from the beginning, but starting from the state $(j_1, k-1)$ up to the instant when the state $(j_2, k-2)$ is reached for the first time, etc., until the instant of reaching the state $(j, 0)$. It can be easily shown that the Markov chain $\{\zeta_m^{(1)}, m \geq 0\}$ passes from the state (i, k) to the absorbing state $(j, 0)$ with the probability given by the matrix formula

$$G_k(1) = G_1^k(1) = G^k(1).$$

It is also obvious that at the first step the transition probability matrices of the Markov chains $\{\zeta_m,\ m \geq 0\}$ and $\{\zeta_m^{(1)},\ m \geq 0\}$ coincide.

The Markov chains $\{\zeta_m^{(2)}, m \geq 0\}$, $\{\zeta_m^{(3)}, m \geq 0\}$, etc., are similarly defined. First, the Markov chain $\{\zeta_m^{(n)}, m \geq 0\}$ passes with the probability $(B_{r_1-k+1})_{is_1}$ from the state (i, k) to the state (s_1, r_1); thereafter, functioning like the Markov chain $\{\zeta_m^{(n-1)},\ m \geq 0\}$, it reaches the state $(j_1, k-1)$. Thereafter, the chain passes with the probability $(B_{r_2-k+2})_{j_1 s_2}$ from the state $(j_1, k-1)$ to the state (s_2, r_2), once again behaving like the Markov chain $\{\zeta_m^{(n-1)}, m \geq 0\}$, it reaches the state $(j_2, k-2)$, etc., until it reaches the absorbing state $(j, 0)$.

The probability that the Markov chain $\{\zeta_m^{(n)}, m \geq 0\}$ passes from the state $(i, 1)$ to the absorbing state $(j, 0)$ is given by the matrix formula

$$G_1(n) = G(n) = \sum_{s=0}^{\infty} B_s G^s(n-1) = \tilde{B}(G(n-1))$$

and the probability of transition from the state (i, k) is

$$G_k(n) = G_1^k(n) = G^k(n).$$

We also observe that at the first n steps the transition probability matrices of the Markov chains $\{\zeta_m, m \geq 0\}$ and $\{\zeta_m^{(n)}, m \geq 0\}$ coincide.

Finally, we observe that if the matrix $G(0)$ is substochastic, then the Markov chains $\{\zeta_m^{(n)},\ m \geq 0\}$ are terminating chains.

Since the embedded Markov chain is ergodic, it necessarily passes from any state to any other state (in particular, one of the states $(i, 0)$, i.e., is absorbing by assumption) in a finite number of steps with probability one. But then, in view of the above remark, as $n \to \infty$

$$G_k(n) = G^k(n) \to G^{(k)} = G^k$$

and, in particular,

$$G(n) = \tilde{B}(G(n-1)) \to G,$$

which proves the third assertion. This completes the proof of the lemma. □

With the use of Lemma 8.1.2, the matrix G can be computed by some iterative algorithm, in which the nth iteration $G(n)$ is computed by the formula

$$G(n) = \tilde{B}(G(n-1)).$$

In general, any stochastic or substochastic matrix $G(0)$ can be used as the zero iteration.

But the first assertion of the lemma is more convenient. It asserts that the matrix generating function $\tilde{B}(A)$ is non-decreasing and a zero matrix can be used as the zero iteration. Then the iterations will increase in number, and computation terminates when the sums $\sum_{j=1}^{l} G(n)_{ij}$ differ from 1 by less than given ε for all i.

8.2. Virtual waiting time

The customer sojourn time is investigated exactly as for the $M/G/1/\infty$ system.

8.2.1. The Takács equation

In this section, as before, $\xi(t)$ denotes the phase of the customer generation process at instant t and $\eta(t)$ is the virtual waiting time at this instant, i.e., the time that is required for the system to become free of all customers at instant t. Let $\zeta(t) = (\xi(t), \eta(t))$. The process $\{\zeta(t), t \geq 0\}$ is a linearwise Markov process with the state set $\mathscr{X} = \{(i, x), i = 1, \dots, l, x \geq 0\}$. Here $\zeta(t) = (i, 0)$ means that the system is empty at instant t, i.e., the virtual waiting time is 0, the customer generation process is in phase i, and $\zeta(t) = (i, x)$, $x > 0$, corresponds to the situation where a time x is required for the system to become free of all customers at instant t; here i, as before, denotes the phase of the customer generation process at this instant.

We set

$$W_i(x, t) = \mathbf{P}\{(\xi(t) = i, \eta(t)) < x\}.$$

As in Section 5.2, it is more convenient to partition the state space $\mathscr{X}$ into two parts: the set $(i, 0)$ of states corresponding to the empty system and the set of states (i, x) for $x \in (0, \infty)$. So, let

$$p_{i0}(t) = W_i(0+, t) = \mathbf{P}\{\xi(t) = i, \eta(t) = 0\}$$

denote the probability that at instant t the system is empty and the customer generation process is in phase i, and

$$U_i(x, t) = W_i(x, t) - p_{i0}(t) = \mathbf{P}\{\xi(t) = i, 0 < \eta(t) < x\}, \qquad x > 0,$$

the probability that at instant t the customer generation process is in phase i, customers exist in the system, and their total service time is less than x.

In order to describe the process $\{\zeta(t), \ t \geq 0\}$ completely, we must specify the initial distribution $W_i(x, 0)$. As before, for the sake of simplicity, we confine ourselves to the case where no customers are in the system at the initial instant, i.e., $U_i(x, 0) \equiv 0$. Then, as in Section 5.2.1, we show that the functions $p_{i0}(t)$ and $U_i(x, t)$ have derivatives $p'_{i0}(t)$, $\partial U_i(x, t)/\partial t$, and $\partial U_i(x, t)/\partial x$.

As usual, we use the vector notation $\boldsymbol{p}_0(t)$, $\boldsymbol{W}(x,t)$, and $\boldsymbol{U}(x,t)$.

The functions $\boldsymbol{p}_0(t)$ and $\boldsymbol{U}(x,t)$ satisfy the Takács equations

$$\boldsymbol{p}_0'^T(t) = \boldsymbol{p}_0^T(t)\Lambda + \frac{\partial}{\partial x}\boldsymbol{U}^T(x,t)\bigg|_{x=0},$$

$$\frac{\partial}{\partial t}\boldsymbol{U}^T(x,t) - \frac{\partial}{\partial x}\boldsymbol{U}^T(x,t) = \boldsymbol{U}^T(x,t)\Lambda - \frac{\partial}{\partial x}\boldsymbol{U}^T(x,t)\bigg|_{x=0} + \int_0^x \boldsymbol{U}^T(x-y,t)N\,dB(y) + \boldsymbol{p}_0^T(t)NB(x) \quad (8.2.1)$$

which are derived exactly as in the scalar case.

8.2.2. Stationary distribution of the virtual waiting time

If the condition $\rho < 1$ is fulfilled, then there exists a limit (stationary) distribution

$$\boldsymbol{W}(x) = \lim_{t\to\infty} \boldsymbol{W}(x,t).$$

As in the preceding section, for the sake of simplicity we assume that

$$\boldsymbol{p}_0 = \boldsymbol{W}(0+) = \lim_{t\to\infty} \boldsymbol{p}_0(t), \qquad \boldsymbol{U}(x) = \boldsymbol{W}(x) - \boldsymbol{p}_0 = \lim_{t\to\infty} \boldsymbol{U}(x,t).$$

Equations for $\boldsymbol{p}_0$ and $\boldsymbol{U}(x)$ are obtained by equating $\boldsymbol{p}_0'(t)$ and $\partial \boldsymbol{U}(x,t)/\partial t$ in (8.2.1) to zero:

$$\mathbf{0}^T = \boldsymbol{p}_0^T\Lambda + \boldsymbol{U}'^T(0),$$

$$-\boldsymbol{U}'^T(x) = \boldsymbol{U}^T(x)\Lambda - \boldsymbol{U}'^T(0) + \int_0^x \boldsymbol{U}^T(x-y)N\,dB(y) + B(x)\boldsymbol{p}_0^T N.$$

To solve this system of equations, we make use of the Laplace–Stieltjes transform

$$\boldsymbol{\omega}(s) = \int_0^\infty e^{-sx}d\boldsymbol{W}(x) = \boldsymbol{p}_0 + \int_0^\infty e^{-sx}d\boldsymbol{U}(x), \qquad \beta(s) = \int_0^\infty e^{-sx}dB(x).$$

We thus obtain

$$\boldsymbol{U}'^T(0) = -\boldsymbol{p}_0^T\Lambda,$$

$$-s[\boldsymbol{\omega}^T(s) - \boldsymbol{p}_0^T] + \boldsymbol{U}'^T(0) = [\boldsymbol{\omega}^T(s) - \boldsymbol{p}_0^T]\Lambda + \boldsymbol{\omega}^T(s)N\beta(s).$$

Hence

$$\boldsymbol{\omega}^T(s)(sI + \Lambda + \beta(s)N) = s\boldsymbol{p}_0^T, \quad (8.2.2)$$

$$\boldsymbol{\omega}^T(s) = s\boldsymbol{p}_0^T(sI + \Lambda + \beta(s)N)^{-1} \quad (8.2.3)$$

(compare formula (8.2.3) with (5.2.1)).

Setting $s = 0$ in (8.2.2), we obtain

$$\boldsymbol{\omega}^T(0)\Lambda^* = \mathbf{0}^T.$$

Hence, by the normalisation condition $\boldsymbol{\omega}^T(0)\mathbf{1} = 1$, we obtain the relation $\boldsymbol{\omega}(0) = \boldsymbol{\pi}$, which has an obvious probabilistic meaning.

It remains to find the vector $\boldsymbol{p}_0$ in (8.2.3). It is easily calculated by the methods used in the preceding sections of this chapter. But analyticity of the function $\boldsymbol{\omega}(s)$ can also be used for this purpose; we will proceed along precisely these lines. Here and in what follows, as in the previous section, we only outline the main idea of the method.

The equation

$$\det(sI + \Lambda + \beta(s)N) = 0$$

has (with account for multiplicity) exactly l roots $s_1, \dots, s_l$ in the right half-plane $\Re s \geq 0$; only one of them (say, s_l) lies on the imaginary axis, and $s_l = 0$. We assume that all these solutions are distinct. Let $\boldsymbol{a}_1, \dots, \boldsymbol{a}_{l-1}$ denote the right eigenvectors of the matrix $sI + \Lambda + \beta(s)N$ corresponding to the zero eigenvalue for $s = s_1, \dots, s_{l-1}$. Multiplying both sides of (8.2.2) by $\boldsymbol{a}_k$, we obtain $l - 1$ equations for $\boldsymbol{p}_0$:

$$\boldsymbol{p}_0^T \boldsymbol{a}_k = 0, \qquad k = 1, \dots, l-1.$$

The vector $\boldsymbol{a}_l = \mathbf{1}$ corresponds to the last solution $s_l = 0$. Multiplying both sides of (8.2.2) by $\mathbf{1}$, we obtain

$$\boldsymbol{\omega}^T(s)(sI + \Lambda^* - (1 - \beta(s))N)\mathbf{1} = s\boldsymbol{\omega}^T(s)\left(\mathbf{1} - \frac{1 - \beta(s)}{s}\boldsymbol{\lambda}\right) = s\boldsymbol{p}_0^T\mathbf{1}.$$

Dividing both sides of this relation by s and making s tend to zero, we obtain

$$p_0 = \boldsymbol{p}_0^T\mathbf{1} = \boldsymbol{\pi}(\mathbf{1} - b\boldsymbol{\lambda}) = 1 - \rho.$$

We derived this relation in the previous section.

The final expression for the Laplace–Stieltjes transform $\tilde{\omega}(s)$ of the stationary distribution of the virtual waiting time is given by the formula

$$\tilde{\omega}(s) = \boldsymbol{\omega}^T(s)\mathbf{1}.$$

Note that, in contrast to the $M/G/1/\infty$ system, in this case the stationary distribution $W(x)$ of the waiting time of a customer does not coincide with the stationary distribution of the virtual waiting time. Its Laplace–Stieltjes transform is given by the formula

$$\omega(s) = \frac{1}{\lambda}\boldsymbol{\omega}^T(s)\boldsymbol{\lambda}. \tag{8.2.4}$$

Formula (8.2.4) is proved in the same way as in Section 7.1.5, where the link between the time-stationary probabilities of the states of the $M/G/1/r$ system and the stationary probabilities of the states of this system at the instants of arrival of customers was established.

Now we can easily find the stationary distribution $V(x)$ of the customer sojourn time in the system. The Laplace–Stieltjes transform of this distribution is

$$\varphi(s) = \omega(s)\beta(s). \tag{8.2.5}$$

As in the case of Poisson input flow, formulas (8.2.4) and (8.2.5) cannot be immediately used to find the stationary distributions of waiting and sojourn times. But these formulas are more convenient for calculating the moments of these characteristics. Using the stationary mean waiting time as an example, we illustrate how this is done.

Differentiating (8.2.2), we obtain

$$\boldsymbol{\omega}'^T(s)(sI + \Lambda + \beta(s)N) + \boldsymbol{\omega}^T(s)(I + \beta'(s)N) = \boldsymbol{p}_0^T. \tag{8.2.6}$$

Setting $s = 0$ in this equality and recalling that $\boldsymbol{\omega}(0) = \boldsymbol{\pi}$, $\beta(0) = 1$, and $\beta'(0) = -b$, we obtain

$$\boldsymbol{\omega}'^T(0)\Lambda^* = \boldsymbol{p}_0^T - \boldsymbol{\pi}^T(I - bN). \tag{8.2.7}$$

It is easily seen that

$$(\boldsymbol{p}_0^T - \boldsymbol{\pi}^T(I - bN))\mathbf{1} = p_0 - (1 - \rho) = 0. \tag{8.2.8}$$

Since Λ^* is an infinitesimal matrix of an irreducible Markov process, its rank is $l - 1$, and the vector $\boldsymbol{\pi}$, which is unique up to a constant, is its left eigenvector corresponding to the zero eigenvalue. The solution of (8.2.7) is determined up to the vector $C\boldsymbol{\pi}$, where C is a constant. To find one more equation for this constant, we differentiate (8.2.6). For the sake of simplicity, we assume that there exists a finite second derivative $\boldsymbol{\omega}''(0)$, though the auxiliary equation (8.2.10) can also be derived without this assumption. Then

$$\boldsymbol{\omega}''^T(s)(sI + \Lambda + \beta(s)N) + 2\boldsymbol{\omega}'^T(s)(I + \beta'(s)N) + \beta''(s)\boldsymbol{\omega}^T(s)N = \mathbf{0}. \tag{8.2.9}$$

Setting $s = 0$ and multiplying (8.2.9) from the right by $\mathbf{1}$, we obtain

$$\boldsymbol{\omega}''^T(0)\Lambda^*\mathbf{1} + 2\boldsymbol{\omega}'^T(0)(I - bN)\mathbf{1} + \beta''(0)\boldsymbol{\omega}^T(0)N\mathbf{1} = 0,$$

or, which is the same,

$$\boldsymbol{\omega}'^T(0)(\mathbf{1} - b\boldsymbol{\lambda}) = -\frac{\lambda b^{(2)}}{2}, \tag{8.2.10}$$

where $b^{(2)}$ is the second moment of the service time. Since $\boldsymbol{\pi}^T(\mathbf{1} - b\boldsymbol{\lambda}) = 1 - \rho > 0$, (8.2.10) is linearly independent of equations (8.2.7).

Thus, the derivative $\boldsymbol{\omega}'(0)$ can be found from equations (8.2.7) and (8.2.10).

In view of (8.2.4) and (8.2.5), the stationary mean times of waiting for service w and sojourn v of a customer in the system are given by the formulas

$$w = -\frac{1}{\lambda}\boldsymbol{\omega}'^T(0)\boldsymbol{\lambda}, \qquad v = w + b.$$

We may also find w from the Little formula

$$\lambda w = Q,$$

where Q is the stationary mean queue length. This formula can be derived by formal repeating of the steps used in deriving the Little formula (7.1.40) for the $MAP/G/1/r$ system. Of course, here we should keep in mind that $r = \infty$ and appropriately modify the expressions used in deriving formula (7.1.40).

8.2.3. Transient distribution of the virtual waiting time

For a system with Markov input flow, as for a system with Poisson flow, the solution of the Takács equation (8.2.1) in the transient case can be expressed in terms of double transform—Laplace and Laplace–Stieltjes transforms. Let

$$\boldsymbol{\omega}(s_1, s_2) = \int_0^\infty \int_0^\infty e^{-s_1 t - s_2 x} \boldsymbol{U}(dx, t)\, dt + \int_0^\infty e^{-s_1 t} \boldsymbol{p}_0(t)\, dt$$

denote the double transform of the virtual waiting time $\eta(t)$ at instant t, and let

$$\boldsymbol{\pi}_0(s) = \int_0^\infty e^{-st} \boldsymbol{p}_0(t) dt.$$

Using the properties of the Laplace and Laplace–Stieltjes transforms, from (8.2.1) under the assumption that the system is empty at the initial instant 0 we obtain

$$\int_0^\infty \frac{\partial}{\partial x} \boldsymbol{U}^T(x, t)\bigg|_{x=0} e^{-st} dt = -\boldsymbol{\pi}_0^T(s)\Lambda + s\boldsymbol{\pi}_0^T(s) - \boldsymbol{p}_0^T(0)$$

and

$$s_1[\boldsymbol{\omega}^T(s_1, s_2) - \boldsymbol{\pi}_0^T(s_1)] - s_2[\boldsymbol{\omega}^T(s_1, s_2) - \boldsymbol{\pi}_0^T(s_1)] + \int_0^\infty \frac{\partial}{\partial x} \boldsymbol{U}^T(x, t)\bigg|_{x=0} e^{-s_1 t} dt$$
$$= [\boldsymbol{\omega}^T(s_1, s_2) - \boldsymbol{\pi}_0^T(s_1)]\Lambda + \beta(s_2)\boldsymbol{\omega}^T(s_1, s_2)N.$$

Hence,

$$\boldsymbol{\omega}^T(s_1, s_2)[s_1 I - s_2 I - \Lambda - \beta(s_2)N] = [\boldsymbol{p}_0^T(0) - s_2\boldsymbol{\pi}_0^T(s_1)], \qquad (8.2.11)$$

or

$$\boldsymbol{\omega}^T(s_1, s_2) = [\boldsymbol{p}_0^T(0) - s_2\boldsymbol{\pi}_0^T(s_1)][s_1 I - s_2 I - \Lambda - \beta(s_2)N]^{-1}. \qquad (8.2.12)$$

In formula (8.2.12), the function $\boldsymbol{\pi}_0(s)$ remains unknown. It can be found by the method used in the previous section. Indeed, for any fixed s_1, $\Re s_1 > 0$, the equation

$$\det(s_1 I - s_2 I - \Lambda - \beta(s_2)N) = 0$$

has (with regard for multiplicity) exactly l roots $s_{21}(s_1), \ldots, s_{2l}(s_1)$ in the right half-plane $\Re s_2 > 0$. Assuming that all solutions are distinct, we denote by $\boldsymbol{a}_1(s_1), \ldots, \boldsymbol{a}_l(s_1)$ the right eigenvectors of the matrix $s_1 I - s_2 I - \Lambda - \beta(s_2)N$ corresponding to the zero eigenvalue for $s_2 = s_{21}(s_1), \ldots, s_{2l}(s_1)$. Multiplying both sides of equality (8.2.11) by $\boldsymbol{a}_k(s_1)$, we obtain l equations for $\boldsymbol{\pi}_0(s_1)$:

$$(\boldsymbol{p}_0^T(0) - s_{2k}(s_1)\boldsymbol{\pi}_0^T(s_1))\boldsymbol{a}_k(s_1) = 0, \qquad k = 1, \ldots, l.$$

But it is not easy to realise the approach described above. Techniques of the theory of semi-Markov renewal processes are more convenient for this purpose.

Let $S(s)$ denote the matrix whose element $S_{ij}(s)$ is the Laplace–Stieltjes transform of a cycle (recall that a cycle is the time between two neighbouring instants when the system is empty) and the probability that the customer generation process at the completion of a cycle is in phase j, provided that it was at the beginning of the cycle in phase i. Then

$$S(s) = (sI - \Lambda)^{-1} N\Gamma(s),$$

where the element $\Gamma_{ij}(s)$ of the matrix $\Gamma(s)$ is the Laplace–Stieltjes transform of the busy period of the system and the probability that the customer generation process at the instant of completion of this busy period is in phase j, provided that it was in phase i at the beginning of this busy period. An expression for the matrix $\Gamma(s)$ will be derived in Section 8.4.

We now introduce a semi-Markov renewal process generated by the instants of termination of cycles, assuming that 0 is such an instant. Now, assuming that at the initial instant the system is empty and the customer generation process is in phase i with the probability $p_{i0}(0)$, we see that the Laplace transform $\boldsymbol{h}(s)$ of the renewal function $\boldsymbol{H}(t)$ (see Section 1.6.1) is of the form

$$\boldsymbol{h}^T(s) = \boldsymbol{p}_0^T(0)(I - (sI - \Lambda)^{-1} N\Gamma(s))^{-1},$$

and the Laplace transform of $\boldsymbol{\pi}_0(s)$ is

$$\boldsymbol{\pi}_0^T(s) = \boldsymbol{h}^T(s)(sI - \Lambda)^{-1}.$$

Turning back to formula (8.2.12), we derive some corollaries.

Assuming that $s_2 = 0$, we obtain

$$\boldsymbol{\omega}^T(s, 0) = \boldsymbol{p}_0^T(0)(sI - \Lambda^*)^{-1}.$$

This expression is simply the Laplace transform of the non-stationary distribution of the customer generation process.

Multiplying both sides of relation (8.2.12) by s_1 and making s_1 tend to zero, we obtain, in view of the limit relation $s\boldsymbol{\pi}_0(s) \underset{s\to 0}{\longrightarrow} \boldsymbol{p}_0$, formula (8.2.3) for the Laplace–Stieltjes transform of the stationary virtual waiting time:

$$Q\boldsymbol{\omega}^T(s) = s\boldsymbol{p}_0^T(sI + \Lambda + \beta(s)N)^{-1}.$$

8.3. Supplementary variables: FCFS discipline

We now consider the method of supplementary variables, using, as in the preceding chapter, the elapsed service time as a supplementary variable under the assumptions given in Section 7.2. Moreover, due to the close analogy with that section, we omit many steps. They can be easily restored by comparison with the computations in Chapter 7.

8.3.1. Stationary state probabilities

In order to find the stationary state probabilities for our system, we consider the linearwise Markov process $\{\zeta(t), t \geq 0\}$ introduced in Section 8.1.4. Using the notation and results of Section 7.2, since the buffer is infinite, we obtain the simultaneous equations

$$\frac{d}{dx}\boldsymbol{q}_k^T(x) = \boldsymbol{q}_k^T(x)\Lambda + u(k-1)\boldsymbol{q}_{k-1}^T(x)N, \qquad k \geq 1, \tag{8.3.1}$$

with the boundary conditions

$$\mathbf{0}^T = \boldsymbol{p}_0^T \Lambda + \int_0^\infty \boldsymbol{q}_1^T(x)\, dB(x), \tag{8.3.2}$$

$$\boldsymbol{q}_k^T(0) = \int_0^\infty \boldsymbol{q}_{k+1}^T(x)\, dB(x) + u(2-k)\boldsymbol{p}_0^T N, \qquad k \geq 1, \tag{8.3.3}$$

where, as before,

$$\boldsymbol{p}_k(x) = [1 - B(x)]\boldsymbol{q}_k(x) \tag{8.3.4}$$

and $u(x)$ is the Heaviside function.

Solving differential equations (8.3.1) with boundary conditions (8.3.2), (8.3.3), we obtain

$$\boldsymbol{q}_k^T(x) = \sum_{j=0}^{k-1} \boldsymbol{q}_{k-j}^T F_j(x), \qquad k \geq 1. \tag{8.3.5}$$

Here $\boldsymbol{q}_k = \boldsymbol{q}_k(0)$ is found from

$$\begin{aligned} \mathbf{0}^T &= \boldsymbol{p}_0^T \Lambda + \boldsymbol{q}_1^T B_0, \\ \boldsymbol{q}_k^T &= \sum_{j=0}^{k} \boldsymbol{q}_{k-j+1}^T B_j + u(2-k)\boldsymbol{p}_0^T N, \qquad k \geq 1, \end{aligned} \tag{8.3.6}$$

i.e., by the formulas

$$\boldsymbol{q}_k^T = \boldsymbol{p}_0^T Q_k, \qquad k \geq 1, \tag{8.3.7}$$

where the matrices Q_k satisfy the recurrence relations

$$\begin{aligned} Q_1 &= -\Lambda B_0^{-1}, \\ Q_k &= \left[Q_{k-1} - \sum_{j=1}^{k-1} Q_{k-j} B_j - u(3-k)N \right] B_0^{-1}, \qquad k \geq 2. \end{aligned} \tag{8.3.8}$$

Finally, the vectors

$$\boldsymbol{p}_k = \int_0^\infty \boldsymbol{p}_k(x)\, dx$$

of stationary state probabilities p_{ik} with no regard for the elapsed service time are expressed in terms of $\boldsymbol{p}_0$ and $\boldsymbol{q}_k$ by the formulas

$$\begin{aligned} \boldsymbol{p}_1^T &= -(\boldsymbol{q}_1^T \Lambda^{-1} + \boldsymbol{p}_0^T), \\ \boldsymbol{p}_k^T &= -\left(\boldsymbol{q}_k^T + \sum_{j=0}^{k-1} \boldsymbol{p}_j^T \Lambda^* \right) \Lambda^{-1}, \qquad k \geq 2. \end{aligned} \tag{8.3.9}$$

Formulas (8.3.5)–(8.3.9) do not differ from the analogous formulas for a system with finite buffer. But they are inadequate for determining the stationary characteristics of the system, because $\boldsymbol{p}_0$ is unknown.

The technique used in Section 7.2.2 cannot help computing $\boldsymbol{p}_0$. Indeed, adding together the equations in (8.3.8), we obtain

$$Q.(I - B_0^*) = \Lambda^*.$$

Hence (7.2.17) reduces to an identity.

8.3.2. Generating functions and Laplace–Stieltjes transforms: application

We introduce the generating functions

$$\boldsymbol{Q}(z) = \sum_{k=1}^{\infty} z^k \boldsymbol{q}_k, \qquad \boldsymbol{P}(z,x) = \sum_{k=1}^{\infty} z^k \boldsymbol{p}_k(x),$$

$$\boldsymbol{P}(z) = \sum_{k=0}^{\infty} z^k \boldsymbol{p}_k = \boldsymbol{p}_0 + \sum_{k=1}^{\infty} \int_0^{\infty} z^k \boldsymbol{p}_k(x)\,dx = \boldsymbol{p}_0 + \int_0^{\infty} \boldsymbol{P}(z,x)\,dx, \qquad |z| \geq 1.$$

Multiplying the kth equation in (8.3.6) by z^k and then summing over $k = 1, 2, \ldots$, by virtue of (8.1.14) and (8.1.15) we obtain

$$\boldsymbol{Q}^T(z) = \frac{1}{z}\boldsymbol{Q}^T(z)\tilde{B}(z) + \boldsymbol{p}_0^T(\Lambda + zN).$$

Hence

$$\boldsymbol{Q}^T(z) = z\boldsymbol{p}_0^T(\Lambda + zN)(zI - \tilde{B}(z))^{-1}. \tag{8.3.10}$$

From relations (8.3.4) and (8.3.5) we obtain

$$\boldsymbol{P}^T(z,x) = (1 - B(x))\boldsymbol{Q}^T(z)F(z,x) = (1 - B(x))\boldsymbol{Q}^T(z)e^{(\Lambda+zN)x}. \tag{8.3.11}$$

Finally, integrating (8.3.11) with respect to x from 0 to ∞, we obtain

$$\boldsymbol{P}^T(z) = \boldsymbol{p}_0^T + \boldsymbol{Q}^T(z)\int_0^{\infty}(1 - B(x))e^{(\Lambda+zN)x}dx$$

or, which is the same,

$$\boldsymbol{P}^T(z) = (1 - z)\boldsymbol{p}_0^T\tilde{B}(z)(\tilde{B}(z) - zI)^{-1}. \tag{8.3.12}$$

Obviously, formula (8.3.12) is the same as (8.1.21).

The analytic properties of the functions $\boldsymbol{Q}(z)$ and $\boldsymbol{P}(z)$ can help in computing the vector $\boldsymbol{p}_0$. This is done exactly as in Section 8.1.5.

We thus obtain the generating function $\boldsymbol{P}^*(z)$ of the stationary distribution of the number of customers for the embedded Markov chain. In analogy with Section 7.2.4, since in a system with infinite buffer the intensity λ_D of the served flow is equal to the intensity λ of the incoming flow, the stationary probabilities $\boldsymbol{p}_k^*$ for the embedded Markov chain are

$$\boldsymbol{p}_k^{*T} = \frac{1}{\lambda} \int_0^\infty \boldsymbol{q}_{k+1}^T(x)\, dB(x), \qquad k \geq 0.$$

Multiplying both sides of this equality by z^k and summing over $k = 0, 1, \ldots$, we obtain

$$\begin{aligned}\boldsymbol{P}^{*T}(z) &= \frac{1}{\lambda z} \boldsymbol{Q}^T(z) \int_0^\infty e^{(\Lambda + zN)x}\, dB(x) \\ &= \frac{1}{\lambda} \boldsymbol{p}_0^T(\Lambda + zN)\tilde{B}(z)(zI - \tilde{B}(z))^{-1} = \frac{1}{\lambda} \boldsymbol{p}_0^T \Lambda(I - zB)\tilde{B}(z)(zI - \tilde{B}(z))^{-1},\end{aligned}$$

where $B = -\Lambda^{-1}N$. Hence, by virtue of the first equality in (8.1.9), we finally obtain

$$\boldsymbol{P}^{*T}(z) = \boldsymbol{p}_0^{*T}(I - zB)\tilde{B}(z)(zI - \tilde{B}(z))^{-1}. \tag{8.3.13}$$

Formula (8.3.13), as should be expected, gives the same expression for $\boldsymbol{P}^{*T}(z)$ as formula (8.1.18).

Finally, we obtain the expression for the Laplace–Stieltjes transform $\delta(s)$ of the stationary distribution of the intervals between the departures of customers of the served flow. In analogy with formula (7.2.34), we obtain

$$\delta(s) = [1 - p_0^* + \boldsymbol{p}_0^{*T}(sI - \Lambda)^{-1}\boldsymbol{\lambda}]\beta(s).$$

Now, since $p_0^* = 1 - \rho$, we obtain the final expression for $\delta(s)$:

$$\delta(s) = [\rho + \boldsymbol{p}_0^{*T}(sI - \Lambda)^{-1}\boldsymbol{\lambda}]\beta(s). \tag{8.3.14}$$

8.4. LCFS discipline

We now study the $MAP/G/1/\infty$ system under the LCFS discipline with service interruption and resumption. The main objective here is not to carry out the in-depth analysis of such a system, but, more exactly, to show how certain results of the above section can be obtained easily with the use of the LCFS discipline.

8.4.1. Busy period

Let $\Gamma(s \mid x)$ denote the matrix whose element $\Gamma_{ij}(s \mid x)$ is the Laplace–Stieltjes transform of the busy period of the system and the probability that the customer generation process at the end of a busy period is in phase j, provided that it was at the beginning of this busy period in phase i, and the busy period is opened by a customer of length x. Let $\Gamma(s)$ denote the analogous matrix but without any additional constraint imposed on the customer length opening the busy period. We recall that in Section 8.2 for the FCFS discipline we used the matrix $\Gamma(s)$ with elements of the same meaning as above. Here, no ambiguity arises because the busy period under the FCFS and LCFS disciplines are the same. (The link between systems under FCFS and LCFS disciplines will be discussed in more detail in Section 8.4.4).

It is immediately seen that

$$\Gamma(s \mid x) = e^{(\Lambda + N\Gamma(s) - sI)x}, \tag{8.4.1}$$

$$\Gamma(s) = \int_0^\infty \Gamma(s \mid x)\, dB(x) = \beta(-\Lambda - N\Gamma(s) + sI). \tag{8.4.2}$$

Formulas (8.4.1) and (8.4.2) are derived exactly as (7.3.4) and (7.3.5) for $\Gamma_{ij}(s \mid x, k)$ and $\Gamma_{ij}(s \mid k)$ for the $MAP/G/1/k$ system, with the only exception that the busy period is the same for all k.

In particular, the matrix $G(x)$ consists of the probabilities $G_{ij}(x)$ that the customer generation process upon completion of a busy period is in phase j, provided that it was at the beginning of this period in phase i and the busy period was opened by a customer of length x. The matrix G consists of the probabilities G_{ij} that are the probabilities $G_{ij}(x)$ but without the condition that the busy period was opened by a customer of length x. These matrices are of the form

$$G(x) = \Gamma(0 \mid x) = e^{(\Lambda + NG)x}, \tag{8.4.3}$$

$$G = \Gamma(0) = \int_0^\infty G(x)\, dB(x) = \beta(-\Lambda - N\Gamma(s)). \tag{8.4.4}$$

Furthermore, we introduce two matrices T and T^*. The element T_{ij} of the matrix T is the mean time during which the customer generation process is in a busy period in phase j, provided that it was in phase i at the beginning of the busy period. Similarly, the element T^*_{ij} of the matrix T^* is the mean time during which the system stays in states in which there are no customers in the queue and the customer generation process is in phase j, provided that it wast the beginning of the busy period in phase i. The matrices $T(x)$ and $T^*(x)$ are of similar meaning, but under the additional condition that the busy period is opened by a customer of length x. These matrices are determined by the formulas

$$T^*(x) = \int_0^x G(y)\, dy, \tag{8.4.5}$$

$$T(x) = T^*(x) + \int_0^x G(y)NT\, dy = T^*(x)(I + NT), \tag{8.4.6}$$

$$T^* = \int_0^\infty (1 - B(x))G(x)\, dx, \tag{8.4.7}$$

$$T = T^* + \int_0^\infty (1 - B(x))G(x)NT\, dx = T^*(I + NT), \tag{8.4.8}$$

which are derived exactly as in Section 7.3 for a system of finite buffer capacity.

8.4.2. Stationary state probabilities

In order to compute the stationary state probabilities, as in Section 7.3, we make use of the supplementary variable method, taking the residual service time as the supplementary variable. We consider the Markov process $\{\zeta(t),\ t \geq 0)$, which differs from the analogous process for a system with finite buffer only in the state set. The state set in our case is $\mathscr{X} = \{(i, 0), i = 1, \dots, l;\ (i, k, x_1, \dots, x_k), x_j \geq 0, i = 1, \dots, l, j = 1, \dots, k,\ k \geq 1\}$.

Introducing, as before, the vectors $\boldsymbol{p}_0$, $\boldsymbol{P}_k(x_1,\dots,x_k)$, and $\boldsymbol{p}_k(x_1,\dots,x_k) = \partial^k \boldsymbol{P}_k(x_1,\dots,x_k)/\partial x_1 \cdots \partial x_k$, $k \geq 1$, we obtain the relations

$$\boldsymbol{p}_0^T(\Lambda + NG) = \mathbf{0}^T, \tag{8.4.9}$$

$$\boldsymbol{p}_k^T(x_1,\dots,x_k) = \boldsymbol{p}_{k-1}^T(x_2,\dots,x_k)N\int_0^\infty G(y)B(x_1+dy). \tag{8.4.10}$$

Formulas (8.4.9) and (8.4.10) virtually do not differ from (7.3.11) and (7.3.14). They allow us to find $\boldsymbol{p}_0$ and $\boldsymbol{p}_k(x_1,\dots,x_k)$ up to a constant, which is then calculated from the normalisation condition given below.

As for a system with finite buffer, for practical application, it suffices to find only $\boldsymbol{p}_k(x) = \partial \boldsymbol{P}_k(x,\infty,\dots,\infty)/\partial x$ or $\boldsymbol{p}_k = \boldsymbol{P}_k(\infty,\dots,\infty)$. These vectors are calculated by the formulas

$$\begin{aligned}
\boldsymbol{p}_k^T(x) &= \boldsymbol{p}_{k-1}^T N\int_0^\infty G(y)B(x+dy), \qquad k \geq 1,\\
\boldsymbol{p}_k^T &= \int_0^\infty \boldsymbol{p}_k^T(x)\,dx\\
&= \boldsymbol{p}_{k-1}^T N\int_0^\infty (1-B(y))G(y)\,dy\\
&= \boldsymbol{p}_{k-1}^T NT^*, \qquad k \geq 1.
\end{aligned} \tag{8.4.11}$$

The unknown constant can be determined from the normalisation condition

$$p_{\cdot} = \sum_{k=0}^{\infty} \boldsymbol{p}_k^T \mathbf{1} = 1.$$

But, as for a single-server system with infinite buffer, the unknown constant can be easily found by a simpler method. For this purpose, we use the obvious equality

$$\sum_{k=0}^{\infty} \boldsymbol{p}_k = \boldsymbol{\pi}.$$

Multiplying this equality by NT^*, in view of formula (8.4.11) we obtain

$$\boldsymbol{\pi}^T NT^* = \boldsymbol{\pi}^T - \boldsymbol{p}_0^T.$$

Since $T_0\mathbf{1} = b\mathbf{1}$, with the use of formulas (8.4.3) and (8.4.7) we obtain

$$p_0 = \boldsymbol{p}_0^T\mathbf{1} = \boldsymbol{\pi}^T\mathbf{1} - \boldsymbol{\pi}^T NT^*\mathbf{1} = 1 - b\boldsymbol{\pi}^T N\mathbf{1} = 1 - \rho.$$

This familiar equality helps to determine uniquely the solution of equilibrium equations (8.4.9).

8.4.3. Stationary distribution of the customer sojourn time. Total work

Since the results of this section repeat the analogous results for a system with finite buffer word for word, we give them without any comments.

The Laplace–Stieltjes transform of the stationary distribution of the sojourn time of a customer of length x in the system is

$$\varphi(s \mid x) = \sum_{k=0}^{\infty} \frac{1}{\lambda} \boldsymbol{p}_k^T N\Gamma(s \mid x)\mathbf{1} = \frac{1}{\lambda}\boldsymbol{\pi}^T N\Gamma(s \mid x)\mathbf{1},$$

and for an arbitrary customer is

$$\varphi(s) = \frac{1}{\lambda}\sum_{k=0}^{\infty} \boldsymbol{p}_k^T N\Gamma(s)\mathbf{1} = \frac{1}{\lambda}\boldsymbol{\pi}^T N\Gamma(s)\mathbf{1}.$$

In particular, the mean sojourn time of a customer of length x in the system is

$$v(x) = \frac{1}{\lambda}\sum_{k=0}^{\infty} \boldsymbol{p}_k^T NT(x)\mathbf{1} = \frac{1}{\lambda}\boldsymbol{\pi}^T NT(x)\mathbf{1},$$

and for an arbitrary customer is

$$v = \frac{1}{\lambda}\sum_{k=0}^{\infty} \boldsymbol{p}_k^T NT\mathbf{1} = \frac{1}{\lambda}\boldsymbol{\pi}^T NT\mathbf{1}.$$

The Laplace–Stieltjes transform $\boldsymbol{\omega}(s)$ of the stationary distribution of the total work and the probability that customer generation process is is phase i is given by the formula

$$\boldsymbol{\omega}(s) = \sum_{k=0}^{\infty} \boldsymbol{\omega}_k(s), \tag{8.4.12}$$

where

$$\boldsymbol{\omega}_k(s) = \boldsymbol{p}_0 R^k(s), \tag{8.4.13}$$

and

$$R(s) = N(\beta(-\Lambda - NG) - \beta(s)I)(sI + \Lambda + NG)^{-1},$$

or, by virtue of (8.4.4),

$$R(s) = N(G - \beta(s)I)(sI + \Lambda + NG)^{-1}. \tag{8.4.14}$$

Substituting (8.4.13) into (8.4.12), we obtain

$$\boldsymbol{\omega}^T(s) = \boldsymbol{p}_0^T \sum_{k=0}^{\infty} R^k(s). \tag{8.4.15}$$

Setting

$$R^*(s) = I + R(s) + R^2(s) + \cdots = I + R(s)R^*(s),$$

in view of (8.4.14) we obtain

$$R^*(s) = (I - R(s))^{-1} = (sI + \Lambda + NG)(sI + \Lambda + \beta(s)N)^{-1}.$$

Hence, with the use of (8.4.15) we obtain

$$\boldsymbol{\omega}^T(s) = \boldsymbol{p}_0^T(sI + \Lambda + NG)(sI + \Lambda + \beta(s)N)^{-1}.$$

Recalling equilibrium equations (8.4.9), we finally obtain

$$\boldsymbol{\omega}^T(s) = s\boldsymbol{p}_0^T(sI + \Lambda + \beta(s)N)^{-1}. \tag{8.4.16}$$

The Laplace–Stieltjes transform $\omega(s)$ of the stationary distribution of the total work is given by the formula

$$\omega(s) = \boldsymbol{\omega}^T(s)\mathbf{1}.$$

8.4.4. Relation between systems with FCFS and LCFS disciplines. Conservative service disciplines

It is immediately seen that formula (8.4.16) for the Laplace–Stieltjes transform of the stationary distribution of the total work in the $MAP/G/1/\infty$ system under the LCFS discipline is the same as formula (8.2.3) for the Laplace–Stieltjes transform of the stationary distribution of the virtual waiting time for the $MAP/G/1/\infty$ system under the FCFS discipline.

This fact is not accidental. Indeed, let us consider a queueing system with infinite buffer, arbitrary input flow, and arbitrary distributions of customer lengths (here by the length of a customer is meant the time during which he is served at unit rate of service). The input flow and customer length may depend on each other in an arbitrary manner, but do not depend on the service discipline. The class of *conservative service disciplines* is defined as follows. At an (arbitrary) instant t, let the system contain n customers. At this moment, the customer with number k is served at rate c_k (i.e., his residual length decreases by c_k per unit time) and quits the system at the instant when his residual length zeroises. Moreover, the service rate c_k may depend not only on the number n of customers in the system, the serial number k of the customer, and the moment t, but also on the past and future behaviour of the customer arrival and service processes. A service discipline is said to be conservative if $\sum_{k=1}^{n} c_k = 1$ at any instant. The FCFS and LCFS disciplines, obviously, belong to the conservative class.

If the system contains at least one customer, the total residual lengths of all customers in the system (total work) for any conservative service discipline decreases at unit rate. Since the input flow and customer length are assumed to be independent of the service discipline, the input work is the same for any, including conservative, service discipline. This means that the total work at any instant in the system is the same for all conservative disciplines, and hence, all characteristics of the system, like the busy period, are the same.

Let us apply this result to the $MAP/G/1/\infty$ system. The total work at any instant is the same for the LCFS and FCFS disciplines. Hence the total work and the virtual waiting time under the FCFS discipline have the same stationary distribution, as observed earlier.

Moreover, since the busy periods for these disciplines are the same, the matrices G defined in Sections 8.1 and 8.4 also coincide. This characteristic can hence be calculated from (8.1.2) or (8.4.4). It is not so easy to demonstrate the equivalence of (8.1.2) and (8.4.4) by direct methods. In the next section, we describe a numerical method to solve (8.1.2) and (8.4.4).

8.4.5. Computation of the matrix G

Equation (8.4.4) can be solved by any iterative method that does not actually differ from the method applied to solve (8.1.2) if the following analogue of Lemma 8.1.2 is used.

LEMMA 8.4.1. *The function $\beta(-\Lambda - NA)$ is a non-decreasing function of stochastic or substochastic matrices A. The matrix G for $\rho < 1$ is stochastic, and the sequence of matrices $G(n) = \beta(-\Lambda - NG(n-1))$, $n = 1, 2, \ldots$, for any stochastic or substochastic matrix $G(0)$ converges to the matrix G.*

PROOF. This lemma is proved as Lemma 8.1.2, but using a different method for constructing the process defining the iteration $G(n)$. This process is conveniently defined by a special n-system, which is defined as follows.

The n-system operates exactly as the initial $MAP/G/1/\infty$ system with the LCFS discipline as long as the system contains less than n customers. But when the number of customers becomes n, the n-system operates by the following rule. No customers arrive at the system, but the customer generation process with intensity $(\Lambda + NG(n-1))_{ij}$ passes from the state i to the state j. Moreover, if the matrix $G(n-1)$ is substochastic, the n-system terminates operation in the state i with intensity $-(\Lambda + NG(n-1))_{ii}$ and its busy period does not terminate. When the number of customers in the n-system becomes less than n once again, the system resumes operation like the initial system.

It is immediately seen that the matrix $G(n)$ of transition probabilities of the customer generation process in a busy period for the n-system is $G(n) = \beta(-\Lambda - NG(n-1))$.

Subsequent steps of the proof are the same as in the proof of Lemma 8.1.2 with regard for the fact that, for $\rho < 1$, a finite number of customers arrive at the initial system with probability one in the busy period. Therefore, the matrix $G(n)$ tends to the matrix G as $n \to \infty$. □

The matrix G can thus be computed by the iteration procedure

$$G(n) = \beta(-\Lambda - NG(n-1)).$$

In general, although any stochastic or substochastic matrix $G(0)$ can be taken as the zero iteration, convergence is easily verified if we take $G(0) = 0$.

9

$MAP/G/1/r$ system: generalisation

This chapter is concerned with certain queueing systems that generalise the $MAP/G/1/r$ system with finite or infinite buffer studied in the previous two chapters.

In Section 9.1, we study a system with batch Markov arrival process and semi-Markov service. We will show how the results obtained in this chapter by the methods described in Section 9.1 for other queueing systems with Markov arrival process and recurrent service can be extended to systems with batch Markov arrivals and semi-Markov service.

Sections 9.2 and 9.3 are devoted to priority systems (for the sake of simplicity, we confine ourselves to two-priority systems) and common buffer of finite capacity; Section 9.2 is concerned with preemptive-priority systems with resumption of interrupted service, whereas Section 9.3 deals with non-preemptive-priority queueing systems. For the sake of diversity, the former system is investigated by the method of the embedded Markov chain, whereas the latter, by the method of supplementary variable. Note that priority systems with Poisson input flow are investigated in several researches, in particular, monographs (Bronstein and Dukhovny, 1976; Gnedenko *et al.*, 1973; Gnedenko and König, 1984; Ivanov, 1973; Jaiswal, 1968; Klimov and Mishkoy, 1979; Volkovinskii and Kabalevskii, 1981).

In Section 9.4, we describe the $MAP/G/1/r$ retrial system with a buffer of finite capacity for primary customers and a buffer (orbit) also of finite capacity for retrial customers and a special discipline for choosing a retrial customer from the orbit.

Section 9.5 is devoted to $MAP/G/1/\infty$ systems with foreground-background processor sharing service discipline.

In Section 9.6, we study the $MAP/G/1/r$ system with the LCFS service interruption discipline and buffer of finite capacity, in which each new arrival carries some random volume (depending on his service time), and the total volume of customers in the system is also bounded by some (non-random) number.

The concluding Section 9.7 deals with the dual of the $MAP/G/1/r$ system, the $G/MSP/1/r$ system with a buffer of finite or infinite capacity, recurrent input flow, and Markov service process (*MSP*)—the generalisation of the $G/M/1/\infty$ system studied in Section 6.4.

The systems investigated in this chapter were chosen not so much for their practical significance as our endeavour to describe certain special techniques used in queueing theory.

9.1. $BMAP/SM/1/r$ system

In order to study queueing systems that generalise the $MAP/G/1/r$ system, we begin with the $BMAP/SM/1/r$, $r \leq \infty$, system with batch Markov arrival process, semi-Markov service process, and a buffer of finite or infinite capacity. As we will show below, this system can be investigated by almost the same methods as those that were applied to study the $MAP/G/1/r$ system in the previous chapters.

Here we restrict ourselves to the computation of the stationary state probabilities of the $BMAP/SM/1/r$ system. Moreover, we study the case of a finite buffer by two methods, namely, the method of supplementary variable and the method of the embedded Markov chain. Both of them pursue the same aim—to show how these methods work for this system. For the case of an infinite buffer, we restrict only to the method of the embedded Markov chain.

9.1.1. System description

The batch Markov arrival process (*BMAP*) is defined as follows. Let a homogeneous Markov process $\{\xi(t), t \geq 0\}$ defined on a finite state set $\{1, 2, \dots, l\}$, $l < \infty$, control the generation of customer batches. As before, the states of this process are called the customer (batch) generation phases. The Markov process $\{\xi(t), t \geq 0\}$ is defined by infinitesimal matrices Λ_k, $0 \leq k \leq \infty$, of order l. The element $(\Lambda_0)_{ij}$, $i \neq j$, of the matrix Λ_0 defines the intensity of transition from the generation phase i to phase j without the arrival of a new customer batch. In turn, the element $(\Lambda_k)_{ij}$, $k \geq 1$, of the matrix Λ_k defines the intensity of transition from the generation phase i to phase j with the arrival of a new customer batch of size k.

We introduce the matrices $\Lambda_k^* = \sum_{s=k}^{\infty} \Lambda_s$. The element $(\Lambda_k^*)_{ij}$, $k \geq 0$, of the matrix Λ_k^* is the intensity of transition from the generation phase i to phase j, with the arrival of a new customer batch of size no less than k. We assume that Λ_0^* is the conservative matrix of intensities of transition of the Markov process $\{\xi(t), t \geq 0\}$ on the generation phase set $\{1, 2, \dots, l\}$ and Λ_1^* is a non-zero matrix.

We assume that the matrix Λ_0^* is indecomposable. In this case, the process $\{\xi(t), t \geq 0\}$ is ergodic and its stationary distribution $\boldsymbol{\pi}^{\mathrm{a}T} = \{\pi_1^{\mathrm{a}}, \pi_2^{\mathrm{a}}, \dots, \pi_l^{\mathrm{a}}\}$ is a unique solution of the equilibrium equations

$$\mathbf{0}^T = \boldsymbol{\pi}^{\mathrm{a}T} \Lambda_0^*$$

under the normalisation condition

$$\boldsymbol{\pi}^{\mathrm{a}T} \mathbf{1} = 1.$$

Now we describe the semi-Markov (*SM*) service. Consider a semi-Markovian service process on a finite set of states—service phases—$\{1, 2, \dots, m\}$, $m < \infty$, with a stochastic control matrix of transition probabilities (the transition probability matrix of the embedded Markov chain) $M = (M_{ij})_{i,j=1,\dots,m}$ and with the semi-Markov kernel $B(x) = (B_{ij}(x))_{i,j=1,\dots,m}$, where M_{ij} is the probability of transition from phase i to phase j and $B_{ij}(x)$, $i, j = 1, \dots, m$, is the conditional distribution function of the time of service in phase i of a customer, provided that the service of the next customer begins in phase j. Let $B_i(x) = \sum_{j=1}^{m} M_{ij} B_{ij}(x)$ denote the unconditional distribution function of the time of service in phase i of a customer.

The service process is described as follows. Let s_k be the instant of completion of service of the kth customer. Assume that the service process at the instant $s_k + 0$ exists in phase i or, briefly, the service phase is i. Then with the probability $M_{ij} B_{ij}(x)$ the service of the $(k+1)$th customer lasts for a time less than x and immediately after the completion of his service (i.e., at the instant $s_{k+1} + 0$) the service phase is j.

We study the case where the service phase remains unchanged in an idle period and this is the phase in which the service process exists upon completion of service of the customer opening this idle period.

Let the matrix M be indecomposable and $b_{ij} = \int_0^\infty t\,dB_{ij}(t) < \infty$ for any $i, j = 1, \dots, m$. Moreover, for the sake of simplicity, we assume that the distribution function $B_{ij}(t)$ is absolutely continuous for any $i, j = 1, \dots, m$.

Let $\boldsymbol{\pi}^{sT} = \{\pi_1^s, \pi_2^s, \dots, \pi_m^s\}$ denote the stationary distribution of the Markov chain with transition probability matrix M—the solution of the equilibrium equations

$$\boldsymbol{\pi}^{sT} = \boldsymbol{\pi}^{sT} M$$

under the normalisation condition

$$\boldsymbol{\pi}^{sT} \mathbf{1} = 1.$$

For the sake of definiteness, we assume that customer batches are served in order of arrival, i.e., according to the FCFS discipline. The service of customers within a batch is random. This means that any customer may find himself in any place in a group of size k with the probability $1/k$.

Finally, for $r < \infty$, we assume that if a batch contains more customers than free places in the buffer, then excessive customers for whom there is no waiting place are lost.

9.1.2. Finite buffer. Introduction of a supplementary variable

The functioning of the queueing system under consideration can be described by a linear-wise process $\{\zeta(t), t \geq 0\}$ with the state set

$$\begin{aligned} \mathcal{X} = \{(i, j, 0), i = 1, \dots, l, j = 1, \dots, m;& \\ (i, j, n, x),\ i = 1, \dots, l, j = 1, \dots, m, n = 1, \dots, r+1, x \geq 0\}.& \end{aligned}$$

The states of the process $\{\zeta(t), t \geq 0\}$ are interpreted as follows:

$\zeta(t) = (i, j, 0)$ is the state in which at instant t the system is empty, the customer generation process is in phase i, and the service process is in phase j; and

$\zeta(t) = (i, j, n, x)$ is the state in which at instant t the system contains n customers, the customer generation process is in phase i, the service process is in phase j, and the time elapsed since the beginning of service of the customer at the server is x.

It is easily seen that the Markov process $\{\zeta(t),\ t \geq 0\}$ introduced here differs from the similar process introduced for the $MAP/G/1/r$ system in Section 7.1.4 in that it contains an additional coordinate defining the service phase of the customer at the server.

Under the above assumptions, the process $\{\zeta(t),\ t \geq 0\}$ is ergodic and its stationary distribution exists and is unique.

Let p_{ij0} denote the stationary probability of the state $(i, j, 0)$, let $p_{ijn}(x)$ be the stationary probability density of the state (i, j, n, x), and let $p_{ijn} = \int_0^\infty p_{ijn}(x)\,dx$ be the stationary probability of the state (i, j, n, x) without regard for the elapsed service time of the customer at the server.

Assuming that

$$q_{ijn}(x) = \frac{p_{ijn}(x)}{1 - B_j(x)},$$

we introduce the vectors

$$\begin{aligned} \boldsymbol{p}_0^T &= (p_{110}, \dots, p_{1m0}, \dots, p_{lm0}), \\ \boldsymbol{p}_n^T(x) &= (p_{11n}(x), \dots, p_{1mn}(x), \dots, p_{lmn}(x)), \\ \boldsymbol{q}_n^T(x) &= (q_{11n}(x), \dots, q_{1mn}(x), \dots, q_{lmn}(x)). \end{aligned}$$

Under such an ordering of vector coordinates, the corresponding vector coordinate for the phases i and j of the customer generation process and service process is $u = (i-1)l + j$. This ordering also will be used in what follows for matrices, which will allow us to use the Kronecker product of matrices.

The stationary distribution of the process $\{\zeta(t),\ t \geq 0\}$ satisfies the differential equations

$$\frac{d}{dx}\boldsymbol{q}_n^T(x) = \sum_{s=1}^{n} \boldsymbol{q}_s^T(x)(\Lambda_{n-s} \otimes I_m), \qquad n = 1, \dots, r, \tag{9.1.1}$$

$$\frac{d}{dx}\boldsymbol{q}_R^T(x) = \sum_{s=1}^{R} \boldsymbol{q}_s^T(x)(\Lambda_{R-s}^* \otimes I_m). \tag{9.1.2}$$

The boundary conditions are defined by the equalities

$$\boldsymbol{0}^T = \boldsymbol{p}_0^T(\Lambda_0 \otimes I_m) + \int_0^\infty \boldsymbol{q}_1^T(x)\, d(I_l \otimes \tilde{B}(x)), \tag{9.1.3}$$

$$\boldsymbol{q}_n^T 0) = \boldsymbol{p}_0^T(\Lambda_n \otimes I_m) + \int_0^\infty \boldsymbol{q}_{n+1}^T(x) d(I_l \otimes \tilde{B}(x)), \qquad n = 1, \dots, r, \tag{9.1.4}$$

$$\boldsymbol{q}_R^T(0) = \boldsymbol{p}_0^T(\Lambda_R^* \otimes I_m). \tag{9.1.5}$$

Here $R = r + 1$, I_k is an identity matrix of order k, $A \otimes B$ is the Kronecker product of the matrices A and B, and $\tilde{B}_{ij}(x) = M_{ij} B_{ij}(x)$.

Furthermore, the following normalisation condition must hold:

$$\boldsymbol{p}_\cdot^T \boldsymbol{1} = 1. \tag{9.1.6}$$

Without going into details of derivation of (9.1.1)–(9.1.5), we only state their difference from (7.2.2)–(7.2.6).

The sums in the right-hand sides of (9.1.1) and (9.1.2) arise as a result of batch arrivals. Indeed, the system at the instant $t + \Delta$ may contain n customers not only in the case where $n - 1$ customers were in the system at the instant t and one more customer had arrived in time Δ, but also in the case where s, $1 \leq s \leq n - 2$, customers were in the system at the instant t and $n-s$ customers had arrived in time Δ. Here the service phase does not change, and this is consistent with the Kronecker product $\Lambda_{n-s} \otimes I_m$ of the matrix Λ_{n-s} and the identity matrix I_m.

Obviously, the Kronecker product $\Lambda_0 \otimes I_m$ in (9.1.3) arises due to the presence of service phases, which, according to the above assumption, do not change in an idle period. The terms $\Lambda_n \otimes I_m$ and $\Lambda_R^* \otimes I_m$ in the right-hand sides of (9.1.4) and (9.1.5) arise because, respectively, n and no less than R customers may arrive concurrently. Finally, the Kronecker product $I_l \otimes \tilde{B}(x)$ in (9.1.3) and (9.1.4) corresponds to the change in the service phase upon completion of service of the routine customer.

In order to solve simultaneous differential equations (9.1.1), (9.1.2), we introduce the following matrix functions $F_n(x)$ of order l:

$$F_0(x) = e^{\Lambda_0 x},$$

$$F_n(x) = \sum_{k=0}^{n-1} \int_0^x F_k(t)\Lambda_{n-k} e^{\Lambda_0(x-t)}dt, \qquad n \geq 1.$$

As in Chapters 7 and 8, the element $(F_n(x))_{jj'}$ of the matrix $F_n(x)$ is the probability that n new customers arrive in time x and the customer generation process at the instant x is in phase j', provided it began at the instant 0 with phase j.

By the definition of the function $F_n(x)$, we have the equalities

$$\frac{d}{dx} F_n(x) = \sum_{k=0}^{n} F_k(x)\Lambda_{n-k}, \qquad n \geq 0. \tag{9.1.7}$$

With the use of the matrices $F_n(x)$, the solution of the simultaneous differential equations (9.1.1), (9.1.2) are of the form

$$\boldsymbol{q}_n^T(x) = \sum_{k=1}^{n} \boldsymbol{q}_k^T(0)(F_{n-k}(x) \otimes I_m), \qquad n = 1, \dots, r, \tag{9.1.8}$$

$$\boldsymbol{q}_R^T(x) = \sum_{k=1}^{R} \boldsymbol{q}_k^T(0)(e^{\Lambda_0^* x} \otimes I_m) - \sum_{k=1}^{r} \boldsymbol{q}_k^T(x). \tag{9.1.9}$$

To prove relations (9.1.8), we substitute them into (9.1.1) and check whether this substitution reduces equations (9.1.1) into identities. Indeed, differentiating (9.1.8), we obtain, by virtue of (9.1.7) and (9.1.8),

$$\frac{d}{dx}\boldsymbol{q}_n^T(x) = \sum_{k=1}^{n} \boldsymbol{q}_k^T(0) \sum_{i=0}^{n-k} (F_i(x)\Lambda_{n-k-i} \otimes I_m)$$

$$= \sum_{k=1}^{n}\sum_{j=1}^{k} \boldsymbol{q}_j^T(0)(F_{k-j}(x)\Lambda_{n-k} \otimes I_m) = \sum_{k=1}^{n} \boldsymbol{q}_k^T(x)(\Lambda_{n-k} \otimes I_m).$$

We thus obtain the right-hand side of (9.1.1).

In order to prove relation (9.1.9), we sum (9.1.1) and (9.1.2) over all values of n and obtain the equation

$$\frac{d}{dx}\boldsymbol{q}_{\cdot}^T(x) = \boldsymbol{q}_{\cdot}^T(x)(\Lambda_0^* \otimes I_m),$$

whose solution is

$$\boldsymbol{q}_{\cdot}^T(x) = \sum_{k=1}^{R} \boldsymbol{q}_k^T(0)(e^{\Lambda_0^* x} \otimes I_m).$$

Hence we obviously arrive at (9.1.9).

Using boundary conditions (9.1.3) and (9.1.4), we find the initial values of $\boldsymbol{q}_n(0)$, $n = 1, \dots, r$, and $\boldsymbol{p}_0$. Let us introduce the matrices

$$B_n = \int_0^\infty F_n(x) \otimes d\tilde{B}(x),$$
$$B_0^* = \int_0^\infty e^{\Lambda_0^* x} \otimes d\tilde{B}(x). \tag{9.1.10}$$

The elements $(B_n)_{uu'}$ and $(B_0^*)_{uu'}$ of the matrices B_n and B_0^* are the conditional probabilities that in the course of service of a customer there arrive n more or at least 0 customers, respectively, and the generation and service processes immediately after the completion of service are in phases i' and j', provided that they were in phase i and j at the commencement of service, where, by our assumption, $u = (i-1)l + j$ and $u' = (i'-1)l + j'$.

Substituting (9.1.8) and (9.1.9) into (9.1.3) and (9.1.4), we obtain

$$\begin{aligned} \boldsymbol{0}^T &= \boldsymbol{p}_0^T(\Lambda_0 \otimes I_m) + \int_0^\infty \boldsymbol{q}_1^T(0)(F_0(x) \otimes I_m)\, d(I_l \otimes \tilde{B}(x)) \\ &= \boldsymbol{p}_0^T(\Lambda_0 \otimes I_m) + \boldsymbol{q}_1(0) B_0, \end{aligned} \tag{9.1.11}$$

$$\begin{aligned} \boldsymbol{q}_n^T(0) &= \boldsymbol{p}_0^T(\Lambda_n \otimes I_m) + \int_0^\infty \sum_{k=1}^{n+1} \boldsymbol{q}_k^T(0)(F_{n+1-k}(x) \otimes I_m)\, d(I_l \otimes \tilde{B}(x)) \\ &= \boldsymbol{p}_0^T(\Lambda_n \otimes I_m) + \sum_{k=1}^{n+1} \boldsymbol{q}_k^T(0) B_{n+1-k}, \qquad n = 1, \dots, r-1; \end{aligned} \tag{9.1.12}$$

$$\begin{aligned} \boldsymbol{q}_r^T(0) &= \boldsymbol{p}_0^T(\Lambda_r \otimes I_m) + \int_0^\infty \sum_{k=1}^{r} \boldsymbol{q}_k^T(0)(e^{\Lambda_0^* x} \otimes I_m)\, d(I_l \otimes \tilde{B}(x)) \\ &\quad - \int_0^\infty \sum_{n=1}^{r}\sum_{k=1}^{n} \boldsymbol{q}_k^T(0)(F_{n-k}(x) \otimes I_m)\, d(I_l \otimes \tilde{B}(x)) \\ &= \boldsymbol{p}_0^T(\Lambda_r \otimes I_m) + \sum_{k=1}^{r} \boldsymbol{q}_k^T(0) B_0^* - \sum_{n=1}^{r}\sum_{k=1}^{n} \boldsymbol{q}_k^T(0) B_{n-k}. \end{aligned}$$

Equations (9.1.11) and (9.1.12) can be rewritten as

$$\mathbf{0}^T = \boldsymbol{p}_0^T(\Lambda_0 \otimes I_m) + \boldsymbol{q}_1^T(0)B_0,$$

$$\boldsymbol{q}_n^T(0) = \boldsymbol{p}_0^T(\Lambda_n \otimes I_m) + \sum_{k=1}^{n+1} \boldsymbol{q}_k^T(0)B_{n+1-k}, \qquad n = 1, \dots, r-1,$$

or in the matrix form as

$$(\boldsymbol{q}_1^T(0), \dots, \boldsymbol{q}_r^T(0)) \begin{pmatrix} B_0 & B_1 - I & B_2 & \dots & B_{r-1} \\ 0 & B_0 & B_1 - I & \dots & B_{r-2} \\ \vdots & \vdots & \vdots & \ddots & \vdots \\ 0 & 0 & 0 & 0 & B_0 \end{pmatrix}$$
$$= -\boldsymbol{p}_0^T(\Lambda_0 \otimes I_m, \Lambda_1 \otimes I_m, \dots, \Lambda_{r-1} \otimes I_m). \quad (9.1.13)$$

It is easily seen that (9.1.13) literally coincides with (7.2.13). Therefore, reasoning as in Section 7.2.2, we obtain (if the matrix B_0 has an inverse) the expression for $\boldsymbol{q}_n(0)$, $n = 1, \dots, r$, in terms of $\boldsymbol{p}_0$:

$$\boldsymbol{q}_n^T(0) = \boldsymbol{p}_0^T Q_n, \quad (9.1.14)$$

where the matrices Q_n are defined by the recurrence relations

$$Q_1 = -(\Lambda_0 \otimes I_m)B_0^{-1},$$

$$Q_n = \left(Q_{n-1} - \sum_{k=1}^{n-1} Q_k B_{n-k} - \Lambda_n \otimes I_m\right) B_0^{-1}, \qquad n = 2, \dots, r.$$

Furthermore, the vector $\boldsymbol{p}_0$ is determined from the equation

$$\boldsymbol{p}_0^T Q = \mathbf{0}^T, \quad (9.1.15)$$

where

$$Q = Q_r - \sum_{k=1}^{r} Q_k B_0^* + \sum_{k=1}^{r}\sum_{j=1}^{k} Q_j B_{k-j} + \Lambda_r \otimes I_m.$$

The solution of (9.1.15) under normalisation condition (9.1.6) is unique.

Equations (9.1.15) can also be solved by the method described in Section 7.1.2.

After finding the stationary probability densities $\boldsymbol{p}_n(x)$, we are able to calculate the stationary probabilities

$$\boldsymbol{p}_n = \int_0^\infty \boldsymbol{p}_n(x)\,dx.$$

To do this, we introduce the diagonal matrix $D(x) = \mathrm{diag}(B_1(x), \dots, B_m(x))$. We observe that

$$\boldsymbol{p}_n^T(x) = \boldsymbol{q}_n^T(x)(I_l \otimes (I_m - D(x))). \quad (9.1.16)$$

From (9.1.16), (9.1.8), and (9.1.9) we immediately obtain the relations

$$p_n^T = \sum_{k=1}^{n} q_k^T(0) H_{n-k}, \qquad n = 1, \dots, r,$$
$$p_R^T = \sum_{k=1}^{R} q_k^T(0) H_0^* - \sum_{k=1}^{r} p_k^T, \tag{9.1.17}$$

where

$$H_n = \int_0^\infty F_n(x) \otimes (I_m - D(x))\, dx, \qquad H_0^* = \int_0^\infty e^{\Lambda_0^* x} \otimes (I_m - D(x))\, dx.$$

In conclusion, let us derive recurrence relations for the matrices $F_n(x)$, $n \geq 0$.

Let $a = \max_i |(\Lambda_0)_{i,i}|$ and $P = I + a^{-1}\Lambda_0$. The matrix P, in view of the properties of the matrices Λ_k, $k \geq 0$, is the transition probability matrix of some terminating Markov chain. We set $S_n = a^{-1}\Lambda_n$.

It is obvious that

$$\int_0^x e^{-ay} \frac{(ay)^n}{n!} e^{\Lambda_0 (x-y)} dy = a^{-1} e^{-ax} \sum_{j=n+1}^{\infty} \frac{(ax)^j}{j!} P^{j-n-1}. \tag{9.1.18}$$

Let us show that the matrices $F_n(x)$ are of the form

$$F_n(x) = e^{-ax} \sum_{\nu=0}^{\infty} \frac{(ax)^\nu}{\nu!} F_{n\nu}, \tag{9.1.19}$$

where $F_{n\nu}$ satisfies the recurrence relations

$$F_{0\nu} = P^\nu,$$
$$F_{n\nu} = \sum_{k=0}^{n-1} \sum_{j=0}^{\nu-1} F_{kj} S_{n-k} P^{\nu-j-1}, \qquad n \geq 1. \tag{9.1.20}$$

Here it is assumed that $\sum_{j=0}^{-1} \cdots = 0$.

We prove this by induction on n. Indeed, for $n = 0$, relations (9.1.19) and (9.1.20) obviously follow from (9.1.18).

Now let (9.1.19) and (9.1.20) be true for all $k = 0, \ldots, n-1$. Let us prove that they hold for n:

$$\begin{aligned} F_n(x) &= \sum_{k=0}^{n-1} \int_0^x F_k(t)\Lambda_{n-k} e^{\Lambda_0(x-t)} dt \\ &= \sum_{k=0}^{n-1} \sum_{\nu=0}^{\infty} F_{k\nu} \Lambda_{n-k} \int_0^x e^{-at} \frac{(at)^\nu}{\nu!} e^{\Lambda_0(x-t)} dt \\ &= \sum_{k=0}^{n-1} \sum_{\nu=0}^{\infty} F_{k\nu} \Lambda_{n-k} a^{-1} e^{-ax} \sum_{j=\nu+1}^{\infty} \frac{(ax)^j}{j!} P^{j-\nu-1} \\ &= e^{-ax} \sum_{\nu=0}^{\infty} \frac{(ax)^\nu}{\nu!} \sum_{k=0}^{n-1} \sum_{j=0}^{\nu-1} F_{kj} S_{n-k} P^{\nu-j-1}. \end{aligned}$$

Using (9.1.19) and (9.1.20), we easily derive formulas for B_n:

$$B_n = \sum_{\nu=0}^{\infty} F_{n\nu} \otimes C_\nu, \qquad n \geq 0,$$

where C_ν is the matrix whose (j, k)-element is of the form

$$(C_\nu)_{jk} = M_{jk} \beta_{\nu jk};$$

here

$$\beta_{\nu jk} = \int_0^\infty e^{-ax} \frac{(ax)^\nu}{\nu!} dB_{jk}(x), \qquad \nu \geq 0,$$

is the ν-order exponential moment of the distribution function $B_{jk}(x)$.

In the particular case where all distribution functions $B_{jk}(x)$ are identical and equal to $B(x)$, we obtain the formulas

$$B_n = \sum_{\nu=n}^{\infty} (F_{n\nu} \otimes M) \beta_\nu,$$

where

$$\beta_\nu = \int_0^\infty e^{-ax} \frac{(ax)^\nu}{\nu!} dB(x), \qquad \nu \geq 0.$$

Now we set $P^* = I + a^{-1}\Lambda_0^*$. We observe that the matrix P^* is a stochastic matrix. As in the preceding section, we obtain

$$e^{\Lambda_0^* x} = e^{-ax} \sum_{\nu=0}^{\infty} \frac{(ax)^\nu}{\nu!} (P^*)^\nu$$

and

$$B_0^* = \sum_{\nu=0}^{\infty} (P^*)^\nu \otimes C_\nu.$$

Using (9.1.19) and (9.1.20), we obtain the formulas for H_n:

$$H_n = \sum_{\nu=0}^{\infty} F_{n\nu} \otimes K_\nu, \quad i = 0, \ldots, r-1, \quad n = 0, \ldots, r,$$

where K_ν is a diagonal matrix and

$$(K_\nu)_{ii} = \int_0^\infty e^{-ax} \frac{(ax)^\nu}{\nu!} (1 - B_i(x))\, dx = a^{-1} \left(1 - \sum_{s=0}^{\nu} \sum_{j=1}^{m} M_{ij} \beta_{sij} \right).$$

Finally, H_0^* is calculated by the formula

$$H_0^* = \sum_{\nu=0}^{\infty} (P^*)^\nu \otimes K_\nu.$$

9.1.3. Finite buffer. Embedded Markov chain

Let us introduce an embedded Markov chain $\{\zeta_k, k \geq 0\}$ at service completion instants with the state set

$$\mathcal{Y} = \{(i, j, n), i = 1, \ldots, l, j = 1, \ldots, m, n = 0, \ldots, r\}.$$

Here $\zeta_k = (i, j, n)$ means that at the instant $s_k + 0$ immediately after the completion of service of the kth customer n customers are in the system and the customer generation process and service process are in phases i and j respectively.

Under these assumptions, the Markov chain $\{\zeta_k, k \geq 0\}$ is ergodic and its stationary distribution exists and is unique.

Let us find the matrix $P = (P_{kn})_{k,n=0,\ldots,r}$ of transition probabilities of the Markov chain $\{\zeta_k, k \geq 0\}$. In order to do this, first introduce the matrices

$$F_n = \int_0^\infty (e^{\Lambda_0 x} \Lambda_n \otimes I_m)\, dx = -\Lambda_0^{-1} \Lambda_n \otimes I_m, \qquad n = 1, \ldots, r,$$

$$F_R = \int_0^\infty (e^{\Lambda_0 x} \Lambda_R^* \otimes I_m)\, dx = -\Lambda_0^{-1} \Lambda_R^* \otimes I_m,$$

where the element $(F_n)_{uu'}$ of the matrix F_n is the probability that the system contains n customers immediately upon arrival of a new batch of customers at an empty system and the customer generation process and the service process pass from phases i and j to phases i' and j' respectively. Here $u = (i-1)l + j$ and $u' = (i'-1)l + j'$. Since the service phase does not change upon arrival of a batch, we see that $(F_n)_{uu'} = 0$ for $j' \neq j$.

We also need the matrices B_n introduced in the preceding section.

Furthermore, we introduce the matrices

$$B_n^* = \int_0^\infty F_n^*(x) \otimes d\tilde{B}(x), \qquad n = 1, \ldots, r,$$

where

$$F_n^*(x) = \sum_{k=0}^{n-1} \int_0^x F_k(t) \Lambda_{n-k}^* e^{\Lambda_0^*(x-t)} dt,$$

whose meaning is obvious. The matrices B_n^* are computed in the same way as the matrices B_n.

Now we write the relations which the matrices P_{kn} satisfy:

$$\begin{aligned} P_{0n} &= \sum_{k=1}^{n+1} F_k B_{n+1-k}, \quad n = 0, \ldots, r-1, \\ P_{0r} &= \sum_{k=1}^{R} F_k B_{R-k}^*, \\ P_{kn} &= B_{n+1-k}, \qquad n = 0, \ldots, r-1, \quad k = 1, \ldots, n+1, \\ P_{kr} &= B_{R-k}^*, \qquad k = 1, \ldots, r. \end{aligned} \tag{9.1.21}$$

Other matrices P_{kn} are zero matrices.

Relations (9.1.21) are derived in the same way as (7.1.8)–(7.1.11).

By virtue of (9.1.21), the transition probability matrix P of the embedded Markov chain $\{\zeta_k, k \geq 0\}$ is

$$P = \begin{pmatrix} P_{00} & P_{01} & P_{02} & \ldots & P_{0,r-1} & P_{0r} \\ B_0 & B_1 & B_2 & \ldots & B_{r-1} & B_r^* \\ 0 & B_0 & B_1 & \ldots & B_{r-2} & B_{r-1}^* \\ \vdots & \vdots & \vdots & \ddots & \vdots & \vdots \\ 0 & 0 & 0 & \ldots & B_0 & B_1^* \end{pmatrix}. \tag{9.1.22}$$

Let

$$\boldsymbol{p}^{*T} = (\boldsymbol{p}_0^{*T}, \ldots, \boldsymbol{p}_r^{*T})$$

denote the vector of stationary probabilities of the embedded Markov chain. Then the equilibrium equations for $\boldsymbol{p}^*$ are of the form

$$\boldsymbol{p}^{*T} P = \boldsymbol{p}^{*T} \tag{9.1.23}$$

with the normalisation condition

$$\boldsymbol{p}^{*T} \mathbf{1} = 1.$$

Since the matrix P is of special structure, equations (9.1.23) can be solved with the use of the method of the preceding section or the method described in Section 7.1.2. Both methods are applied in the same way as earlier.

Knowing the stationary distribution of the embedded Markov chain, the stationary distribution of the linearwise process $\{\zeta(t), t \geq 0\}$ introduced in Section 9.1.2, can be easily calculated with the use of renewal theory apparatus. (see Section 7.1.3 and 7.1.4.

We introduce two vectors $\boldsymbol{t}_0$ and $\boldsymbol{t}_1$ whose components are (with regard for the phases of the customer generation and service processes) the mean lengths of the intervals between successive changes in the state of the embedded Markov chain. These intervals consist of an idle period and the succeeding time of service of a customer for the vector $\boldsymbol{t}_0$ or they may consist only of the service time of one customer for the vector $\boldsymbol{t}_1$. Here, as usual, the subscripts i and j of the components of the vectors $\boldsymbol{t}_0$ and $\boldsymbol{t}_1$ denote the phases of the customer generation and service processes after the first of two successive changes of state of the embedded Markov chain. Hence

$$\boldsymbol{t}_0 = -(\Lambda_0^{-1}\mathbf{1}) \otimes \mathbf{1}_m + \mathbf{1}_l \otimes \boldsymbol{b},$$
$$\boldsymbol{t}_1 = \mathbf{1}_l \otimes \boldsymbol{b},$$

where $\boldsymbol{b}$ is the vector with coordinates

$$b_i = \int_0^\infty t\,dB_i(t).$$

The mean time T between successive changes in the state of the embedded Markov chain in steady state is given by the formula

$$T = \boldsymbol{p}_0^{*T}\boldsymbol{t}_0 + \sum_{n=1}^{r} \boldsymbol{p}_n^{*T}\boldsymbol{t}_1.$$

The time-stationary probabilities $\boldsymbol{p}_0$ and stationary probability densities $\boldsymbol{p}_n(x)$ introduced in Section 9.1.2, are expressed via the stationary distribution $\boldsymbol{p}^*$ of the embedded Markov chain as

$$\boldsymbol{p}_0^T = -\frac{1}{T}\boldsymbol{p}_0^{*T}(\Lambda_0^{-1} \otimes I_m),$$

$$\boldsymbol{p}_n^T(x) = \frac{1}{T}\left(-\boldsymbol{p}_0^{*T}\left(\sum_{k=1}^{n}\left(\Lambda_0^{-1}\Lambda_k F_{n-k}(x)\right) \otimes (I - D(x))\right)\right.$$
$$\left. + \sum_{k=1}^{n} \boldsymbol{p}_k^{*T}\left(F_{n-k}(x) \otimes (I - D(x))\right)\right), \qquad n = 1, \ldots, r,$$

$$\boldsymbol{p}_R^T(x) = \frac{1}{T}\left(-\boldsymbol{p}_0^{*T}\left(\sum_{k=1}^{r}(\Lambda_0^{-1}\Lambda_k F_{R-k}^*(x)) \otimes (I - D(x))\right.\right.$$
$$\left.\left. + (\Lambda_0^{-1}\Lambda_R^* F_0^*(x)) \otimes (I - D(x))\right) + \sum_{k=1}^{r} \boldsymbol{p}_k^{*T}\left(F_{R-k}^*(x) \otimes (I - D(x))\right)\right),$$

where $D(x)$ is the matrix defined in Section 9.1.2. These formulas are different from (7.1.17)–(7.1.19), as before, in that they contain the Kronecker product and an additional sum pertaining to the semi-Markov service and non-ordinary input flow.

Other characteristics of the system are investigated in the same fashion, with the use of supplementary variables.

In conclusion, let us study, in brief, certain performance indices of the queueing system.

We first find λ, λ_A, λ_L, and λ_D—the intensities of input flow, accepted flow, lost input flow, and served flow respectively. In order to evaluate them, we introduce the vectors

$$\boldsymbol{\lambda}_k = \Lambda_k \mathbf{1}, \qquad k \geq 1,$$
$$\boldsymbol{\lambda}_k^* = \Lambda_k^* \mathbf{1}, \qquad k \geq 1,$$
$$\boldsymbol{a}_k = \sum_{l=k}^{\infty} l \boldsymbol{\lambda}_l, \qquad k \geq 1.$$

The jth element of these vectors is interpreted as follows:

$(\boldsymbol{\lambda}_k)_j$ is the intensity of arrival of a batch of k customers, provided that its generation terminates in phase j;

$(\boldsymbol{\lambda}_k^*)_j$ is the intensity of arrival of a batch of size equal to or greater than k customers, provided its generation terminates in phase j; and

$(\boldsymbol{a}_k)_j$ is the intensity of arrival of a customer in a group of size equal to or greater than k, provided that its generation terminates in phase j.

Assuming that the vector $\boldsymbol{a}_1$ is bounded, we write out the formulas for the intensities of these flows:

$$\begin{aligned}
\lambda &= \boldsymbol{\pi}^{\mathrm{a}T} \boldsymbol{a}_1, \\
\lambda_A &= \sum_{l=0}^{R} \boldsymbol{p}_l^T \left(\left(\sum_{k=0}^{R-l} k \boldsymbol{\lambda}_k + (R-l) \boldsymbol{\lambda}_{R-l+1}^* \right) \otimes \mathbf{1} \right), \\
\lambda_D &= \lambda_A, \\
\lambda_L &= \sum_{l=0}^{R} \boldsymbol{p}_l^T ((\boldsymbol{a}_{R-l+1} - (R-l) \boldsymbol{\lambda}_{R-l+1}^*) \otimes \mathbf{1}).
\end{aligned} \tag{9.1.24}$$

Certain other performance characteristics are computed by the following formulas:

the probability P_0 that the system is idle is calculated by the formula

$$P_0 = \boldsymbol{p}_0 (\mathbf{1} \otimes \mathbf{1});$$

the probability π that a customer is lost is calculated by the formula

$$\pi = \frac{\lambda_L}{\lambda} = \frac{1}{\lambda} \sum_{l=0}^{R} \boldsymbol{p}_l^T ((\boldsymbol{a}_{R-l+1} - (R-l) \boldsymbol{\lambda}_{R-l+1}^*) \otimes \mathbf{1}),$$

the mean number of customers in the system $\bar{N}$ is calculated by the formula

$$\bar{N} = \sum_{k=1}^{R} k \boldsymbol{p}_k^T (\mathbf{1} \otimes \mathbf{1}).$$

We observe that in distinction to a system with (ordinary) Markov arrival process, a stationary queue distribution exists in a system with batch Markov arrival process even if the input flow intensity λ is infinite. This becomes clear if we recall that no more than R customers can simultaneously arrive at a system with a buffer of finite capacity r. Therefore, from the viewpoint of stationary state probabilities, we can straightaway assume that any arriving customer batch contains no more than R customers; hence the input flow intensity must necessarily be finite.

Finally, we observe that the mean stationary customer sojourn time v is determined by the Little formula

$$\lambda_A v = \bar{N},$$

which can be proved by simple qualitative reasoning.

9.1.4. Infinite buffer

For the $MAP/SM/1/r$ system, no difficulties arise while passing from a finite buffer to an infinite buffer with the use of results of Chapter 8. So, in this section we only briefly mention the key points of analysis of the $MAP/SM/1/\infty$ system.

Using the technique of Chapter 8, we see that the necessary and sufficient condition for the $MAP/SM/1/\infty$ system to have a steady state is $\rho < 1$, where ρ is the traffic intensity of the system defined by the formula $\rho = \lambda b$. Here the intensity λ of the input flow is given by formula (9.1.24) and the mean service time b in the steady state is given by the expression $b = \boldsymbol{\pi}^{sT}\boldsymbol{b}$.

As for the $MAP/G/1/\infty$ system, the main difficulty in passing to the infinite buffer is the calculation of the vector $\boldsymbol{p}_0$ of the stationary probabilities that there is no customer (with account for the generation and service process phases) in the system. But all those methods that were applied for this purpose in Chapter 8 are applicable without any change to the $MAP/SM/1/\infty$ system.

Indeed, let us introduce the matrix G whose element $G_{uu'}$ is the probability that the phases of the customer generation and service processes at the end of a busy period opened by the only one customer in the system are i' and j', provided that they were i and j, respectively, at the beginning of the busy period. Here $u = (i-1)l + j$, $u' = (i'-1)l + j'$.

The matrix G can be found from the equation

$$G = \hat{B}(G) \tag{9.1.25}$$

where

$$\hat{B}(A) = \sum_{k=0}^{\infty} B_k A^k,$$

and the matrix B_k is defined by formulas (9.1.10). Equation (9.1.25) does not differ from (8.1.2) and hardly requires any explanation.

The matrix G can also be found from the equation

$$G = \int_0^{\infty} d(I_l \otimes \tilde{B}(x)) e^{x \sum_{k=0}^{\infty} (\Lambda_k \otimes I_m) G^k}, \tag{9.1.26}$$

which is an analogue of (8.4.4) and is derived as follows. Let the system at the initial instant begin to serve the only (tagged) customer. Moreover, his service time (length) is x, the service phase at the initial instant is i, and at the final instant is j. If other customers arrive in the course of service of this customer, the service discipline is changed as follows. The newly arrived customers interrupt the service of the (only) tagged customer and are served according to the LCFS discipline, but without interruption. The service of the customer arriving first at the server begins with service phase j, that of the second customer begins with the phase at which the service process was after the completion of service of the first customer, etc. Obviously, such a change of discipline has no influence on the distribution of the busy period, and the busy period terminates at the instant of completion of service of the tagged customer. Since the phases of the customer generation and service processes change during the (real) service time x (i.e., with account for interruptions) of the tagged customer according to the matrix expression $e^{x\sum_{k=0}^{\infty}(\Lambda_k \otimes I_m)G^k}$, we arrive at relation (9.1.26) with the help of the total probability formula.

Now, as on Section 8.1.2, introducing the matrices

$$\hat{B}_k = -(\Lambda_0^{-1}\Lambda_k) \otimes I_m, \qquad k \geq 1,$$

$$\tilde{B}_k = \sum_{i=0}^{\infty} B_{k+i}G^i, \qquad k \geq 0,$$

$$P_{0k}^{(k)} = \sum_{i=1}^{k+1} \hat{B}_k \tilde{B}_{k+1-i}, \qquad k \geq 0,$$

we find the stationary probabilities $\boldsymbol{p}_k^*$, $k \geq 0$, of the embedded Markov chain $\{\zeta_n, n \geq 0\}$ up to the normalisation constant from the equations

$$\boldsymbol{p}_0^{*T} = \boldsymbol{p}_0^{*T} P_{00}^{(0)},$$

$$\boldsymbol{p}_k^{*T} = \left(\boldsymbol{p}_0^{*T} P_{0k}^{(k)} + \sum_{m=1}^{k-1} \boldsymbol{p}_m^{*T} \tilde{B}_{k-m+1}\right)(I - \tilde{B}_1)^{-1}, \qquad k \geq 1.$$

The time-stationary probability densities $\boldsymbol{p}_k(x)$, $k \geq 1$, and probabilities $\boldsymbol{p}_k$, $k \geq 0$, can be found from formulas (9.1.8), (9.1.14), and (9.1.17). For this purpose, we can use the results of Section 9.1.3.

The stationary probabilities of the embedded Markov chain and the time-stationary state probabilities can also be calculated by means of the generating function apparatus. We will not do this here, because the key ideas are the same as those in Chapter 8.

The last point that we concern here is the virtual waiting time. It is more convenient to define the virtual waiting time for our system by the linearwise Markov process $\{\zeta(t),\ t \geq 0\}$ with the state set $\mathcal{X} = \{(i, j, x), i = 1, \dots, l, j = 1, \dots, m, x \geq 0\}$. Here $\zeta(t) = (i, j, 0)$ means that the system is empty at the instant t, i.e., the virtual waiting time is zero and the customer generation and service processes are in phases i and j respectively, and $\zeta(t) = (i, j, x)$, $x > 0$, means that a time x is required for the system to get rid of the customers that existed at the instant t. Here i, as before, stand for the phase of the customer generation process at this instant and j is the phase of the customer service process at the instant of completion of service of the last customer in the system at the instant t.

As usual, we use the vectors $\boldsymbol{p}_0(t)$, $\boldsymbol{W}(x, t)$, and $\boldsymbol{U}(x, t)$. The functions $\boldsymbol{p}_0(t)$, $\boldsymbol{W}(x, t)$, and $\boldsymbol{U}(x, t)$ have the same meaning as in Section 8.2, but their coordinates $p_{u0}(t)$, $W_u(x, t)$,

$U_u(x,t)$, where $u = (i-1)l + j$, take into account the customer generation phase i and service phase j.

The Takács equation for this case is of the form

$$p_0'^T(t) = p_0^T(t)\Lambda + \frac{\partial}{\partial x}U^T(x,t)\Big|_{x=0},$$

$$\frac{\partial}{\partial t}U^T(x,t) - \frac{\partial}{\partial x}U^T(x,t) = U^T(x,t)\Lambda - \frac{\partial}{\partial x}U^T(x,t)\Big|_{x=0} + \int_0^x U^T(x-y,t)\,dN(y) + p_0^T(t)N(x). \qquad (9.1.27)$$

Here

$$\Lambda = \Lambda_0 \otimes I_m, \qquad N(x) = \sum_{k=1}^{\infty} \Lambda_k \otimes \tilde{B}^{(k*)}(x).$$

Equation (9.1.27), in essence, coincides with (8.2.1). This is clear if we recall that the matrix $N(x)$ in (9.1.27) defines the total service time of all customers in a batch with regard for the phase changes of the customer generation and service processes.

We now consider the steady state of the queueing system. Let $W(x) = \lim_{t\to\infty} W(x,t)$ and let $\omega(s)$ be the corresponding Laplace–Stieltjes transform. Then in the steady state

$$\omega^T(s)(sI + \Lambda + \tilde{N}(s)) = s p_0^T, \qquad (9.1.28)$$

where

$$\tilde{N}(s) = \int_0^{\infty} e^{-sx}\,dN(x) = \sum_{k=1}^{\infty} \Lambda_k \otimes \beta^k(s) \qquad (9.1.29)$$

and $\tilde{\beta}(s)$ is the Laplace–Stieltjes transform of $\tilde{B}(x)$.

Let us set $s = 0$ in formula (9.1.28). Since $\omega(0) = p_. = \pi$ is the vector of the stationary distribution of the customer generation/service process and

$$\Lambda + \tilde{N}(0) = \sum_{k=0}^{\infty} \Lambda_k \otimes M^k, \qquad (9.1.30)$$

we obtain

$$\pi^T \sum_{k=0}^{\infty} \Lambda_k \otimes M^k = 0. \qquad (9.1.31)$$

The matrix $\sum_{k=0}^{\infty} \Lambda_k \otimes M^k$ is the infinitesimal matrix of the irreducible Markov process. Therefore, (9.1.31) has a unique solution satisfying the normalisation condition. Since the vector π^{s} satisfies the equation

$$\pi^{\mathrm{s}T} M^k = \pi^{\mathrm{s}T}$$

and the vector π^{a}, the equation

$$\pi^{\mathrm{a}T} \sum_{k=0}^{\infty} \Lambda_k = \pi^{\mathrm{a}T} \Lambda_0^* = \mathbf{0}^T,$$

for all $k \geq 0$, this solution is

$$\pi = \pi^{\mathrm{a}} \otimes \pi^{\mathrm{s}}. \tag{9.1.32}$$

This result means that in the steady state the phases of the customer generation process (at any instant) and of the service process (at the instant when the service of all customers in the system at an arbitrary instant is completed) are independent random variables and their distributions coincide with the stationary distributions of the customer generation process and embedded chain of the semi-Markov service process.

Differentiating formula (9.1.28) with respect to s and taking $s = 0$, we obtain

$$\omega'^T(0)(\Lambda + \tilde{N}(0)) = p_0^T - \omega^T(0)(I + \tilde{N}'(0)).$$

Multiplying the last equality by $\mathbf{1}$, in view of (9.1.30) and since the product of an infinitesimal matrix and $\mathbf{1}$ is the zero vector, we see that

$$0 = p_0^T \mathbf{1} - \omega^T(0)(I + \tilde{N}'(0))\mathbf{1};$$

hence, by virtue of relations (9.1.29) and (9.1.32), we obtain

$$p_0 = 1 - (\pi^{\mathrm{a}T} \otimes \pi^{\mathrm{s}T}) \sum_{k=1}^{\infty} \sum_{n=1}^{k} (\Lambda_k \otimes M^{n-1} \bar{B} M^{k-n})(\mathbf{1}_l \otimes \mathbf{1}_m),$$

where $p_0 = p_{\cdot,\cdot,0}$ is the time-stationary probability that no customers exist in the system and $\bar{B}$ is the matrix consisting of the elements $M_{jj'} b_{jj'}$.

Therefore,

$$\begin{aligned} p_0 &= 1 - \sum_{k=1}^{\infty} \pi^{\mathrm{a}T} \Lambda_k \mathbf{1}_l \sum_{n=1}^{k} \pi^{\mathrm{s}T} (M^{n-1} \bar{B} M^{k-n}) \mathbf{1}_m \\ &= 1 - \sum_{k=1}^{\infty} \pi^{\mathrm{a}T} \boldsymbol{\lambda}_k \sum_{n=1}^{k} \pi^{\mathrm{s}T} \boldsymbol{b} = 1 - \pi^{\mathrm{a}T} \sum_{k=1}^{\infty} k \boldsymbol{\lambda}_k b = 1 - \lambda b, \end{aligned}$$

i.e., the time-stationary probability that no customers exist in the system is given by the standard formula $p_0 = 1 - \rho$, which, in particular, corroborates the necessity of the condition $\rho < 1$ for the existence of steady state.

9.2. $MAP/G_2/1/r$ system with preemptive priority

In this and next sections, we consider priority queueing systems with Markov input flow. Furthermore, we only find the stationary distribution of the number of customers in a two-priority system with a common buffer of finite capacity. Due to the priorities, customers of the category of higher urgency or importance are preferred in service by worsening the service quality to others.

In this section, we study a preemptive-priority system, using the method of the embedded Markov chain.

9.2.1. System description

We consider a single-server queueing system which two types of customers arrive at. The service time of a first-type customer has some distribution function $B_1(x)$ with mean $b_1 = \int_0^\infty x dB_1(x) < \infty$, whereas that of the second-type customer has a distribution function $B_2(x)$ with mean $b_2 = \int_0^\infty x dB_2(x) < \infty$. For the sake of simplicity, we assume that the densities $b_k(x) = B_k'(x)$ of $B_k(x)$, $k = 1, 2$, exist.

The first-type customers, whom we refer to as priority customers, have preemptive priority over the second-type customers (non-priority customers). This means that a priority customer arriving at a system containing no priority customers is immediately taken up for service, interrupting the service of the non-priority customer (if any). Customers of each type are served according to the order of arrival (the FCFS discipline). The interrupted service of non-priority customers is resumed.

Customers of both types form a common queue. The number of waiting places in the system is finite, equal to r. This means that a customer of any type arriving at an instant when the system is full (the server is busy serving a customer and all r waiting place are occupied) is lost. To avoid the trivial case, we assume that $r \geq 1$.

The Markov flow of customers arriving at the system is described as follows. As before, the matrix Λ with elements Λ_{ij}, $i, j = 1, \ldots, l$, $i \neq j$, describes the changes in the customer generation phase without any new arrival. But now there are two matrices N_1 and N_2 corresponding to the changes in the customer generation phase with an arrival of a new customer. The first matrix corresponds to the arrival of a priority customer, whereas the second matrix, to the arrival of a non-priority one. As usual, let $\Lambda^* = \Lambda + N_1 + N_2$ denote the infinitesimal matrix of the customer generation process. We also assume that the matrix Λ^* is indecomposable and the matrices N_1 and N_2 are non-zero.

This model generalises the $MAP_2/G_2/1/r$ queueing system with two independent Markov flows and preemptive priority. Indeed, the $MAP_2/G_2/1/r$ system reduces to the above system if

$$\Lambda = \Lambda_1 \otimes I_{l_2} + I_{l_1} \otimes \Lambda_2, \quad N_1 = \tilde{N}_1 \otimes I_{l_2}, \quad N_2 = I_{l_1} \otimes \tilde{N}_2,$$

where Λ_1 and Λ_2, and $\tilde{N}_1$ and $\tilde{N}_2$ are the matrices of intensities of changes in the customer generation phases without and with an arrival for the Markov arrival flows of priority and non-priority customers, l_1 and l_2 are the numbers of generation phases of the Markov arrival flows of priority and non-priority customers, and $l = l_1 l_2$.

9.2.2. Embedded Markov chain

We consider

- the instants of completion of service of non-priority customers, and
- the instants at which the system is free of priority customers, provided that the system was empty at the preceding instant and then a priority customer arrived at the empty system.

The sequence of such instants is denoted by $\{\tau_k, k \geq 1\}$. Obviously, there are no priority customers in the system immediately after the instants τ_k.

The embedded Markov chain, with which we investigate the system, is formed by the phase of the customer generation process and the number of (non-priority) customers in the system immediately after the instants τ_k. Obviously, the state set of the embedded Markov chain is $\mathscr{X} = \{(i, m),\ i = 1, \dots, l,\ m = 0, \dots, r\}$, where i denotes the customer generation phase and m is the number of non-priority customers in the system.

Let $P = (P(m_1, m_2))_{m_1, m_2 = 0, \dots, r}$ be the transition probability matrix of the embedded Markov chain. Obviously, the elements $P(m_1, m_2)$ of the matrix P are square matrices of order l.

Before computing the matrix $P(m_1, m_2)$, we consider certain auxiliary matrices related to the behaviour of the system during the service of a customer or in the period when the system is busy serving priority customers.

In what follows, by an m-system, $m = 0, \dots, r$, is meant a system that differs from the initial system only in the number m of waiting places. The r-system is, obviously, identical with the initial system.

We introduce the matrix $F(k_1, k_2 \mid x)$ whose element $(F(k_1, k_2 \mid x))_{ij}$ is the probability that in time x at a system with an infinite number of waiting places (or, which is the same, at an m-system, where $m > k_1 + k_2$, containing only one priority customer at the initial instant whose service time is greater than x) k_1 priority and k_2 non-priority customers arrive and the customer generation process passes from phase i to phase j.

The matrices $F(k_1, k_2 \mid x)$ are found from the recurrence relations

$$F(0, 0 \mid x) = e^{\Lambda x}, \tag{9.2.1}$$

$$F(k, 0 \mid x) = \int_0^x F(k-1, 0 \mid y) N_1 F(0, 0 \mid x-y)\, dy, \qquad k \geq 1, \tag{9.2.2}$$

$$F(0, k \mid x) = \int_0^x F(0, k-1 \mid y) N_2 F(0, 0 \mid x-y)\, dy, \qquad k \geq 1, \tag{9.2.3}$$

$$\begin{aligned} F(k_1, k_2 \mid x) = & \int_0^x F(k_1 - 1, k_2 \mid y) N_1 F(0, 0 \mid x-y)\, dy \\ & + \int_0^x F(k_1, k_2 - 1 \mid y) N_2 F(0, 0 \mid x-y)\, dy, \qquad k_1, k_2 \geq 1. \end{aligned} \tag{9.2.4}$$

Since relations (9.2.1)–(9.2.4) are derived in the same way as relations (7.1.2) and (7.1.3), we give only a brief note concerning (9.2.4). Indeed, k_1 priority and k_2 non-priority customers arrive by the instant x, if $k_1 - 1$ priority and k_2 non-priority customers arrive by some intermediate instant y and then a priority customer arrives in time dy, and finally no more customers arrive in time $x - y$, or, vice versa, $k_2 - 1$ non-priority and k_1 priority customers arrive by the instant y and then one non-priority customer arrives in time dy and finally no more customers arrive in time $x - y$.

Let us now assume that the $(k_1 + k_2)$-system at the initial instant contains only one priority customer (at the server) whose service is not completed by the instant x. Let $F^*(k_1, k_2 \mid x)$ denote the matrix whose elements are the probabilities (with regard for the change in customer generation phase) that at the instant x the system contains k_1 priority and k_2 non-priority customers (a greater number of customers may arrive, but the customers that arrive when the system is full are lost). Then

$$F^*(0,0 \mid x) = e^{\Lambda^* x}, \tag{9.2.5}$$

$$F^*(k,0 \mid x) = \int_0^x F(k-1,0 \mid y) N_1 F^*(0,0 \mid x-y)\, dy, \qquad k \geq 1, \tag{9.2.6}$$

$$F^*(0,k \mid x) = \int_0^x F(0,k-1 \mid y) N_2 F^*(0,0 \mid x-y)\, dy, \qquad k \geq 1, \tag{9.2.7}$$

$$F^*(k_1,k_2 \mid x) = \int_0^x F(k_1-1,k_2 \mid y) N_1 F^*(0,0 \mid x-y)\, dy + \int_0^x F(k_1,k_2-1 \mid y) N_2 F^*(0,0 \mid x-y)\, dy, \quad k_1,k_2 \geq 1. \tag{9.2.8}$$

Relations (9.2.5)–(9.2.8) are derived in the same manner as (7.1.3), (7.1.4) with account for the remark made while deriving (9.2.1)–(9.2.4).

The matrices $B_1(k_1,k_2)$ and $B_1^*(k_1,k_2)$ describe the probabilities (with regard for the change in customer generation phase) that upon completion of the service of the priority customer (the only customer at the initial instant), k_1 priority and k_2 non-priority customers remain in the m-system. Here we must consider two cases separately: the case where the total number (k_1+k_2) of remaining customers is less than the number m of free waiting places and the case where the total number (k_1+k_2) of remaining customer is equal to the number m of free waiting places.

In the first case,

$$B_1(k_1,k_2) = \int_0^\infty F(k_1,k_2 \mid x)\, dB_1(x), \qquad k_1,k_2 \geq 0.$$

In the second case,

$$B_1^*(k_1,k_2) = \int_0^\infty F^*(k_1,k_2 \mid x)\, dB_1(x), \qquad k_1,k_2 \geq 0.$$

Let us consider the busy period of the m-system, $m \geq 0$, opened by $(i+1)$ priority customers (i.e., at the beginning of the busy period the system contains, besides the priority customer at the server, i more priority customers in the queue). Let $A(k \mid m,i)$, $k = 0,\dots,m$, $i = 0,\dots,\max\{0,m-1\}$, denote the matrix whose elements are the probabilities (with regard for the change in customer generation phase) that upon completion of this busy period the m-system contains k non-priority customers.

The probabilities $A(k \mid m,i)$ satisfy the relations

$$A(0 \mid 0,0) = B_1^*(0,0), \tag{9.2.9}$$

$$A(0 \mid 1,0) = B_1(0,0) + B_1^*(1,0) A(0 \mid 1,0), \tag{9.2.10}$$

$$A(1 \mid 1,0) = B_1^*(0,1) + B_1^*(1,0) A(1 \mid 1,0), \tag{9.2.11}$$

$$\begin{aligned}A(k \mid m, 0) = B_1(0, k) &+ \Bigg(B_1(1, 0) + \sum_{i=2}^{m-1} B_1(i, 0) A(0 \mid m-1, i-2) \\ &+ B_1^*(m, 0) A(0 \mid m-1, m-2) \Bigg) A(k \mid m, 0) \\ &+ \sum_{j=1}^{k} \Bigg(\sum_{i=2}^{m-1} B_1(i, 0) A(j \mid m-1, i-2) \\ &+ B_1^*(m, 0) A(j \mid m-1, m-2) \Bigg) A(k-j \mid m-j, 0) \\ &+ \sum_{j=1}^{k} \sum_{i=1}^{m-j-1} B_1(i, j) A(k-j \mid m-j, i-1) \\ &+ \sum_{j=1}^{k} B_1^*(m-j, j) A(k-j \mid m-j, m-j-1), \\ &\qquad m \geq 2, \quad k = 0, \dots, m-1, \end{aligned} \tag{9.2.12}$$

$$\begin{aligned}A(m \mid m, 0) = B_1^*(0, m) &+ \Bigg(B_1(1, 0) + \sum_{i=2}^{m-1} B_1(i, 0) A(0 \mid m-1, i-2) \\ &+ B_1^*(m, 0) A(0 \mid m-1, m-2) \Bigg) A(m \mid m, 0) \\ &+ \sum_{j=1}^{m-1} \Bigg(\sum_{i=2}^{m-1} B_1(i, 0) A(j \mid m-1, i-2) \\ &+ B_1^*(m, 0) A(j \mid m-1, m-2) \Bigg) A(m-j \mid m-j, 0) \\ &+ \sum_{j=1}^{m-2} \sum_{i=1}^{m-j-1} B_1(i, j) A(m-j \mid m-j, i-1) \\ &+ \sum_{j=1}^{m-1} B_1^*(m-j, j) A(m-j \mid m-j, m-j-1), \\ &\qquad m \geq 2, \end{aligned} \tag{9.2.13}$$

$$\begin{aligned}A(k \mid m, i) = \sum_{j=0}^{k} A(j \mid m-1, i-1) A(k-j \mid m-j, 0), \\ m \geq 2, \quad k = 0, \dots, m-1, \quad i = 1, \dots, m-1, \end{aligned} \tag{9.2.14}$$

$$A(m \mid m, i) = \sum_{j=0}^{m-1} A(j \mid m-1, i-1) A(m-j \mid m-j, 0),$$

$$m \geq 2, \quad i = 1, \ldots, m-1. \tag{9.2.15}$$

We now show how formulas (9.2.9)–(9.2.15) are derived, using relations (9.2.12) and (9.2.15) as examples.

Upon completion of the busy period in which the m-system is busy serving priority customers, there exist exactly k non-priority customers if

- during the service of the first customer (here and in what follows, by the first customer is meant the priority customer that opens the busy period), no priority customer arrives, and k non-priority customers arrive;
- during the service of the first customer, one priority and 0 non-priority customers arrive, and then after the completion of the busy period (with priority customers) opened by a priority customer, the m-system contains k non-priority customers;
- during the service of the first customer, i, $i = 2, \ldots, m-1$, priority and 0 non-priority customers arrive; then after the completion of the busy period of the $(m-1)$-system opened by $i-1$ priority customers, no non-priority customer is in the system, and, finally, after the completion of the busy period of the m-system, k non-priority customers are in the system;
- after the completion of service of the first customer, m priority and 0 non-priority customers remain in the system; then after the completion of the busy period of the $(m-1)$-system opened by $m-1$ priority customers, no non-priority customer exists in the system, and, finally, after the completion of the busy period of the m-system k non-priority customers are in the system;
- during the service of the first customer, i, $i = 2, \ldots, m-1$, priority and 0 non-priority customers arrive; then after the completion of the busy period of the $(m-1)$-system opened by $i-1$ priority customers, j, $j = 1, \ldots, k$, non-priority customers are in the system; and, finally, after the completion of the busy period of the $(m-j)$-system $k-j$ non-priority customers remain in the system;
- after the completion of service of the first customer m priority and 0 non-priority customers remain in the system; then after the completion of the busy period of the $(m-1)$-system opened by the leading $m-1$ priority customers, j, $j = 1, \ldots, k$, non-priority customers remain in the system; and, finally, after the completion of the busy period of the $(m-j)$-system, $k-j$ non-priority customers exist in the system;
- during the service of the first customer, j, $j = 1, \ldots, k$, non-priority and i, $i = 1, \ldots, m-j-1$, priority customers arrive, and then after the completion of the busy period of the $(m-j)$-system opened by i priority customers, $k-j$ non-priority customers exist in the system;
- after the completion of service of the first customer, j, $j = 1, \ldots, k$, non-priority and $m-j$ priority customers exist, and then after the completion of the busy period of the

$(m-j)$-system opened by $m-j$ priority customers, $k-j$ non-priority customers are in the system.

We thus arrive at formula (9.2.12).

While deriving relation (9.2.15), we must keep in mind the following: after the completion of the busy period of the m-system opened by $i+1$ priority customers, there will be m non-priority customers, provided that after the completion of the busy period of the $(m-1)$-system opened by i priority customers j, $j=0,\ldots,m-1$, non-priority customers exist in the system, and then after the completion of the busy period of the $(m-j)$-system $m-j$ non-priority customers are in the system.

The matrices $A(k \mid m, i)$ are calculated recursively in m.

We take $A(k \mid m) = A(k \mid m, 0)$.

Let $B_2(k \mid m)$, $m \geq 0$, $k = 0, \ldots, m$, denote the matrix whose elements are the probabilities (with regard for the change in customer generation phase) that after the completion of service of a non-priority customer (possibly, with interruptions for serving priority customers) k other non-priority customers remain in the m-system. Let the elements of the matrix $F_2(k \mid m, x)$ denote the same probabilities, provided that the time of service (without regard for interruptions by priority customers) of this non-priority customer is x. Then

$$F_2(0 \mid 0, x) = e^{\Lambda^* x}, \tag{9.2.16}$$

$$F_2(0 \mid m, x) = e^{(N_1 A(0|m-1)+\Lambda)x}, \qquad m \geq 1, \tag{9.2.17}$$

$$F_2(k \mid m, x) = \int_0^x \left(F_2(k-1 \mid m, y) N_2 + \sum_{j=0}^{k-1} F_2(j \mid m, y) N_1 A(k-j \mid m-j-1) \right) F_2(0 \mid m-k, x-y)\, dy,$$
$$m \geq 1, \quad k = 1, \ldots, m-1, \tag{9.2.18}$$

$$F_2(m \mid m, x) = \int_0^x F_2(m-1 \mid m, y) N_2 F_2(0 \mid 0, x-y)\, dy, \qquad m \geq 1. \tag{9.2.19}$$

Let us explain the derivation of formulas (9.2.16)–(9.2.19) using (9.2.18) and (9.2.19) as examples. Consider the served length y of some tagged non-priority customer of total length x, when the m-system first contains k other non-priority customers (then during the service of the remaining length $x-y$ of this customer, no non-priority customer must arrive at the $(m-k)$-system, which corresponds to the last factor in (9.2.18)). By this instant (corresponding to the length y of the tagged customer) either $(k-1)$ non-priority customers must be in the m-system and one more non-priority customer must arrive at this instant, or j, $j=0,\ldots,k-1$, non-priority customers must be in the m-system and one more priority customer must arrive at this instant, and after the completion of the busy period of priority customers of the $(m-j-1)$-system, $k-j$ more non-priority customers must remain in the system.

As concerns formula (9.2.19), m non-priority customers may exist in the m-system after the completion of service of a non-priority customer if and only if the last customer accepted for service in the m-system is a non-priority customer.

Obviously,

$$B_2(k \mid m) = \int_0^\infty F_2(k \mid m, x)\, dB_2(x), \quad m \geq 1, \quad k = 0, \dots, m.$$

In what follows, we also need two matrices B_1 and B_2. The element $(B_1)_{ij}$ of the matrix B_1 ($(B_2)_{ij}$ of the matrix B_2) is the probability that after the completion of the idle period a priority (a non-priority) customer arrives and the customer generation process is in phase j, provided that it was at the beginning of the idle period in phase i. These matrices are defined by the formula

$$B_k = \int_0^\infty e^{\Lambda x} N_k\, dx = -\Lambda^{-1} N_k, \qquad k = 1, 2.$$

Now we are able to calculate the transition probability matrix P of the embedded Markov chain.

The embedded Markov chain passes from the state $(i, 0)$ to the state (j, m) if after the completion of the idle period the customer generation process passes first to an intermediate phase k and finally to phase j in two different situations:

- if a priority customer arrives, then after the completion of the busy period of priority customers opened by this customer; or
- if a non-priority customer arrives, then after the completion of service of this non-priority customer m non-priority customers remain there.

Hence,

$$P(0, m) = B_1 A(m \mid r) + B_2 B_2(m \mid r), \qquad m = 0, \dots, r. \tag{9.2.20}$$

Thereafter, the embedded Markov chain passes from the state with m_1, $m_1 = 1, \dots, r$, customers to the state with m_2, $m_2 = m_1 - 1, \dots, r$, customers if after the completion of service of a non-priority customer in the system in which $m_1 - 1$ non-priority customers were in the queue, $m_2 - m_1 + 1$ more non-priority customers arrive. Since the probability of this event coincides with the probability that the $(r - m_1 + 1)$-system after the completion of service of a non-priority customer contains $m_2 - m_1 + 1$ non-priority customers, we obtain

$$P(m_1, m_2) = B_2(m_2 - m_1 + 1 \mid r - m_1 + 1), \quad m_1 = 1, \dots, r, \quad m_2 = m_1 - 1, \dots, r. \tag{9.2.21}$$

Finally, after the completion of service of a non-priority customer the number of non-priority customers cannot decrease by more than one. Therefore,

$$P(m_1, m_2) = (0), \quad m_1 = 2, \dots, r, \quad m_2 = 0, \dots, m_1 - 2. \tag{9.2.22}$$

Hence, the matrix P is of the form

$$P = \begin{pmatrix} P(0,0) & P(0,1) & \cdots & P(0,r-1) & P(0,r) \\ P(1,0) & P(1,1) & \cdots & P(1,r-1) & P(1,r) \\ 0 & P(2,1) & \cdots & P(2,r-1) & P(2,r) \\ \vdots & \vdots & \ddots & \vdots & \vdots \\ 0 & 0 & \cdots & P(r-1,r-1) & P(r-1,r) \\ 0 & 0 & \cdots & P(r,r-1) & P(r,r) \end{pmatrix}, \tag{9.2.23}$$

where the matrices $P(m_1, m_2)$ are defined by formulas (9.2.20)–(9.2.22).

Denoting the vector of stationary state probabilities of the embedded Markov chain by $\boldsymbol{p}^* = (\boldsymbol{p}_0^{*T}, \ldots, \boldsymbol{p}_r^{*T})^T$, we obtain the simultaneous equilibrium equations for this vector

$$\boldsymbol{p}^{*T} = \boldsymbol{p}^{*T} P$$

with the normalisation condition

$$\sum_{m=0}^{r} \boldsymbol{p}_m^{*T} \mathbf{1} = 1.$$

Methods to solve them for a matrix P of type (9.2.23) are discussed in Chapter 7.

9.2.3. Time-stationary state probabilities

Let us consider the time of service of a priority customer in a system with an infinite number of waiting places. Let $T_1(k_1, k_2)$, $k_1, k_2 \geq 0$, denote the matrix whose elements are the mean lengths of the periods in this time interval in which exactly k_1 priority and k_2 non-priority customers are in the system. Here the element $(T_1(k_1, k_2))_{ij}$ of the matrix $T_1(k_1, k_2)$ means that at the initial instant the customer generation process was in phase i and at the instant when k_1 priority and k_2 non-priority customers are in the queue, the process stays in phase j. The elements of the matrix $T_1^*(k_1, k_2)$, $k_1, k_2 \geq 0$, are of similar meaning, but for the $(k_1 + k_2)$-system. Obviously,

$$T_1(k_1, k_2) = \int_0^\infty dB_1(x) \int_0^x F(k_1, k_2 \mid t)\, dt = \int_0^\infty F(k_1, k_2 \mid t)(1 - B_1(t))\, dt,$$
$$T_1^*(k_1, k_2) = \int_0^\infty dB_1(x) \int_0^x F^*(k_1, k_2 \mid t)\, dt = \int_0^\infty F^*(k_1, k_2 \mid t)(1 - B_1(t))\, dt.$$

We now consider a busy period of the m-system, $m \geq 0$, opened by n priority customers. Let $S(k_1, k_2 \mid m, n)$, $0 \leq k_1 + k_2 \leq m$, $n = 0, \ldots, \max\{0, m-1\}$, denote the matrix whose elements are the mean lengths of the intervals in this busy period when exactly k_1 priority and k_2 non-priority customers are in the queue. The (i, j)th element of the matrix $S(k_1, k_2 \mid m, n)$ is interpreted in the same way as above.

The matrices $S(k_1, k_2 \mid m, n)$ satisfy the relations

$$S(0, 0 \mid 0, 0) = T_1^*(0, 0), \tag{9.2.24}$$
$$S(0, 0 \mid 1, 0) = T_1(0, 0) + B_1^*(1, 0) S(0, 0 \mid 1, 0), \tag{9.2.25}$$
$$S(k_1, k_2 \mid 1, 0) = T_1^*(k_1, k_2) + B_1^*(1, 0) S(k_1, k_2 \mid 1, 0), \quad k_1, k_2 \geq 0, \quad k_1 + k_2 = 1, \tag{9.2.26}$$

$$
\begin{aligned}
S(0,k \mid m,0) = T_1(0,k) + \Bigg(& B_1(1,0) + \sum_{i=2}^{m-1} B_1(i,0)A(0 \mid m-1, i-2) \\
& + B_1^*(m,0)A(0 \mid m-1, m-2) \Bigg) S(0,k \mid m,0) \\
& + \sum_{j=1}^{k} \Bigg(\sum_{i=2}^{m-1} B_1(i,0)A(j \mid m-1, i-2) \\
& + B_1^*(m,0)A(j \mid m-1, m-2) \Bigg) S(0,k-j \mid m-j,0) \\
& + \sum_{j=1}^{k} \sum_{i=1}^{m-j-1} B_1(i,j)S(0,k-j \mid m-j, i-1) \\
& + \sum_{j=1}^{k} B_1^*(m-j,j)S(0,k-j \mid m-j, m-j-1), \\
& \qquad m \geq 2, \quad k = 0, \ldots, m-1,
\end{aligned} \tag{9.2.27}
$$

$$
\begin{aligned}
S(0,m \mid m,0) = T_1^*(0,m) + \Bigg(& B_1(1,0) + \sum_{i=2}^{m-1} B_1(i,0)A(0 \mid m-1, i-2) \\
& + B_1^*(m,0)A(0 \mid m-1, m-2) \Bigg) S(0,m \mid m,0) \\
& + \sum_{j=1}^{m-1} \Bigg(\sum_{i=2}^{m-1} B_1(i,0)A(j \mid m-1, i-2) \\
& + B_1^*(m,0)A(j \mid m-1, m-2) \Bigg) (0,m-j \mid m-j,0) \\
& + \sum_{j=1}^{m-2} \sum_{i=1}^{m-j-1} B_1(i,j)S(0,m-j \mid m-j, i-1) \\
& + \sum_{j=1}^{m-1} B_1^*(m-j,j)S(0,m-j \mid m-j, m-j-1), \qquad m \geq 2,
\end{aligned} \tag{9.2.28}
$$

$$
\begin{aligned}
S(k_1,k_2 \mid m,0) = T_1(k_1,k_2) & + \sum_{i=2}^{m-1} B_1(i,0)S(k_1-1,k_2 \mid m-1, i-2) \\
& + B_1^*(m,0)S(k_1-1,k_2 \mid m-1, m-2)
\end{aligned}
$$

$$
\begin{aligned}
&+\left(B_1(1,0)+\sum_{i=2}^{m-1} B_1(i,0)A(0\mid m-1,i-2)\right.\\
&\left.+\ B_1^*(m,0)A(0\mid m-1,m-2)\right)S(k_1,k_2\mid m,0)\\
&+\sum_{j=1}^{k_2}\left(\sum_{i=2}^{m-1} B_1(i,0)A(j\mid m-1,i-2)\right.\\
&\left.+\ B_1^*(m,0)A(j\mid m-1,m-2)\right)S(k_1,k_2-j\mid m-j,0)\\
&+\sum_{j=1}^{k_2}\sum_{i=1}^{m-j-1} B_1(i,j)S(k_1,k_2-j\mid m-j,i-1)\\
&+\sum_{j=1}^{k_2} B_1^*(m-j,j)S(k_1,k_2-j\mid m-j,m-j-1),\\
&\quad m\geq 2,\quad k_1\geq 1,\quad k_2\geq 0,\quad k_1+k_2\leq m-1,
\end{aligned}
\tag{9.2.29}
$$

$$
\begin{aligned}
S(k_1,k_2\mid m,0)={}&T_1^*(k_1,k_2)+\sum_{i=2}^{m-1} B_1(i,0)S(k_1-1,k_2\mid m-1,i-2)\\
&+\ B_1^*(m,0)S(k_1-1,k_2\mid m-1,m-2)\\
&+\left(B_1(1,0)+\sum_{i=2}^{m-1} B_1(i,0)A(0\mid m-1,i-2)\right.\\
&\left.+\ B_1^*(m,0)A(0\mid m-1,m-2)\right)S(k_1,k_2\mid m,0)\\
&+\sum_{j=1}^{k_2}\left(\sum_{i=2}^{m-1} B_1(i,0)A(j\mid m-1,i-2)\right.\\
&\left.+\ B_1^*(m,0)A(j\mid m-1,m-2)\right)S(k_1,k_2-j\mid m-j,0)\\
&+\sum_{j=1}^{k_2}\sum_{i=1}^{m-j-1} B_1(i,j)S(k_1,k_2-j\mid m-j,i-1)\\
&+\sum_{j=1}^{k_2} B_1^*(m-j,j)S(k_1,k_2-j\mid m-j,m-j-1),\\
&\quad m\geq 2,\quad k_1\geq 1,\quad k_2\geq 0,\quad k_1+k_2=m,
\end{aligned}
\tag{9.2.30}
$$

$$S(0,k \mid m,n) = \sum_{j=0}^{k} A(j \mid m-1, n-1) S(0, k-j \mid m-j, 0),$$
$$m \geq 2, \quad k = 0, \ldots, m-1, \quad n = 1, \ldots, m-1, \tag{9.2.31}$$

$$S(0,m \mid m,n) = \sum_{j=0}^{m-1} A(j \mid m-1, n-1) S(0, m-j \mid m-j, 0),$$
$$m \geq 2, \quad n = 1, \ldots, m-1, \tag{9.2.32}$$

$$S(k_1,k_2 \mid m,n) = S(k_1-1, k_2 \mid m-1, n-1)$$
$$+ \sum_{j=0}^{k_2} A(j \mid m-1, n-1) S(k_1, k_2-j \mid m-j, 0),$$
$$m \geq 2, \quad k_1 \geq 1, \quad k_2 \geq 0, \quad k_1+k_2 \leq m,$$
$$n = 1, \ldots, m-1. \tag{9.2.33}$$

Since relations (9.2.24)–(9.2.33) are derived by the total expectation formula according to the one and same rule, we discuss relation (9.2.29) only. Our probabilistic explanation is given in matrix form without any resort to generation phase changes which are hidden in the matrix expression.

The mean time $S(k_1, k_2 \mid m, 0)$ necessarily contains the mean time $T_1(k_1, k_2)$ spent by the m-system in states with k_1 priority and k_2 non-priority customers in the queue during the service of the (priority) customer that opens the busy period (the first term in the right-hand side of (9.2.29)).

Now (if at least one new customer arrives during the service time of the first customer) two cases are possible.

First, after the completion of service of the first customer the system may contain i, $i \geq 1$, priority and j, $j \geq 1$, non-priority customers (with the probability $B_1(i,j)$ for $i+j < m$ and $B_1^*(m-j, j)$ for $i+j = m$). Then the mean time $S(k_1, k_2-j \mid m-j, i-1)$ spent by the $(m-j)$-system in states with k_1 priority and k_2-j non-priority customers in the queue in the busy period of priority customers must also be included (the ninth or tenth term in the right-hand side of (9.2.29)).

Second, after the completion of service of the first customer the system may contain i, $i = 1, \ldots, m$, priority and 0 non-priority customers (with the probability $B_1(i, 0)$ for $i < m$ and $B_1^*(m, 0)$ for $i = m$). Then we divide the busy period that remains after the completion of service of the first customer into two parts. The first part is the busy period of the $(m-1)$-system that is opened by $i-1$ priority customers (for $i = 1$, this busy period is absent). Of course, the mean time $S(k_1, k_2 \mid m, 0)$ must be added to the mean time $S(k_1-1, k_2 \mid m-1, i-2)$ spent by the $(m-1)$-system in states with k_1-1 priority and k_2 non-priority customers in the queue in the busy period of priority customers (the second or third term in the right-hand side of (9.2.29)).

After the completion of the first part, the system contains one priority customer and may contain j, $j = 0, \ldots, k_2$, non-priority customers (with the probability $A(j \mid m-1, i-2)$), whereas the second part is the busy period of an $(m-j)$-system. Therefore, the mean time $S(k_1, k_2-j \mid m-j, 0)$ spent by the $(m-j)$-system in states with k_1 priority and k_2-j non-priority customers in the system in the busy period of priority customers must be included. Thus, we obtain the fourth (for $i = 1$), fifth (for $i = 2, \dots, m-1$, $j = 0$), sixth (for $i = m$, $j = 0$), seventh (for $i = 2, \dots, m-1$, $j = 1, \dots, k_2$), and eighth (for $i = m$, $j = 1, \dots, k_2$) terms in the right-hand side of (9.2.29).

Relations (9.2.24)–(9.2.33) help in recursive finding of the matrices $S(k_1, k_2 \mid m, n)$ over m.

Now we set $S(k_1, k_2 \mid m) = S(k_1, k_2 \mid m, 0)$.

Let us assume that at the initial instant the m-system begins to serve the only non-priority customer. Let us consider the actual service time of this customer. It consists of his service time itself and the total time during which his service is interrupted by priority customers. Let $T_2(k_1, k_2 \mid m)$, $m \geq 0$, $k_1 + k_2 = 0, \dots, m$, denote the matrix whose elements are the mean lengths of these parts of the actual service time of this non-priority customer when k_1 priority and $(k_2 + 1)$ non-priority customers were in the m-system. The element $(T_2(k_1, k_2 \mid m))_{ij}$ of the matrix $T_2(k_1, k_2 \mid m)$ means that at the initial instant the process was in phase i and at the instant when k_1 priority and $(k_2 + 1)$ non-priority customers are in the system the process stays in phase j.

The matrices $T_2(k_1, k_2 \mid m)$ satisfy the relations

$$
\begin{aligned}
T_2(0, k \mid m) &= \int_0^\infty dB_2(x) \int_0^x F_2(k \mid m, y)\, dy \\
&= \int_0^\infty F_2(k \mid m, x)(1 - B_2(x))\, dt, \qquad k = 0, \dots, m, \\
T_2(k_1, k_2 \mid m) &= \int_0^\infty dB_2(x) \int_0^x \sum_{j=0}^{k_2} F_2(j \mid m, y) N_1 S(k_1 - 1, k_2 - j \mid m - j - 1)\, dy \\
&= \sum_{j=0}^{k_2} \int_0^\infty F_2(j \mid m, x)(1 - B_2(t))\, dx N_1 S(k_1 - 1, k_2 - j \mid m - j - 1), \\
&\qquad k_2 = 0, \dots, m-1, \quad k_1 = 1, \dots, m - k_2.
\end{aligned}
$$

Their proof is left to the reader as an exercise.

The last matrix we need is the matrix T^* whose element T^*_{ij} is the mean sojourn time of the customer generation process in phase i during an idle period, provided it was at the beginning of this idle period in phase i. As shown in Section 7.1.3, this matrix is of the form

$$T^* = -\Lambda^{-1}.$$

Let $\boldsymbol{s}(r)$ denote the vector whose coordinates are the mean lengths of the busy period of priority customers of the initial system (r-system) in which priority customers are served, and let $\boldsymbol{t}_2(m)$, $m = 0, \dots, r$, be the vector whose coordinates are the mean (actual) service times (i.e., with account for interruptions) of a non-priority customer in the m-system. The ith coordinates of the vectors $\boldsymbol{s}(r)$ and $\boldsymbol{t}_2(m)$ mean that the busy period of priority customers or service of a non-priority customer begins with phase i of the customer generation process. Obviously,

$$\boldsymbol{s}(r) = \sum_{k_1+k_2 \leq r} S(k_1, k_2 \mid r)\mathbf{1}, \qquad \boldsymbol{t}_2(m) = \sum_{k_1+k_2 \leq m} T_2(k_1, k_2 \mid m)\mathbf{1}.$$

Furthermore, let $\boldsymbol{t}^*$ be the vector whose coordinates are the mean times between the instant when the embedded Markov chain reaches the state 0 and the next instant of change of state of the chain. The coordinates of the vector $\boldsymbol{t}^*$ has the same meaning as before. Then

$$\boldsymbol{t}^* = T^*\mathbf{1} + B_1\boldsymbol{s}(r) + B_2\boldsymbol{t}_2(r).$$

Finally, let T denote the mean time between the instants of change of state of the embedded Markov chain in the stationary state. Obviously,

$$T = \boldsymbol{p}_0^{*T}\boldsymbol{t}^* + \sum_{m=1}^{r} \boldsymbol{p}_m^{*T}\boldsymbol{t}_2(r-m+1).$$

Now we are able to find the time-stationary probabilities of the states $\boldsymbol{p}_0$ and $\boldsymbol{p}_{k_1k_2}$, where the coordinates of the vector $\boldsymbol{p}_0$ are the probabilities that there no customers exist in the system and the coordinates of the vector $\boldsymbol{p}_{k_1k_2}$, $k_1+k_2 = 1,\dots,r+1$, are the probabilities that the system contains k_1 priority and k_2 non-priority customers, where the number of the coordinate of the respective vector denotes the number of the phase of the customer generation process. Indeed, by the general theorem on the relationship between the time-stationary probabilities of a semi-Markov process and stationary probabilities for embedded Markov chain, we see that

$$\boldsymbol{p}_0^T = \frac{1}{T}\boldsymbol{p}_0^{*T}T^*,$$

$$\boldsymbol{p}_{k0}^T = \frac{1}{T}\boldsymbol{p}_0^{*T}B_1S(k_1-1,0\mid r), \qquad k=1,\dots,r+1,$$

$$\boldsymbol{p}_{0k}^T = \frac{1}{T}\left(\boldsymbol{p}_0^{*T}B_2T_2(0,k-1\mid r) + \sum_{m=1}^{k}\boldsymbol{p}_m^{*T}T_2(0,k-m\mid r-m+1)\right), \qquad k=1,\dots,r,$$

$$\boldsymbol{p}_{0,r+1}^T = \frac{1}{T}\left(\boldsymbol{p}_0^{*T}B_2T_2(0,r\mid r) + \sum_{m=1}^{r}\boldsymbol{p}_m^{*T}T_2(0,r+1-m\mid r-m+1)\right),$$

$$\boldsymbol{p}_{k_1k_2}^T = \frac{1}{T}\Bigg(\boldsymbol{p}_0^{*T}(B_1S(k_1-1,k_2\mid r) + B_2T_2(k_1,k_2-1\mid r)) + \sum_{m=1}^{k_2}\boldsymbol{p}_m^{*T}T_2(k_1,k_2-m\mid r-m+1)\Bigg), \qquad k_1,k_2\ge 1, \quad k_1+k_2\le r+1.$$

9.2.4. Performance indices

Since the main performance indices of the system are computed in the same way as those for a system without any priority, here we consider only the intensities $\lambda^{(s)}$, $\lambda_L^{(s)}$, and $\lambda_D^{(s)}$, $s=1,2$, of the flow of customers arriving at the system, the flow of customers lost at the input of the system, and the flows of served priority customers ($s=1$) and non-priority customers ($s=2$), and the probabilities $\pi^{(1)}$ and $\pi^{(2)}$ of losses of priority and non-priority customers.

The intensity $\lambda^{(s)}$ of arriving flow of customers is given by the formula

$$\lambda^{(s)} = \boldsymbol{\pi}^T N_s \mathbf{1},$$

where, as usual, $\boldsymbol{\pi}$ is the vector of stationary probabilities of the states of the customer generation process.

To calculate the intensity of the lost flow, it suffices to observe that, according to the service discipline for customers of different priorities, both priority and non-priority customers are lost only if the total number of customers in the buffer is r. Therefore,

$$\lambda_L^{(s)} = \sum_{s=1}^{2} \sum_{k_1+k_2=r} \boldsymbol{p}_{k_1 k_2}^T N_s \mathbf{1}. \tag{9.2.34}$$

The intensity $\lambda_D^{(s)}$ of the served flow is easily found from the expression

$$\lambda_D^{(s)} = \lambda^{(s)} - \lambda_L^{(s)}.$$

Finally, the probability $\pi^{(s)}$ that a customer is lost is given by the formula

$$\pi^{(s)} = \frac{\lambda_L^{(s)}}{\lambda^{(s)}},$$

which hardly needs any explanation.

9.3. $MAP/G_2/1/r$ system with non-preemptive priority

The system we study in this section differs from the system studied in the preceding section only in that the preemptive priority is replaced by a non-preemptive priority, in which a newly arriving customer cannot interrupt the service of the non-priority customer at the server, but is served ahead of all non-priority customers in the queue. Moreover, we apply the method of supplementary variables to study the $MAP/G_2/1/r$ system with non-preemptive priority.

Using the notation and assumptions of Section 9.2 for the parameters of the Markov flow and distribution function $B_s(x)$, we introduce the following additional technical conditions. We assume that $B_s(x) < 1$ for all x and $\beta_s(-\sigma_i) \neq 0$ for all eigenvalues σ_i of the matrix Λ, where $\beta_s(y)$ is the Laplace–Stieltjes transform of the distribution function $B_s(x)$, $s = 1, 2$.

9.3.1. Differential equations

The functioning of the system is described by the linearwise Markov process $\{\gamma(t), t \geq 0\}$ with state set $\mathcal{X} = \{(i, 0), i = 1, \dots, l;\ (s, i, n, m, x), s = 1, 2, i = 1, \dots, l, 0 \leq n + m \leq r, x \geq 0\}$. The states of the process $\{\gamma(t), t \geq 0\}$ are interpreted as follows:

$(i, 0)$ denotes the state in which the system is empty and the customer generation process is in phase i, and

(s, i, n, m, x) denotes the situation in which a priority customer (for $s = 1$) or a non-priority customer (for $s = 2$) is being served at the server, the customer generation process is in phase i, the buffer contains n priority and m non-priority customers, and the time elapsed since the beginning of the customer at the server is x.

Let p_{i0} be the stationary probability of the state $(i, 0)$, and let $p_{sinm}(x)$ be the stationary probability density of the state (s, i, n, m, x). Let

$$\boldsymbol{p}_0 = (p_{10}, p_{20}, \dots, p_{l0})^T,$$
$$\boldsymbol{p}_{snm}(x) = (p_{s1nm}(x), p_{s2nm}(x), \dots, p_{slnm}(x))^T.$$

Representing the vector $\boldsymbol{p}_{snm}(x)$ as

$$\boldsymbol{p}_{snm}(x) = [1 - B_s(x)]\boldsymbol{q}_{snm}(x) \tag{9.3.1}$$

and reasoning as in Section 7.2, we easily derive the differential equations for $\boldsymbol{q}_{snm}(x)$, $s = 1, 2, 0 \le n + m \le r$:

$$\frac{d}{dx}\boldsymbol{q}^T_{snm}(x) = \boldsymbol{q}^T_{snm}(x)\Lambda + u(n)\boldsymbol{q}^T_{s,n-1,m}(x)N_1 + u(m)\boldsymbol{q}^T_{s,n,m-1}(x)N_2, \qquad 0 \le n + m \le r - 1, \tag{9.3.2}$$

$$\frac{d}{dx}\boldsymbol{q}^T_{s,r-m,m}(x) = \boldsymbol{q}^T_{s,r-m,m}(x)\Lambda^* + u(r - m)\boldsymbol{q}^T_{s,r-m-1,m}(x)N_1 + u(m)\boldsymbol{q}^T_{s,r-m,m-1}(x)N_2, \qquad m = 0, \dots, r, \tag{9.3.3}$$

with the boundary conditions

$$\boldsymbol{0}^T = \boldsymbol{p}_0^T\Lambda + \sum_{s=1}^{2}\int_0^\infty \boldsymbol{q}^T_{s00}(x)\,dB_s(x), \tag{9.3.4}$$

$$\boldsymbol{q}^T_{1nm} = \sum_{s=1}^{2}\int_0^\infty \boldsymbol{q}^T_{s,n+1,m}(x)\,dB_s(x) + u(1 - n - m)\boldsymbol{p}_0^T N_1, 0 \le n + m \le r - 1, \tag{9.3.5}$$

$$\boldsymbol{q}^T_{20m} = \sum_{s=1}^{2}\int_0^\infty \boldsymbol{q}^T_{s,0,m+1}(x)\,dB_s(x) + u(1 - m)\boldsymbol{p}_0^T N_2, \qquad m = 0, \dots, r - 1, \tag{9.3.6}$$

$$\boldsymbol{q}_{1nm} = \boldsymbol{0}, \qquad n + m = r, \tag{9.3.7}$$

$$\boldsymbol{q}_{2nm} = \boldsymbol{0}, \qquad m = 0, \dots, r - 1, \quad n = 1, \dots, r - m, \tag{9.3.8}$$

$$\boldsymbol{q}_{20r} = \boldsymbol{0}, \tag{9.3.9}$$

where $u(x)$ is the Heaviside function and $\boldsymbol{q}_{snm} \equiv \boldsymbol{q}_{snm}(0)$.

Let $p_{sinm} = \int_0^\infty p_{sinm}(x)\,dx$ denote the stationary probability of the states (s, i, n, m, x) without account for the elapsed service time x. Let $\boldsymbol{p}_{snm} = (p_{s1nm}, \dots, p_{slnm})^T$. Then the normalisation condition is as follows:

$$\boldsymbol{p}_0^T\boldsymbol{1} + \boldsymbol{p}^T_{\cdot,\cdot,\cdot}\boldsymbol{1} = 1. \tag{9.3.10}$$

9.3.2. Solution of the differential equations

Now we need the matrix functions $F(n, m \mid x)$ and $F^*(n, m \mid x)$ introduced in Section 9.2.2. It is easily seen that these functions satisfy the relations

$$\begin{aligned} \frac{d}{dx} F(0,0 \mid x) &= F(0,0 \mid x)\Lambda, \\ \frac{d}{dx} F(n,m \mid x) &= F(n,m \mid x)\Lambda + u(n)F(n-1,m \mid x)N_1 \\ &\quad + u(m)F(n,m-1 \mid x)N_2, \qquad n+m > 0, \end{aligned} \tag{9.3.11}$$

and

$$\begin{aligned} \frac{d}{dx} F^*(0,0 \mid x) &= F^*(0,0 \mid x)\Lambda^*, \\ \frac{d}{dx} F^*(n,m \mid x) &= F^*(n,m \mid x)\Lambda^* + u(n)F(n-1,m \mid x)N_1 \\ &\quad + u(m)F(n,m-1 \mid x)N_2, \qquad n+m > 0. \end{aligned} \tag{9.3.12}$$

The solution of differential equations (9.3.2) and (9.3.3) is of the form

$$q^T_{snm}(x) = \sum_{k=0}^{n} \sum_{l=0}^{m} q^T_{s,n-k,m-l} F(k,l \mid x), \qquad 0 \le n+m \le r-1, \tag{9.3.13}$$

$$q^T_{s,r-m,m}(x) = \sum_{k=0}^{r-m} \sum_{l=0}^{m} q^T_{s,r-m-k,m-l} F^*(k,l \mid x), \qquad m = 0, \dots, r. \tag{9.3.14}$$

Formulas (9.3.13) and (9.3.14) can be easily proved with recourse to the probabilistic meaning of the matrix functions $F(n, m \mid x)$ and $F^*(n, m \mid x)$ and vectors $q_{snm} = p_{snm}(0)$. Validity of (9.3.13) and (9.3.14) can also be established by substituting them into differential equations (9.3.2) and (9.3.3), which, in view of (9.3.11) and (9.3.12), transform them to identities.

In formulas (9.3.13) and (9.3.14), the vectors q_{skl}, $s = 1, 2$, $k = 0, \dots, r$, $l = 0, \dots, r-k$ are unknown yet. In order to find these vectors, we substitute expressions (9.3.13) and (9.3.14) into the equations for the boundary conditions. We thus arrive at the linear algebraic equations in unknowns p_0 and q_{snm}

$$-p_0^T \Lambda = \sum_{s=1}^{2} q^T_{s00} B_{s00}, \tag{9.3.15}$$

$$\begin{aligned} q^T_{1nm} &= \sum_{k=0}^{n+1} \sum_{l=0}^{m} q^T_{1,n+1-k,m-l} B_{1kl} + \sum_{k=0}^{m} q^T_{2,0,m-k} B_{2,n+1,k} \\ &\quad + u(1-n-m) p_0^T N_1, \qquad m = 0, \dots, r-2, \quad n = 0, \dots, r-2-m, \end{aligned} \tag{9.3.16}$$

$$q^T_{1,r-1-m,m} = \sum_{k=0}^{r-m} \sum_{l=0}^{m} q^T_{1,r-m-k,m-l} B^*_{1kl} + \sum_{k=0}^{m} q^T_{2,0,m-k} B^*_{2,r-m,k}, \qquad m = 0, \dots, r-1, \tag{9.3.17}$$

$$q_{20m}^T = \sum_{s=1}^{2}\sum_{k=0}^{m+1} q_{s,0,m+1-k}^T B_{s0k} + u(1-m)p_0^T N_2, \qquad m = 0, \dots, r-2, \tag{9.3.18}$$

$$q_{2,0,r-1}^T = \sum_{s=1}^{2}\sum_{k=0}^{r} q_{s,0,r-k}^T B_{s0k}^*, \tag{9.3.19}$$

where

$$B_{skl} = \int_0^\infty F(k, l \mid x)\, dB_s(x), \qquad B_{skl}^* = \int_0^\infty F^*(k, l \mid x)\, dB_s(x).$$

The elements of the matrix B_{skl} (with account for the changes in the customer generation phases) are the probabilities that k priority and l non-priority customers arrive during the time of service of a priority (for $s = 1$) or a non-priority (for $s = 2$) customer, and the elements of the matrix B_{skl}^* are the probabilities that after the completion of service of a priority (for $s = 1$) or a non-priority (for $s = 2$) customer k priority and l non-priority more customers appear in the queue, provided that $r - k - l$ customers were at the beginning of service.

Before finding the boundary vectors q_{skl} from equations (9.3.15)–(9.3.19), we rearrange them. We introduce

$$V_n = \begin{pmatrix} B_{100} & B_{110} - I & B_{120} & \dots & B_{1n0} \\ 0 & B_{100} & B_{110} - I & \dots & B_{1,n-1,0} \\ 0 & 0 & B_{100} & \dots & B_{1,n-2,0} \\ \vdots & \vdots & \vdots & \ddots & \vdots \\ 0 & 0 & 0 & \dots & B_{100} \end{pmatrix},$$

$$A_{lm} = \begin{pmatrix} B_{10m} & B_{11m} & B_{12m} & \dots & B_{1lm} \\ 0 & B_{10m} & B_{11m} & \dots & B_{1,l-1,m} \\ 0 & 0 & B_{10m} & \dots & B_{1,l-2,m} \\ \vdots & \vdots & \vdots & \ddots & \vdots \\ 0 & 0 & 0 & \dots & B_{10m} \end{pmatrix},$$

$$q_{lm} = (q_{10m}^T, \dots, q_{1lm}^T)^T,$$

$$q_1 = (q_{r-1,0}^T, q_{r-2,1}^T, \dots, q_{0,r-1}^T)^T.$$

By re-arranging, equations (9.3.15), (9.3.16), and (9.3.18) reduce to the form

$$q_1^T W = d^T,$$

where W is the upper triangular block matrix

$$W = \begin{pmatrix} V_{r-1} & A_{r-2,1} & A_{r-3,2} & \dots & A_{0,r-1} \\ 0 & V_{r-2} & A_{r-3,1} & \dots & A_{0,r-2} \\ 0 & 0 & V_{r-3} & \dots & A_{0,r-3} \\ \vdots & \vdots & \vdots & \ddots & \vdots \\ 0 & 0 & 0 & \dots & V_0 \end{pmatrix}$$

and $\boldsymbol{d}$ is a vector.

We observe that the matrices W and V_n have triangular block structure. Since the matrix B_{100} is invertible (which follows from the additional assumption made at the beginning of this section), we express all unknowns $\boldsymbol{q}_{1nm}$, $0 \le n+m \le r-1$, in terms of $\boldsymbol{q}_{20j}$ and $\boldsymbol{p}_0$.

Thus, the vectors $\boldsymbol{q}_{1nm}$ admit the representation

$$\boldsymbol{q}_{1nm}^T = \sum_{j=0}^{m} \boldsymbol{q}_{20j}^T \chi_{n,m-j} + \boldsymbol{p}_0^T \sigma_{nm}, \qquad 0 \le n+m \le r-1, \tag{9.3.20}$$

where χ_{nm} and σ_{nm} are some matrices.

In order to determine the matrices χ_{nm} and σ_{nm} in formula (9.3.20), we make use of the method of indefinite coefficients. To this end, we substitute (9.3.20) into (9.3.15), (9.3.16), and (9.3.18). Since the vectors $\boldsymbol{q}_{20j}$ and $\boldsymbol{p}_0$ are arbitrary and their coordinates are non-negative, equating their coefficients in the right-hand and left-hand sides of equalities obtained after substitution, we arrive at the recurrence relations

- for the matrices χ_{nm},

$$\chi_{00} = -B_{200}B_{100}^{-1}, \tag{9.3.21}$$

$$\chi_{nm} = \left(\chi_{n-1,m} - \sum_{l=0}^{m-1} \chi_{n,m-1-l}B_{1,0,l+1} - \sum_{k=0}^{n-1}\sum_{l=0}^{m} \chi_{n-1-k,m-l}B_{1,k+1,l} - B_{2nm}\right)B_{100}^{-1},$$
$$m = 0,\dots,r-2, \qquad n = 1,\dots,r-1-m, \tag{9.3.22}$$

$$\chi_{0m} = \left(\delta_{1m}I - B_{20m} - \sum_{l=0}^{m-1} \chi_{0l}B_{1,0,m-l}\right)B_{100}^{-1}, \qquad m = 1,\dots,r-1; \tag{9.3.23}$$

- for the matrices σ_{nm},

$$\sigma_{00} = -\Lambda B_{100}^{-1}, \tag{9.3.24}$$

$$\sigma_{nm} = \left[\sigma_{n-1,m} - \sum_{l=0}^{m-1} \sigma_{n,m-1-l}B_{1,0,l+1} - \sum_{k=0}^{n-1}\sum_{l=0}^{m} \sigma_{n-1-k,m-l}B_{1,k+1,l} - u(2-n-m)N_1\right]B_{100}^{-1},$$
$$m = 0,\dots,r-2, \qquad n = 1,\dots,r-1-m, \tag{9.3.25}$$

$$\sigma_{0m} = -\left[\sum_{l=0}^{m-1} \sigma_{0,m-1-l}B_{1,0,l+1} + u(2-m)N_2\right]B_{100}^{-1}, \qquad m = 1,\dots,r-1, \tag{9.3.26}$$

where δ_{ij} is the Kronecker delta.

Thus, using formulas (9.3.21)–(9.3.26) and knowing the vectors $\boldsymbol{q}_{20j}$ and $\boldsymbol{p}_0$, we are able to compute the vectors $\boldsymbol{q}_{1nm}$, $0 \le n+m \le r-1$, by formula (9.3.20).

Let us find the vectors $\boldsymbol{q}_{20j}$ and $\boldsymbol{p}_0$. Note that (9.3.17) was not used to find the vectors $\boldsymbol{q}_{1nm}$. Substituting expression (9.3.20) of the vectors $\boldsymbol{q}_{1nm}$ into the right-hand side of (9.3.17), we obtain

$$\sum_{j=0}^{m} \boldsymbol{q}_{20j}^T \varphi_{mj} = \boldsymbol{p}_0^T \psi_m, \qquad m = 0, \ldots, r-1, \tag{9.3.27}$$

where

$$\varphi_{mj} = \chi_{r-1-m,m-j} - \sum_{l=j}^{m} \sum_{k=0}^{r-1-m} \chi_{k,l-j} B_{1,r-m-k,m-l}^* - u(m-j) \sum_{l=j}^{m-1} \chi_{r-m,l-j} B_{1,0,m-l}^*$$
$$- B_{2,r-m,m-j}^*, \quad m = 0, \ldots, r-1, \quad j = 0, \ldots, m,$$
$$\psi_m = \sum_{l=0}^{m} \sum_{k=0}^{r-1-m} \sigma_{kl} B_{1,r-m-k,m-l}^* + \sum_{l=0}^{m-1} \sigma_{r-m,l} B_{1,0,m-l}^*$$
$$- \sigma_{r-1-m,m}, \qquad m = 0, \ldots, r-1.$$

Equations (9.3.27) in unknowns $\boldsymbol{q}_{20j}$, $j = 0, \ldots, r-1$, are derived from equations (9.3.15)–(9.3.18) where $\boldsymbol{q}_{1nm}$, $0 \le n+m \le r-1$, and $\boldsymbol{q}_{20m}$, $m = 0, \ldots, r-1$, are assumed to be unknown with a non-degenerate coefficient matrix. In fact, equations (9.3.27) are derived from (9.3.15)–(9.3.18) by means of elimination of the unknowns $\boldsymbol{q}_{1nm}$, $0 \le n+m \le r-1$. Hence the matrix of coefficients of (9.3.27) must also be non-degenerate. Moreover, this matrix is triangular. Therefore, we will seek for the solution in the form

$$\boldsymbol{q}_{20j}^T = \boldsymbol{p}_0^T Q_{20j}, \qquad j = 0, \ldots, r-1, \tag{9.3.28}$$

where Q_{20j} are some matrices. In order to determine them, we once again apply the method of indefinite coefficients. Substituting (9.3.28) into equalities (9.3.27), we obtain

$$Q_{20m} = \left(\psi_m - \sum_{j=0}^{m-1} Q_{20j} \varphi_{mj} \right) \varphi_{mm}^{-1}, \qquad m = 0, \ldots, r-1.$$

The invertibility of the matrices φ_{mm} is implied in the non-degeneracy of the matrix of coefficients (9.3.27) and its triangular form.

From (9.3.20) and (9.3.28) we immediately obtain the expression of $\boldsymbol{q}_{1nm}$

$$\boldsymbol{q}_{1nm}^T = \boldsymbol{p}_0^T Q_{1nm}, \qquad 0 \le n+m \le r-1, \tag{9.3.29}$$

where the matrices Q_{1nm} are determined from the recurrence relations

$$Q_{1nm} = \sum_{j=0}^{m} Q_{20j} \chi_{n,m-j} + \sigma_{nm}, \qquad 0 \le n+m \le r-1.$$

Thus, we have found all vectors $\boldsymbol{q}_{snm}$ in terms of the vector $\boldsymbol{p}_0$. Now we must find $\boldsymbol{p}_0$. This vector can be found from the thus far unused (9.3.19) by substituting expression (9.3.20) for $\boldsymbol{q}_{1nm}$ and expression (9.3.28) for $\boldsymbol{q}_{20m}$. Since the unknowns $\boldsymbol{p}_0$, $\boldsymbol{q}_{1nm}$, and $\boldsymbol{q}_{20m}$ are determined from equations (9.3.15)–(9.3.19) uniquely up to a constant, the matrix of coefficients of the equations derived from (9.3.19) for $\boldsymbol{p}_0$ is degenerate and its rank is $l-1$, i.e., one less than the number of unknowns p_{i0}, $i = 1, \ldots, l$. But to find the vector $\boldsymbol{p}_0$, we use not these equations, but equivalent ones derived below. For this purpose, we turn back to equations (9.3.2) and (9.3.3) and sum them over all possible n and m. Solving the differential equations obtained, we arrive at

$$\boldsymbol{q}_{s,\cdot,\cdot}^T(x) = \boldsymbol{q}_{s,\cdot,\cdot}^T e^{\Lambda^* x}, \qquad s = 1, 2. \tag{9.3.30}$$

Hence it follows that

$$\boldsymbol{q}_{\cdot,\cdot,\cdot}^T(x) = \boldsymbol{q}_{\cdot,\cdot,\cdot}^T e^{\Lambda^* x}. \tag{9.3.31}$$

Now summing (9.3.4)–(9.3.9) over all possible s, n, and m, and using (9.3.30) and (9.3.31), we obtain

$$\boldsymbol{q}_{\cdot,\cdot,\cdot}^T = \sum_{s=1}^{2} \boldsymbol{q}_{s,\cdot,\cdot}^T \int_0^\infty e^{\Lambda^* x} dB_s(x) + \boldsymbol{p}_0^T \Lambda^*.$$

Hence, by virtue of (9.3.29), we obtain the equality

$$\boldsymbol{p}_0^T \Lambda^* = \boldsymbol{p}_0^T \sum_{s=1}^{2} Q_{s,\cdot,\cdot}(I - B_{s00}^*).$$

Introducing the matrix

$$Q = \Lambda^* - \sum_{s=1}^{2} Q_{s,\cdot,\cdot}(I - B_{s00}^*),$$

we rewrite the last equality as

$$\boldsymbol{p}_0^T Q = \boldsymbol{0}^T. \tag{9.3.32}$$

The vector $\boldsymbol{p}_0$ is determined by linear algebraic equations (9.3.32) up to a constant which is found, after computing the stationary probabilities $\boldsymbol{p}_{snm}$, also up to a constant, from normalisation condition (9.3.10).

9.3.3. Stationary probabilities

After computing $\boldsymbol{p}_0$ and $\boldsymbol{q}_{snm}(x)$, we easily find the stationary probabilities of the states of the system under consideration.

Indeed, the vectors $p_{sinm}(x)$ of stationary probability densities of the states (s, i, n, m, x) are determined by formula (9.3.1), whereas the vectors p_{sinm} of stationary probabilities of the states (s, i, n, m, x) are defined, without any regard for the elapsed waiting time x, as follows from (9.3.13) and (9.3.14), by the expressions

$$p^T_{snm} = \sum_{k=0}^{n}\sum_{l=0}^{m} q^T_{s,n-k,m-l} T_{skl}, \qquad 0 \le n+m \le r-1,$$

$$p^T_{s,r-m,m} = \sum_{k=0}^{r-m}\sum_{l=0}^{m} q^T_{s,r-m-k,m-l} T^*_{skl}, \qquad m = 0, \dots, r,$$

where

$$T_{skl} = \int_0^\infty [1 - B_s(x)] F(k, l \mid x)\, dx, \qquad 0 \le k+l \le r-1,$$

$$T^*_{skl} = \int_0^\infty [1 - B_s(x)] F^*(k, l \mid x)\, dx, \qquad 0 \le k+l \le r.$$

We recall that the stationary state probabilities are calculated up to a constant, which is then found from the normalisation condition (9.3.10).

9.3.4. Performance indices

The main performance indices of the $MAP_2/G_2/1/r$ queueing system with non-preemptive priority are determined by the same formulas as for the system with preemptive priority. To do so, in formula (9.2.34) we set

$$p_{k_1k_2} = \sum_{s=1}^{2} p_{sk_1k_2}.$$

9.4. $MAP/G/1/r$ retrial system

In this section, we once again turn to a single-server queueing system with Markov input flow, a buffer of finite capacity, and single-type customers. But, along with the customers arriving at the system from outside, retrial customers may also circulate in the system. Let us recall (see Section 3.7) that a primary customer becomes a retrial customer if upon his arrival the server and all waiting places are occupied, and a primary customer differs from a retrial customer in that the retrial customer repeats his attempts to get service from the queue of retrial customers after certain (random) time intervals.

Here for the $MAP/G/1/r$ system with one of the possible service disciplines for primary and retrial customers, we will find the stationary distributions of the number of primary and retrial customers in the system by means of analysing the embedded Markov chain generated by the instants when the server becomes free.

9.4.1. Description of the system

We consider a $MAP/G/1/r$ system with buffer of finite capacity r which a Markov flow of customers arrives at, and the customer generation process, as usual, is defined by matrices Λ and N of order l. Each customer is served for a random time with some distribution function $B(x)$.

We assume that all conditions that were imposed on Λ, N, and $B(x)$ in Chapter 7 are fulfilled. Furthermore, we use many of the symbols of Chapter 7. So, $\Lambda^* = \Lambda + N$ denotes the infinitesimal matrix of the customer generation process and $b = \int_0^\infty x dB(x) < \infty$ is the mean service time.

The system contains separate queues for primary and retrial customers. Moreover, primary and retrial customers are served according to the following service disciplines.

A newly arriving customer first goes to the server (if it is free) or joins the queue of primary customers if there is any free place. The queue of primary customers can hold at most r, $r \geq 1$, customers. A customer for service is chosen from the queue of primary customers according to the standard FCFS discipline.

If the server and all waiting places for primary customers are occupied, then a newly arriving customer becomes a retrial customer and joins the queue for retrial customers, called the orbit. The number of retrial customers in the orbit is also bounded by a number m. If the orbit contains m retrial customers, a customer is lost if he arrives at an instant when the queue for primary customers is full.

Each retrial customer staying at instant t in the orbit along with $n-1$, $1 \leq n \leq m$, other retrial customers, irrespective of the process of operation of the system up to this instant t and, in particular, irrespective of the number of his previous retrials in the time interval $(t, t+\Delta)$, with the probability $(\gamma_n/n)\Delta + o(\Delta)$ repeats his attempt to get service. If the server is busy, the customer cannot join the queue for primary customers and remains in the orbit. Thus, a retrial customer repeats his attempts to get service until he is accepted by the server for service. The total intensity γ_n of retrial customers may depend on the number n of retrial customers in the system at an instant, though. For the sake of uniformity, we set $\gamma_0 = 0$.

Such a service discipline for retrial customers implies that primary customers are served with a non-preemptive priority.

9.4.2. Auxiliary results

In what follows, we need some auxiliary matrices. They can be described in terms of the usual $MAP/G/1/\infty$ system with the FCFS discipline; this is what we do in this section.

Let B_k and B_k^*, $k \geq 0$, denote, respectively, the matrices whose elements are the probabilities that during the service of a customer k and at least k customers arrive and the customer generation process passes to phase j, provided that at the initial instant it was in phase i. Let $F_k(x)$ and $F_k^*(x)$ denote similar matrices, but under the additional condition that the service of a customer lasts for a time x. Recall that these matrices are defined by formulas (7.1.1)–(7.1.6).

Furthermore, we need three vectors $\boldsymbol{t}_k$, $\boldsymbol{t}_k^*$, and $\boldsymbol{t}_k^{**}$, whose coordinates t_{ik}, t_{ik}^*, and t_{ik}^{**}, $i = 1, \dots, l$, define the mean sojourn times of a customer in states when k, at least k, and no more than k, respectively, customers had arrived during the service of a customer, provided that the customer generation process was in phase i at the service beginning. It is easily seen that these vectors, according to Section 7.1.3, are defined by the formulas

$$\boldsymbol{t}_k = \int_0^\infty (1 - B(x)) F_k(x) \mathbf{1}\, dx, \qquad k \geq 0,$$

$$\boldsymbol{t}_k^{**} = \sum_{n=0}^{k} \boldsymbol{t}_n, \qquad k \geq 0,$$

$$\boldsymbol{t}_k^{*} = \int_0^\infty (1 - B(x)) F_k^*(x) \mathbf{1}\, dx, \qquad k \geq 0.$$

Now we consider the busy period of an ordinary $MAP/G/1/n$ system without retrial customers, $n = 1, \dots, r$.

We introduce two matrices $L_k(n, s)$ and $L_k^*(n, s)$, $k \geq 0$, $n \geq 1$, $s = 0, \dots, n-1$, whose elements $(L_k(n, s))_{ij}$ and $(L_k^*(n, s))_{ij}$, $i, j = 1, \dots, l$, are the probabilities that k and at least k customers, respectively, are lost in a busy period and the customer generation process passes to phase j when the busy period terminates, provided that at the beginning of the busy period the customer generation process was in phase i and the system contained, besides the customer at the server, s more customers in the queue.

The matrices $L_k(n, s)$ and $L_k^*(n, s)$ satisfy the relations

$$L_0(n, 0) = F_0 + \left(F_1 + \sum_{i=2}^{n} F_i L_0(n-1, i-2) \right) L_0(n, 0), \tag{9.4.1}$$

$$\begin{aligned} L_k(n, 0) = {} & \left(F_1 + \sum_{i=2}^{n} F_i L_0(n-1, i-2) \right) L_k(n, 0) \\ & + \sum_{j=1}^{k} \sum_{i=2}^{n} F_i L_j(n-1, i-2) L_{k-j}(n, 0) \\ & + \sum_{j=1}^{k} F_{n+j} L_{k-j}(n, n-1), \qquad k \geq 1, \end{aligned} \tag{9.4.2}$$

$$L_k(n, s) = \sum_{j=0}^{k} L_j(n-1, s-1) L_{k-j}(n, 0), \qquad k \geq 1, \; s = 1, \dots, n-1, \tag{9.4.3}$$

$$L_0^*(1, 0) = F_0 + F_1^* L_0^*(1, 0), \tag{9.4.4}$$

$$\begin{aligned} L_0^*(n, 0) = F_0 + \Bigg(& F_1 + \sum_{i=2}^{n-1} F_i L_0^*(n-1, i-2) \\ & + F_n^* L_0^*(n-1, n-2) \Bigg) L_0^*(n, 0), \qquad n \geq 2, \end{aligned} \tag{9.4.5}$$

$$\begin{aligned}
L_k^*(n,0) &= \left(F_1 + \sum_{i=2}^{n} F_i L_0(n-1,i-2)\right) L_k^*(n,0) \\
&\quad + \sum_{j=1}^{k-1}\sum_{i=2}^{n} F_i L_j(n-1,i-2) L_{k-j}^*(n,0) \\
&\quad + \sum_{i=2}^{n} F_i L_k^*(n-1,i-2) L_0^*(n,0) \\
&\quad + \sum_{j=1}^{k-1} F_{n+j} L_{k-j}^*(n,n-1) \\
&\quad + F_{n+k}^* L_0^*(n,n-1), \qquad k \geq 1, \qquad (9.4.6)
\end{aligned}$$

$$\begin{aligned}
L_k^*(n,s) &= \sum_{j=0}^{k-1} L_j(n-1,s-1) L_{k-j}^*(n,0) \\
&\quad + L_k^*(n-1,s-1) L_0^*(n,0), \qquad k \geq 1, \\
&\qquad s = 1,\dots,n-1. \qquad (9.4.7)
\end{aligned}$$

We now briefly outline the derivation of formulas (9.4.1)–(9.4.4). It must be recalled that the matrix product describes the changes in the customer general phases.

In the $MAP/G/1/n$ system, no customer is lost in a busy period if

(1) no customer arrives during the service of the first customer;

(2) exactly one customer arrives during the service of the first customer and no customer is lost in the busy period opened by this new customer; and

(3) i, $i = 2,\dots,n$, new customers arrive during the service of the first customer and no customer is lost in the busy period of the $MAP/G/1/n-1$ system opened by the customer at the server and with $i-2$ customers in the queue, and, finally, no customer is lost in the busy period of the $MAP/G/1/n$ system opened by the sole customer in the system.

We thus prove relation (9.4.1).

Relation (9.4.2) is derived as follows:

- during the service of the first customer, exactly one new customer arrives, and then exactly k customers are lost during the busy period of the $MAP/G/1/n$ system opened by this new customer;

- during the service of the first customer, i, $i = 2,\dots,n$, new customers arrive, and no customer is lost in the busy period of the $MAP/G/1/n-1$ system opened by the customer at the server and with $i-2$ customers in the queue, and k customers are lost in the busy period of the $MAP/G/1/n$ system opened by the sole customer in the system;

- during the service of the first customer, i, $i = 2, \ldots, n$, new customers arrive, and then j, $j = 1, \ldots, k$, customers are lost in the busy period of the $MAP/G/1/n-1$ system opened by the customer at the server and with $i-2$ customers in the queue, and, finally, $k-j$ customers are lost in the busy period of the $MAP/G/1/n$ system opened by the sole customer in the system;

- $j, j = 1, \ldots, k$, customers are lost during the service of the first customer and $k-j$ customers are lost in the busy period of the $MAP/G/1/n-1$ system opened by the customer at the server and with $(n-1)$ customers in the queue.

Relation (9.4.3) follows from the fact that k customers are lost in the busy period of the $MAP/G/1/n$ system opened by the customer at the server and with s customers in the queue only if $j, j = 0, \ldots, k$, customers are lost in the busy period of the $MAP/G/1/n-1$ opened by the customer at the server and with $s-1$ customers in the queue and then $k-j$ customers are lost in the busy period of the $MAP/G/1/n$ system opened by the sole customer in the system.

To derive relation (9.4.4), it suffices to observe that at least 0 customer is lost in the busy period of the $MAP/G/1/1$ system if

- no customer arrives during the service of the first customer;

- at least one new customer arrives during the service of the first customer and at least 0 customer is lost from the $MAP/G/1/1$ system in the busy period opened by this new customer.

Formulas (9.4.5)–(9.4.7) are derived in the same manner, and their derivation is left to the reader as an exercise.

From relations (9.4.1)–(9.4.7) it follows that the matrices $L_k(n, s)$ and $L_k^*(n, s)$ can be determined recursively in n. Invertibility of the matrices in recursive computations is guaranteed by the irreducibility of the customer generation process.

The element $(L_0^*(n, 0))_{ij}$ of the matrix $L_0^*(n, 0)$ is the probability that the customer generation process passes from phase i to phase j in a busy period of the $MAP/G/1/n$ system.

For the sake of brevity, we set $L_k = L_k(r, 0)$ and $L_k^* = L_k^*(r, 0)$; these matrices are needed in what follows.

Let us consider the $MAP/G/1/n$, $n = 1, \ldots, r$, system once again, and introduce two vectors $\mathbf{g}_k(n, s)$ and $\mathbf{g}_k^*(n, s)$, $k \geq 0, n \geq 1, s = 0, \ldots, n-1$, whose coordinates $g_{ik}(n, s)$ and $g_{ik}^*(n, s)$, $i = 1, \ldots, l$, are the mean times spent by the system in the busy period in states in which k customers and at least k customers are lost, provided that at the beginning of the busy period the customer generation process was in phase i and the system contained, besides the customer at the server, s more customers in the queue.

In order to find $\mathbf{g}_0(n, 0)$, we represent the time spent by the $MAP/G/1/n$ in a busy period in states in which no customer is lost as the sum of three terms (second and third terms may be zero):

- the part of service time of the first customer during which no customer was lost;

- the part of the busy period of the $MAP/G/1/n-1$ system opened by $i-1$, $i = 1, \ldots, n$, customers (one customer at the server and $i-2$ customers in the queue out of a total number i of customers that arrived during the time of service of the first customer) in which no customer was lost; here no customer must be lost in the course of service of the first customer, which is consistent with the constraint $i \leq n$;

- the part of the busy period of the $MAP/G/1/n$ system opened by the last customer of i customers that had arrived (during the service of the first customer) in which no customer was lost; here the condition that there were no losses in the previous part also holds.

Now, with the use of the total expectation formula, we obtain

$$\mathbf{g}_0(n,0) = \mathbf{t}_n^{**} + \sum_{i=2}^{n} F_i \mathbf{g}_0(n-1, i-2) + \left(F_1 + \sum_{i=2}^{n} F_i L_0(n-1, i-2) \right) \mathbf{g}_0(n,0).$$

Similarly we derive the relations

$$\begin{aligned}
\mathbf{g}_k(n,0) = \mathbf{t}_{n+k} &+ \sum_{i=2}^{n} F_i \mathbf{g}_k(n-1, i-2) + \left(F_1 + \sum_{i=2}^{n} F_i L_0(n-1, i-2) \right) \mathbf{g}_k(n,0) \\
&+ \sum_{j=1}^{k} \sum_{i=2}^{n} F_i L_j(n-1, i-2) \mathbf{g}_{k-j}(n,0) \\
&+ \sum_{j=1}^{k} F_{n+j} \mathbf{g}_{k-j}(n, n-1), \quad k \geq 1,
\end{aligned}$$

$$\mathbf{g}_k(n,s) = \mathbf{g}_k(n-1, s-1) + \sum_{j=0}^{k} L_j(n-1, s-1) \mathbf{g}_{k-j}(n,0), \ k \geq 0, \ s = 1, \ldots, n-1,$$

$$\mathbf{g}_0^*(1,0) = b\mathbf{1} + F_1^* \mathbf{g}*_0(1,0),$$

$$\begin{aligned}
\mathbf{g}_0^*(n,0) = b\mathbf{1} &+ \sum_{i=2}^{n-1} F_i \mathbf{g}_0^*(n-1, i-2) + F_n^* \mathbf{g}_0^*(n-1, n-2) \\
&+ \left(F_1 + \sum_{i=2}^{n-1} F_i L_0^*(n-1, i-2) + F_n^* L_0^*(n-1, n-2) \right) \mathbf{g}_0^*(n,0), \ n \geq 2,
\end{aligned}$$

$$\begin{aligned}
\mathbf{g}_k^*(n,0) = \mathbf{t}_{n+k}^* &+ \sum_{i=2}^{n} F_i \mathbf{g}_k^*(n-1, i-2) + \left(F_1 + \sum_{i=2}^{n} F_i L_0(n-1, i-2) \right) \mathbf{g}_k^*(n,0) \\
&+ \sum_{j=1}^{k-1} \sum_{i=2}^{n} F_i L_j(n-1, i-2) \mathbf{g}_{k-j}^*(n,0) + \sum_{i=2}^{n} F_i L_k^*(n-1, i-2) \mathbf{g}_0^*(n,0) \\
&+ \sum_{j=1}^{k-1} F_{n+j} \mathbf{g}_{k-j}^*(n, n-1) + F_{n+k}^* \mathbf{g}_0^*(n, n-1), \quad k \geq 1,
\end{aligned}$$

$$g_k^*(n,s) = g_k^*(n-1,s-1) + \sum_{j=0}^{k-1} L_j(n-1,s-1)g_{k-j}^*(n,0)$$
$$+ L_k^*(n-1,s-1)g_0^*(n,0), \quad k \geq 0, \quad s = 1,\dots,n-1.$$

Like the matrices $L_k(n,s)$ and $L_k^*(n,s)$, the vectors $g_k(n,s)$ and $g_k^*(n,s)$ are also computed recursively in n.

Obviously, the vector g_0^* describes the mean lengths of the busy period of the $MAP/G/1/n$ system and the coordinate g_{i0}^* takes account of the condition that the busy period is opened when the customer generation process is in phase i.

In what follows, we only need the vectors $g_k = g_k(r,0)$ and $g_k^* = g_k^*(r,0)$.

Finally, for the $MAP/G/1/n$, $n = 1,\dots,r$, system, we introduce the vectors $s_k(n,s)$, $n \geq 1$, $k = 0,\dots,n$, $s = 0,\dots,n-1$, where the coordinate $s_{ik}(n,s)$, $i = 1,\dots,l$, of the vector $s_k(n,s)$ is the mean time spent by the system in the states when k customers are in the queue in a busy period, provided that at the beginning of the busy period the customer generation process was in phase i and the system, besides the customer at the server, contained s more customers in the queue. Once again applying the total expectation formula, we obtain the relations for the vectors $s_k(n,s)$:

$$s_0(1,0) = t_0 + F_1^* s_0(1,0),$$
$$s_0(n,0) = t_0 + \left(F_1 + \sum_{i=2}^{n-1} F_i L_0^*(n-1,i-2) + F_n^* L_0^*(n-1,n-2)\right) s_0(n,0), \quad n \geq 2,$$
$$s_k(n,0) = t_k + \sum_{i=2}^{n-1} F_i s_{k-1}(n-1,i-2) + F_n^* s_{k-1}(n-1,n-2)$$
$$+ \left(F_1 + \sum_{i=2}^{n-1} F_i L_0^*(n-1,i-2) + F_n^* L_0^*(n-1,n-2)\right) s_k(n,0),$$
$$k = 1,\dots,n-1,$$
$$s_1(1,0) = t_1^* + F_1^* s_1(1,0),$$
$$s_n(n,0) = t_n^* + \sum_{i=2}^{n-1} F_i s_{n-1}(n-1,i-2) + F_n^* s_{n-1}(n-1,n-2)$$
$$+ \left(F_1 + \sum_{i=2}^{n-1} F_i L_0^*(n-1,i-2) + F_n^* L_0^*(n-1,n-2)\right) s_n(n,0), \quad n \geq 2,$$
$$s_0(n,s) = L_0^*(n-1,s-1)s_0(n,0), \quad s = 1,\dots,n-1,$$
$$s_k(n,s) = s_{k-1}(n-1,s-1) + L_0^*(n-1,s-1)s_k(n,0),$$
$$k = 1,\dots,n, \qquad s = 1,\dots,n-1.$$

The vectors $s_k(n,s)$ are also computed by a recursive algorithm.

In what follows, we only need the vectors $s_k = s_k(r,0)$.

9.4.3. Embedded Markov chain

Let us turn back to the initial $MAP/G/1/r$ retrial system.

The time interval from the instant of arrival of a customer at a free server (the system may contain retrial customers) to the instant when the server becomes free is called the busy period of the system. Similarly, the idle period is the time interval from the instant when the server becomes free (of the customers in the queue of primary customers) up to the instant of beginning of service of the next customer (who had arrived from outside or who is chosen from orbit). For our system, the busy and idle periods depend not only on the phase of the customer generation process but also on the number of retrial customers in the orbit at the beginning of the respective period.

Let us consider the instants of completion of successive busy periods. Obviously, the number of retrial customers in the system (and customer generation phases) at these instants form a homogeneous Markov chain.

We introduce the matrices

P_{kn}, $k, n = 0, \dots, m$, is the matrix whose element $(P_{kn})_{ij}$, $i, j = 1, \dots, l$, is the probability that at the instant of completion of a busy period the system contains n retrial customers and customer generation process is in phase j, provided that at the end of the preceding busy period there were k retrial customers and the customer generation process was in phase i;

D_n^-, $n = 0, \dots, m$, is the matrix whose element $(D_n^-)_{ij}$, $i, j = 1, \dots, l$, is the probability that at the instant of completion of an idle period, a retrial customer is taken up for service and the customer generation process passes to phase j, provided that at the beginning of this idle period n retrial customers were in the orbit and the customer generation process was in phase i; and

D_n^+, $n = 0, \dots, m$, is the matrix whose element $(D_n^+)_{ij}$, $i, j = 1, \dots, l$, is the probability that at the instant of completion of an idle period a new (primary) customer is taken up for service and the customer generation process is in phase j, provided that at the beginning of this idle period n retrial customers were in the orbit and the customer generation process was in phase i.

In order to compute the matrix D_n^-, it suffices to recall that with the probability $e^{(\Lambda-\gamma_n I)t}$ the idle period is not completed by the instant t, and then with the probability $\gamma_n\, dt$ one of n retrial customers in the orbit is taken up by the server in the time interval $(t, t+dt)$ without changing the phase of the customer generation process. Therefore,

$$D_n^- = \int_0^\infty \gamma_n e^{(\Lambda-\gamma_n I)t}\, dt = \gamma_n(\gamma_n I - \Lambda)^{-1}.$$

The matrix D_n^+ is calculated similarly:

$$D_n^+ = \int_0^\infty e^{(\Lambda-\gamma_n I)t} N\, dt = (\gamma_n I - \Lambda)^{-1} N.$$

We can find the matrix P_{kn} with the use of the total probability formula, recalling that at the completion of an idle period the server takes up either a retrial customer (thereby reducing the number of retrial customers in the orbit by one) or a new customer (leaving the queue of retrial customers unchanged) and in the busy period the queue of retrial customers increases by an appropriate number of customers. Hence

$$\begin{aligned}
P_{kn} &= D_k^+ L_{n-k} + D_k^- L_{n-k+1}, & k &= 0,\dots,m-1, \quad n = k,\dots,m-1,\\
P_{k,k-1} &= D_k^- L_0, & k &= 1,\dots,m,\\
P_{km} &= D_k^+ L_{m-k}^* + D_k^- L_{m-k+1}^*, & k &= 0,\dots,m,\\
P_{kn} &= 0, & k &= 2,\dots,m, \quad\quad n = 0,\dots,k-2.
\end{aligned}$$

Thus, the transition probability matrix P of the embedded Markov chain is of the form

$$P = \begin{pmatrix} P_{00} & P_{01} & P_{02} & \dots & P_{0,m-1} & P_{0m} \\ P_{10} & P_{11} & P_{12} & \dots & P_{1,m-1} & P_{1m} \\ 0 & P_{21} & P_{22} & \dots & P_{2,m-1} & P_{2m} \\ \dots & \dots & \dots & \dots & \dots & \dots \\ 0 & 0 & 0 & \dots & P_{m,m-1} & P_{mm} \end{pmatrix}.$$

Introducing $\boldsymbol{p}^- = (\boldsymbol{p}_0^{-T}, \boldsymbol{p}_1^{-T}, \dots, \boldsymbol{p}_m^{-T})^T$, where the coordinate p_{ik}^-, $i = 1,\dots,l$, of the vector $\boldsymbol{p}_k^-$ is the stationary probability of the state (i,k) of the embedded Markov chain, we arrive at the equilibrium equations

$$\boldsymbol{p}^{-T} = \boldsymbol{p}^{-T} P \tag{9.4.8}$$

with the normalisation condition

$$\sum_{k=0}^{m} \boldsymbol{p}_k^{-T} \mathbf{1} = 1.$$

These equilibrium equations are solved by methods described earlier.

Knowing the solution $\boldsymbol{p}^-$ of (9.4.8), we easily find the stationary probabilities $\boldsymbol{p}^+ = (\boldsymbol{p}_0^{+T}, \boldsymbol{p}_1^{+T}, \dots, \boldsymbol{p}_m^{+T})^T$ of the embedded Markov chain generated by the instants of commencement of busy periods:

$$\boldsymbol{p}_k^{+T} = \boldsymbol{p}_k^{-T} D_k^+ + \boldsymbol{p}_{k+1}^{-T} D_{k+1}^-.$$

9.4.4. Time-stationary distributions

Let us find the time-stationary distributions of the number of retrial and primary customers in the queueing system. To do this, we, as usual, use elements of the theory of semi-Markov renewal processes. Here we only find the marginal distributions of these characteristics, because their joint distributions are rather unwieldy.

Let $\boldsymbol{T}_k$, $k = 0,\dots,m$, denote the vector whose coordinate T_{ik}, $i = 1,\dots,l$, is the mean length of the idle period, provided that at the beginning of this period k retrial customers were in the system and the customer generation process was in phase i. Then

$$\boldsymbol{T}_k = \int_0^\infty e^{(\Lambda - \gamma_k I)t} \mathbf{1}\, dt = (\gamma_k I - \Lambda)^{-1} \mathbf{1}.$$

Let T^- and T^+ denote the mean length of the idle period and the mean length of the busy period of the system in steady state respectively. Obviously,

$$T^- = \sum_{k=0}^{m} \boldsymbol{p}_k^{-T} \boldsymbol{T}_k, \qquad T^+ = \sum_{k=0}^{m} \boldsymbol{p}_k^{+T} \boldsymbol{g}_0^*.$$

Let

$$T_k = \boldsymbol{p}_k^{-T} \boldsymbol{T}_k, \qquad k = 0, \dots, m,$$

be the mean time in which k retrial customers are in the orbit in an idle period of the system in steady state. Similarly,

$$T_k^{(1)} = \sum_{n=0}^{k} \boldsymbol{p}_n^{+T} \boldsymbol{g}_{k-n}, \quad k = 0, \dots, m-1, \quad T_m^{(1)} = \sum_{n=0}^{m} \boldsymbol{p}_n^{+T} \boldsymbol{g}_{m-n}^*$$

and

$$T_n^{(2)} = \sum_{k=0}^{m} \boldsymbol{p}_k^{+T} \boldsymbol{s}_{n-1}, \qquad n = 1, \dots, r+1,$$

denote the mean times during which k retrial customers exist in the orbit and n primary customers do in the system in a busy period in stationary state respectively.

Finally, let $p_k^{(1)}$, $k = 0, \dots, m$, and $p_n^{(2)}$, $n = 0, \dots, r+1$, be the stationary probabilities that the system contains k retrial and n primary customers respectively. Then

$$p_k^{(1)} = \frac{1}{T^- + T^+}(T_k + T_k^{(1)}), \qquad k = 0, \dots, m,$$

$$p_0^{(2)} = \frac{1}{T^- + T^+} \sum_{k=0}^{m} T_k, \quad p_n^{(2)} = \frac{1}{T^- + T^+} T_n^{(2)}, \qquad n = 1, \dots, r+1.$$

The last expressions define the time-stationary distributions of the number of retrial customers in the orbit and the number of primary customers in the system respectively.

9.5. $MAP/G/1/\infty$ system with foreground–background processor sharing discipline

We once again turn back to the $MAP/G/1/\infty$ system with infinite buffer. The above notation and constraints on the initial parameters of the system hold for this system. We assume that the traffic intensity of the system is $\rho = \lambda b < 1$. Here, as usual, $\lambda = \boldsymbol{\pi}^T N \mathbf{1}$ is the input flow intensity, $\boldsymbol{\pi}$ is the vector of stationary probabilities of the states of the customer generation process, and b is the mean service time. Further in this section it is more convenient to refer to the service time as the customer length.

The foreground–background processor sharing discipline consists of serving the customer whose served length is minimal (i.e., the customer that spent less time at the server than any other customer). If there are i such customers, then each is served at rate $1/i$.

For this system, we will find stationary distributions of sojourn time and number of customers in the system.

9.5.1. Auxiliary system

In what follows, we require an x-system, which is also a $MAP/G/1/\infty$ system, but with the FCFS discipline. Besides, the customer length in the x-system has the distribution function

$$B^{(x)}(y) = \begin{cases} B(y), & y \le x; \\ 1, & y > x. \end{cases}$$

In other words, all customers of length exceeding x arriving at the initial system have in the x-system a length equal to x. Characteristics pertaining to the x-system are superscripted with (x).

Let $G^{(x)}(u \mid y)$ denote the matrix whose element $G_{ij}^{(x)}(u \mid y)$ is the probability that the busy period of the x-system is less than u and the customer generation process at the instant of completion of this busy period is in phase j, provided that the total length of customers at the beginning of the busy period was y and the customer generation process was in phase i. Then the Laplace–Stieltjes transform $\Gamma^{(x)}(s \mid y)$ of the matrix $G^{(x)}(u \mid y)$, according to Section 8.4.1, satisfies the relation

$$\Gamma_{ij}^{(x)}(s \mid y) = e^{(\Lambda + N\Gamma^{(x)}(s) - sI)y},$$

where the matrix $\Gamma^{(x)}(s)$ is defined by the equation

$$\Gamma^{(x)}(s) = \int_0^\infty e^{(\Lambda + N\Gamma^{(x)}(s) - sI)y} dB^{(x)}(y).$$

Let $\boldsymbol{w}^{(x)}(y)$ denote the vector whose coordinate $w_i^{(x)}(y)$ is the probability that in the x-system in steady state, the virtual waiting time is less than y and the customer generation phase is i. As shown from Section 8.2.2, the Laplace–Stieltjes transform $\boldsymbol{\omega}^{(x)}(s)$ of the vector $\boldsymbol{w}^{(x)}(y)$ is defined by the formula

$$\boldsymbol{\omega}^{(x)T}(s) = s\boldsymbol{p}_0^{(x)T}(sI + \Lambda + \beta^{(x)}(s)N)^{-1},$$

where the coordinate $p_{i0}^{(x)}$ of the vector $\boldsymbol{p}_0^{(x)}$ is the stationary probability that no customers exist in the x-system and the customer generation process is in phase i, and $\beta^{(x)}(s)$ is the Laplace–Stieltjes transform of the length of a customer in the x-system. Methods to find the vector $\boldsymbol{p}_0^{(x)}$ are discussed in sufficient detail in Chapter 8.

Let $P_k^{(x)}$, $k \ge 0$, denote the matrix consisting of elements $(P_k^{(x)})_{ij}$, where $(P_k^{(x)})_{ij}$ is the probability that k customers of length x are served in a busy period of the x-system and the customer generation process exists in phase j at the instant of completion of this busy period, provided that the customer generation process was in phase i at the beginning of this busy period. Let $\tilde{P}^{(x)}(z)$ denote the generating function of this matrix.

In the x-system, using the reverse order of service with interruption and resumption and reasoning as in Section 8.4.1, we find that the matrix $\tilde{P}^{(x)}(z)$ satisfies the relation

$$\tilde{P}^{(x)}(z) = \int_0^{x-} e^{(\Lambda + N\tilde{P}^{(x)}(z))y} dB(y) + z(1 - B(x))e^{(\Lambda + N\tilde{P}^{(x)}(z))x}.$$

We observe that the first term in the right-hand side of this formula describes the situation in the busy period opened by a customer of length $y < x$, whereas the second term describes the situation in the busy period opened by a customer of length x.

These relations define the functions $G^{(x)}(u \mid y)$, $\boldsymbol{w}^{(x)}(y)$, and $P_k^{(x)}$, which will be used for analysing the initial systems in what follows.

9.5.2. Stationary distribution of the sojourn time

Let us turn back to the initial system.

Let $\boldsymbol{v}(u \mid x)$ denote the vector whose coordinate $v_i(u \mid x)$ is the stationary probability that the sojourn time of a customer of length x in the system is less than u and the customer generation process at the instant of completion of his service is in phase i.

In order to find the vector $\boldsymbol{v}(u \mid x)$, we tag a customer of length x arriving at the system. We observe that customers of served length greater than x have no influence on the service of the tagged customer. Moreover, customers of (total) length less than x that exist in the system at the instant of arrival of the tagged customer and those that arrive during the service of this tagged customer necessarily quit the system earlier than the tagged customer. Finally, customers of (total) length greater than x whose served length at the instant of arrival of the tagged customer was less than x or the customers that had arrived during the service of the tagged customer will have a served length x at instant when the tagged customer quits the system. Therefore, the sojourn time of the tagged customer is equal to the time in which the system gets rid of all customers of served length less than x or, in other words, ends at the instant of completion of the busy period of the x-system.

Note that in steady state of the x-system with the probability $w_i^{(x)}(y)$ the virtual waiting time is less than y and customer generation process is in phase i. The customer arrival intensity, depending on the state, is determined by the matrix N, and the total input flow intensity is λ. Since after the arrival of a customer of length x the total time required for the x-system to get rid of all customer (including the customers of total length y that were in the system earlier and the newly arrived customer of length x) coincides with the busy period of the x-system opened by customers of total length $y + x$, and its distribution is defined by the matrix $G^{(x)}(u \mid y + x)$, we obtain

$$\boldsymbol{v}^T(u \mid x) = \frac{1}{\lambda} \int_0^\infty \boldsymbol{w}^{(x)T}(dy) N G^{(x)}(u \mid y + x),$$

or, in terms of the Laplace–Stieltjes transform,

$$\boldsymbol{\varphi}^T(s \mid x) = \frac{1}{\lambda} \int_0^\infty \boldsymbol{w}^{(x)T}(dy) N \Gamma^{(x)}(s \mid y + x). \tag{9.5.1}$$

The vector $\boldsymbol{v}(u)$ of stationary probabilities $v_i(u)$ that the sojourn time of a customer in the initial system is less than u and the customer generation process at the instant of completion of service of this customer is in phase i is defined in terms of Laplace-Stieltjes transform by the formula

$$\boldsymbol{\varphi}(s) = \int_0^\infty \boldsymbol{\varphi}(s \mid x)\, dB(x). \tag{9.5.2}$$

Finally, the Laplace–Stieltjes transforms $\varphi(s \mid x)$ and $\varphi(s)$ of the stationary distributions $v(u \mid x)$ of sojourn times of customers of length x and $v(u)$, respectively, are of the form

$$\varphi(s \mid x) = \boldsymbol{\varphi}^T(s \mid x)\mathbf{1}, \tag{9.5.3}$$

$$\varphi(s) = \boldsymbol{\varphi}^T(s)\mathbf{1}. \tag{9.5.4}$$

Formulas (9.5.1)–(9.5.4) define the steady-state characteristics associated with the sojourn time of a customer in the system.

9.5.3. Stationary distribution of the number of customers in the system

Let us consider a busy period of the initial system and on this period take the instant τ_x at which either the maximum served length attains a value x, or the busy period ends if all customers that had arrived in this busy period have a (total) length less than x. It is immediately seen that if the system contains customers at the instant $\tau_x + 0$, then they all will be of a (served) length x. Furthermore, the distribution of the number of customers of length x at the instant τ_x is given by the matrix $P_k^{(x)}$ (with regard for the customer generation process at the beginning of the busy period and at the instant τ_x).

Let $Q(x, n)$, $n \geq 1$, denote the matrix of elements $Q_{ij}(x, n)$ where $Q_{ij}(x, n)$ is the mean time spent by the system up to the instant τ_x in a state in which n customers were in the system and the customer generation process was in phase j, provided that the customer generation process at the beginning of the busy period was in phase i.

Increasing x by Δ, we find that $Q(x, n)$ increases to $o(\Delta)$:

- first, by $nP_n^{(x)}\Delta$ due to the fact that n customers were in the system with the probability $P_n^{(x)}$ at the instant τ_x and $Q(x, n)$ increases by $n\Delta$ by the instant $\tau_{x+\Delta}$;
- second, by $kP_k^{(x)}NQ(x, n-k)\Delta$ due to the fact that there were k, $k = 1, \ldots, n-1$, customers with the probability $P_k^{(x)}$ in the system at the instant τ_x, and in the time period $k\Delta$ in which x increases by Δ, a customer arrives with the probability $kN\Delta$ and by the instant when the maximal served length attains x the mean time spent by the system in the state with $n-k$ customers in the new busy period opened by this new customer is $Q(x, n-k)$.

Thus, $Q(x, n)$ is defined by the equations

$$Q'(x, n) = nP_n^{(x)} + \sum_{k=1}^{n-1} kP_k^{(x)}NQ(x, n-k).$$

Obviously, the initial condition is

$$Q(0, n) = 0.$$

Passing to the generating function

$$\tilde{Q}(x, z) = \sum_{n=1}^{\infty} z^n Q(x, n),$$

we find that it is defined by the differential equations

$$\frac{\partial \tilde{Q}(x,z)}{\partial x} = z\frac{\partial \tilde{P}^{(x)}(z)}{\partial z}(I + N\tilde{Q}(x,z)) \tag{9.5.5}$$

with the initial condition

$$\tilde{Q}(0,z) = 0. \tag{9.5.6}$$

Note that equations (9.5.5) are, in fact, linear ordinary differential equations and, by virtue of initial condition (9.5.6), they can be easily integrated numerically.

Now consider a busy period of the system, denoting by G the matrix whose element G_{ij} is the probability that the customer generation process at the end of this busy period is in phase j, provided that it was in phase i at the beginning of the busy period. The matrix G is related to the matrix $\tilde{P}^{(x)}(z)$ by the formula

$$G = \tilde{P}^{(\infty)}(1).$$

Furthermore, since the busy period of our system coincides with the busy period of a similar system but with the FCFS discipline, the matrix G can be found by any of the methods described in Chapter 8, Sections 8.1 and 8.4.

Similarly, let us consider the idle period and denote by B the matrix with elements B_{ij}, where B_{ij} is the probability that the customer generation process at the beginning of a busy period that follows an idle period is in phase j, provided that it was at the beginning of the idle period in phase i. Recalling formula (7.1.7), we find that

$$B = \int_0^\infty e^{\Lambda x} N\,dx = -\Lambda^{-1}N.$$

Now we set

$$R^{(1)} = GB, \qquad R^{(2)} = BG.$$

Obviously, $R^{(1)}$ and $R^{(2)}$ are the transition probability matrices of the embedded Markov chains defined by the customer generation phases at the instants immediately after the beginning of successive busy periods and at the instants of completion of successive busy periods.

Denoting the vector of stationary probabilities of the first and second chain by $\boldsymbol{p}^{(1)}$ and $\boldsymbol{p}^{(2)}$, respectively, we see that these vectors are defined by the equilibrium equations

$$\boldsymbol{p}^{(1)T}R^{(1)} = \boldsymbol{p}^{(1)T}, \qquad \boldsymbol{p}^{(2)T}R^{(2)} = \boldsymbol{p}^{(2)T}$$

with the normalisation conditions

$$\boldsymbol{p}^{(1)T}\mathbf{1} = 1, \qquad \boldsymbol{p}^{(2)T}\mathbf{1} = 1;$$

obviously, the vectors $\boldsymbol{p}^{(1)}$ and $\boldsymbol{p}^{(2)}$ are related by the formulas

$$\boldsymbol{p}^{(1)T}G = \boldsymbol{p}^{(2)T}, \qquad \boldsymbol{p}^{(2)T}B = \boldsymbol{p}^{(1)T},$$

which are helpful in finding only one of the vectors in numerical computations.

In what follows, we need the vectors $\boldsymbol{t}_k$, where the coordinate t_{ik} of the vector $\boldsymbol{t}_k$ is the mean time spent by the system in a cycle (i.e., a busy period followed by the succeeding idle period) in steady mode in the state with k customers and customer generation phase i.

Since t_{i0} in fact is the mean time spent by the system in steady mode in the state with customer generation phase i only in an idle period, we see that

$$\boldsymbol{t}_0^T = \boldsymbol{p}^{(2)T} \int_0^\infty e^{\Lambda x} dx = -\boldsymbol{p}^{(2)T} \Lambda^{-1}.$$

Similarly, t_{ik}, $k \geq 1$, is the mean time spent by the system in steady mode in the state with k customers and customer generation phase i only in a busy period, and

$$\boldsymbol{t}_k^T = \boldsymbol{p}^{(1)T} Q(\infty, k), \qquad k \geq 1.$$

Finally, the mean length T of a cycle of the system in steady mode is given by the formula

$$T = \sum_{k=0}^{\infty} \boldsymbol{t}_k^T \mathbf{1}.$$

We introduce the vectors $\boldsymbol{p}_k$, where p_{ik} is the (time-) stationary probability that k customers exist in the system and the customer generation process is in phase i at any instant. Now using the semi-Markov process theory apparatus, we obtain

$$\boldsymbol{p}_k = \frac{1}{T} \boldsymbol{t}_k.$$

In particular, the stationary probabilities p_k that k customers exist in the system are given by the formula

$$p_k = \boldsymbol{p}_k^T \mathbf{1}.$$

These results are conveniently expressed in terms of generating functions. Assuming that

$$\boldsymbol{p}(z) = \sum_{k=1}^{\infty} z^k \boldsymbol{p}_k, \qquad p(z) = \sum_{k=1}^{\infty} z^k p_k,$$

we obtain

$$\boldsymbol{p}^T(z) = \frac{1}{T}(\boldsymbol{p}^{(1)T} \tilde{Q}(\infty, z) - \boldsymbol{p}^{(2)T} \Lambda^{-1}), \qquad p(z) = \boldsymbol{p}^T(z)\mathbf{1},$$

where

$$T = (\boldsymbol{p}^{(1)T} \tilde{Q}(\infty, 1) - \boldsymbol{p}^{(2)T} \Lambda^{-1})\mathbf{1}.$$

The formulas derived in this section are useful in finding steady-state characteristics associated with the number of customers in the system.

9.6. $MAP/G/1/r$ system with LCFS discipline and bounded total volume of customers

In this section, we study the $MAP/G/1/n$ queueing system with reverse order of service and service interruption, in which every customer has, besides service time (length), also a random volume, which, in general, depends on the customer length. The total volume of customers and the total number of waiting places (to avoid confusion between the length and volume of a customer, we use the term 'number of waiting places' instead of 'buffer capacity') in the system are bounded by some (deterministic) numbers. Such a queueing system is helpful in more adequate describing many real technical systems, though the reverse order of service with interruption is rather artificial. Although the method of this section can, in principle, be applied to study systems with the FCFS discipline, serious technical complications may arise.

9.6.1. System description

We consider the

$$MAP/G/1/r$$

system with r, $0 \leq r \leq \infty$, waiting places. As usual, we set $R = r + 1$.

The input for the system is a Markov flow of customers, for which we use the usual notation and assumptions.

Each customer arriving at the system has, besides a random service time (which in what follows is referred to as the service length), also a random (positive) volume. The joint distribution of the length and volume is defined by some function $B(x, y)$. For the sake of simplicity, we assume that the joint distribution density $b(x, y) = \partial^2 B(x, y)/(\partial x \partial y)$ exists. The total volume of customers in the system is bounded by a (non-random) number Y, $0 < Y < \infty$. If the sum of the volume of an arriving customer plus the volumes of other customers in the system exceeds Y or there are r customers in the queue, the newly arrived customer is lost.

We use the following notation:

$b_2(y) = \int_0^\infty b(x, y)\, dx$ is the probability density of the customer volume;

$B_2(y) = B(\infty, y) = \int_0^y b_2(u)\, du$ is the distribution function of the customer volume;

$b(x \mid y) = b(x, y)/b_2(y)$ is the conditional probability density of the customer length under the condition that the customer volume is y; and

$B(x \mid y) = \int_0^x b(u \mid y)\, du$ is the conditional distribution function of the customer length under the condition that the customer volume is y.

The service discipline for the system is the reverse order of service with interruption (i.e., the LCFS discipline with interruption), in which every accepted customer is immediately taken up for service, supplanting the customer at the server to the first place in the queue. The interrupted service of a customer is resumed later.

We assume that the steady state of the system exists. Here we will not study conditions for existence of such a mode; they, in general, are rather complicated.

For our system, we will find the (time-) stationary state probabilities and stationary distribution of the sojourn time of a customer in the system at any instant.

9.6.2. Auxiliary system

In what follows, we need the (m, V)-system, $m = 0, \dots, r$, which is exactly the same as the initial system, but the number of waiting places is $r - m$ and the total volume of customers is bounded by a number V. Obviously, the $(0, Y)$-system coincides with the initial system. The characteristics pertaining to the (m, V)-system are superscripted with (m, V).

Let $G^{(m,V)}(u)$ be the matrix whose element $G_{ij}^{(m,V)}(u)$ is the probability that the busy period of the (m, V)-system is less than u and the customer generation process at the instant of completion of this busy period is in phase j, provided that it was at the beginning of the busy period in phase i. Moreover, we assume that the busy period is zero if a customer of length exceeding V arrives at the (m, V)-system. Then the Laplace–Stieltjes transform $\Gamma^{(m,V)}(s)$ of the matrix $G^{(m,V)}(u)$ satisfies the relation

$$\Gamma^{(m,V)}(s) = [1 - B_2(V)]I + \int_0^V b_2(y)\,dy \int_0^\infty e^{(\Lambda+N\Gamma^{(m+1,V-y)}(s)-sI)x} b(x \mid y)\,dx. \tag{9.6.1}$$

This relation is derived as follows. First, the volume of the customer that opens a busy period of the (m, V)-system may with the probability $1 - B_2(V)$ be greater than V. Then the busy period is equal to zero. Second, a busy period with the probability $b_2(y)\,dy$, $0 < y < V$, may be opened by a customer of intermediate volume $(y, y + dy)$ and the length of the customer opening this busy period lies with the probability $b(x \mid y)\,dx$ in the interval $(x, x + dx)$. Moreover, the service of this customer may be interrupted and each such interruption is a busy period of the $(m + 1, V - y)$-system.

Relation (9.6.1) can be used to compute $\Gamma^{(m,V)}(s)$ recursively from $m = r$ to $m = 0$. Here we must set $\Gamma^{(R,V)}(s) = I$.

We introduce the matrix

$$\Lambda^{(m,V)} = \Lambda + N\Gamma^{(m,V)}(0).$$

It is the infinitesimal matrix of the customer generation process in idle periods, i.e., the period obtained by discarding all those time intervals in which the (m, V)-system is busy.

9.6.3. Stationary distribution of the number of customers in the system

Let us turn back to the initial system.

Let $\boldsymbol{p}_0$ denote the vector whose coordinates p_{i0} are the stationary probabilities that no customers exist in the system and the customer generation process is in phase i; let $\boldsymbol{p}_1(x, y)$ denote the vector of stationary probabilities $p_{i1}(x, y)$ that the system contains only one customer of (served) length x and volume y and the customer generation process is in phase i; and let $\boldsymbol{p}_m(x, y, v)$, $m = 2, \dots, R$, denote the vector of stationary densities $p_{im}(x, y, v)$ of probabilities that m customers exist in the system and the (served) length and volume of the customer at the server are x and y respectively, the total volume of other customers in the system is v, and the customer generation process is in phase i.

Obviously, the total volume of customers in the system, by the formulation of the problem, lies between 0 and Y; hence $p_{i1}(x, y) = 0$ for $y > Y$ and $p_{im}(x, y, v) = 0$ for $y + v > Y$. Therefore, in what follows, without any stipulation of this fact, we take $y < Y$ or $y + v < Y$.

We will find the vectors $\boldsymbol{p}_0$, $\boldsymbol{p}_1(x, y)$, and $\boldsymbol{p}_m(x, y, v)$, $m = 2, \ldots, R$, using Lemma 1.4.1 successively for systems with bounded total number of customers $0, 1, \ldots, R$.

First, we discard those instants at which customers exist in the system and glue the remaining intervals. Then the phases of the customer generation process thus obtained form an irreducible continuous-time Markov process with l states and infinitesimal matrix $\Lambda^{(0,Y)}$. Hence the vector $\boldsymbol{p}_0$ can be found from the simultaneous algebraic equations

$$\boldsymbol{p}_0^T \Lambda^{(0,Y)} = \boldsymbol{0}^T$$

under the normalisation condition

$$\boldsymbol{p}_0^T \boldsymbol{1} = p_0,$$

where p_0 is the stationary probability of absence of customers in the system, which can be determined from the normalisation condition after computing all other probabilities.

In order to compute the vector $\boldsymbol{p}_1(x, y)$, we consider a system with no waiting place, which differs from the initial system in that if a customer of volume y is in the system, then the matrix of arrival (lost) intensities is $\Lambda^{(1,Y-y)}$. By virtue of Lemma 1.4.1, the stationary state probabilities of this system are, up to a normalisation constant, the corresponding stationary state probabilities of the initial system. Hence, for the vector $\boldsymbol{p}_1(x, y)$, we easily derive the equation

$$(\boldsymbol{p}_1^T(x, y))'_x = \boldsymbol{p}_1^T(x, y) \left(\Lambda^{(1,Y-y)} - \frac{b(x \mid y)}{1 - B(x \mid y)} I \right).$$

Solving this equation, we obtain

$$\boldsymbol{p}_1^T(x, y) = [1 - B(x \mid y)] \boldsymbol{p}_1^T(0, y) e^{\Lambda^{(1,Y-y)} x}. \tag{9.6.2}$$

Similarly, for the vector $\boldsymbol{p}_m(x, y, v)$, $m = 2, \ldots, R$, we obtain the equation

$$(\boldsymbol{p}_m^T(x, y, v))'_x = \boldsymbol{p}_m^T(x, y, v) \left(\Lambda^{(m,Y-y-v)} - \frac{b(x \mid y)}{1 - B(x \mid y)} I \right)$$

whose solution is

$$\boldsymbol{p}_m^T(x, y, v) = [1 - B(x \mid y)] \boldsymbol{p}_m^T(0, y, v) e^{\Lambda^{(m,Y-y-v)} x}. \tag{9.6.3}$$

For the initial conditions $\boldsymbol{p}_1(0, y)$ and $\boldsymbol{p}_m(0, y, v)$, it is easy to derive the relations

$$\begin{aligned}
\boldsymbol{p}_1^T(0, y) &= b_2(y) \boldsymbol{p}_0^T N, \\
\boldsymbol{p}_2^T(0, y, v) &= b_2(y) \int_0^\infty \boldsymbol{p}_1^T(x, v)\, dx N, \\
\boldsymbol{p}_m^T(0, y, v) &= b_2(y) \int_0^v dz \int_0^\infty \boldsymbol{p}_{m-1}^T(x, z, v - z)\, dx N, \qquad m = 3, \ldots, R.
\end{aligned}$$

Substituting $\boldsymbol{p}_1(x, v)$ and $\boldsymbol{p}_m(x, z, v - z)$ from formulas (9.6.2) and (9.6.3) into the last equality, after we obtain the expressions for the vectors of stationary state probability densities

$$\boldsymbol{p}_1^T(x, y) = [1 - B(x \mid y)] b_2(y) \boldsymbol{p}_0^T N e^{\Lambda^{(1,Y-y)} x}, \tag{9.6.4}$$

$$\boldsymbol{p}_m^T(x, y, v) = [1 - B(x \mid y)] b_2(y) \boldsymbol{c}_m^T(v) e^{\Lambda^{(m,Y-y-v)} x}, \qquad m = 2, \ldots, R, \tag{9.6.5}$$

where the vectors $\boldsymbol{c}_m(v)$ are found from the recurrence relations

$$\boldsymbol{c}_2^T(v) = b_2(v)\boldsymbol{p}_0^T N B_1(v, Y - v), \tag{9.6.6}$$

$$\boldsymbol{c}_m^T(v) = \int_0^v b_2(z)\boldsymbol{c}_{m-1}^T(v - z)B_{m-1}(z, Y - v)\,dz, \qquad m = 2, \dots, R, \tag{9.6.7}$$

and the matrices $B_m(y, v)$ are given by the formulas

$$B_m(y, v) = \int_0^\infty [1 - B(x \mid y)]e^{\Lambda^{(m,v)}x}dx N, \qquad m = 1, \dots, R. \tag{9.6.8}$$

In what follows, we also need the vectors $\boldsymbol{p}_m(v)$, $m = 1, \dots, R$, of stationary densities $p_{im}(v)$ of probabilities that the system contains m customers, the total volume of customers in the system is v, and the customer generation process is in phase i. Then

$$\boldsymbol{p}_1(v) = \int_0^\infty \boldsymbol{p}_1(x, v)\,dx,$$

$$\boldsymbol{p}_m(v) = \int_0^v dz \int_0^\infty \boldsymbol{p}_m(x, z, v - z)\,dx, \qquad m = 2, \dots, R,$$

or, after integration,

$$\boldsymbol{p}_m(v) = \boldsymbol{c}_{m+1}(v), \qquad m = 1, \dots, R.$$

Finally, the vector $\boldsymbol{p}_m$, $m = 1, \dots, R$, of stationary probabilities p_{im} that the system contains m customers and the customer generation process exists in phase i is given by the formula

$$\boldsymbol{p}_m = \int_0^Y \boldsymbol{c}_{m+1}(v)\,dv, \qquad m = 1, \dots, R,$$

and the stationary probability p_m that the system contains m customers is defined by the formula

$$p_m = \boldsymbol{p}_m^T \mathbf{1}, \qquad m = 1, \dots, R.$$

The stationary probability p_0 that no customers exist in the system is found from the normalisation condition

$$\sum_{m=0}^{R} p_m = 1.$$

In conclusion, we write out the expression for the probability π that a customer is lost:

$$\pi = \frac{1}{\lambda}\left([1 - B_2(Y)]\boldsymbol{p}_0^T + \sum_{m=1}^{r}\int_0^Y [1 - B_2(v)]\boldsymbol{p}_m^T(v)\,dv + \boldsymbol{p}_R^T\right) N\mathbf{1}.$$

9.6.4. Stationary distribution of the customer sojourn time

Let us find the stationary distribution of the time of sojourn of a customer in the system. First, we introduce the matrix $\widetilde{W}^{(m,V)}(s \mid y)$ whose element $\widetilde{W}_{ij}^{(m,V)}(s \mid y)$ is the Laplace–Stieltjes transform of the sojourn time of a customer of volume y, $y < V$, in the (m, V)-system, $m = 0, \dots, r$, and the probability that at the instant of completion of service of this customer the customer generation process is in phase j, provided that this customer arrived at the initial instant at an empty (m, V)-system and the customer generation process was in phase i. Since the sojourn time of this customer is the sum of the service time itself and the total time of interruptions of his service, and each interruption is a busy period of the $(m+1, V-y)$-system, we find that $\widetilde{W}^{(m,V)}(s \mid y)$ is given by the relation

$$\widetilde{W}^{(m,V)}(s \mid y) = \int_0^\infty e^{(\Lambda + N\Gamma^{(m+1,V-y)}(s) - sI)x} b(x \mid y)\, dx. \tag{9.6.9}$$

Let $\boldsymbol{\omega}(s)$ denote the vector whose coordinate $\omega_i(s)$ is the Laplace–Stieltjes transform of the time of sojourn of a customer in the system and the probability that at the instant of completion of service of this customer the customer generation process is in phase i in the steady state of the system. For the sake of definiteness, we assume that the sojourn time of lost customers is zero. Hence the vector $\boldsymbol{\omega}(s)$ is defined by the formula

$$\begin{aligned}\boldsymbol{\omega}^T(s) = \frac{1}{\lambda}\Bigg\{&\left([1 - B_2(Y)]\boldsymbol{p}_0^T + \sum_{m=1}^r \int_0^Y [1 - B_2(v)]\boldsymbol{p}_m^T(v)\, dv + \boldsymbol{p}_R^T\right) N \\ &+ \int_0^Y \boldsymbol{p}_0^T N \widetilde{W}^{(0,Y)}(s \mid y) b_2(y)\, dy \\ &+ \sum_{m=1}^r \int_0^Y \boldsymbol{p}_m^T(v) N\, dv \int_0^{Y-v} \widetilde{W}^{(m,Y-v)}(s \mid y) b_2(y)\, dy\Bigg\}.\end{aligned}$$

Finally, the Laplace–Stieltjes transform of the stationary distribution of the sojourn time is

$$\omega(s) = \boldsymbol{\omega}^T(s)\mathbf{1}.$$

9.6.5. Particular cases

We now consider a few particular cases of our results.

1. Independent length and volume of a customer. If the length and volume of a customer are independent, then in all formulas we must set $b(x \mid y) = b(x)$ and $B(x \mid y) = B(x)$, where $b(x)$ and $B(x)$ are the probability density and distribution function of the customer length.

Using the matrix Laplace–Stieltjes transform

$$\beta(U) = \int_0^\infty e^{-Ux} b(x)\, dx$$

of the customer length, we rewrite relations (9.6.1), (9.6.8), and (9.6.9) as

$$\begin{aligned}\Gamma^{(m,V)}(s) &= [1 - B_2(V)]I + \int_0^V b_2(y)\beta(sI - \Lambda - N\Gamma^{(m+1,V-y)}(s))\, dy, \\ B_m(y, v) &= (\Lambda^{(m,v)})^{-1}[I - \beta(\Lambda^{(m,v)})]N, \\ \widetilde{W}^{(m,V)}(s \mid y) &= \beta(sI - \Lambda - N\Gamma^{(m+1,V-y)}(s)).\end{aligned}$$

2. System with unbounded number of waiting places. In this case, $r = \infty$ and the matrices $\Gamma^{(m,V)}(s) = \Gamma^{(V)}(s)$, $\Lambda^{(m,v)} = \Lambda^{(v)}$, and $B_m(y, v) = B(y, v)$ do not depend on m.

Relation (9.6.1) reduces to the integral equation

$$\Gamma^{(V)}(s) = [1 - B_2(V)]I + \int_0^V b_2(y)\,dy \int_0^\infty e^{(\Lambda + N\Gamma^{(V-y)}(s) - sI)x} b(x \mid y)\,dx,$$

which can be used to find $\Gamma^{(V)}(s)$.

We introduce the generating function

$$\boldsymbol{c}(z, v) = \sum_{m=2}^{\infty} z^m \boldsymbol{c}_m(v).$$

Then the vector $\boldsymbol{c}_m(v)$ can be found in terms of the generating function $\boldsymbol{c}(z, v)$ from the integral equation

$$\boldsymbol{c}^T(z, v) = z \int_0^v b_2(y) \boldsymbol{c}^T(z, v - y) B(y, Y - v)\,dy + z^2 b_2(v) \boldsymbol{p}_0^T N B(v, Y - v).$$

Knowing the generating function $\boldsymbol{c}(z, v)$, we easily calculate other characteristics of the system in terms of this generating function. In particular, the generating function

$$\boldsymbol{p}(z, v) = \sum_{m=1}^{\infty} z^m \boldsymbol{p}_m(v)$$

is of the form

$$\boldsymbol{p}(z, v) = \boldsymbol{c}(z, v)/z.$$

3. Systems with unbounded total volume. In this case, $Y = \infty$, the matrices $\Gamma^{(m,V)}(s) = \Gamma^{(m)}(s)$, $\Lambda^{(m,V)} = \Lambda^{(m)}$, and $B_m(y, V) = B_m(y)$ do not depend on V.

If the volume of a customer is disregarded, then the system is transformed into a non-priority $MAP/G/1/r$ system with LCFS service discipline and interruption. In particular, introducing the vector $\boldsymbol{p}_m^*(x)$ whose coordinate $p_{im}^*(x)$ is the stationary probability density of the (served) length of the customer at the server, and the probability that the system contains m customers and the customer generation process is in phase i, from formulas (9.6.4)–(9.6.8) we obtain

$$\boldsymbol{p}_m^{*T}(x) = [1 - B(x)] \boldsymbol{r}_m^T N e^{\Lambda^{(m)} x},$$

where the vectors $\boldsymbol{r}_m$ are defined by the recurrence relations

$$\boldsymbol{r}_1 = \boldsymbol{p}_0,$$
$$\boldsymbol{r}_m^T = \boldsymbol{r}_{m-1}^T B_{m-1}, \qquad m = 2, \ldots, R,$$

and the matrix B_m is of the form

$$B_m = N \int_0^\infty [1 - B(x)] e^{\Lambda^{(m)} x}\,dx, \qquad m = 1, \ldots, R.$$

Here the vector $\boldsymbol{p}_m$ of stationary state probabilities is

$$\boldsymbol{p}_m^T = \boldsymbol{p}_0^T B_1 \cdots B_m, \qquad m = 1, \ldots, R,$$

which coincides with formula (7.3.15) to within notation.

9.7. $G/MSP/1/r$ system

In this section, we study a queueing system that is slightly different from the systems considered earlier in this and two previous chapters. Namely, we assume that the input flow is recurrent and the service process is Markov. Such a system, denoted by $G/MSP/1/r$ (*MSP* stands for Markov service process), is reasonably called the dual to the $MAP/G/1/r$ system. Cases of both finite and infinite buffers are investigated. The stationary probabilities of the states of the system for an embedded Markov chain, as well as the time-stationary probabilities, are derived. The Laplace-Stieltjes transforms of the steady-state distributions of waiting and sojourn times of a customer are also given.

9.7.1. System description

Let us consider the $G/MSP/1/r$ ($r \leq \infty$) system with a recurrent input flow, in which the inter-arrival times have some distribution function $A(x)$. We assume that the mean inter-arrival time $a = \int_0^\infty x\,dA(x)$ is finite and, additionally, if the queue distribution is time-stationary, the inter-arrival time cannot take the values jt, where t is a positive number and $j = 0, 1, \ldots$

The Markov service process is defined as follows. There is a Markov process with continuous time and a finite number l of states (service phases). Now, if at some instant a customer is under service and the service is in phase i, $i = 1, \ldots, l$, then in a small time Δ the phase i changes with the probability $\Lambda_{ij}\Delta + o(\Delta)$ to phase j, $j = 1, \ldots, l$, $j \neq i$, and the service of the customer continues, or the phase with the probability $N_{ij}\Delta + o(\Delta)$ changes to phase j, $j = 1, \ldots, l$, but the service is completed and the customer quits the system. As usual, the matrices consisting of Λ_{ij} and N_{ij} are denoted by Λ and N respectively. We introduce the matrix $\Lambda^* = \Lambda + N$, assuming that the matrix Λ^* is indecomposable and the matrix N is non-zero. We also assume that the service phase does not change in an idle period. Let π denote the vector of stationary probabilities of the Markov service process (i.e., the Markov process with an infinitesimal matrix Λ^*).

A customer that arrives at the instant when the buffer is full is lost.

The service discipline is FCFS.

We only apply the method of the embedded Markov chain to analysing this system.

We use the above notation $R = r + 1$.

Since the Markov service process generalises a phase-type service process, the results of this section are applicable to the $G/PH/1/r$ system.

9.7.2. Finite buffer

We consider the successive instants τ_n, $n \geq 0$, of arrival of customers at the system.

Let $\xi(t)$ be the phase of service at instant t and let $\nu(t)$ be the number of customers in the system at this instant.

Let the random variables $\xi_n = \xi(\tau_n + 0)$ and $\nu_n = \nu(\tau_n)$ define the service phase and the number of customers in the system immediately after the arrival of the nth customer. Moreover, assume that $\zeta_n = (\xi_n, \nu_n)$. Then the sequence $\{\zeta_n,\ n \geq 0\}$ forms a homogeneous Markov chain called the embedded chain. Obviously, the state set $\mathcal{X}$ of the embedded Markov chain is $\mathcal{X} = \{(i, k), i = 1, \ldots, l, k = 1, \ldots, R\}$, where i and k denote the service phase and number of customers in the system immediately after the arrival of a customer.

Let us find the matrix of transition probabilities of the embedded Markov chain $\{\zeta_n,\ n > 0\}$. For this purpose, first we must find

- the matrix $F_k(x)$ whose element $(F_k(x))_{ij}$ is the probability that k customers are served in time x and the service process passes to phase j, provided that more than k customers, along with the customer at the server, were in the system at the initial instant, the service process was in phase i, and the service of the customer at the server was not completed in time x;
- the matrix A_k whose element $(A_k)_{ij}$ is the probability that k customers are served in an inter-arrival period and the service process passes to phase j, provided that more than k customers were in the system at the initial instant and the service process was in phase i; and
- the matrices $F_k^*(x)$ and A_k^*, which are similar to the matrices $F_k(x)$ and A_k, but under the condition that exactly k customers were in the system at the initial instant.

The matrices $F_k(x)$ and $F_k^*(x)$ satisfy the recurrence relations

$$F_0(x) = e^{\Lambda x}, \tag{9.7.1}$$

$$F_k(x) = \int_0^x F_{k-1}(y) N F_0(x-y)\,dy, \qquad k \geq 1, \tag{9.7.2}$$

$$F_0^*(x) = I, \tag{9.7.3}$$

$$F_k^*(x) = \int_0^x F_{k-1}(y) N F_0^*(x-y)\,dy, \qquad k \geq 1, \tag{9.7.4}$$

and the matrices A_k and A_k^* are determined by the formulas

$$A_k = \int_0^\infty F_k(x)\,dA(x), \qquad k \geq 0, \tag{9.7.5}$$

$$A_k^* = \int_0^\infty F_k^*(x)\,dA(x), \qquad k \geq 0. \tag{9.7.6}$$

Relations (9.7.1)–(9.7.6) are derived in the same way as (7.1.1)–(7.1.6) for the $MAP/G/1/r$ system, except for formula (9.7.3), which follows from the fact that the service phase does not change in an idle period.

The embedded Markov chain may pass from a state with i, $i = 1, \ldots, R$, customers to only one of the states with j, $j = 1, \ldots, \min(i+1, R)$, customers. Furthermore, transition from a state with i, $i = 1, \ldots, r$, customers to a state with j, $j = 2, \ldots, i+1$, customers occurs only if exactly $j-i-1$ customers are served in an inter-arrival period and transition to a state with one customer occurs only if all i customers in the system are served in an inter-arrival period. Similarly we determine transitions from a state with R customers, except for the transition to a state with R customers, too; this transition occurs not only when only one customer is served, but also if no customer is served and the newly arrived customer is lost.

Thus, the transition probability matrix P of the embedded Markov chain in the block form $P = (P_{km})_{k,m=1,\dots,R}$ is

$$P = \begin{pmatrix} A_1^* & A_0 & 0 & \dots & 0 & 0 \\ A_2^* & A_1 & A_0 & \dots & 0 & 0 \\ A_3^* & A_2 & A_1 & \dots & 0 & 0 \\ \vdots & \vdots & \vdots & \ddots & \vdots & \vdots \\ A_r^* & A_{r-1} & A_{r-2} & \dots & A_1 & A_0 \\ A_R^* & A_r & A_{r-1} & \dots & A_2 & A_1 + A_0 \end{pmatrix}.$$

It is easily seen that the matrix P, within order of numeration, is of the same form as for the $MAP/G/1/r$ system.

Let p_{ik}^*, $i = 1, \dots, l$, $k = 1, \dots, R$, denote the stationary probability (for the embedded Markov chain) that the system contains k customers and the service process is in phase i. Let $\boldsymbol{p}_k^* = (p_{1k}^*, \dots, p_{lk}^*)^T$ and $\boldsymbol{p}^* = (\boldsymbol{p}_1^{*T}, \dots, \boldsymbol{p}_R^{*T})^T$. Then $\boldsymbol{p}^*$ satisfies the equilibrium equations

$$\boldsymbol{p}^{*T} = \boldsymbol{p}^{*T} P \tag{9.7.7}$$

under the normalisation condition

$$p_{\cdot,\cdot}^* = 1.$$

In view of the structure of the matrix P, equations (9.7.7) can be solved by methods described in the two opening sections of Chapter 7. Here we must recall that reverse numeration is used for the states and the transformation of the matrix P must begin with the first state if the method of Section 7.1 is used, but with the last state if the method of Section 7.2 (if the matrix A_0 is invertible) is applied.

Knowing the stationary probabilities of the states of the embedded Markov chain, we easily find other steady-state characteristics of the system.

First, let us find the vector $\boldsymbol{p}_k^- = (p_{1k}^-, \dots, p_{lk}^-)^T$, $k = 1, \dots, R$, of stationary probabilities p_{ik}^- that at the instant (immediately before) of arrival of a customer k other customers are in the system and the Markov service process is in phase i. For this purpose, we observe that a customer upon arrival finds k other customers in the system only if $k + 1$ customers exist in the system after his arrival. Since at the instant of arrival of a customer the phase of the service process does not change, we obtain

$$\boldsymbol{p}_k^- = \boldsymbol{p}_{k+1}^*, \qquad k = 0, \dots, r - 1.$$

It is easily seen that for $k = r$, R,

$$\boldsymbol{p}_k^{-T} = \sum_{j=k}^{R} \boldsymbol{p}_j^{*T} A_{j-k}, \qquad k = r, R.$$

In particular, the stationary probability π that a customer is lost is

$$\pi = \boldsymbol{p}_R^{-T} \mathbf{1} = \boldsymbol{p}_R^{*T} A_0 \mathbf{1}.$$

In order to find the time-stationary probabilities of states, we introduce two matrices T_k and T_k^* whose elements $(T_k)_{ij}$ and $(T_k^*)_{ij}$, $i, j = 1, \dots, l$, are the mean sojourn time of the $G/MSP/1/m$ system (with the buffer of capacity m) in the state $(j, M-k)$ between two successive instants of arrival of customers, provided that the number of customers in the system was $M = m+1$ after the arrival of the first customer and the service process was in phase i. In the first case, it is assumed that $M > k$, whereas in the second case, $M = k$.

The matrices T_k and T_k^* are determined by the relations

$$T_k = \int_0^\infty (1 - A(x)) F_k(x)\, dx, \tag{9.7.8}$$

$$T_k^* = \int_0^\infty (1 - A(x)) F_k^*(x)\, dx, \tag{9.7.9}$$

which are derived exactly as the analogous relations in Section 7.1.3.

Proceeding as in Section 7.1.3, we see that the vectors $\boldsymbol{p}_k$, $k = 0, \dots, R$, of time-stationary state probabilities of the queueing system under consideration are given by the formulas

$$\boldsymbol{p}_0^T = \frac{1}{T} \sum_{m=1}^{R} \boldsymbol{p}_m^{*T} T_m^*, \tag{9.7.10}$$

$$\boldsymbol{p}_k^T = \frac{1}{T} \sum_{m=k}^{R} \boldsymbol{p}_m^{*T} T_{m-k}, \qquad k = 1, \dots, R, \tag{9.7.11}$$

where T is the mean time between changes of the state of the embedded Markov chain. The mean time for our system coincides with the mean inter-arrival time a, i.e., $T = a$.

Let us write out a relation between the vectors $\boldsymbol{p}_k$ and $\boldsymbol{\pi}$ that will be of help in what follows. To this end, we first observe that the matrix functions $F_k(x)$ satisfy differential equation (7.2.8). Integrating by parts, from formulas (9.7.5) and (9.7.6) with regard for (7.2.8), (9.7.1)–(9.7.4), (9.7.8), and (9.7.9), we obtain

$$\begin{aligned} A_0 &= -(1 - A(x))\, F_0(x)\big|_0^\infty + \int_0^\infty (1 - A(x)) F_0(x)\, dx\, \Lambda \\ &= I + T_0 \Lambda, \end{aligned} \tag{9.7.12}$$

$$\begin{aligned} A_k &= -(1 - A(x))\, F_k(x)\big|_0^\infty + \int_0^\infty (1 - A(x))(F_k(x)\Lambda + F_{k-1}(x) N)\, dx \\ &= T_k \Lambda + T_{k-1} N, \qquad k \ge 1, \end{aligned} \tag{9.7.13}$$

$$\begin{aligned} A_k^* &= -\left.(1 - A(x)) \int_0^x F_{k-1}(y)\, dy N\right|_0^\infty + \int_0^\infty (1 - A(x)) F_{k-1}(x)\, dx N \\ &= T_{k-1} N, \qquad\qquad k \ge 1. \end{aligned} \tag{9.7.14}$$

Writing out equilibrium equations (9.7.7) as

$$p_1^{*T} = \sum_{m=1}^{R} p_m^{*T} A_m^*,$$

$$p_k^{*T} = \sum_{m=k-1}^{R} p_m^{*T} A_{m-k+1}, \qquad k = 2, \ldots, r,$$

$$p_R^{*T} = p_r^{*T} A_0 + p_R^{*T} (A_0 + A_1),$$

summing these equations over k from 1 to R, and substituting the expressions of A_k and A_k^* from formulas (9.7.12)–(9.7.14), we obtain

$$p_{\cdot}^{*T} = p_{\cdot}^{*T} + \sum_{k=1}^{R} \sum_{m=k}^{R} p_m^{*T} T_{m-k} \Lambda^*$$

or, in view of (9.7.11),

$$p_{\cdot}^{*T} = p_{\cdot}^{*T} + T \sum_{k=1}^{R} p_k^T \Lambda^*.$$

From this relation we immediately find that

$$\sum_{k=1}^{R} p_k^T \Lambda^* = \mathbf{0}^T.$$

Hence, in turn, we obtain

$$\sum_{k=1}^{R} p_k = c\pi, \tag{9.7.15}$$

where c is the normalisation constant which can expressed as

$$c = 1 - p_0. \tag{9.7.16}$$

Here $p_k = p_{\cdot,k}$, $k = 0, \ldots, R$, is the (time-) stationary probability that k customers exist in the system.

This result has a clear physical meaning. It shows that the conditional (time-) stationary probability that the Markov service process is in phase i, provided that customers exist in the system, coincides with the unconditional stationary probability that the Markov service process is in phase i. But, in contrast to systems with Markov input flow, here we are concerned with the conditional probability that customers exist in the system. This becomes clear if we recall that the service phase does not change in an idle period.

Finally, we find certain characteristics related to the customer sojourn time in the system. To do this, we let $V_k(x)$ denote the matrix whose element $(V_k(x))_{ij}$ is the probability that at least k customers will be served within the time x and the service process upon completion of service passes to phase j, provided that k customers were in the system and the service process was in phase i at the beginning of service. Let $\Phi_k(s)$ denote the Laplace–Stieltjes transform of $V_k(x)$. Since the probability that the service of a customer is completed in the time interval $[x, x+dx)$ is given by the matrix formula $e^{\Lambda x} N\,dx$ and the total service time of k customers is the sum of their service times, we obtain

$$\Phi_1(s) = \int_0^\infty e^{-sx} e^{\Lambda x} N\,dx = (sI - \Lambda)^{-1} N,$$
$$\Phi_k(s) = \Phi_1^k(s), \qquad k \geq 2.$$

Hence we easily find the Laplace–Stieltjes transforms $\omega(s)$ and $\varphi(s)$ of stationary distributions of the waiting and sojourn times of an arbitrary customer provided he was not lost:

$$\omega(s) = \frac{1}{1-\pi} \sum_{k=0}^{r} \boldsymbol{p}_k^{-T} \Phi_1^k(s) \mathbf{1},$$
$$\varphi(s) = \frac{1}{1-\pi} \sum_{k=0}^{r} \boldsymbol{p}_k^{-T} \Phi_1^{k+1}(s) \mathbf{1}.$$

In particular, differentiating these formulas at the point $s = 0$, we see that the mean waiting time w for service and the mean sojourn time v of an arbitrary customer accepted for service in the steady state are

$$w = -\frac{1}{1-\pi} \sum_{k=1}^{r} \boldsymbol{p}_k^{-T} \sum_{j=0}^{k-1} (-\Lambda^{-1} N)^j \Lambda^{-1} \mathbf{1}$$
$$= -\frac{1}{1-\pi} \sum_{j=0}^{r-1} \left(\sum_{k=j+1}^{r} \boldsymbol{p}_k^{-T} \right) (-\Lambda^{-1} N)^j \Lambda^{-1} \mathbf{1},$$
$$v = -\frac{1}{1-\pi} \sum_{k=0}^{r} \boldsymbol{p}_k^{-T} \sum_{j=0}^{k} (-\Lambda^{-1} N)^j \Lambda^{-1} \mathbf{1}$$
$$= -\frac{1}{1-\pi} \sum_{j=0}^{r} \left(\sum_{k=j}^{r} \boldsymbol{p}_k^{-T} \right) (-\Lambda^{-1} N)^j \Lambda^{-1} \mathbf{1}.$$

In numerical computations, it is convenient to use the formula

$$V_k(x)\mathbf{1} = 1 - \sum_{i=0}^{k-1} F_i(x) \mathbf{1}, \qquad k \geq 1,$$

This formula yields the following relation for $W(x)$ and $V(x)$:

$$W(x) = 1 - \frac{1}{1-\pi} \sum_{i=0}^{r-1} \left(\sum_{k=i+1}^{r} \boldsymbol{p}_k^{-T} \right) F_i(x) \mathbf{1},$$
$$V(x) = 1 - \frac{1}{1-\pi} \sum_{i=0}^{r} \left(\sum_{k=i}^{r} \boldsymbol{p}_k^{-T} \right) F_i(x) \mathbf{1}.$$

9.7.3. Infinite buffer

We now consider a system with infinite buffer ($r = \infty$).

We can show that the necessary and sufficient condition for existence of a steady state for a system with infinite buffer is the standard condition $\rho < 1$, where $\rho = (a\mu)^{-1}$ is the traffic intensity of the system and $\mu = \boldsymbol{\pi}^T N\mathbf{1}$ is the stationary service intensity.

As for a system with finite buffer, we consider an embedded Markov chain with the state set $\mathscr{X} = \{(i,k), i = 1,\dots,l, k \geq 1\}$ which now is countable. As above, introducing the vectors $\boldsymbol{p}_k^*$, $k \geq 1$, and $\boldsymbol{p}^* = (\boldsymbol{p}_1^{*T}, \boldsymbol{p}_2^{*T}, \dots)^T$ of stationary state probabilities of the embedded Markov chain, we obtain (9.7.7) for $\boldsymbol{p}^*$, where the transition probability matrix P is of the form

$$P = \begin{pmatrix} A_1^* & A_0 & 0 & 0 & \dots \\ A_2^* & A_1 & A_0 & 0 & \dots \\ A_3^* & A_2 & A_1 & A_0 & \dots \\ A_4^* & A_3 & A_2 & A_1 & \dots \\ \vdots & \vdots & \vdots & \vdots & \ddots \end{pmatrix}.$$

Let us write out equilibrium equations (9.7.7) in full:

$$\boldsymbol{p}_1^{*T} = \sum_{m=1}^{\infty} \boldsymbol{p}_m^{*T} A_m^*, \tag{9.7.17}$$

$$\boldsymbol{p}_k^{*T} = \sum_{m=k-1}^{\infty} \boldsymbol{p}_m^{*T} A_{m-k+1}, \qquad k \geq 2. \tag{9.7.18}$$

Before solving equations (9.7.17)–(9.7.18), we introduce the vectors $\boldsymbol{p}_k$, $k \geq 0$, of time-stationary state probabilities and derive certain useful relations.

First, for $\boldsymbol{p}_k$, we obtain the formulas

$$\boldsymbol{p}_0^T = \frac{1}{T} \sum_{m=1}^{\infty} \boldsymbol{p}_m^{*T} T_m^*, \tag{9.7.19}$$

$$\boldsymbol{p}_k^T = \frac{1}{T} \sum_{m=k}^{\infty} \boldsymbol{p}_m^{*T} T_{m-k}, \qquad k \geq 1, \tag{9.7.20}$$

which are analogues of formulas (9.7.10) and (9.7.11).

Second, the vectors $\boldsymbol{p}_k$ satisfy the relation

$$\sum_{k=1}^{\infty} \boldsymbol{p}_k = (1 - p_0)\boldsymbol{\pi}, \tag{9.7.21}$$

which can be easily derived using the above proof of formula (9.7.15).

In order to derive the next relation, we recall the equality

$$\Lambda\mathbf{1} = -N\mathbf{1}. \tag{9.7.22}$$

This equality and formulas (9.7.12)–(9.7.14) yield

$$T_{k-1}N\mathbf{1} = A_k^*\mathbf{1}, \qquad k \geq 1, \tag{9.7.23}$$

$$T_0 N\mathbf{1} = (I - A_0)\mathbf{1}, \tag{9.7.24}$$

$$T_k N\mathbf{1} = (T_{k-1}N - A_k)\mathbf{1}, \qquad k \geq 1. \tag{9.7.25}$$

Now we consider the product $(1 - p_0)\rho^{-1}$, which, by virtue of formula (9.7.21), admits the representation

$$(1 - p_0)\rho^{-1} = T \sum_{k=1}^{\infty} \boldsymbol{p}_k^T N\mathbf{1}.$$

Substituting the expressions of the time-stationary probabilities in terms of the stationary probabilities of the embedded Markov chain defined by formula (9.7.20), we obtain

$$\begin{aligned}(1 - p_0)\rho^{-1} &= \sum_{k=1}^{\infty}\sum_{m=k}^{\infty} \boldsymbol{p}_m^{*T} T_{m-k} N\mathbf{1} \\ &= \sum_{m=1}^{\infty} \boldsymbol{p}_m^{*T} T_{m-1} N\mathbf{1} + \sum_{k=2}^{\infty}\sum_{m=0}^{\infty} \boldsymbol{p}_{k+m}^{*T} T_m N\mathbf{1}.\end{aligned} \tag{9.7.26}$$

Now from formulas (9.7.23) and (9.7.17) we obtain

$$\sum_{m=1}^{\infty} \boldsymbol{p}_m^{*T} T_{m-1} N\mathbf{1} = \sum_{m=1}^{\infty} \boldsymbol{p}_m^{*T} A_m^*\mathbf{1} = \boldsymbol{p}_1^{*T}\mathbf{1} = p_1^*, \tag{9.7.27}$$

where $p_k^* = \boldsymbol{p}_k^{*T}\mathbf{1}$ denotes the stationary probability (for the embedded Markov chain) that k customers exist in the system.

Similarly, for $k \geq 2$ formulas (9.7.24), (9.7.25), and (9.7.18) yield

$$\begin{aligned}\sum_{m=0}^{\infty} \boldsymbol{p}_{k+m}^{*T} T_m N\mathbf{1} &= \boldsymbol{p}_k^{*T}\mathbf{1} - \sum_{m=k}^{\infty} \boldsymbol{p}_m^{*T} A_{m-k}\mathbf{1} + \sum_{m=0}^{\infty} \boldsymbol{p}_{k+1+m}^{*T} T_m N\mathbf{1} \\ &= p_k^* - p_{k+1}^* + \sum_{m=0}^{\infty} \boldsymbol{p}_{k+1+m}^{*T} T_m N\mathbf{1} \\ &= p_k^* - p_{k+1}^* + p_{k+1}^* - p_{k+2}^* + \sum_{m=0}^{\infty} \boldsymbol{p}_{k+2+m}^{*T} T_m N\mathbf{1} \\ &= \cdots = p_k^*.\end{aligned} \tag{9.7.28}$$

From formulas (9.7.26)–(9.7.28) we finally obtain

$$(1 - p_0)\rho^{-1} = \sum_{k=1}^{\infty} p_k^* = 1,$$

i.e., the standard expression

$$\rho = 1 - p_0, \tag{9.7.29}$$

interrelating the traffic intensity of the system and the probability that there are no customers in the system.

We can now solve equilibrium equations (9.7.17)–(9.7.18). We will seek for the solution in the form

$$\boldsymbol{p}_k^{*T} = \boldsymbol{p}_1^{*T} G^{k-1}, \qquad k \geq 1, \tag{9.7.30}$$

where G solves the equation

$$G = \sum_{k=0}^{\infty} G^k A_k = \tilde{A}(G). \tag{9.7.31}$$

LEMMA 9.7.1. *For $\rho < 1$, equation* (9.7.31) *has a unique solution in the class of matrices whose all eigenvalues are less than one in absolute value. This solution is a matrix whose all elements are positive, and the iterations $G^{(n)} = \tilde{A}(G^{(n-1)})$, $n \geq 1$, converge to the solution as $n \to \infty$ if the initial iteration $G^{(0)}$ is a matrix with eigenvalues less than one 1 in absolute value.*

Validity of the lemma immediately follows from findings in (Neuts, 1981).

In numerical computations, it is more convenient to take $G^{(0)}$ to be a zero matrix so that the sequence $G^{(n)}$, $n \geq 0$, may monotonically converge to G.

It is easily seen that for any $\boldsymbol{p}_1^*$ the sequence of vectors $\boldsymbol{p}_k^*$ defined by formula (9.7.30) where G solves (9.7.31) satisfies all equations in (9.7.18), and moreover,

$$\sum_{k=1}^{\infty} \sum_{i=1}^{l} |p_{ik}^*| < \infty, \tag{9.7.32}$$

because each eigenvalue of the matrix G is less than one in absolute value.

With the use of (9.7.17), we will find the unknown vector $\boldsymbol{p}_1^*$. For this purpose, we first turn back to equation (9.7.31) and substitute the expressions of A_k from formulas (9.7.12) and (9.7.13). We thus obtain

$$G = I + U\Lambda + GUN, \tag{9.7.33}$$

where

$$U = \sum_{k=0}^{\infty} G^k T_k \tag{9.7.34}$$

(we can show that the elements of the matrix U are positive). Multiplying both sides of (9.7.33) by $\mathbf{1}$, by virtue of (9.7.22) we obtain

$$(I - G)\mathbf{1} = (I - G)UN\mathbf{1}.$$

Since the matrix $(I - G)$ is non-degenerate, the last equality is equivalent to the equality

$$UN\mathbf{1} = \mathbf{1},$$

which, in its turn, implies that the matrix UN is stochastic, because its elements are non-negative.

From the nature of the matrices U and N it also follows that all columns of the matrix UN are either strictly positive or zero. The state set of the Markov chain with the transition probability matrix UN hence consists of two subsets; the first subset consists of communicating states only, whereas the second subset consists of unessential states only. Hence, in turn, it follows that the equation

$$\boldsymbol{p}_1^{*T} = \boldsymbol{p}_1^{*T} UN \tag{9.7.35}$$

has a solution which is unique within a constant.

Substituting the expressions for A_m^* from formula (9.7.14) and $\boldsymbol{p}_m^*$ from (9.7.30) into (9.7.17), we arrive at (9.7.35), which, as shown above, has a solution being unique within a constant, which is found from the normalisation condition easily reduced to

$$\boldsymbol{p}_1^{*T}(I-G)^{-1}\mathbf{1} = 1. \tag{9.7.36}$$

Finally, applying the Foster theorem 1.4.3, we see that, by virtue of (9.7.32), formula (9.7.30) yields the (unique) solution of equilibrium equations (9.7.17)–(9.7.18), where the matrix G and the vector $\boldsymbol{p}_1^*$ are found from (9.7.31), (9.7.35), and (9.7.36), and the matrix U is defined by (9.7.34).

If the matrix U has an inverse, then the vector $\boldsymbol{p}_1^*$ can be conveniently computed with the use of relations (9.7.20), (9.7.21), and (9.7.29). Substituting the vectors $\boldsymbol{p}_k^*$ from (9.7.30) into (9.7.20), we obtain, by virtue of (9.7.21) and (9.7.29),

$$\begin{aligned}\frac{1}{T}\sum_{k=1}^{\infty}\sum_{m=k}^{\infty}\boldsymbol{p}_1^{*T}G^{m-1}T_{m-k} &= \frac{1}{T}\boldsymbol{p}_1^{*T}\sum_{m=0}^{\infty}G^m\sum_{k=0}^{\infty}G^kT_k \\ &= \frac{1}{T}\boldsymbol{p}_1^{*T}(I-G)^{-1}U = \rho\boldsymbol{\pi}^T,\end{aligned}$$

which yields the expression of $\boldsymbol{p}_1^*$

$$\boldsymbol{p}_1^{*T} = \rho T\boldsymbol{\pi}^T U^{-1}(I-G).$$

Other aspects of the $G/MSP/1/\infty$ system can be investigated along the same lines as for the $G/MSP/1/r$ system with a finite buffer.

Thus, the following formulas hold:

- for the vectors $\boldsymbol{p}_k^-$, $k \geq 0$, of stationary probabilities that at the instant of arrival of a customer k other customers are in the system,

$$\boldsymbol{p}_k^{-T} = \boldsymbol{p}_{k+1}^{*T} = \boldsymbol{p}_1^{*T}G^k;$$

- for the vectors $\boldsymbol{p}_k$, $k \geq 0$, of time-stationary probabilities,

$$\boldsymbol{p}_0^T = \frac{1}{T}\sum_{m=1}^{\infty}\boldsymbol{p}_1^{*T}G^{m-1}T_m^*,$$

$$\boldsymbol{p}_k^T = \frac{1}{T}\sum_{m=k}^{\infty}\boldsymbol{p}_1^{*T}G^{m-1}T_{m-k} = \frac{1}{T}\boldsymbol{p}_1^{*T}G^{k-1}U, \qquad k \geq 1;$$

– for the Laplace-Stieltjes transforms $\omega(s)$ and $\varphi(s)$ of stationary distributions of the waiting and sojourn times of a customer,

$$\omega(s) = \sum_{k=0}^{\infty} \mathbf{p}_1^{*T} G^k \Phi_1^k(s)\mathbf{1},$$

$$\varphi(s) = \sum_{k=0}^{\infty} \mathbf{p}_1^{*T} G^k \Phi_1^{k+1}(s)\mathbf{1};$$

– for the mean waiting time w for service and the mean sojourn time v of an arbitrary customer,

$$\begin{aligned} w &= -\mathbf{p}_1^{*T} \sum_{j=0}^{\infty} \left(\sum_{k=j+1}^{\infty} G^k \right) (-\Lambda^{-1}N)^j \Lambda^{-1}\mathbf{1} \\ &= -\mathbf{p}_1^{*T} G(I-G)^{-1} \sum_{j=0}^{\infty} G^j (-\Lambda^{-1}N)^j \Lambda^{-1}\mathbf{1}, \\ v &= -\mathbf{p}_1^{*T} \sum_{j=0}^{\infty} \left(\sum_{k=j}^{\infty} G^k \right) (-\Lambda^{-1}N)^j \Lambda^{-1}\mathbf{1} \\ &= -\mathbf{p}_1^{*T} (I-G)^{-1} \sum_{j=0}^{\infty} G^j (-\Lambda^{-1}N)^j \Lambda^{-1}\mathbf{1}. \end{aligned}$$

Let us consider in more detail the computation of distribution functions $W(x)$ and $V(x)$, which for the queueing system with infinite capacity can be represented in the form

$$\begin{aligned} W(x) &= 1 - \mathbf{p}_1^{*T} \sum_{i=0}^{\infty} \left(\sum_{k=i+1}^{\infty} G^k \right) F_i(x)\mathbf{1} \\ &= 1 - \mathbf{p}_1^{*T} G(I-G)^{-1} \sum_{i=0}^{\infty} G^i F_i(x)\mathbf{1}, \end{aligned} \tag{9.7.37}$$

$$\begin{aligned} V(x) &= 1 - \mathbf{p}_1^{*T} \sum_{i=0}^{\infty} \left(\sum_{k=i}^{\infty} G^k \right) F_i(x)\mathbf{1} \\ &= 1 - \mathbf{p}_1^{*T} (I-G)^{-1} \sum_{i=0}^{\infty} G^i F_i(x)\mathbf{1}. \end{aligned} \tag{9.7.38}$$

We set

$$R(x) = \sum_{i=0}^{\infty} G^i F_i(x) \tag{9.7.39}$$

and denote by a the absolute value of the minimal diagonal element of the matrix Λ. We also set $Q = I + \Lambda/a$.

We seek for the matrix function $R(x)$ in the form

$$R(x) = e^{-ax} \sum_{i=0}^{\infty} \frac{(ax)^i}{i!} R_i. \tag{9.7.40}$$

Since

$$F_0(x) = e^{\Lambda x} = e^{-ax} \sum_{i=0}^{\infty} \frac{(ax)^i}{i!} Q^i,$$

relations (9.7.2), (9.7.39), and (9.7.40) yield

$$\begin{aligned}
e^{-ax} \sum_{i=0}^{\infty} \frac{(ax)^i}{i!} R_i &= e^{-ax} \sum_{i=0}^{\infty} \frac{(ax)^i}{i!} Q^i \\
&\quad + G \int_0^x e^{-ay} \sum_{i=0}^{\infty} \frac{(ay)^i}{i!} R_i N e^{-a(x-y)} \sum_{j=0}^{\infty} \frac{(a(x-y))^j}{j!} Q^j \, dy \\
&= e^{-ax} \sum_{i=0}^{\infty} \frac{(ax)^i}{i!} Q^i \\
&\quad + e^{-ax} \sum_{i=0}^{\infty} a^i \sum_{j=0}^{i} G R_j N Q^{i-j} \int_0^x \frac{y^j (x-y)^{i-j}}{j!(i-j)!} dy \\
&= e^{-ax} \sum_{i=0}^{\infty} \frac{(ax)^i}{i!} Q^i \\
&\quad + e^{-ax} \sum_{i=1}^{\infty} \frac{(ax)^i}{i!} \sum_{j=0}^{i-1} \frac{1}{a} G R_j N Q^{i-1-j}.
\end{aligned}$$

We find the matrices R_i from the recurrence relation

$$\begin{aligned}
R_0 &= I, \\
R_i &= Q^i + \frac{1}{a} \sum_{j=0}^{i-1} G R_j N Q^{i-1-j} \\
&= \left(Q^{i-1} + \frac{1}{a} \sum_{j=0}^{i-2} G R_j N Q^{i-2-j} \right) Q + \frac{1}{a} G R_{i-1} N, \qquad i \geq 1,
\end{aligned}$$

or

$$R_0 = I, \tag{9.7.41}$$

$$R_i = R_{i-1} Q + \frac{1}{a} G R_{i-1} N, \qquad i \geq 1. \tag{9.7.42}$$

Thus, from formulas (9.7.37)–(9.7.39) it follows that the distribution functions $W(x)$ and $V(x)$ are

$$W(x) = 1 - \boldsymbol{p}_1^{*T} G(I - G)^{-1} R(x)\mathbf{1},$$
$$V(x) = 1 - \boldsymbol{p}_1^{*T} (I - G)^{-1} R(x)\mathbf{1},$$

where the matrix function $R(x)$ is determined by relations (9.7.40)–(9.7.42).

10

Queueing networks

In Chapter 2, we have touched on the concept of a queueing network. The theory of queueing networks is well developed nowadays, and its results are published in numerous monographs and papers. Of course, here we can only briefly outline what has been done in this field. In this chapter, we concern the theory of exponential queueing networks with a single customer class and present its main results, namely, the Jackson and Gordon–Newell theorems. According to these theorems, the multivariate stationary distribution of the number of customers at network nodes can be represented in a multiplicative form so that the computation of network performance indices can be radically simplified. Without exaggeration, these theorems are at the heart of the theory of queueing networks.

10.1. Network classes

We begin the study of queueing networks with a detailed description of open and closed exponential networks. Here we confine ourselves only to networks with single-type customers. As mentioned in Chapter 2, open queueing networks differ from closed networks in that customers of an open network arrive from outside and quit after service at nodes chosen at random, whereas the same customers constantly circulate in a closed network.

In this section, we derive equations for the intensities of flows at individual nodes. These equations also hold for non-exponential queueing networks.

10.1.1. Open queueing networks

First we study open queueing networks.

An open queueing network consists of M nodes which customers arrive at from an external source (node 0). A node i is assumed to be a multi-server queueing system with n_i identical servers and a common buffer of infinite capacity, i.e., $r_i = \infty$. The input flow from the external source is Poisson with intensity λ_0. Each customer arriving with the probability θ_{0i}, $\sum_{i=1}^{M} \theta_{0i} = 1$, does not depend on the state and prehistory of the network, and is directed to some node i, $i = 1, \dots, M$. For our study, it is convenient to assume that $\theta_{00} = 0$. According to the Poisson flow rarefaction theorem 2.1.1, arrivals from an external source at node i form a Poisson flow of intensity $\lambda_{0i} = \lambda_0\theta_{0i}$, $i = 1, \dots, M$.

The service time of any customer at any server of the node i is a random variable, does not depend on the state and prehistory of the network (and on its service at other network nodes if it had arrived at the node i not from an external source but from another node), and is exponentially distributed with parameter μ_i, $i = 1, \dots, M$.

Finally, we must specify a discipline by which customers are chosen for service at the network nodes. For the sake of definiteness, we restrict to the FCFS discipline, though the results stated below are equally applicable to other disciplines as well.

Now we describe the route of customers in the network. A customer upon completion of service at node i with the probability θ_{ij}, which does not depend on the state and prehistory of the network, instantaneously passes to node j, $i, j = 1, \dots, M$, and with the probability $\theta_{i0} = 1 - \sum_{j=1}^{M} \theta_{ij}$, $i = 1, \dots, M$, quits the network. The matrix $\Theta_0 = (\theta_{ij})_{i,j=0,\dots,M}$ is stochastic and is called the matrix of transitions on the set $\mathcal{M}_0 = \{0, 1, \dots, M\}$ of network nodes or the *route matrix*. In what follows, we assume that the matrix Θ_0 is indecomposable. Physically, indecomposability of the matrix Θ_0 implies that a customer arriving at the network with a non-zero probability after a certain number of transitions may join any node. Moreover, since the zero column of the indecomposable matrix Θ_0 contains at least one non-zero element, each customer will necessarily quit the network.

The network thus defined is denoted by

$$\mathcal{N}_o(M) = \langle M, \Theta_0, \lambda_0, \mu_i, n_i, r_i = \infty, \text{FCFS}, i = 1, \dots, M \rangle.$$

Knowing the initial data for the network, that is, the input flow, service times at nodes, and transition laws (route matrices), we study the flows circulating in the network.

Let λ_{ij} be the intensity of flow of customers from node i to node j, $i, j = 0, \dots, M$, $\lambda_{00} = 0$. Let

$$\lambda_i = \lambda_{\cdot,i} = \sum_{j=0}^{M} \lambda_{ji} \qquad \lambda_{i,\cdot} = \sum_{j=0}^{M} \lambda_{ij}$$

be the intensities of the total input and output flows to and from node i, respectively. We consider the steady (equilibrium) state of the network. Later we will describe a condition for existence of a steady state of the network. In the steady state, the input flow intensity at node i must be equal to its output flow intensity. Therefore, we denote these intensities by λ_i, i.e., $\lambda_{i,\cdot} = \lambda_{\cdot,i} = \lambda_i$, $i = 0, \dots, M$. Hence

$$\lambda_{ij} = \lambda_i \theta_{ij}, \qquad i, j = 0, \dots, M. \tag{10.1.1}$$

Summing equality (10.1.1) over $i = 0, 1, \dots, M$, we obtain equilibrium equations for the intensities of flows in the network:

$$\lambda_j = \sum_{i=0}^{M} \lambda_i \theta_{ij}, \qquad j = 0, \dots, M. \tag{10.1.2}$$

Since Θ_0 is a stochastic indecomposable matrix, simultaneous homogeneous equations (10.1.2) in unknowns λ_j, $j = 0, \dots, M$, are of order $M + 1$ and rank M (equations (10.1.2) coincide with the equilibrium equations for the Markov chain generated by the transition matrix Θ_0 and the state set $\mathcal{M}_0 = \{0, 1, \dots, M\}$ to within normalisation factor). Therefore, all unknowns in (10.1.2) can be expressed in terms of one of them, say, λ_0. Therefore, for a given external input intensity λ_0, the intensities λ_j, $j = 1, \ldots, M$, are strictly positive and uniquely defined by (10.1.2).

It must be noted that equilibrium equations (10.1.2) also hold in a more general case where the external input to node i is a recurrent flow with distribution function $A_i(x)$, $i = 1, \ldots, M$, and the service time at each server of node i has an arbitrary distribution function $B_i(x)$, $i = 1, \ldots, M$. In this sense, equilibrium equations (10.1.2) are invariant to the distribution functions $A_i(x)$ and $B_i(x)$ for fixed mean inter-arrival time a_i and mean service time b_i at node i.

Let us turn back to the transition matrix Θ_0. Along with this matrix, we also need its submatrix $\Theta = (\theta_{ij})_{i,j=1,\ldots,M}$.

Now we study the random walk of a customer in the set $\mathcal{M} = \{1, 2, \ldots, M\}$ of network nodes, beginning from the instant of his arrival at the network to the instant of departure from the network, considering only those node-to-node transitions that occur in unit time (in one step). Let η_n be the label of the node at which the customer stays at the nth step. Then $\{\eta_n, n \geq 0\}$ is a homogeneous absorbing Markov chain with transition matrix Θ_0 and state set $\mathcal{M}_0 = \{0, 1, 2, \ldots, M\}$, where the state 0 is an absorbing state, i.e., a customer reaches the state 0 means that he leaves the network (though it is more appropriate to distinguish two states, namely, the source 0_a and the sink 0_b, we disregard such details and take $0_a \equiv 0_b \equiv 0$, i.e., each customer arrives at the network from outside and returns to outside on leaving the network).

Let N_{ij} denote the mean number of visits of a customer at node j before quitting the network (absorption at node 0), provided that the random walk of the customer started from node i, $i, j = 1, \ldots, M$. Clearly, the probability $\mathbf{P}\{\eta_0 = i\}$ of this condition is θ_{0i}, $i = 1, \ldots, M$.

The number N_{ij} satisfies the relations

$$N_{ij} = \delta_{ij} + \sum_{k=1}^{M} \theta_{ik} N_{kj}, \qquad i, j = 1, \ldots, M. \tag{10.1.3}$$

Indeed, to find N_{ij}, the contribution of the initial state of the network to the number of visits of a customer at node j must be added to the contributions of succeeding steps. But the contribution of the initial state is non-zero and equals 1 only if $i = j$. This contribution is defined by the first term in the right-hand side. In the next step, the customer passes to node k. Therefore, the contribution of this state, by virtue of the Markov property, is N_{kj} and does not depend on the contribution of the previous step. Summing over all states $k = 1, 2, \ldots, M$, we arrive at (10.1.3).

We express (10.1.3) in the matrix form as

$$N = I + \Theta N. \tag{10.1.4}$$

Since Θ is the principal submatrix of the indecomposable stochastic matrix Θ_0, the matrix $(I - \Theta)$ is invertible (Kemeny and Snell, 1960) and

$$(I - \Theta)^{-1} = I + \Theta + \Theta^2 + \cdots \tag{10.1.5}$$

Therefore, from (10.1.4) we obtain

$$N = (I - \Theta)^{-1}. \tag{10.1.6}$$

The matrix N is called the *fundamental matrix* of the absorbing Markov chain $\{\eta_n, n \geq 0\}$.

From (10.1.4)–(10.1.6) we obtain the equality $\Theta N = N\Theta$. By virtue of this equality, relations (10.1.3) can rewritten as

$$N_{ij} = \delta_{ij} + \sum_{k=1}^{M} N_{ik}\theta_{kj}, \qquad i, j = 1, \ldots, M. \tag{10.1.7}$$

Let

$$h_j = \sum_{i=1}^{M} \theta_{0i} N_{ij}, \qquad j = 1, \ldots, M,$$

stand for the mean number of visits of a customer at node j before quitting the network. Multiplying both sides of (10.1.7) by θ_{0i} and then summing over $i = 1, 2, \ldots, M$, we find that h_j, $j = 1, \ldots, M$, satisfies the simultaneous equations

$$h_j = \theta_{0j} + \sum_{i=1}^{M} h_i \theta_{ij}, \qquad j = 1, \ldots, M. \tag{10.1.8}$$

Using the vectors $\boldsymbol{\lambda}^T = (\lambda_1, \ldots, \lambda_M)$, $\boldsymbol{\theta}_0^T = (\theta_{01}, \ldots, \theta_{0M})$, $\boldsymbol{\lambda}_0 = \lambda_0 \boldsymbol{\theta}_0$, and $\boldsymbol{h}^T = (h_1, \ldots, h_M)$, we rewrite (10.1.2) and (10.1.8) in the matrix form as

$$\boldsymbol{\lambda}^T (I - \Theta) = \boldsymbol{\lambda}_0^T,$$
$$\boldsymbol{h}^T (I - \Theta) = \boldsymbol{\theta}_0^T.$$

Hence

$$\boldsymbol{\lambda}^T = \boldsymbol{\lambda}_0^T (I - \Theta)^{-1} = \boldsymbol{\lambda}_0^T N, \tag{10.1.9}$$

$$\boldsymbol{h}^T = \boldsymbol{\theta}_0^T (I - \Theta)^{-1} = \boldsymbol{\theta}_0^T N. \tag{10.1.10}$$

Comparing (10.1.9) and (10.1.10), we obtain

$$\lambda_j = \lambda_0 h_j, \qquad j = 1, \ldots, M. \tag{10.1.11}$$

Since $\lambda_j > 0$, $j = 1, \ldots, M$, and $\lambda_0 > 0$, from (10.1.11) we see that $h_j > 0$, $j = 1, \ldots, M$.

Formula (10.1.11) can also be obtained from the following qualitative considerations. The mean number of visits of a customer at node j in time t, on the one hand, is $\lambda_j t$ and, on the other hand, is equal to the mean number $\lambda_0 t$ of arrivals in time t multiplied by the mean number h_j of visits at node j of each customer arriving at the network.

Let h_0 denote the mean number of visits of a customer at node 0 after quitting the network. On the one hand, $h_0 = 1$, and, on the other hand, substituting (10.1.11) into (10.1.2) for $j = 0$, we obtain

$$h_0 = \sum_{i=1}^{M} h_i \theta_{i0} = 1. \tag{10.1.12}$$

Combining (10.1.8) and (10.1.12), we obtain

$$h_j = \sum_{i=0}^{M} h_i \theta_{ij}, \qquad j = 0, \dots, M. \tag{10.1.13}$$

Solving (10.1.13), we should keep in mind that $h_0 = 1$.

10.1.2. Closed queueing networks

A network in which K customers (they neither leave the network nor are joined by any customer from outside) circulate in $\mathcal{M} = \{1, 2, \dots, M\}$ nodes is said to be closed.

As for an open network, we assume that any node i of the closed network is a multi-server queueing system with n_i identical servers, waiting capacity $r_i \geq K - 1$, and FCFS service discipline. We also assume that the distribution function of the service time at any server of node i is exponential with parameter μ_i, $i = 1, \dots, M$. The route of customers in the network, as before, is defined by the transition probabilities θ_{ij}, $i, j = 1, \dots, M$. Since no customer quits the network, the equality $\sum_{j=1}^{M} \theta_{ij} = 1$, $i = 1, \dots, M$, holds. We also assume that the routing stochastic matrix $\Theta = (\theta_{ij})_{i,j=1,\dots,M}$ is indecomposable. A closed network with such parameters is denoted by

$$\mathcal{N}_c(M, K) = \langle M, K, \Theta, \mu_i, n_i, r_i \geq K - 1, \text{FCFS}, i = 1, \dots, M \rangle.$$

By analogy with open networks, taking a labelled node i^*, let us consider the random walk of a customer beginning from the instant of transition from node i^* to some node i to the instant of returning to node i^*. This random walk is described by the homogeneous absorbing Markov chain $\{\eta_n^*, n \geq 0\}$ on the state set $\mathcal{M}$ with absorbing state i^*. Taking $\mathcal{M}^* = \mathcal{M} \setminus \{i^*\}$, we find that the role of the matrix Θ for the open network is now played by the matrix $\Theta^* = (\theta_{ij})_{i,j \in \mathcal{M}^*}$, and that of θ_{i0}, by the probability θ_{ii^*}. Now let h_j denote the mean number of visits of a customer at node j, $j \in \mathcal{M}^*$, in the time interval between two successive visits at node i^*. Obviously, if the absorbing state i^* is taken into consideration, then we must take $h_{i^*} = 1$. Reasoning as in the case of an open network, we find that h_j, $j = 1, \dots, M$, satisfies the equations (the analogy of (10.1.13)):

$$h_j = \sum_{i=1}^{M} h_i \theta_{ij}, \qquad j = 1, \dots, M. \tag{10.1.14}$$

Since the node i^* is an arbitrary node of the set $\mathcal{M}$, equations (10.1.14) in the general case possess a solution $h_j > 0$, $j = 1, \dots, M$, which is defined up to a constant. For a fixed node i^*, as demonstrated earlier, we have $h_{i^*} = 1$ and the solution of (10.1.14) is unique.

10.2. Open exponential networks

The main objective of this section is to prove the Jackson theorem, which asserts that the stationary distribution of the number of customers at the nodes of an open exponential network is the product of marginal distributions of these numbers and the marginal distributions are defined by the same formulas as for an ordinary $M/M/n/\infty$ system.

10.2.1. Product form of the stationary distribution of the number of customers at network nodes

For the open exponential network described in Section 10.1.1, let $\nu_i(t)$, $i = 1, \ldots, M$, denote the number of customers at node i at instant t and set $\boldsymbol{\nu}(t) = (\nu_1(t), \ldots, \nu_M(t))$.

Let us study the random process $\{\boldsymbol{\nu}(t), t \geq 0\}$. Its state set is

$$\mathscr{X} = \{\boldsymbol{k} = (k_1, \ldots, k_M), k_i \geq 0, i = 1, \ldots, M\}.$$

The states of the set $\mathscr{X}$ are interpreted as follows: if $\boldsymbol{\nu}(t) = (k_1, \ldots, k_M)$ at instant t, then at this instant there are $k_1, k_2 \ldots, k_M$ customers at nodes $1, 2, \ldots, M$, respectively (note that the notation of the vectors $\boldsymbol{\nu}(t)$ and $\boldsymbol{k}$ differ from our previous notation for row vectors; the vector expressions $\boldsymbol{\nu}(t)$ and $\boldsymbol{k}$ are purely symbolic here).

Since we are studying an exponential network, the process $\{\boldsymbol{\nu}(t), t \geq 0\}$ is Markov and homogeneous. So, in order to find the network performance indices, we can use the apparatus of continuous-time Markov processes with discrete state set. Here we restrict ourselves to the stationary state of the network, because the Jackson theorem holds *only* in this case.

We assume that the limit (stationary) probabilities

$$p(k_1, \ldots, k_M) = \lim_{t\to\infty} \mathbf{P}\{\boldsymbol{\nu}(t) = (k_1, \ldots, k_M)\}$$

exist. Later we define the ergodicity condition for existence of these probabilities. For the sake of brevity, we use also the expression $p(\boldsymbol{k}) = p(k_1, \ldots, k_M)$ in what follows.

To derive the equilibrium equations for the stationary probabilities $p(\boldsymbol{k})$, $\boldsymbol{k} \in \mathscr{X}$, we need some additional network operation parameters. Let $\rho_i = \lambda_i/\mu_i$, $\gamma_i(k_i) = \min(k_i, n_i)$, and $\mu_i(k_i) = \mu_i\gamma_i(k_i)$. Here ρ_i is the traffic intensity of a server at node i, $\gamma_i(k_i)$ is the number of concurrently operating servers at node i at instant t under the condition that at this instant there are k_i customers at node i and $\mu_i(k_i)$ has the meaning of intensity of service at node i, provided that at instant t there are k_i customers at node i (see Section 3.2).

Now we are in a position to derive the equilibrium equations. To derive them, we use the global balance principle by equating the probability flow emanating from each state $\boldsymbol{k} \in \mathscr{X}$ and the probability flow entering into this state.

Departure from the state $\boldsymbol{k} = (k_1, \ldots, k_i, \ldots, k_M)$ is possible either due to the arrival of a customer from outside (with the intensity λ_0) or due to the completion of service of the customer at node i and transition of this customer from the node to some other node or his exit from the network (with the intensity $\mu_i(k_i)(1 - \theta_{ii})$). Then the total probability flow emanating from the state $\boldsymbol{k}$ is $p(\boldsymbol{k})\left[\lambda_0 + \sum_{i=1}^{M} \mu_i(k_i)(1 - \theta_{ii})\right]$.

Now let us consider how to enter the state $\boldsymbol{k}$. This can be achieved due to the arrival of a customer from outside at node i (with the intensity $\lambda_0\theta_{0i}$), $i = 1, \ldots, M$, provided that $k_i - 1 \geq 0$ customers were at this node. The total probability flow in this case is $\lambda_0 \sum_{i=1}^{M} p(\boldsymbol{k} - \boldsymbol{e}_i)\theta_{0i}u(k_i)$. Here, as before, $u(x)$ is the Heaviside function and $\boldsymbol{e}_i$ is an M-dimensional vector whose all components are 0, except its ith component which is equal to 1. The state $\boldsymbol{k}$ can also be attained if a customer is being served at node i and quits the network, provided that $k_i + 1$ customers were at node i (this occurs with the intensity $\mu_i(k_i+1)\theta_{i0}$). Then the total probability flow due to the exit of a customer from the network is $\sum_{i=1}^{M} p(\boldsymbol{k} + \boldsymbol{e}_i)\mu_i(k_i + 1)\theta_{i0}$. Finally, the state $\boldsymbol{k}$ can be attained if a customer upon completion of service at node i jumps to node j, provided that k_i+1 and k_j-1 customers were at nodes i and j respectively (such a transition occurs with the intensity $\mu_i(k_i+1)\theta_{ij}$). In this case, the total probability flow due to the transition of a customer from one node to another is $\sum_{i=1}^{M}\sum_{\substack{j=1\\ j\neq i}}^{M} p(\boldsymbol{k}+\boldsymbol{e}_i-\boldsymbol{e}_j)\mu_i(k_i+1)\theta_{ij}u(k_j)$.

Equating the total probability flows entering into the state $\boldsymbol{k}$ and the total probability flow emanating from this state, we obtain the equilibrium equations

$$p(\boldsymbol{k})\left[\lambda_0+\sum_{i=1}^{M}\mu_i(k_i)(1-\theta_{ii})\right]=\lambda_0\sum_{i=1}^{M}p(\boldsymbol{k}-\boldsymbol{e}_i)\theta_{0i}u(k_i)+\sum_{i=1}^{M}p(\boldsymbol{k}+\boldsymbol{e}_i)\mu_i(k_i+1)\theta_{i0}$$
$$+\sum_{i=1}^{M}\sum_{\substack{j=1\\ j\neq i}}^{M}p(\boldsymbol{k}+\boldsymbol{e}_i-\boldsymbol{e}_j)\mu_i(k_i+1)\theta_{ij}u(k_j),\qquad \boldsymbol{k}\in\mathcal{X}. \quad (10.2.1)$$

Let us now solve these equilibrium equations. Its solution is determined by the Jackson theorem.

THEOREM 10.2.1 (Jackson theorem). *In an open homogeneous exponential network*

$$\mathcal{N}_o(M)=\langle M,\Theta_0,\lambda_0,\mu_i,n_i,r_i=\infty,\mathrm{FCFS},i=1,\dots,M\rangle,$$

if the matrix Θ is indecomposable and the condition $\rho_i<n_i$, $i=1,\dots,M$, holds, then the process $\{\boldsymbol{\nu}(t),t\geq 0\}$ is ergodic and its stationary distribution $p(\boldsymbol{k})$, $\boldsymbol{k}\in\mathcal{X}$, admits the product form

$$p(\boldsymbol{k})=p(k_1,\dots,k_M)=\prod_{i=1}^{M}p_i(k_i), \quad (10.2.2)$$

where

$$p_i(k_i)=\begin{cases}\dfrac{\rho_i^{k_i}}{k_i!}p_i(0), & k_i=0,\dots,n_i-1;\\[2ex] \dfrac{\rho_i^{k_i}}{n_i!\,n_i^{k_i-n_i}}p_i(0), & k_i\geq n_i.\end{cases} \quad (10.2.3)$$

Moreover,

$$\sum_{k_i=0}^{\infty}p_i(k_i)=1,\qquad i=1,\dots,M. \quad (10.2.4)$$

PROOF. To prove the theorem, we make use of the concept of the partial balance (recall that unlike local and global balances, which always hold, partial balance is to be established in each case individually).

We assume that two types of partial balances hold for our network. For the first-type partial balance, we assume that the probability flows for each state $\boldsymbol{k}\in\mathcal{X}$ are equal: emanating flow from the state $\boldsymbol{k}$ due to the arrival of customers from outside is equal to the entering flow at the state $\boldsymbol{k}$ due to the completion of service at network nodes and exit of customers from the network. As a result of such a partial balance, we obtain the equations

$$p(\boldsymbol{k})\lambda_0 = \sum_{i=1}^{M} p(\boldsymbol{k}+\boldsymbol{e}_i)\mu_i(k_i+1)\theta_{i0}, \qquad \boldsymbol{k} \in \mathscr{X}. \tag{10.2.5}$$

According to the second type of partial balance, the probability flows for each state $\boldsymbol{k} = (k_1, \ldots, k_i, \ldots, k_M) \in \mathscr{X}$ and each node i, $i = 1, \ldots, M$, are equal: the emanating flow from the state $\boldsymbol{k}$ due to the completion of service at node i is equal to the entering flow to state $\boldsymbol{k}$ due to the arrival of a customer from outside to node i and due to the completion of service of a customer at node j and his transition to node i. In this case, the partial balance equations take the form

$$\begin{aligned} p(\boldsymbol{k})\mu_i(k_i)(1-\theta_{ii}) &= p(\boldsymbol{k}-\boldsymbol{e}_i)\lambda_0\theta_{0i}u(k_i) \\ &+ \sum_{\substack{j=1 \\ j\neq i}}^{M} p(\boldsymbol{k}+\boldsymbol{e}_j-\boldsymbol{e}_i)\mu_j(k_j+1)\theta_{ji}u(k_i), \quad i = 1, \ldots, M, \quad \boldsymbol{k} \in \mathscr{X}. \end{aligned} \tag{10.2.6}$$

Summing (10.2.6) over $i = 1, 2, \ldots, M$, and then adding them to (10.2.5), we obtain the initial equilibrium equations (10.2.1). Therefore, if solutions are found for each of equations (10.2.5) and (10.2.6) (partial sets of equilibrium equations) and they are consistent, then they will also be solutions of equilibrium equations (10.2.1).

In general, for the solutions of partial sets of equilibrium equations to be consistent, it is sufficient that the global balance equations have a solution. But this condition is not a necessary condition.

Therefore, our problem is to find consistent solutions of (10.2.5) and (10.2.6). Precisely these are defined by formulas (10.2.2)–(10.2.4), as shown below.

First we observe that for a fixed $i = 1, \ldots, M$, formulas (10.2.3) and (10.2.4) define the stationary distribution of the number of customers in the $M/M/n_i/\infty$ system with Poisson input of intensity λ_i and exponentially distributed service time with parameter μ_i, which is the same for any server. According to Section 3.2, the local balance equations for such a queueing system are

$$\lambda_i p_i(k_i) = \mu_i(k_i+1)p_i(k_i+1), \quad k_i \geq 0, \quad i = 1, \ldots, M. \tag{10.2.7}$$

In view of the assumption that $p(\boldsymbol{k}) > 0$, $\boldsymbol{k} \in \mathscr{X}$, from (10.2.7) we obtain

$$\begin{aligned} \frac{p(\boldsymbol{k}-\boldsymbol{e}_i)}{p(\boldsymbol{k})} &= \frac{\mu_i(k_i)}{\lambda_i}, && k_i \geq 1, && i = 1, \ldots, M, \\ \frac{p(\boldsymbol{k}+\boldsymbol{e}_i)}{p(\boldsymbol{k})} &= \frac{\lambda_i}{\mu_i(k_i+1)}, && && i = 1, \ldots, M, \\ \frac{p(\boldsymbol{k}+\boldsymbol{e}_j-\boldsymbol{e}_i)}{p(\boldsymbol{k})} &= \frac{\lambda_j}{\lambda_i}\frac{\mu_i(k_i)}{\mu_j(k_j+1)}, && k_i \geq 1, \quad i \neq j, && i, j = 1, \ldots, M. \end{aligned} \tag{10.2.8}$$

Dividing both sides of (10.2.5) and (10.2.6) by $p(\boldsymbol{k})$ and using relations (10.2.8), we obtain the equalities

$$\lambda_0 = \sum_{i=1}^{M} \lambda_i \theta_{i0}, \tag{10.2.9}$$

$$\mu_i(k_i)(1-\theta_{ii}) = \frac{\mu_i(k_i)}{\lambda_i}\lambda_0\theta_{0i} + \sum_{\substack{j=1\\ j\neq i}}^{M} \frac{\lambda_j}{\lambda_i}\mu_i(k_i)\theta_{ji}, \quad \boldsymbol{k} \in \mathscr{X}, \quad i = 1,\dots,M. \tag{10.2.10}$$

For $k_i = 0$, since $\mu_i(0) = 0$, equalities (10.2.10) become identities. For $k_i > 0$, we see that $\mu_i(k_i) > 0$. Therefore, multiplying both sides of (10.2.10) by $\lambda_i/\mu_i(k_i)$, we reduce them to the relations

$$\lambda_i = \lambda_0\theta_{0i} + \sum_{j=1}^{M} \lambda_j\theta_{ji}, \qquad i = 1,\dots,M,$$

which, like relation (10.2.9), are mere equations for the stationary intensities of flows circulating in the network. Thus, substituting (10.2.3) and (10.2.4) into partial sets of equilibrium equations (10.2.5) and (10.2.6), we arrive at identities. Therefore, (10.2.3) and (10.2.4) also solve equilibrium equations (10.2.1).

Note that formulas (10.2.2) and (10.2.3) give a positive solution of (10.2.1). This solution is bounded if the series $\sum_{k_i=0}^{\infty} p_i(k_i)$, $i = 1,\dots,M$, converge. They converge if $\rho_i < n_i$, $i = 1,\dots,M$ (see Section 3.2). Since all states of the network communicate with each other, by virtue of Theorem 1.5.4 the process $\{\boldsymbol{\nu}(t), t \geq 0\}$ is ergodic. This completes the proof of the theorem. □

From the Jackson theorem, at first glance, we may conclude that in the stationary state the networks nodes operate as independent $M/M/n_i/\infty$ systems with Poisson input flows of intensities $\lambda_i, i = 1,\dots,M$. In general, this is not true. The probabilistic characteristics of nodes, such as sojourn times at individual nodes, may be dependent; furthermore, the flows in the network may be non-Poisson. Therefore, the independence of network nodes in such a case is purely apparent. Here we will not touch upon this topic, and details can be found in special literature (e.g., (Warland, 1988)).

10.2.2. Network performance indices

Since the stationary distribution $p(\boldsymbol{k})$ is represented in product form (10.2.2), it is relatively easy to calculate the performance indices of individual nodes and certain characteristics of the network. In particular, the probability that a network node i is idle is $p_i(0)$, the mean number N_i of customers at node i is given by formula (3.2.6) (the parameters λ, μ, and n are to be prefixed with index i), and the mean sojourn time v_i of a customer at node i is given by the Little formula

$$\lambda_i v_i = N_i, \qquad i = 1,\dots,M. \tag{10.2.11}$$

Validity of this formula can be proved by simple qualitative considerations.

Obviously, the total mean number N of customers in the network can be found from the formula

$$N = \sum_{i=1}^{M} N_i. \tag{10.2.12}$$

Therefore, the total mean sojourn time v of a customer in the network (mean time of response) can also be found from the Little formula

$$\lambda_0 v = N. \tag{10.2.13}$$

Its physical meaning is self-evident. Its proof is given in (Warland, 1988).

10.3. Closed exponential networks

In this section, we prove the Gordon–Newell theorem, which is the analogue of the Jackson theorem for a closed exponential network. While computing the performance indices of a closed network, the most difficult problem from the computational viewpoint is to find the normalisation factor. So, we will give one possible method to compute this factor.

10.3.1. Product form of the stationary distribution of the number of customers at network nodes

For the closed exponential queueing network described above, let us study the queueing processes at the network nodes. The notation of Section 10.2 is retained for $\nu_i(t)$, $\boldsymbol{\nu}(t)$, $\boldsymbol{k}$, and other entities. The random process $\{\boldsymbol{\nu}(t),\ t \geq 0\}$ describing the operation of the closed network

$$\mathcal{N}_c(M, K) = \langle M, K, \Theta, \mu_i, n_i, r_i \geq K - 1, \text{FCFS}, i = 1, \dots, M \rangle,$$

is a homogeneous Markov process defined on the state set

$$\mathscr{X}(M, K) = \{\boldsymbol{k} = (k_1, \dots, k_M), k_i = 0, \dots, K, i = 1, \dots, M, k. = K\}. \tag{10.3.1}$$

Physically, the state $\boldsymbol{k} = (k_1, \dots, k_M)$ is interpreted as in Section 10.2.

The number of states in (10.3.1) is finite. Furthermore, by the assumption, the matrix Θ is indecomposable, and hence, all states in (10.3.1) communicate with each other. Therefore, the Markov process $\{\boldsymbol{\nu}(t), t \geq 0\}$ is ergodic. Hence the stationary probabilities $p(\boldsymbol{k}) > 0$, $\boldsymbol{k} \in \mathscr{X}(M, K)$, exist and satisfy the equilibrium equations

$$p(\boldsymbol{k})\sum_{j=1}^{M}\mu_j(k_j)(1-\theta_{jj}) = \sum_{i=1}^{M}\sum_{\substack{j=1 \\ j\neq i}}^{M} p(\boldsymbol{k}+\boldsymbol{e}_i-\boldsymbol{e}_j)\mu_i(k_i+1)\theta_{ij}u(k_j), \qquad \boldsymbol{k} \in \mathscr{X}(M, K), \tag{10.3.2}$$

under the normalisation condition

$$\sum_{k \in \mathscr{X}(M,K)} p(\boldsymbol{k}) = 1. \tag{10.3.3}$$

It is easy to derive equilibrium equations (10.3.2). They are similar to those for an open network, except for the difference that there is no need to specify the probability flows related to arrivals from outside and their departures since there are no such flows in a closed network.

We now formulate the fundamental result of the theory of closed homogeneous exponential networks—the cornerstone in the development of this theory.

THEOREM 10.3.1 (Gordon–Newell theorem). *In the closed homogeneous exponential network*

$$\mathcal{N}_c(M, K) = \langle M, K, \Theta, \mu_i, n_i, r_i \geq K-1, \text{FCFS}, i = 1, \ldots, M \rangle,$$

if the transition matrix Θ *is indecomposable, then the stationary distribution* $p(\boldsymbol{k})$, $\boldsymbol{k} \in \mathscr{X}(M, K)$, *admits the product form*

$$p(\boldsymbol{k}) = G^{-1}(M, K) \prod_{i=1}^{M} \frac{d_i^{k_i}}{\beta_i(k_i)}, \qquad \boldsymbol{k} \in \mathscr{X}(M, K), \tag{10.3.4}$$

where $d_i = h_i/\mu_i$, h_i, $i = 1, \ldots, M$, *is a positive solution of equations* (10.1.14),

$$\beta_i(k) = \prod_{j=1}^{k} \gamma_i(j), \quad k \geq 1, \quad \beta_i(0) = 1,$$

and the normalisation constant is defined by the condition

$$G(M, K) = \sum_{k \in \mathscr{X}(M,K)} \prod_{i=1}^{M} \frac{d_i^{k_i}}{\beta_i(k_i)}. \tag{10.3.5}$$

For the sake of brevity, we write $\mathscr{X}$ for $\mathscr{X}(M, K)$, though the context in many cases may require that the expression be written in full form.

PROOF. As already noted, the expressions of the equilibrium equations and their solution for closed networks are somewhat analogous to those for open networks. Naturally, an idea arises that this analogy can be extended to the proof as well. In other words, assume that partial balance between probability flows also holds for a closed network, i.e., the emanating flow from node j due to the completion of service is equal to the entering flow at node j due to the completion of service of the customer at node i and its transition to node $j, i = 1, \ldots, M, i \neq j$. Equality of these flows is defined by the partial balance equation

$$p(\boldsymbol{k})\mu_j(k_j)(1-\theta_{jj}) = \sum_{\substack{i=1 \\ j \neq i}}^{M} p(\boldsymbol{k} + \boldsymbol{e}_i - \boldsymbol{e}_j)\mu_i(k_i+1)\theta_{ij}u(k_j), \quad j = 1, \ldots, M, \quad \boldsymbol{k} \in \mathscr{X}. \tag{10.3.6}$$

Obviously, summing both sides of (6) for $j = 1, 2, \ldots, M$, we obtain system (2). Therefore, if substitution of (4) into (6) yields an identity, then (2) is also converted into an identity.

Dividing both sides of (10.3.6) by $p(\boldsymbol{k})$ and then substituting (10.3.4) into (10.3.6), we obtain

$$\mu_j(k_j)(1-\theta_{jj}) = \sum_{\substack{i=1\\ j\neq i}}^{M} \frac{d_i\gamma_j(k_j)}{\gamma_i(k_i+1)d_j}\mu_i(k_i+1)\theta_{ij}u(k_j), \quad j=1,\ldots,M, \quad \boldsymbol{k}\in\mathscr{X}.$$

Since $\mu_i(k_i) = \mu_i\gamma_i(k_i)$ and $\gamma_i(k_i) = \min(k_i, n_i)$, for $k_j = 0$ we arrive at an identity. If $k_j > 0$, then, since $d_i = h_i/\mu_i$ and $h_i > 0$, the last equalities reduce to

$$h_j(1-\theta_{jj}) = \sum_{\substack{i=1\\ j\neq i}}^{M} h_i\theta_{ij}, \quad j=1,\ldots,M, \quad \boldsymbol{k}\in\mathscr{X}.$$

Hence we obviously obtain equalities (10.1.14). Therefore, we obtain an identity. This completes the proof of the theorem. □

Simple but elegant formulas (10.3.4) and (10.3.5) have certain demerits. First, before carrying out computations by these formulas, we solve linear algebraic equations (10.1.14), which may be of high order if the dimensionality of the network is large. But this difficulty can be overcome in some cases due to the simple structure of the matrix Θ. A far more difficult problem arises in computing the normalisation factor $G(M, K)$. The number of states in the set $\mathscr{X}(M, K)$ is $\binom{M+K-1}{M-1}$ and it rapidly increases with M and K. Even for $M = 5$ and $K = 10$, this number is 1000 and for $K = 100$ and $M = 5$ the number is $5\cdot 10^6$. Clearly, there is no sense in computing $G(M, K)$ directly from formula (10.3.5). Therefore, simple efficient algorithms must be designed to compute the normalisation factor in order to save memory and processing time. Such algorithms are suggested in numerous publications (Basharin *et al.*, 1989; Gnedenko and König, 1984; Zhozhikashvili and Vishnevskii, 1988; Warland, 1988). Below we describe one approach to computing the normalisation factor, which, in a sense, is a classical method.

10.3.2. The normalisation factor: the convolution method of computation

First we study a closed queueing network with single-server nodes, i.e., $n_i = 1$, $i = 1, \ldots, M$. In this case, $\gamma_i(k_i) = 0$ at $k_i = 0$, and $\gamma_i(k_i) = 1$ at $k_i \geq 1$. Therefore, $\beta_i(k) = 1$ for $k \geq 0$. Hence, by virtue of the Gordon–Newell theorem,

$$p(\boldsymbol{k}) = G^{-1}(M, K)\prod_{i=1}^{M} d_i^{k_i}, \qquad \boldsymbol{k}\in\mathscr{X}(M, K), \tag{10.3.7}$$

where the normalisation factor is defined by the condition

$$G(M, K) = \sum_{\boldsymbol{k}\in\mathscr{X}(M,K)}\prod_{i=1}^{M} d_i^{k_i}. \tag{10.3.8}$$

We introduce the auxiliary sets $\mathcal{X}(m, k) = \{\boldsymbol{k} = (k_1, \dots, k_m), k_i \geq 0, i = 1, \dots, m, k. = k\}$, $m = 1, \dots, M$, $k = 1, \dots, K$, and the functions

$$g(m, k) = \sum_{\boldsymbol{k} \in \mathcal{X}(m,k)} \prod_{i=1}^{m} d_i^{k_i}, \quad m = 1, \dots, M, \quad k = 1, \dots, K. \tag{10.3.9}$$

Obviously, $g(m, k)$ is the normalisation factor for the network $N_c(m, k)$ with m nodes and k customers. We also need boundary conditions for $g(m, k)$ at $m = 0$ or $k = 0$. If $k = 0$, then $\mathcal{X}(m, 0)$, $m > 0$, consists of only one zero vector $\boldsymbol{k} = (0, \dots, 0)$ of dimensionality m. Therefore, from (10.3.9) we obtain

$$g(m, 0) = 1, \qquad m = 1, \dots, M. \tag{10.3.10}$$

If $m = 0$, then $\mathcal{X}(0, k)$ does not contain any vector. In this case, we must set

$$g(0, k) = 0, \qquad k = 0, \dots, K. \tag{10.3.11}$$

Now we represent the set $\mathcal{X}(m, k)$ for fixed m and k as the union of two disjoint sets

$$\mathcal{X}_0(m, k) = \{\boldsymbol{k} = (k_1, \dots, k_m), k_i \geq 0, i = 1, \dots, m-1, k_m = 0k. = k\},$$
$$\mathcal{X}_1(m, k) = \{\boldsymbol{k} = (k_1, \dots, k_m), k_i \geq 0, i = 1, \dots, m-1, k_m \geq 1, k. = k\}.$$

Then (10.3.9) is expressed as

$$g(m, k) = \sum_{\boldsymbol{k} \in \mathcal{X}_0(m,k)} \prod_{i=1}^{m} d_i^{k_i} + \sum_{\boldsymbol{k} \in \mathcal{X}_1(m,k)} \prod_{i=1}^{m} d_i^{k_i}. \tag{10.3.12}$$

Since the vectors $\boldsymbol{k}$ in the first term of (10.3.12) are of the form $\boldsymbol{k} = (k_1, \dots, k_{m-1}, 0)$, we see that $\prod_{i=1}^{m} d_i^{k_i} = \prod_{i=1}^{m-1} d_i^{k_i}$, and hence, it can be expressed as $\sum_{\boldsymbol{k} \in \mathcal{X}(m-1,k)} \prod_{i=1}^{m-1} d_i^{k_i}$, which is merely $g(m-1, k)$. Now we consider the second term in (10.3.12). Applying the substitutions $k'_m = k_m - 1$ and $k_m = k'_m$, we obtain $\mathcal{X}_1(m, k) = \mathcal{X}(m, k-1)$. Thus the second term reduces to $d_m \sum_{\boldsymbol{k} \in \mathcal{X}(m,k-1)} \prod_{i=1}^{m} d_i^{k_i}$. By the definition of $g(m, k)$, the second term is equal to $d_m g(m, k-1)$.

We thus obtain

$$g(m, k) = g(m-1, k) + d_m g(m, k-1), \quad m = 1, \dots, M, \quad k = 1, \dots, K. \tag{10.3.13}$$

Since $g(M, K) = G(M, K)$, formula (10.3.13), along with initial conditions (10.3.10) and (10.3.11), defines a recurrent algorithm for computing the normalisation factor $G(M, K)$. This algorithm is known as the convolution algorithm or the Buzen algorithm (Basharin *et al.*, 1989; Gnedenko and König, 1984; Zhozhikashvili and Vishnevskii, 1988; Warland, 1988).

While computing the normalisation factor $G(M, K)$, this algorithm uses MK multiplications and MK additions. It suffices to store just K values: $(K - k)$ values of $g(m-1, j)$, $j = k+1, \dots, K$, and $g(m, j)$, $j = 1, \dots, k$.

Let us now study multi-server nodes, i.e., $n_i \geq 1$, $i = 1, \dots, M$. We introduce the auxiliary functions

$$f_i(k) = \frac{d_i^k}{\beta_i(k)}, \quad i = 1, \dots, M, \quad k \geq 0.$$

Moreover, by analogy with single-server node, let us consider the sets $\mathcal{X}(m, k)$ and the functions

$$g(m,k) = \sum_{\boldsymbol{k}\in\mathcal{X}(m,k)} \prod_{i=1}^{m} f_i(k_i), \quad m = 1,\ldots,M, \quad k = 1,\ldots,K. \tag{10.3.14}$$

The boundary conditions $g(m, 0)$ and $g(0, k)$ for these functions are, obviously, defined by formulas (10.3.10) and (10.3.11).

For fixed m and k, let us represent the set $\mathcal{X}(m, k)$ as the union of $(k + 1)$ disjoint sets

$$\mathcal{X}_l = \{\boldsymbol{k} = (k_1,\ldots,k_m), k_i \geq 0, i = 1,\ldots,m-1, k_m = l, k. = k\}, \qquad l = 0,\ldots,k.$$

Applying the substitution $k'_m = k_m - l$ and using the notation $k_m = k'_m$, we obtain $\mathcal{X}_l = \mathcal{X}(m-1, k-l)$. Therefore, (10.3.14) takes the form

$$g(m,k) = \sum_{l=0}^{k} f_m(l) \sum_{\boldsymbol{k}\in\mathcal{X}(m-1,k-l)} \prod_{i=1}^{m-1} f_i(k_i).$$

But, by the definition,

$$\sum_{\boldsymbol{k}\in\mathcal{X}(m-1,k-l)} \prod_{i=1}^{m-1} f_i(k_i) = g(m-1, k-l).$$

We thus obtain

$$g(m,k) = \sum_{l=0}^{k} f_m(l) g(m-1, k-l), \quad m = 1,\ldots,M, \quad k = 1,\ldots,K. \tag{10.3.15}$$

Since $g(M, K) = G(M, K)$, formula (10.3.15), together with initial conditions (10.3.10) and (10.3.11), defines a recurrent algorithm to compute the normalisation factor $G(M, K)$ for a network with multi-server nodes. We can show that the total number of arithmetic operations needed to compute $G(M, K)$ is of order $(M-1)(K+2)^2$. This is left to the reader as an exercise.

10.3.3. Performance indices of network nodes

As in the case of open networks, representation of the stationary distribution $p(\boldsymbol{k})$, $\boldsymbol{k} \in \mathcal{X}(M, K)$, in the product form greatly simplifies computation of various steady-state characteristics of network nodes. In distinction to an open network, for which certain node characteristics can be computed by formulas known for multi-server queueing systems, for a closed network we must first derive formulas for computing the characteristics of its nodes. Below we derive these formulas. First we introduce a multidimensional random variable $(\xi_1,\ldots,\xi_M)$ with probability distribution $p(\boldsymbol{k})$, $\boldsymbol{k} \in \mathcal{X}(M, K)$. To compute the steady-state performance indices of node i, we must know the marginal distribution $p_i(k) = \mathbf{P}\{\xi_i = k\}$, $k = 0,\ldots,K$.

First we study single-server nodes and consider the probability

$$\mathsf{P}\{\xi_i \geq k\} = G^{-1}(M, K) \sum_{\substack{k \in \mathscr{X}(M,K) \\ k_i \geq k}} \prod_{i=1}^{M} d_i^{k_i}.$$

Setting $k_i' = k_i - k$ and $k_i = k_i'$, as in the preceding section, we obtain

$$\mathsf{P}\{\xi_i \geq k\} = G^{-1}(M, K) d_i^k \sum_{k \in \mathscr{X}(M,K-k)} \prod_{j=1}^{M} d_j^{k_j}.$$

Here the sum in the right-hand side, in view of (10.3.9), is $g(M, K-k)$. Therefore,

$$\mathsf{P}\{\xi_i \geq k\} = \frac{d_i^k g(M, K-k)}{G(M, K)}, \quad k \geq 0, \quad i = 1, \ldots, M. \tag{10.3.16}$$

Since

$$\begin{aligned} &\mathsf{P}\{\xi_i = k\} = \mathsf{P}\{\xi_i \geq k\} - \mathsf{P}\{\xi_i \geq k+1\}, \qquad k = 0, \ldots, K-1, \\ &\mathsf{P}\{\xi_i = K\} = \mathsf{P}\{\xi_i \geq K\}, \end{aligned}$$

we finally obtain

$$p_i(k) = \mathsf{P}\{\xi_i = k\} = \frac{d_i^k [g(M, K-k) - u(K-k) d_i g(M, K-k-1)]}{G(M, K)}, \qquad k = 0, \ldots, K, \quad i = 1, \ldots, M, \tag{10.3.17}$$

where $g(M, 0) = 1$.

Now the mean number $N_i = \mathsf{E}\xi_i$ of customers at node i is

$$\begin{aligned} \mathsf{E}\xi_i = \sum_{k=1}^{K} k p_i(k) = \frac{1}{G(M, K)} [&d_i g(M, K-1) - d_i^2 g(M, K-2) + 2 d_i^2 g(M, K-2) \\ &- 2 d_i^3 G(M, K-3) + \cdots + (K-1) d_i^{K-1} g(M, 1) - (K-1) d_i^K + K d_i^K]. \end{aligned}$$

Hence we obtain

$$N_i = \mathsf{E}\xi_i = \frac{1}{G(M, K)} \sum_{k=1}^{K} d_i^k g(M, K-k), \qquad i = 1, \ldots, M. \tag{10.3.18}$$

Let us find the throughput of node i, which is merely the intensity λ_i of output flow from node i. Note that λ_i, $i = 1, \ldots, M$, satisfies the equilibrium equations for the steady-state intensities of flows in the network (see Section 10.1). These equations can be formally derived from (10.1.2) for $\theta_{i0} = 0$, $i = 1, \ldots, M$. But these equations are helpless for finding the intensity λ_i, because their solution is determined up to a constant. Only after finding the marginal distribution $p_i(k)$, $k = 0, \ldots, K$, we can calculate λ_i, but from other considerations that are similar to those used in finding the output flow intensity of an ordinary queueing system. Indeed, the intensity λ_i of the output flow from node i is equal to the intensity μ_i of service at node i multiplied by the stationary probability $1 - p_i(0)$, which is equal to the fraction of time during which the server of node i is busy. Thus,

$$\lambda_i = \mu_i[1 - p_i(0)], \qquad i = 1, \dots, M.$$

Hence, by virtue of formula (10.3.16) and the equality $d_i = h_i/\mu_i$, we finally obtain

$$\lambda_i = h_i \frac{g(M, K-1)}{G(M, K)}, \qquad i = 1, \dots, M. \tag{10.3.19}$$

Now we can find the mean sojourn time v_i of a customer at node i from the Little formula:

$$\lambda_i v_i = N_i, \qquad i = 1, \dots, M. \tag{10.3.20}$$

It is easy to prove qualitatively the Little formula for this case, though there is a rigorous proof.

Now we consider the general case of multi-server nodes and derive the expression for $p_M(k) = \mathbf{P}\{\xi_M = k\}$. Reasoning as in the preceding section, we obtain

$$\mathbf{P}\{\xi_M = k\} = \frac{1}{G(M, K)} \sum_{\substack{k \in \mathcal{X}(M,K) \\ k_M = k}} \prod_{i=1}^{M} f_i(k_i) = \frac{f_M(k)}{G(M, K)} \sum_{k \in \mathcal{X}(M-1, K-k)} \prod_{i=1}^{M-1} f_i(k_i).$$

Hence, since the sum in the right-hand side of the last equality is equal to $g(M-1, K-k)$, we obtain the final expression for the unknown probability

$$\mathbf{P}\{\xi_M = k\} = \frac{f_M(k) g(M-1, K-k)}{G(M, K)}, \qquad k = 0, \dots, K. \tag{10.3.21}$$

In order to find the probability $\mathbf{P}\{\xi_i = k\}$ for $i \neq M$, we can re-order the nodes ($i \leftrightarrow M$) and then solve this problem again, which is indeed a problem for a different network. There are other methods to find marginal distributions for multi-server nodes, which are omitted here. An interested reader may refer, for example, to (Basharin *et al.*, 1989).

Bibliography

Basharin, G. P. (1965). A single server with a finite queue and items of several types. *Theory Probab. Appl.* **10**, 261–274.

Basharin, G. P. (1967). On the service of two flows with non-preemptive priority in the system of full accessibility ith limited waiting room. *Izv. USSR Acad. Sci.. Ser. Tech. Cybern.* #2, 72–86 (in Russian).

Basharin, G. P., Kharkevich, A. D., and Shneps, M. A. (1968). *Queues in Telephone Systems.* Nauka, Moscow (in Russian).

Basharin, G. P., Bocharov, P. P., and Kogan, Ya. A. (1989). *Queueing Analysis for Computer Networks: Theory and Computational methods.* Nauka, Moscow (in Russian).

Bellman, R. (1960). *Introduction to Matrix Analysis.* McGraw–Hill, New York.

Belyaev, Yu. K. (1962). Piecewise Markov processes and its applications to reliability problems. In: *Proc 6 All-Union Conf. Theory Probab. Math. Statist.* Vilnius, pp. 309–323 (in Russian).

Berman, A., and Plemmons, P. (1979). *Nonnegative Matrices in the Mathematical Sciences.* Academic Press, New York.

Bhat, N. U., and Basawa, I.V. (1992). *An Introduction to Queueing Theory.* Clarendon Press, Oxford.

Blondia, C. (1992). Performance evaluation of $M/1$-stage in an ATM switching element. *Performance Evaluation* **15**, 1–20.

Blondia, C. (1998). The $N/G/1$ finite capacity queue. *Stochastic Models* **5**, 273–294.

Bocharov, P. P. (1985). *Single-Server Queueing Systems of Finite Capacity.* Peoples' Friendship Univ. Press, Moscow (in Russian).

Bocharov, P. P. (1985). Analysis of single-server queueing system of finite capacity with customers of several types and FIFO discipline. In: *Methods of Teletraffic Theory in Communication Systems and Computers.* Nauka, Moscow, pp. 51–61 (in Russian).

Bocharov, P. P. (1993). Sobre algunas propiedades de *PH*-distribuciones. In: *Revista del Seminario de Ensenanza y Titulación* **IX**, #75, pp. 1–16.

Bocharov, P. P. (1995). Analysis of the $MAP/G/1/r$ queueing system of finite capacity. *Vestn. Ross. Univ. Druzh. Nar., Ser. Prikl. Mat. Inf.* #1, 52–67 (in Russian).

Bocharov, P. P. (1996). Stationary distribution of a finite queue with recurrent input and Markov service. *Autom. Remote Control* **57**, 1274–1283.

Bocharov, P. P., and Phong, N. H. (1996). Analysis of the $MAP_2/G_2/1/r$ queueing system with non-preemptive priority. *Vestn. Ross. Univ. Druzh. Nar., Ser. Prikl. Mat. Inf.* #2, 67–84 (in Russian).

Bocharov, P. P., and Naumov, V. A. (1977). On some finite capacity queueing systems. *Probl. Inform. Transm.* **13**, 96–104 (in Russian).

Bocharov, P. P., and Naumov, V. A. (1986). Matrix-geometric stationary distribution for the $PH/PH/1/r$ queue. *Elektron. Informationsverarbeitung Kybernetik* **22**, 179–186.

Bocharov, P. P. and Pechinkin, V. A. (1995). *Queueing Theory.* Peoples' Friendship Univ. Press, Moscow (in Russian).

Bocharov, P. P., and Spesivov, S. S. (1986). *Tables of Steady-State Characteristics of Single-Server Queueing Systems of Finite Capacity.* Peoples' Friendship Univ. Press, Moscow (in Russian).

Bocharov, P. P., Pechinkin, V. A., and Albores, F. J. (1997). *Teoria de Colas.* Tlaxcala Univ. Press, Tlaxcala.

Bocharov, P. P., Pechinkin, A. V., and Phong, N. H. (2000). Stationary probabilities of the states of the $MAP/G/1/r$ retrial system with priority servicing of primary customers. *Autom. Remote Control* **61**, 1300–1309.

Bocharov, P. P., D'Apice, C., Pechinkin, A. V., and Phong, N. H. (2001). Single-server queueing system of finite capacity with batch arrivals and semi-Markovian service. *Vestn. Ross. Univ. Druzh. Nar., Ser. Prikl. Mat. Inf.* #1, 64–79 (in Russian).

Bolch, G., Greiner, S., de Meer, H., and Trivedi, K. (1998). *Queueing Networks and Markov Chains. Modeling and Performance Evaluation with Computer Science Applications.* Wiley, Chichester.

Borovkov, A. A. (1976). *Stochastic Processes in Queueing Theory.* Springer, New York.

Borovkov, A. A. (1984). *Asymptotic Methods in Queueing Theory.* Wiley, New York.

Bose, S. K. (2001). *An Introduction to Queueing Systems.* Kluwer, New York.

Brémaud, P. (1981). *Point Processes and Queues.* Springer, New York.

Brodi, S. M., and Pogosian, M. A. (1973). *Embedded Stochastic Processes in Queueing Theory.* Naukova Dumka, Kiev (in Russian).

Bronstein, O. I., and Dukhovny, I. M. (1976). *Models of Priority Service in Information Computer Networks.* Nauka, Moscow (in Russian).

Bunday, B. D. (1996). *An Introduction to Queueing Theory.* Arnold, London.

Chao, X., Miyazawa, M., and Pinedo, M. (1999). *Queueing Networks. Customers, Signals, and Product Form Solutions.* Wiley, Chichester.

Chung, K. L. (1967). *Markov Chains with Stationary Transition Probabilities.* Springer, New York.

Cohen, J. W. (1969). *The Single Server Queue.* North-Holland, Amsterdam.

Cooper, R. B. (1981). *Introduction to Queueing Theory.* Arnold, London.

Cox, D. R., and Smith, W. L. (1961). *Queues.* Methuen, London, 1961.

Dshalalow, J. H., Ed. (1995). *Advances in Queueing. Theory, Methods, and Open Problems.* CRC, New York.

Dudin, A. N., and Klimenok, V. I. (2000). *Queueing Systems with Correlated Flows.* Byelorussian State Univ. Press, Minsk (in Russian).

Erlang, A. K. (1917). Solution of some problems in the theory of probabilities of significance in automatic telephone exchanges. *The Post Office Electrical Eng. J.* **10**, 189–197.

Falin, G. I., and Templeton, J. G. C. (1997). *Retrial Queues.* CRC, London.

Feller, W. (1966). *An Introduction to Probability Theory and its Applications.* Wiley, New York.

Franken, P., König, D., Arndt, U., Schmidt, V. (1982). *Queues and Point Processes.* Wiley, Chichester.

Gantmacher, F. R. (1959). *The Theory of Matrices.* Chelsea, New York.

Gelenbe, E. and Pujolle, G. (1998). *Introduction to Queueing Networks.* Wiley, New York.

Gikhman, I. I., and Skorokhod, A. V. (1966). *Introduction to the theory of random processes.* Nauka, Moscow (in Russian).

Gnedenko, B. W., and König, D., Eds. (1983). *Handbuch der Bedienungstheorie I. Grundlagen und Methoden.* Akademie, Berlin.

Gnedenko, B. W., and König, D., Eds. (1984). *Handbuch der Bedienungstheorie II. Formeln und andere Ergebnisse.* Akademie, Berlin.

Gnedenko, B. V., and Kovalenko, I.N. (1991). *Introduction to Queueing Theory.* Birkhäuser, Boston.

Gnedenko, B. V., Danielian, E. A., Dimitrov, B. N., Klimov, G. P., and Matveev, V. F. (1973). *Priority Queueing Systems.* Moscow State Univ. Press, Moscow (in Russian).

Graham, A. (1981). *Kronecker Products and Matrix Calculus: With Applications.* Ellis Horwood, Chichester.

Grigelionis, B. (1962a). On accuracy of approximation of renewal processes composition by a Poisson process. *Liet. Mat. Rink.* **11**, 135–143 (in Russian).

Grigelionis, B. (1962b). Accuracy estimation of multivariate limit theorem on the convergence to Poisson law. *Liet. Mat. Rink.* **11**, 143–148 (in Russian).

Gross, D., and Harris, C. M. (1974). *Fundamentals of Queueing Theory.* Wiley, New York.

Ivanov, G. A. (1973). *Priority Queues in Transient Mode.* Moscow State Univ. Press, Moscow (in Russian).

Ivchenko, G. I., Kashtanov, V. A., and Kovalenko, I. N. (1982). *Theory of Mass Service.* Vysshaya Shkola, Moscow (in Russian).

Jaiswal, N. (1968). *Priority Queues.* Academic Press, New York.

Kaj, I. (2002). *Stochastic Modelling in Broadband Communications Systems.* SIAM, S-UUPS, Uppsala.

Kalashnikov, V. V. (1994). *Mathematical Methods in Queueing Theory.* Kluwer, Dordrecht.

Karlin, S., and McGregor, J. (1957a). The differential exudations of birth and death processes and Stieltjes Moment Problem. *Trans. AMS* **85**, 458–546.

Karlin, S., and McGregor, J. (1957b). The classification of birth and death processes. *Trans. AMS* **86**, 366–400.

Kato, T. (1966). *Perturbation Theory for Linear Operators.* Springer, Berlin.

Kaufmann, A., and Cruon, R. (1961). *Les phenomenes d'attente.* Dunod, Paris.

Kelly, F. P. (1979). *Reversibility and Stochastic Networks.* Wiley, New York.

Kemeny, J. G., and Snell, J. L. (1960). *Finite Markov Chains.* Van Nostrand, Princeton, NJ.

Khintchine, A. Y. (1960). *Mathematical Methods in the Theory of Queueing.* Griffin, London.

Kleinrock, L. (1984). *Queueing Systems. Vol.. I: Theory.* Wiley, New York.

Kleinrock, L., and Gall, R. (1996). *Queueing Systems: Problems and Solutions.* Wiley, New York.

Klimow, G. P. (1979). *Bedienungsprozesse.* Birkhäuser, Basel.

Klimov, G. P., and Mishkoy, G. K. (1979). *Priority Service Systems with Orientation.* Moscow State Univ. Press, Moscow (in Russian).

Koliadenkova, L. G., Pechinkin, A. V., and Trishechkin, S. I. (2000). The $MAP_2/G_2/1/n$ system with preemptive priority and common queue. *Vestn. Ross. Univ. Druzh. Nar., Ser. Prikl. Mat. Inf.* #1, 72–90 (in Russian).

Korolyuk, V. S., and Turbin, A. F. (1976). *Semi-Markov Processes and their Applications.* Naukova Dumka, Kiev (in Russian).

Latouche, G., and Ramaswami, V. (1993). A logarithmic reduction algorithm for quasi-birth-and-death processes. *J. Appl. Prob.* **30**, 650–674.

Lavenberg, S. S. (1978). Stability and maximum departure rate of certain open queueing networks having finite capacity constraints. *Rev. Francaise Automat. Informat. Rech. Operat., Ser. Bleue* **12**, 331–347.

Little, J. D. (1961). A proof of the queueing formula $L = \lambda W$. *Oper. Res.* **9**, 383–387.

Lucantoni, D. M. (1991). New results on single server with a batch markovian queue. *Stochastic Models* **7**, 1–46.

Lucantoni, D. M., and Neuts, M. F. (1994). Some steady-state distributions for the $MAP/SM/1$ queue. *Stochastic Models* **10**. 575–598.

Lucantoni, D. M., Meier-Hellstern, K. S., and Neuts, M. F. (1990). A single-server queue with server vacations and a class of non-renewal processes. *Adv. Appl. Prob.* **22**, 676–705.

Matveev, V. F., and Ushakov, V. G. (1975) *Queueing Systems.* Moscow State Univ. Press, Moscow (in Russian).

Medhi, J. (2003). *Stochastic Models in Queueing Theory.* Academic Press, London.

Morse, P. M. (1958). *Queues, Inventories and Maintenance.* Wiley, New York.

Naumov, V. A. (1985). *Numerical Methods of Analysis of Markov Systems.* Peoples' Friendship Univ. Press, Moscow (in Russian).

Naumov, V. A. (1987). Markov models of customer flows. In: *Queueing Systems and Informatics.* Peoples' Friendship Univ. Press, Moscow, 67–73 (in Russian).

Naumov, V. A. (1995). Matrix-multiplicative approach to quasi-birth-and-death processes analysis. In: *Proc. 1st Internat. Conf. on Matrix-Analytic Methods in Stochastic Models.* Detroit, August 24–25, 1995.

Nelson, R. (1995). *Probability, Stochastic Processes, and Queueing Theory: the Mathematics of Computer Performance Modeling.* Springer, New York.

Neuts, M. F. (1979). A versatile Markov point process. *J. Appl. Prob.* **16**, 764–779.

Neuts, M. F. (1981). *Matrix-Geometric Solutions in Stochastic Models. An Algorithmic Approach.* The Johns Hopkins Univ. Press, London.

Neuts, M. F. (1989). *Structured Stochastic Matrices of* $M/G/1$ *type and their Applications.* Marcel Decker, New York.

Newell, G. (1971). *Applications of Queueing Theory.* Chapman & Hall, London.

Ng Chee Hock (1997). *Queueing Modelling Fundamentals.* Wiley, New York.

Pechinkin, A. V. (1983). On an invariant mass service system. *Math. Operationsforsch. Stat., Ser. Optimization* **14**, 433–444.

Pechinkin, A. V. (1999a). The $BMAP/G/1/\infty$ queueing system with foreground–background processor sharing. *Autom. Remote Control* #10, 108–114 (in Russian).

Pechinkin, A. V. (1999b). The $MAP/G/1/n$ queueing system under LIFO discipline with interruptions and constraints on the common customer volume. *Autom. Remote Control* #12, 114–120 (in Russian).

Prabhu, N. U. (1980). *Stochastic Storage Processes. Queues, Insurance Risk and Dams.* Springer, New York.

Prabhu, N. U. (1997). *Foundations of Queueing Theory.* Kluwer, Boston.

Riordan, J. (1962). *Stochastic Service Systems.* Wiley, New York.

Saaty, T. L. (1961). *Elements of Queueing Theory with Applications.* McGraw–Hill, New York.

Sevastyanov, B. A. (1957). An ergodic theorem for Markov processes and its application to telephone systems with refusals. *Theory Probab. Appl.* **2**, 106–116 (in Russian).

Simpson, A. (1970). Inversion of a class of matrices occurring in control system theory by an m-step procedure allied to Gaussian elimination. *Electron. Lett.* **6**, 500–501.

Stoyan, D. (1983). *Comparison Methods for Queues and Other Stochastic Models.* Wiley, New York.

Takács, L. (1962). *Introduction to the Theory of Queues.* Oxford Univ. Press, New York.

Takagi, H. (1994a). *Queueing Analysis: Discrete-Time Systems.* Elsevier, Amsterdam.

Takagi, H. (1994b). *Queueing Analysis: Finite Systems.* Elsevier, Amsterdam.

Tanner, M. (1995). *Practical Queueing Analysis.* McGrow–Hill Cambridge.

Tijms, H. (1994). *Stochastic Models: An Algorithmic Approach.* Wiley, Chichester.

Tikhonenko, O. M. (1990). *Queueing Models in Information Processing Systems.* Universitetskoe, Minsk (in Russian).

van Dijk, N. M. (1993). *Queueing Networks and Product Forms: A System Approach.* Wiley, Chichester.

Volkovinskii, M. I., and Kabalevskii, A. N. (1981). *Analysis of Priority Queues with Switching Time.* Energoizdat, Moscow (in Russian).

Warland, J. (1988). *An Introduction to Queueing Networks.* Prentice Hall, NJ.

Yashkov, S. F. (1989). *Queueing Analysis in Computers.* Radio i Svyaz, Moscow (in Russian).

Zozhikashvili, V. A., and Vishnevskii, V. M. (1988). *Queueing Networks. Theory and Applications to Computer Networks.* Radio i Svyaz, Moscow (in Russian).

Index